MAJOR CHEMICAL HAZARDS

Ellis Horwood Series in
CHEMICAL ENGINEERING
Series Editor: Dr. DAVID SHARP, OBE, former General Secretary, Society of Chemical Industry

***THERMODYNAMICS FOR CHEMISTS AND CHEMICAL ENGINEERS**
M.H. CARDEW, University of Birmingham
***ENVIRONMENTAL BIOTECHNOLOGY**
F.C. FORSTER and D.A.J. WASE, University of Birmingham
MAJOR CHEMICAL HAZARDS
V. C. MARSHALL, University of Bradford
BIOLOGICAL TREATMENT OF WASTE-WATER
M. WINKLER, University of Surrey

Ellis Horwood Limited also publishes related titles in
APPLIED SCIENCE AND INDUSTRIAL TECHNOLOGY
INDUSTRIAL CHEMISTRY
BIOLOGICAL SCIENCES

* *In preparation*

MAJOR CHEMICAL HAZARDS

V. C. MARSHALL
M.Phil., Ph.D., C.Eng., F.I.Chem.E., F.Inst.E., F.S.R.S., M.I.O.S.H., F.R.S.A.
Honorary Fellow in Industrial Technology,
formerly Director of Safety Services and sometime Senior Lecturer in Chemical Engineering, in the University of Bradford

ELLIS HORWOOD LIMITED
Publishers · Chichester

Halsted Press: a division of
JOHN WILEY & SONS
New York · Chichester · Brisbane · Toronto

First published in 1987 by
ELLIS HORWOOD LIMITED
Market Cross House, Cooper Street,
Chichester, West Sussex, PO19 1EB, England
The publisher's colophon is reproduced from James Gillison's drawing of the ancient Market Cross, Chichester.

Distributors:
Australia and New Zealand:
JACARANDA WILEY LIMITED
GPO Box 859, Brisbane, Queensland 4001, Australia
Canada:
JOHN WILEY & SONS CANADA LIMITED
22 Worcester Road, Rexdale, Ontario, Canada
Europe and Africa:
JOHN WILEY & SONS LIMITED
Baffins Lane, Chichester, West Sussex, England
North and South America and the rest of the world:
Halsted Press: a division of
JOHN WILEY & SONS
605 Third Avenue, New York, NY 10158, USA

British Library Cataloguing in Publication Data
Marshall, V.C.
Major chemical hazards. —
(Ellis Horwood series in chemical engineering)
1. Hazardous substances
2. Chemical industries — Safety measures
I. Title
660.2′804 TP149

Library of Congress Card No. 86–27611

ISBN 0–85312–969–X (Ellis Horwood Limited)
ISBN 0–470–20813–9 (Halsted Press)

Printed in Great Britain by Unwin Bros. of Woking

CONTENTS

DEDICATION.

This book is dedicated to the University of Bradford and to the Institution of Chemical Engineers without whose assistance, in many diverse ways, this book could not have been written.

"Oh yet we trust that somehow good will be the final goal of ill"

Alfred, Lord Tennyson.

ACKNOWLEDGEMENTS.

Thanks are expressed to the Institution of Chemical Engineers for permission to reproduce terms from its book "Nomenclature for Hazard and Risk Assessment in the Process Industries".

Permission is gratefully acknowledged to the BASF for Figures 11.1, 11.2, 13.4, 13.5, 13.6, 13.7 and 13.8, the Controller of Her Majesty's Stationery Office and the Health and Safety Executive for 13.12, 13.13 and 13.14, the National Institute for Occupational Safety and Health for Figure 14.1. The management of the former Nypro Company for Figures 21.5, 21.6, 21.7, 21.8 and 21.9 and the cover illustration. The Scotsman Newspaper for 9.16. Sim Chem Ltd for 12.5 and 13.15. Technica Ltd for 15.12 and 21.15. TNO (Netherlands) for 9.15. The Safety and Reliability Directorate for 15.4, 15.5 and 15.6. Mr. G.W. Watkins of Bath University for finding 17.1.

Finally, thanks are recorded to Mr. H. Coulton for his assistance with some of the illustrations.

Old fashioned and highly hazardous technique for manufacturing high explosives. (See page 454)
Photo courtesy Nobel's Explosives Company Limited

FOREWORD

Bryan Harvey C.B.E., M.A.(Oxon), S.M.(Harvard), F.S.A.

[Formerly H.M. Chief Inspector of Factories and Deputy Director General of the Health & Safety Executive. Latterly Chairman of the Advisory Committee on Major Hazards.]

Dr. Marshall, the author of this book, was a member of the Major Hazards Advisory Committee and served with it from its inception in 1975 until it completed its work in 1983. He took a full part in all the work of the Committee but as Chairman I believe that his great contribution lay in the research that he did and in the series of written papers which he produced and which provided such important background material for the deliberations of the Committee.

It is therefore a valuable development that he should build on the basis of his wide experience and extensive enquiries in this field to produce a comprehensive text book on the subject.

Chief Inspectors of Factories (myself included) had been warning of the possibility of disaster from 1967 onwards, but alas it took a Flixborough to concentrate industry's, and for that matter the public's mind, on the problems involved and to force significant changes in the design, construction, and above all the operation of high technology plants. After Flixborough it might have been expected that the lessons would have been learned on a world-wide basis, but Mexico City and Bhopal, both on a scale to dwarf Flixborough, suggest that there is still much to be learned if advanced technology is to regain the confidence of the public. Vital to improving safety is the provision of information. Emotional reactions prompted by media coverage do not solve the problems though exposure in the media may do good in drawing attention to the need for vigilance.

One of the difficulties in preventing disasters is that memories of past events fade and perhaps within a decade many of those whose memories of past failures remain fresh retire from active roles and what could be described a group experience is lost. A new generation of managers and technicians may be tempted to make the same mistakes again, even if in a different context. There is therefore a continuing need to retain a pool of experience upon which each new group of managers, designers and technicians can call.

While it is true that the reports of the Advisory Committee did something to satisfy this need and that the subsequent European Directive and the national Regulations based upon it paint a broad picture of how the difficulties can be tackled, it is detailed information which is needed and that is where this publication will prove invaluable. Indeed it is something of a milestone that the author should have brought together so much detailed information to form what is clearly the most up to date and authoritative work on major chemical hazards.

In this connection it is worth stressing that not only will the reader be able to refer to the theoretical considerations involved, but will find that these are backed by comprehensive analysis of the circumstances of past disasters, where theory has, with such dire results, been proved in practice.

Dr. Marshall is to be congratulated on this book and I hope and believe that it will become a standard work of reference. I wish it every success.

Bryan Harvey.

FOREWORD

Chapter 1

INTRODUCTION

1.1. THE AIM OF THE BOOK.

During the last sixty or so years there have been a number of events which have occurred in the chemical and process industries and which have given rise to grave public concern. These events have been explosions, cataclysmic fires or releases of toxic substances. Some of them have led to hundreds, and in one case at least, to thousands of deaths.

Sometimes the places at which they have occurred, and which have given their name to the disastrous event, had previously been unknown to the world at large, as for example, Flixborough and Seveso.

Such events have become associated with the term "Major Hazard" or with the more precise term "Major Chemical Hazard" (MCH). The meaning of these terms will be discussed below.

The shock produced by these disasters has led to demands for controls to be exercised over the type of installations which are deemed to be capable of giving rise to such events. There has been pressure for legislation and for other forms of state intervention; this pressure has resulted in many countries adopting legislation; in recent years there has been action by the European Community.

At the same time there has been a rapid development of new managerial techniques to identify hazards as a step towards their effective control. Such techniques are now being used at the design stage of new plant either to eliminate the hazards at source or to reduce them to a minimum.

The aim of the book is:

(1) to identify those hazards which possess the potential for major harm.
(2) to examine the circumstances in which this potential may be realised.
(3) to identify the consequences which are likely to ensue if the potential be realised.
(4) to examine those methods of control which will minimise the probability of such events occurring.
(5) to survey the measures which may be taken to mitigate the consequences in the event that there is a failure of the preventative measures.

In pursuit of these aims, it is intended that these analyses should be conducted in a rigorous manner and with the use of a self-consistent terminology.

1.2. NOMENCLATURE.

The question of the meanings to be attached to certain terms used in the book is very important and one which, in some cases, cannot be resolved by consulting a dictionary. This is especially true of the words "hazard" and "risk" which are frequently used in the book and for which the dictionary meanings are interchangeable. The importance of these definitions is such that Chapter 4 is largely devoted to this subject. They are collated in Appendix 1.

1.3. PERSONAL CONTRIBUTION TO THE SUBJECT.

1.3.1. ACTIVITIES.

This book embodies the results of twelve years of study in the field covered by its title. The author commenced the study in June 1974 when he was called upon by two trade unions, the Transport and General Workers Union (T&GWU) and the Association of Scientific, Technical and Managerial Staffs (ASTMS) to give them technical advice as to the causes of the Flixborough disaster in which members of these Unions had been killed and injured in the explosion at the Nypro Works near Flixborough. This entailed conducting an internal inquiry attended by members of these Unions employed at the Nypro Works and giving evidence to the Official Court of Inquiry into the causes of the disaster.

Subsequently he was called upon by ASTMS to advise them in the matter of the Whiddy Island, Bantry Bay, disaster of January 1979 and he gave evidence to the Irish Government's Official Court of Inquiry in December of that year.

In January 1975 he was appointed to membership of the Health and Safety Commission's Advisory Committee on Major Hazards (ACMH) which completed its work in 1983. Since then he has been appointed to membership of the Commission's Advisory Committee on the Major Hazards Aspects of Transport.

He has been a member of the Loss Prevention Panel of the Institution of Chemical Engineers since 1976. He has served as an Assessor on matters relating to Health and Safety at two Public Inquiries. The first, at Cheltenham in 1981, was concerned with the possible hazard to the public arising from the storage of pressurised North Sea gas. The second, at Thornton Cleveleys, Lancashire in 1982, was concerned with the possible hazard to the public arising from the storage and handling of liquid chlorine at a nearby works.

1.3.2. PUBLICATIONS.

In part this book is based upon papers which were prepared and submitted to the ACMH. Some of the conclusions in these papers have become incorporated in the First and Second Reports of the Committee, [ACMH (1976) and ACMH (1979)]. There have been, in addition, more than fifty papers by the author on the subject which have been read to conferences or otherwise published.

1.4. CONVENTIONS ADOPTED.

1.4.1. UNITS.

SI units are used throughout, with the exception of the unit of mass. Because it is convenient to do so in the field covered by this work, the unit of mass generally used is the tonne, (1000 kilograms or 1 megagram). Within the limits of accuracy of most of the calculations in the book, the tonne may be taken as equivalent to the imperial ton (1 tonne = 0.984 ton). However, this does not necessarily hold true for the U S short ton of 2000 lbs (1 tonne = 1.102 short tons).

1.4.2. REFERENCES.

These are given in the form "Smith (1976)" when it is the first time that the book is referring to a work by that author published in 1976. Subsequent references to further papers by that author, published in the same year, will give rise to references of the form "Smith (1976a)", "Smith (1976b)" and so on. But 1976, 1976a, 1976b are in the order in which the papers are cited, not the chronological order of their publication in 1976. For the convenience of the reader, the number of the page on which the reference is cited, is given.

1.5. ABBREVIATIONS AND SYMBOLS.

All terms subject to abbreviation are set out in full on first citation and for the convenience of the reader are further collated in Appendix 2.

The abbreviation UK needs comment. It is an abbreviation for the United Kingdom of Great Britain and Northern Ireland, Great Britain being England, Wales and Scotland. Northern Ireland is, however, legally self-governing and legislation on occupational health and safety by the Westminster Parliament does not apply to Northern Ireland. The expression UK will be used as a convenient abbreviation but its use will not necessarily imply that Northern Ireland is included.

The abbreviation US will be used for the United States of America.

All mathematical or other symbols used are identified in the immediate vicinity of any equation in which they appear and the author accordingly has not considered it necessary to list them separately.

Interpolations by the present author into quotations from others are identified by the initials "VM".

1.6. ACKNOWLEDGEMENTS.

This book obviously draws upon the work of many others in the field and this work is acknowledged in references in the text.

The author benefited greatly from the stimulation derived from his membership of the Advisory Committee on Major Hazards and many of his conclusions have been tempered by the constructive criticism of his colleagues on that Committee. In their final form, his conclusions contain within them, as a vital element, the fruits of this constructive criticism and dialogue.

1.7. ADDENDUM.

In a world of rapid change the author considered it likely that during the course of writing this book events may occur which may render amendments to the text desirable. Further, new material may be published, or otherwise come to the author's attention, in the course of writing.

As the book has been produced as a "camera-ready" typescript it is not convenient, in general, to incorporate such amendments in the text as they arise. So, accordingly, the book has been provided with an Addendum (Chapter 25), in which any amendments are assembled in order of page number. The reader is advised to consult the Addendum first in order to ascertain whether any particular section of the text requires amendment.

1.8. THANKS.

The author's best thanks are due to Mrs. Pauline Brier who typed the text with such great care and who pointed out textual errors. Any remaining errors and inconsistencies are entirely the author's responsibility.

Chapter 2

THE NATURE OF THE PROBLEM

2.1. HISTORICAL ORIGINS OF MAJOR CHEMICAL HAZARDS.

2.1.1. CATACLYSMIC FIRES.

Nash (1976) records city fires dating back before the beginning of the present era and it is clear from his accounts that some of the world's oldest cities such as London have been destroyed wholly or partially several times in their history.

An early record of a chemical fire is given in Annual Register (1759) where it is recorded that the Worcester Wagon was burned out as the result of the spillage of a bottle of aqua fortis (nitric acid). The earliest recorded incident of a major fire arising from hydrocarbon gases may be that at Cleveland, Ohio, USA 1944, in which there was serious loss of containment of liquefied natural gas (LNG) which killed about 130 people and did extensive damage to property [ACMH (1979)]. The first major liquefied petroleum gas (LPG) fire, perhaps was that at Meldrin, Georgia, USA 1959, which killed 23 people [ACMH (1979)].

2.1.2. EXPLOSIONS.

The accidental explosion of "black powder" (which later became called "gunpowder" after the invention of the gun) must go back to the time of the invention of the substance which took place at least a thousand years ago. The cases given below, therefore, cannot be regarded as constituting the earliest occurrences; they are merely examples which have happened to be recorded.

For example, James II of Scotland was killed as the result of the bursting of a cannon at Roxburgh in 1460. The earliest record of an explosion quoted in Nash (1976) is that of a gunpowder explosion in Boston in 1645 which killed 3 people and destroyed one-third of the city. Biasutti (1985) quotes the explosion of 85 tons of black powder at Brescia, Italy in 1769, which destroyed 190 houses. It is perhaps sufficient to say, at this point, that the invention of gunpowder brought about a new phenomenon, the chemical explosion. Subsequent inventions in the nineteenth century brought into existence explosives which were more powerful than gunpowder.

The first recorded explosion involving the new generation of explosives took place in Heleneburg, Sweden in 1864. This killed 5 persons, one of them was Emil Nobel, brother of the inventor of dynamite [Biasutti (1985)].

Ammonium nitrate which is used mainly as a fertiliser, rather than as a high explosive, has been involved in a number of serious explosions. Clancey (1963) records 16 explosions, of which the first, involving 7 kg, occurred in Kensington, London in 1896.

Gas explosions must have followed shortly after the invention of coal gas by Murdock in 1795, but there seems to be little information on the subject. For example, Nash (1976) does not give any examples which are clearly recognisable as early gas explosions. Hake & MacNab (1892) state "the clouds formed by petroleum vapours and other volatile hydrocarbons have also given rise to similar (i.e. similar to explosions in clouds of dust, VM) explosions in cellars and magazines, or even in the open air . . ." This latter also foreshadows the discussion of vapour cloud explosions (VCEs) which follows below.

The first clearly recognisable VCE (the nature of VCEs is discussed in Chapter 12) was that which occurred at Ludwigshafen in 1943. Information on it is scanty because of wartime censorship and comes mainly from Stahl et al (1949). Dust explosions have long been recognised. Palmer (1973) quotes an explosion in a flour mill at Turin, Italy in 1785. National Fire Protection Association (1957) quotes the occurrence of 1,120 major dust explosions in the United States and Canada in the period 1860-1956.

2.1.3. TOXIC RELEASES.

It is recorded that a mixture of pitch and sulphur was used during the siege of Plataea in 429 BC, [Encyclopaedia Britannica (1951), (Vol.5, page 350)] which presumably gave rise to clouds of sulphur dioxide. Apart from this there does not seem to be any record of a major toxic release before the First World War. Though an act of war and not strictly comparable to an industrial release, the gas attack at Ypres on 22nd April 1915 which killed 5,000 soldiers, demonstrated the potential dangers of chlorine. The first major non-military release of chlorine on record was that at Wyandotte, Michigan, USA 1917, in which 17 tonnes of the gas were spilled causing one fatality [ACMH (1979)].

2.2. THE GROWTH OF THE CHEMICAL AND PROCESS INDUSTRIES.

2.2.1. THE NATURE OF THE INDUSTRIES.

Manufacturing industry may be divided into two distinct branches, mechanical manufacturing and process manufacturing. Mechanical manufacturing includes all those activities in which the products are of a defined shape and dimensions and in which the operations of the industry consist of shaping parts and/or fitting them together. Two examples are a weaving shed or a car assembly plant.

Process manufacturing on the other hand includes all processes whose end-products are liquids or gases and also some solids where these are a starting point for mechanical manufacture. Examples are the bottled gas industry, the sulphuric acid industry and the industry for producing nylon chips. An alternative definition is that the process industries are those concerned with changing the physical and/or chemical state of matter. Strictly speaking the process industries include the chemical industry but the term "process industry" will be used below to refer to that portion of it which is not commonly recognised as being the chemical industry as, for example, petroleum refining.

Excluded from the definition for the purposes of this book are extractive industries such as coal mining, the combustion of fuel to raise steam, and the smelting of metals. The manufacturing and re-processing of fuel elements for nuclear power is a process industry but one which is excluded from discussion in this book because of its associated radioactive hazards.

Perhaps the most important common feature that the chemical and process industries possess is that in them increased output can be achieved at will by "scaling up" whereas, in manufacturing industry, increasing output usually entails replication of production units. This property of the process industries will be discussed further below. It is of crucial importance to the major chemical hazards problem.

2.2.2. GENERAL HISTORY.

2.2.2.1. The growth of the chemical and process industries.

In world terms the chemical and process industries have achieved, in the last three decades, a growth rate which has been appreciably higher than that of industrial production as a whole. However, the rate has markedly slackened in recent years in Western Europe.

The process industries were cradled in Europe and, in spite of enormous growth of the chemical industry in North America, Europe today still produces more than half the world's chemicals, [Telfer (1980)]. The hydrocarbon industries which handle crude oil and process it to yield fuels and petrochemicals and which extract and distribute natural gas have a turnover at least as large in Europe as a whole (including Eastern Europe) as they have in North America. Features which are recognised as typical of the North American scene such as the creation of a vast network of pipelines to handle liquid and gaseous hydrocarbons also exist in Europe though they are mainly to be found in the USSR and in Eastern Europe.

It seems useful to distinguish four phases of development in the chemical industry.

2.2.2.2. Early beginnings.

The first phase was ushered in at the end of the 18th century and early 19th century by the rapid growth of the textile industry which brought demands for detergents, bleaches, mordants and dyestuffs. These demands were met by inorganic chemicals, such as sulphuric acid, sulphur dioxide, salt cake, soda and chlorine. At the same time there was a growth in the manufacture of coal gas, which though organic in origin was not as yet a source of chemicals, the tar being an unwanted, and difficult to dispose of, by-product. The problems of the coal gas industry were paralleled by those of the coke making industry to which it closely corresponded. Coal gas and coke oven gas and other nineteenth century fuel gases were highly toxic, containing between 4% and 40% of carbon monoxide.

Boiler explosions became an everyday occurrence throughout industry as virtually every factory had to install a boiler to provide power and heat. These boilers, of the fire tube type, often lacked safety valves and low level alarms. Water treatment was in its infancy and in hard water areas boilers rapidly furred up. They did, however, operate at pressures much lower than those current today.

The period produced many human casualties. A combination of ruthless commercial pressure, laissez faire politics, and crass ignorance produced the most serious abuses. Air, water and land was polluted and the health of work people assailed. Eventually legislation could no longer be avoided. The UK Alkali Works Act was passed in 1863 primarily to control the atmospheric pollution caused by the Le Blanc process which in South Lancashire alone was discharging tens of tons a day of HCl into the atmosphere. In the workplace industrial diseases became recognised and became the subject of legislation. The operation of steam boilers became regulated by law and came under the surveillance of insurance companies.

But major chemical hazards, as we know them today, were in their infancy except for a few incidents involving the manufacture and storage of explosives. The reason for this was the small scale on which chemicals were produced and the low pressures which were technically feasible in operating chemical plant.

2.2.2.3. Chemicals from coal.

The second phase also owed its origin to the growth of the textile industry. The demand for dyestuffs was outstripping the supply of vegetable dyes and the discovery of the first aniline dye by Perkin in 1856 was followed by the establishment of the aniline dyestuffs industry which took place principally in Germany. This industry was based upon the coal tar which had formerly been a nuisance but which now became a source of hundreds of different organic chemicals. These developments also intensified the demand for nitric acid as the aniline and its derivatives on which dyestuffs industry was based was produced by the reduction of nitrobenzene and other aromatic nitro-compounds. Parallel with this there grew the need for nitrates as fertilizers.

There were considerable developments in the explosives industry in the nineteenth century which resulted in gunpowder becoming obsolescent as a militare explosive except for specialised purposes. Nitroglycerine and nitrocellulose (gun cotton) were discovered before 1850; Nobel invented dynamite in 1862. These discoveries led to numerous explosions which will be discussed in Chapter 10.

All of these developments enormously increased the demand for nitric acid and created the "nitrogen fixation" problem. It was the solution of this problem, which arose from the impending exhaustion of the Chile saltpetre deposits, which ushered in the third phase.

2.2.2.4. The fixation of nitrogen.

This third phase commenced in the years immediately prior to the First World War with the commercial success of the Haber-Bosch process for the manufacture of ammonia. It may be argued that this process represented the first real marriage between engineering and chemistry as this was a process using both a high temperature (ca 500°C) and a high pressure (100-250 bars). It is interesting to compare the equipment constructed at the Ludwigshafen works of the Badischen Aniline und Soda Fabrik (BASF) to carry out this process with the kind of apparatus illustrated in George Davis's "Handbook of Chemical Engineering", [Davis (1901)]. Much of this latter apparatus was made by carpenters and blacksmiths and would not have seemed out of place in a farmyard. Yet it represented the ultimate achievement in nineteenth-century chemical engineering.

It was the creation of this new and powerful industry which gave rise to an event which may be said to have marked the beginning of the era of major chemical hazards. This event was the Oppau, Ludwigshafen, explosion of 21st September 1921 in which a store of some 4,000 tonnes of an ammonium nitrate mixture exploded, killing over 500 people. This was in no sense an ammunition dump, the ammonium nitrate was used as a fertiliser.

It might, however, be noted that during the preceding First World War some 17 cases were recorded, in Europe alone, of munitions works and ammunition dumps exploding in which, in some individual cases, the death toll exceeded 1,000.

What characterised this third era? Firstly, the scale of chemical plant was increasing as it was found that there was no obvious upper limit to the throughput of chemical plant, particularly if the process were continuous rather than batchwise. Secondly, it marked the beginning of the importance of petroleum as a product of the process industries though, to begin with, it was not a raw material for the chemical industry. This arose as the consequence of the rapid growth of the motor car industry, particularly in the United States where Henry Ford had started selling his "Model T" in 1908. This led to increasing demands for motor spirit and eventually led to the invention of cracking and reforming processes. These were developments of enormous importance as they freed the petroleum refiner from some of the constraints imposed by the composition of crude oil by making possible the manufacture of petroleum spirit from heavy hydrocarbons. Later these developments were to lead to the present-day petrochemicals industry.

The most serious chemical accidents in the period between the wars, apart from the disaster at Oppau, involved chlorine. Today the record of the chlorine industry is very much better, both in the USA and in Europe.

2.2.2.5. Chemicals from oil.

The fourth era is the post Second World War era characterised by the extremely rapid development of the petrochemicals industry and especially of the techniques of handling liquefied gases.

The petrochemicals industry was born in the 1930s. Researchers in America, Russia and the United Kingdom showed the feasibility of producing olefines from gas oil [Ruhemann (1949)]. The present author worked on the first UK petrochemicals plants which were started up at the British Celanese Works, Spondon, Derbyshire, in the early years of the Second World War.

The trend away from coal had now started. However, the peculiar national position of Germany in the thirties with no indigenous oil and a perceived need to prepare for war-time blockade diverted research in this important centre of the chemical industry into "oil from coal" rather than "chemicals from oil".

The scale of handling of chemicals has grown enormously in the last three decades including the transport, by ship, road, rail and pipeline of crude and refined petroleum and other products. The number of sites and of transport vehicles with inventories sufficient, should containment be lost, of producing a vapour cloud sufficiently large as to give rise to a serious fire or explosion has increased very rapidly indeed. Though most of the incidents involving liquefied gases, for example, have occurred in the USA there have, nevertheless, been seven major disasters in Europe involving the handling of liquid hydrocarbons together with a number of incidents involving ships carrying these materials, of which the most serious was the Bantry Bay disaster in Eire in January 1979 in which 50 persons were killed.

The present author, in Marshall (1979) has pointed out that, in recent years, C_3 and C_4 hydrocarbons have been involved in some 40% of all serious incidents involving flammable liquids and gases and that propane appears to be about 30 times as dangerous as petrol.

2.2.3. SUMMARY.

The rapid growth of the chemical and process industries has occurred in a number of phases, only the latter of which have given rise to major chemical hazards.

Figure 2.1. expresses in graphical form the salient developments of major chemical, and other related hazards, during the present century.

2.3. THE GROWTH IN SCALE OF PROCESS UNITS.

2.3.1. THE GROWTH IN DEMAND.

The previous sub-section has referred to the continuous growth of the chemical and process industries, both as regards the variety of their products and as regards the quantities produced. It would seem useful to provide some data on this growth before proceeding to examine the reasons for the growth in the size of the process units which make up chemical plant. (For definitions of "process unit" and "plant" see Appendix I).

Figure 2.2. is a graph showing how the production of sulphuric acid and

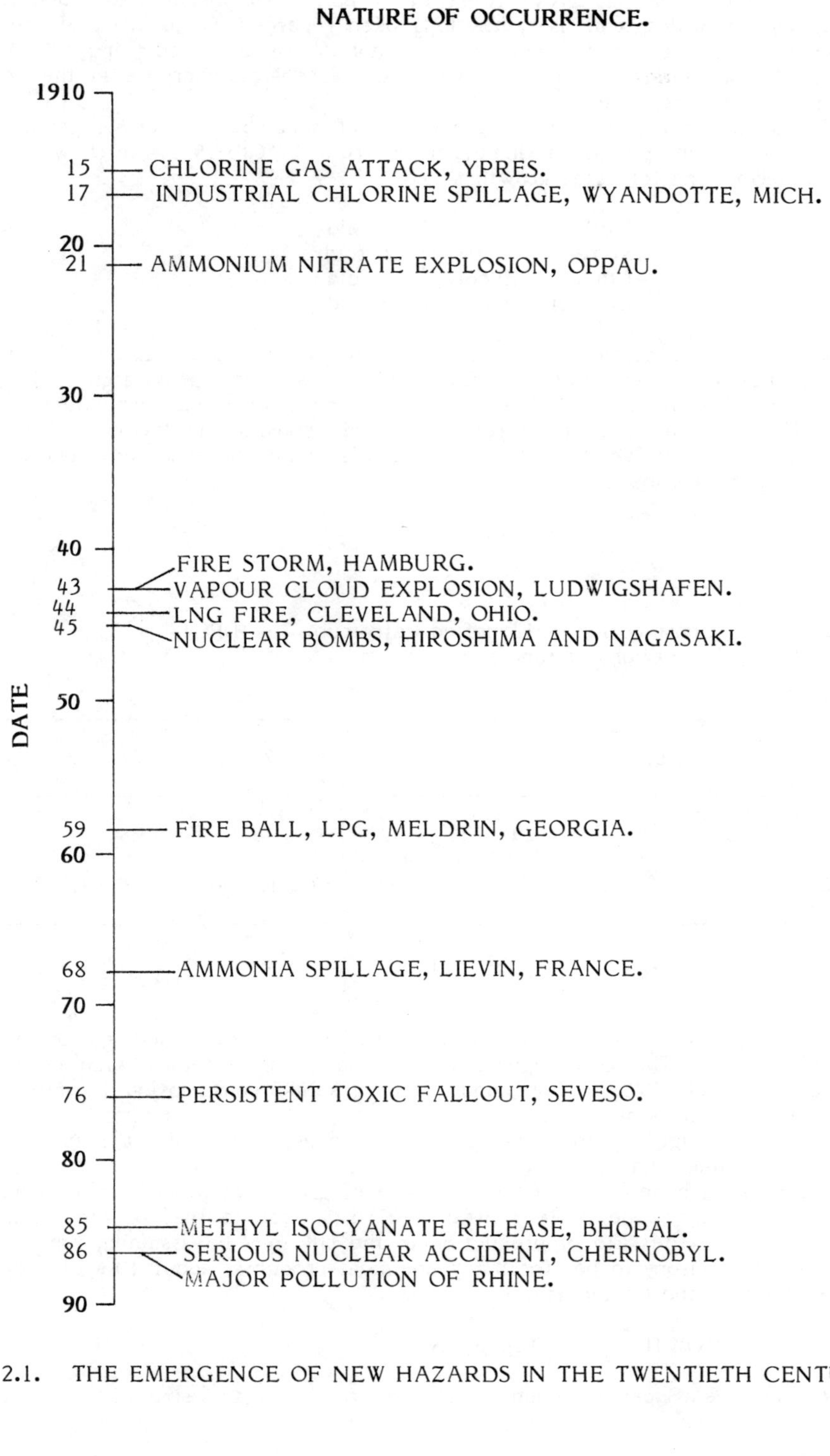

Fig. 2.1. THE EMERGENCE OF NEW HAZARDS IN THE TWENTIETH CENTURY.

chlorine in the U S has grown with the years. Sulphuric acid manufacture, because of its widespread uses, has long been regarded as an index of the level of general chemical production. The graph shows a ten-fold increase in the demand for sulphuric acid in forty years and a ten-fold increase in the demand for chlorine in thirty years.

Figure 2.3. is a graph of the production of a number of basic organic chemicals derived from petroleum in the USA from 1950 to 1975. It shows, over a 25-year period, the following growths:

Methanol	6 fold
Styrene	10 fold
Ethylene Glycol	7 fold
Ethanol	3 fold

An area which is particularly relevant to major chemical hazards is that of liquefied petroleum gas (LPG). Table 2.1. gives consumption figures for LPG both for the US and the UK. It shows that consumption has risen fairly rapidly in the UK but that sales have remained fairly constant in the USA during the seventies. Most LPG in the USA is used for home heating and most of it is transported by pipeline.

TABLE 2.1. Consumption of Liquefied Petroleum Gas (LPG) in millions of tonnes per annum.

Country	1925	65	70	71	72	73	74	75	76	77	78	79
U S	.0007	-	29	29	32	32	31	29	30	28	-	-
U K	Nil?	0.23	0.56	0.75	1.1	1.33	1.33	1.2	1.2	1.3	1.25	1.44

For the U S total consumption figures are shown i.e. including consumption by gas works. The U K figures exclude such consumption which ceased, in any case, around 1970. For comparison, gasoline consumption in the U S is ca 350 million tonnes per annum.

The table suggests that the per capita consumption of LPG in the U S is about five times that of the U K

The paragraph demonstrates the pattern of growth of production in a number of typical areas. The pattern in the UK is similar to that of the US but at a lower level. The rate of production in the UK may be assumed, for purposes of rough comparison, to be, for many chemical products, about 10% of the level of production of the United States.

2.3.2. THE GROWTH OF PROCESS UNITS.

It does not seem open to doubt that the growth in the size of process units

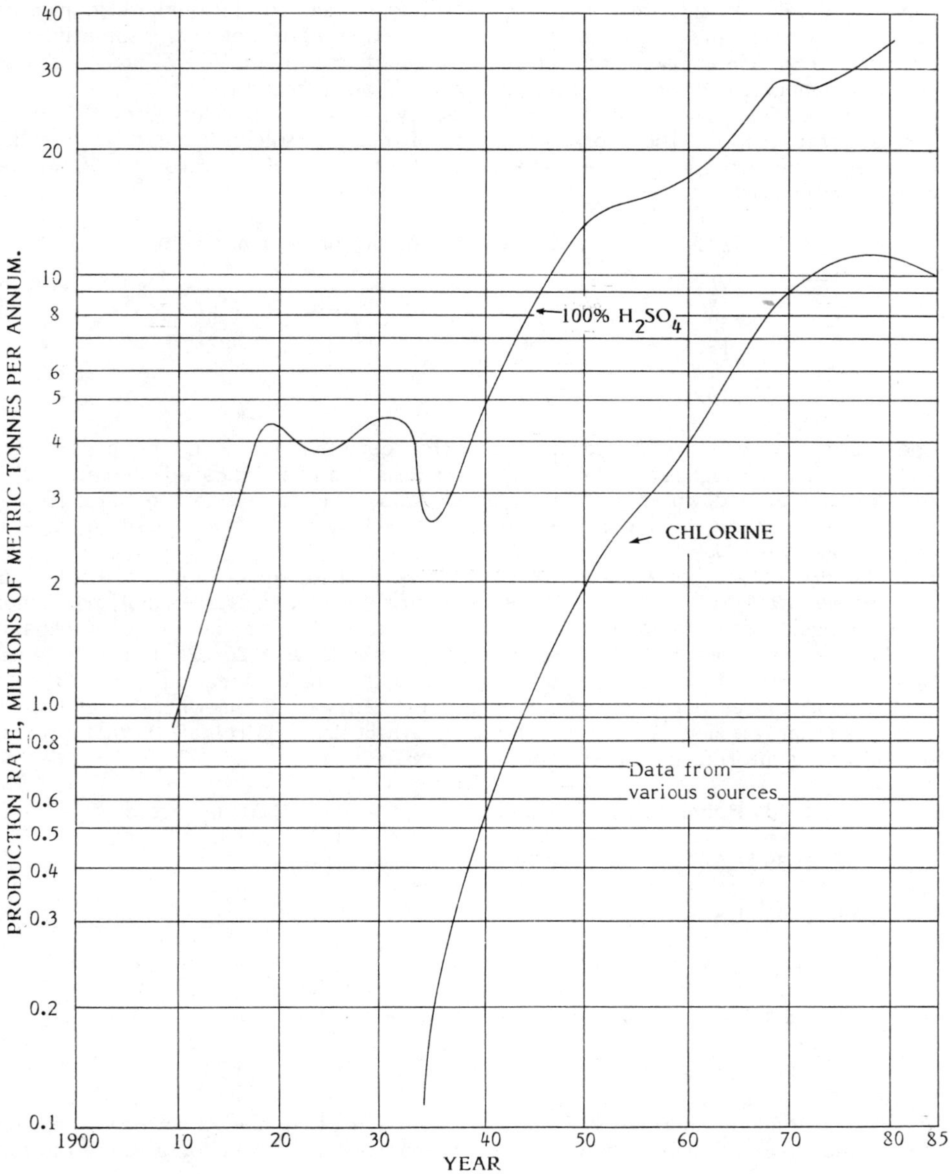

Fig. 2.2. GROWTH IN THE PRODUCTION RATES OF SULPHURIC ACID AND CHLORINE IN THE US.

has matched or more than matched the growth in the output of the various branches of the chemical industry. However, finding data to demonstrate the growth of process unit size in anything other than the most general terms is not easy as this is seldom treated by authors in the field of chemical technology. The present author has been unable to find any book or paper devoted to this specific theme, though reference is sometimes made to the subject in prefaces to books on chemical technology. Information can, however, sometimes be obtained from chemical plant contractors about the plants they have built, but this is not necessarily an adequate reflection of world trends.

Firth (1977) gives detailed information available on the Northbrook Chemical Works, Bradford. The works commenced making sulphuric acid in the late eighteenth century and figures are available from 1850 onwards. Production rates were approximately:

1850	1 1/4	tonnes per day of chamber acid
1871	6	" " " " " "
1908	16	" " " " " "
1926	36	" " " " " "

The works, however, included at the latter period more than one process unit and thus the throughput of each unit was less than is indicated above. When the works closed down in 1972 the business was transferred to the parent company which operates a single stream contact sulphuric acid unit with a capacity of 600 tonnes per day.

The present author worked for Simon Carves Ltd, (now SimChem Ltd) during the period 1954-59. This is a leading firm of contractors supplying, among other things, contact sulphuric acid plants. At that time the range of capacity of their single stream contact sulphuric acid plants was 20-220 tonnes per day. However in recent years the company has built single stream process units of up to 2,000 tonnes per day and larger process units exist elsewhere in the world.

A similar history applies to ammonia synthesis. Strelzoff (1981) provides the following information on stream capacities:

1913	Haber-Bosch, Germany.	20	tonnes per day
1939	USA	18-45	" " "
1939-1945	USA	80	" " "
ca 1960	USA	150-200	" " "
1963	Texas City	600	" " "
1980	Chiba, Japan	1,500	" " "

Imperial Chemical Industries PLC (ICI) have supplied the "knowhow" to build the ammonia plants listed below. The capacities are those of a single stream.

1966	Immingham, UK	570	tonnes per day
1969	Oulu, Finland	750	" " "
1969	Billingham	950	" " "

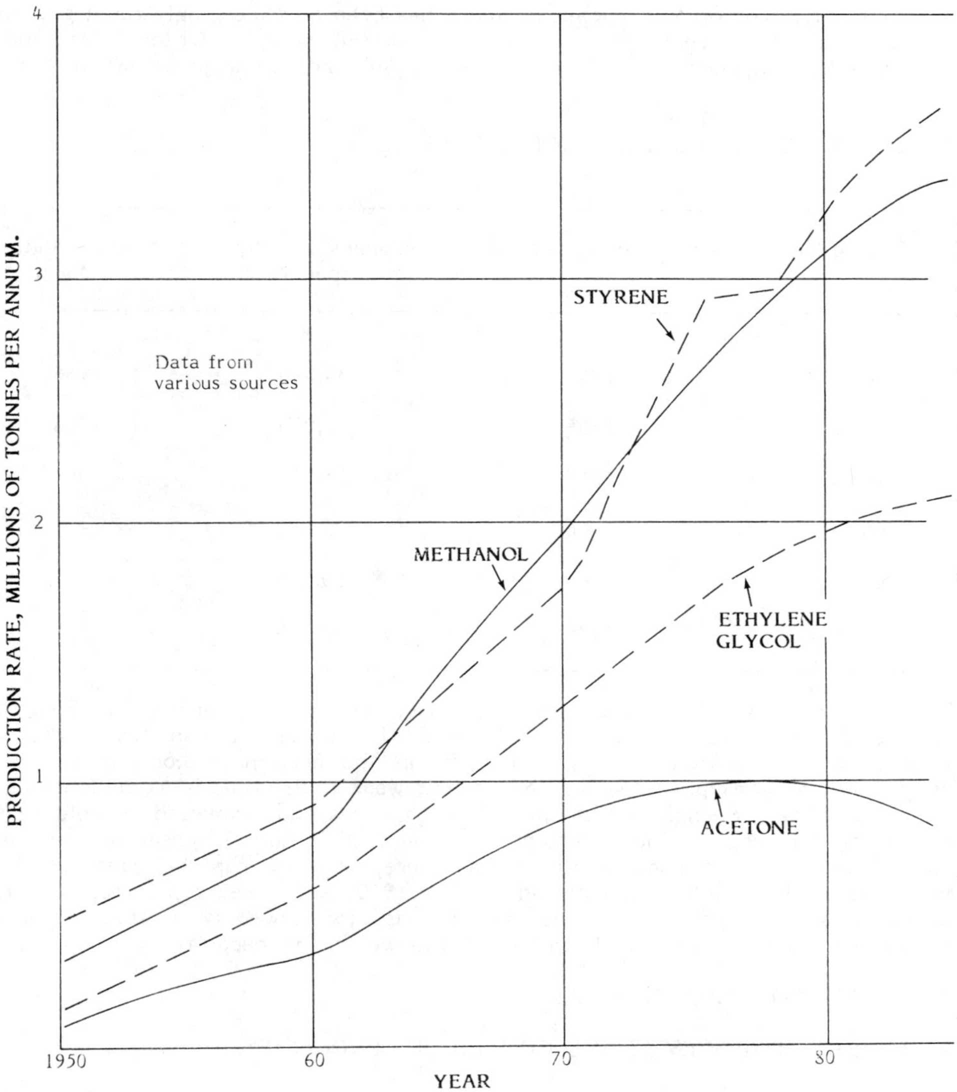

Fig. 2.3. PRODUCTION RATES FOR PETROCHEMICALS US.

1975	Tulicarin, India	1,100 tonnes per day
1977	Shiraz, Iran	1,200 " " "

ICI (1979), from which the above data is drawn, claims that the ICI "lozenge" converter can be designed to give single stream capacities of up to 2,500 tonnes per day.

In the field of organic chemicals there has been an exceedingly rapid growth in the capacity of process units. Data on olefine units at Wilton, Cleveland, UK, have been made available in Mitchell (1980) and are summarised in Table 2.2.

TABLE 2.2. Stream Capacities, ICI Olefine Units.

Olefine Unit	Date Commissioned	Tonnes of Ethylene per Year	Date Shut Down
No. 1	1951	35,000	1966
No. 2	1956	40,000	1966
No. 3	1959	55,000	1969
No. 4	1966	200,000	-
No. 5	1969	450,000	-
No. 6	1979	500,000	-

This table is of special interest as it demonstrates that as units 4 and 5 came on stream they replaced units 1, 2 and 3 which had become too small in capacity to be economically viable. A document in the present author's possession, of unknown authorship, states that No. 5 unit was, at the time of commissioning, the world's largest single stream plant though it was not expected to hold that distinction for long. The document also claims that unit 5 consumed annually 1.5×10^6 tonnes of naphtha which was more than the naphtha consumed by the entire British petrochemical industry in 1960, i.e. 9 years previously. It might be said, in parenthesis, that growth has now slowed down and, in some recent years, these units have been operated at well below capacity.

2.4. THE ECONOMIES OF SCALE.

2.4.1. BATCH PROCESSES AND CONTINUOUS PROCESSES.

No reference has been made in the subsections above to the manner in which the processes discussed are conducted, but they are, in fact, all continuous processes.

The distinction between batch and continuous processes may be summed up as follows. In a batch process, at any point in the process, such properties of the system as the state of matter (solid, liquid or gas), chemical composition, temperature or pressure, are subject to a cyclic variation. In a continuous process, at any point in the process, these properties remain more or less constant. In fact it is the aim of control systems to reduce variations in them to a minimum.

In the field of process technology, for example, fermentation is generally conducted as a batch process whereas the operation of modern steam raising plant is an example of a continuous process in which fuel and water are fed in and combustion products, steam and blowdown are continuously taken out.

The chemical and process plants which are objects of public apprehension are mostly continuous process plants. Today batch-wise processes are mainly to be found in the relatively small scale production of pharmaceuticals, dyestuffs, pesticides, paint, varnish and resins. However where highly toxic substances are processed, batch processes may still pose major toxic hazards. Some varnish and resin manufacture using batch processes is on sufficiently large a scale as to bring it into the major chemical hazards category on account of possible fires or vapour cloud explosions.

2.4.2. WHY IT PAYS TO BE BIG; CAPITAL COSTS.

The basic law which governs the capital cost of chemical and process units and plant is:

$$C_x = C_y \times R^z$$

Where C_x is the cost of a plant or process unit with a capacity of x tonnes in unit time and C_y is the cost of a plant or a process unit of similar pattern with a capacity of y tonnes in unit time, R is the ratio x/y and z is an exponent. Provided that $z \leqslant 1$, and this is generally the case, there are economies in capital cost to be gained from increased scale of the individual units. In practice the value of z lies somewhere between 0.6 and 0.66 for whole chemical plants, but for individual process units the range may be somewhat wider. The values just quoted have led to the rule being known variously as the "six-tenths rule" or the "two-thirds rule".

The factors which give rise to these rules are well known to chemical engineers but are perhaps less well known to those qualified in other disciplines. In view of the importance of this rule it may be useful to discuss the rationale behind it. For example, the first task involved in the design of process units may be viewed, in very broad terms, as seeking to determine two leading dimensions. These are length and path width. Path length is a function of the "duty", examples of duty being the provision of suitable equipment for cooling or heating a fluid over a given temperature range or for the reduction of the concentration of a soluble constituent of a gas by scrubbing it with a solvent. In broad terms this duty will determine the length of a heat exchanger tube or the height of a scrubbing tower and these will be independent of the rate of throughput. The length of path being determined, it will then be found that the path width is a function of the rate of throughput. This path width will determine the number of tubes in a heat exchanger or the cross-sectional area of a scrubbing tower. Thus the difference between two plants differing only in capacity will be characterised by the diameter of the vessels and not by their length or height.

The implications of this are that for the cylindrical portion of the outer shell of a heat exchanger or a scrubbing tower the cost (assuming that the wall thickness does not vary) will be given by:

$$C_x = C_y \times R^{0.5}$$

For the ends of the column, however, the cost will be proportional to the cross-sectional area i.e:

$$C_x = C_y \times R^{1.0}$$

There is also the cost of the tubes or column packing to be considered and these will have a value for z near to 1.0. The cost of fabrication of the process unit will be the sum of a number of components with values of z lying between 0.5 and 1.0. It should be borne in mind that the cost of process units is only one element in the cost of a whole plant. This includes site purchase and preparation, design, procurement and erection, the provision of buildings, roads, pipework services etc. All of these items have their own characteristic value for z which may, in many cases, be near to zero.

For example, given that the larger plant has the same flow sheet as the smaller, design costs will be invariant with respect to scale i.e. z = 0. For many instruments for measuring plant variables and for the provision of a control building the scale of the throughput is of little influence, hence here too z = 0. On the other hand, large electric motors have a value for z which is close to 1. The application of this rule and the differing values of the exponent z for the various items which make up a complex plant are treated in detail in section 25 of Perry (1973).

When the different elements are summed up the value of z for the whole plant will be likely to be between 0.6 and 0.7. To give an example, if the value of z is circa 0.66 then a plant of capacity 2y will cost about 1.60 times the cost of a plant of capacity y and a plant of capacity 4y will cost about 2.50 times the cost of a plant of capacity y. It is thus much cheaper to build one plant with a capacity of 2y than two plants with a capacity of y. This is further illustrated in Figure 2.4. which shows the relative costs for ethylene plants [Kniel et al (1980)].

However, the economies of scale-up are not necessarily optimised by simply reproducting the flow sheet of a small plant in a large plant. The economies of scale are sometimes most fully realised in a larger plant by greater sophistication and this could take the form, for example, of more elaborate automatic control.

2.4.3. WHY IT PAYS TO BE BIG; RUNNING COSTS.

Even greater savings can be made in labour costs by scaling up plant. Process supervision costs are invariant with regard to stream capacity i.e. z = 0. The cost of maintenance labour is likely also to have an exponent close to zero.

Where the operation entails heat losses, as most do, the heat loss per unit of throughput is lessened by scaling up.

2.4.4. EFFECTS OF SCALE-UP ON SAFETY SURVEILLANCE.

As with process supervision in the general sense, safety supervision in the particular sense might also be expected to be more or less invariant with respect to size, i.e. $z \geqslant 0$. Certainly this would be true for Hazop studies (for definition see Appendix I) as the cost of conducting them relates only to the complexity of the flow sheet and not to the throughput of the plant. For such activities as non-destructive testing of pressure vessels the cost may be a function of the surface area of the vessel in which case z = 0.5 to 0.60.

Broadly, therefore, it may be said that the cost of safety surveillance per unit of output to achieve the same standards is largely independent of scale. On the other hand, in a competitive world, the owner of two plants of capacity y will be able to afford less money for safety surveillance than the owner of one 2y plant.

FIGURE 2.4. COST VERSUS CAPACITY (and Feedstock) U S ETHYLENE PLANTS.

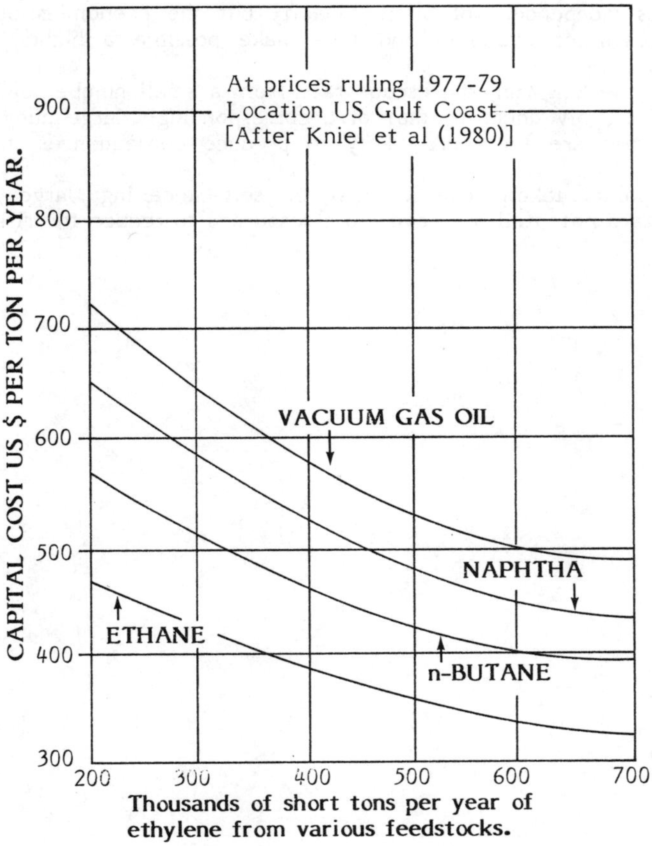

2.4.5. THE EFFECT OF SCALE-UP ON INVENTORY.

In broad terms, inventory, the hold-up of process materials in process units, may be taken as being a linear function of capacity, i.e.:

$$I_x = I_y \times R^{1.0}$$

where Ix and Iy are capacities for throughputs x and y and R is the ratio x/y.

It follows, therefore, that the consequences of loss of containment are going to be roughly double if there is a catastrophic failure of a process unit in a plant where the capacity ratio = 2. On the other hand the probability of failure of a 2y unit is probably no higher than that of a y unit. Over a sufficiently long time the total material loss from one 2y plant is not much greater (if it is greater) than from two y plants. However, as will be discussed later, the public may still rate the consequences of one large escape as more serious than an equivalent number of small ones if only because the one large escape may have consequences beyond the boundary fence which may not occur with the small escapes.

2.4.6. SUMMARY.

There can be no doubt that scaling up produces considerable savings in both capital and running costs. The cost of safety surveillance to a given standard is more or less independent of plant capacity but the economies of scale may increase the resources available and thus make possible a higher standard of safety surveillance.

Though, in the long run, the escape rate from a small number of large plants is likely to be roughly equal to that of a correspondingly large number of small plants, the former are the more likely to produce consequences in the public domain.

The two points taken together may be self-cancelling; large plants have more major hazard potential but have more resources to reduce the risk.

Chapter 3

THE REPORTING OF MAJOR CHEMICAL ACCIDENTS

3.1. SOURCES OF INFORMATION.

3.1.1. THE SCOPE OF REPORTING.

To obtain a comprehensive overview of all major chemical accidents from the early years of the chemical industry to the present for all countries of the world and in full detail is impossible. Judgement, therefore, has to be exercised as to whether the partial lists which have been compiled, which provide sometimes scanty information, represent a valid sample of the whole. The author's view is that, broadly speaking, they represent a sample which is biased somewhat in the direction of making the situation look worse than it is. This is because incidents which led to heavy loss of life or to severe damage are more likely to be reported than accidents which failed to realise their full potential.

Figure 3.1., which is taken from Marshall (1979), is a Venn diagram which attempts to sum up, in a notional form, the extent to which the sum total of major chemical accidents has achieved the record. The main conclusion which the diagram is intended to illustrate is that the greater the size of release the more likely it is that it will, firstly, be locally reported and that, secondly, it will also be internationally reported. It follows that the relatively few major chemical accidents have a much greater probability of having been reported internationally, than the much greater number of minor accidents.

3.1.2. A GENERAL COMPILATION.

It would be impossible for an individual investigator to attempt to gather together from the original sources a compilation of all accidents which are of interest in the field. Indeed, it is unnecessary to do so as it is possible to draw on a number of very useful existing compilations. Of these, the most general in scope is that of Nash (1976). This compilation of world-wide disasters is unique. Griffiths & Fryer (1979) have this to say of it: "This situation (i.e. the earlier situation faced by investigators) has been greatly transformed for the better by the recent publication of the book "Darkest Hours" by J.R. Nash which includes an encyclopaedic listing of occurrences of multiple fatality disasters for both natural and man-made causes". However, they go on to say: "Although this new compilation is a great improvement . . . . it is not by any means totally satisfactory. We find we cannot resolve the problems presented by conflicting accounts of the same incidents, nor can we arrive at any reliable estimate of how near the data base is to being complete; and, of course, this lamentable catalogue of misfortune is added to day by day". To this it may

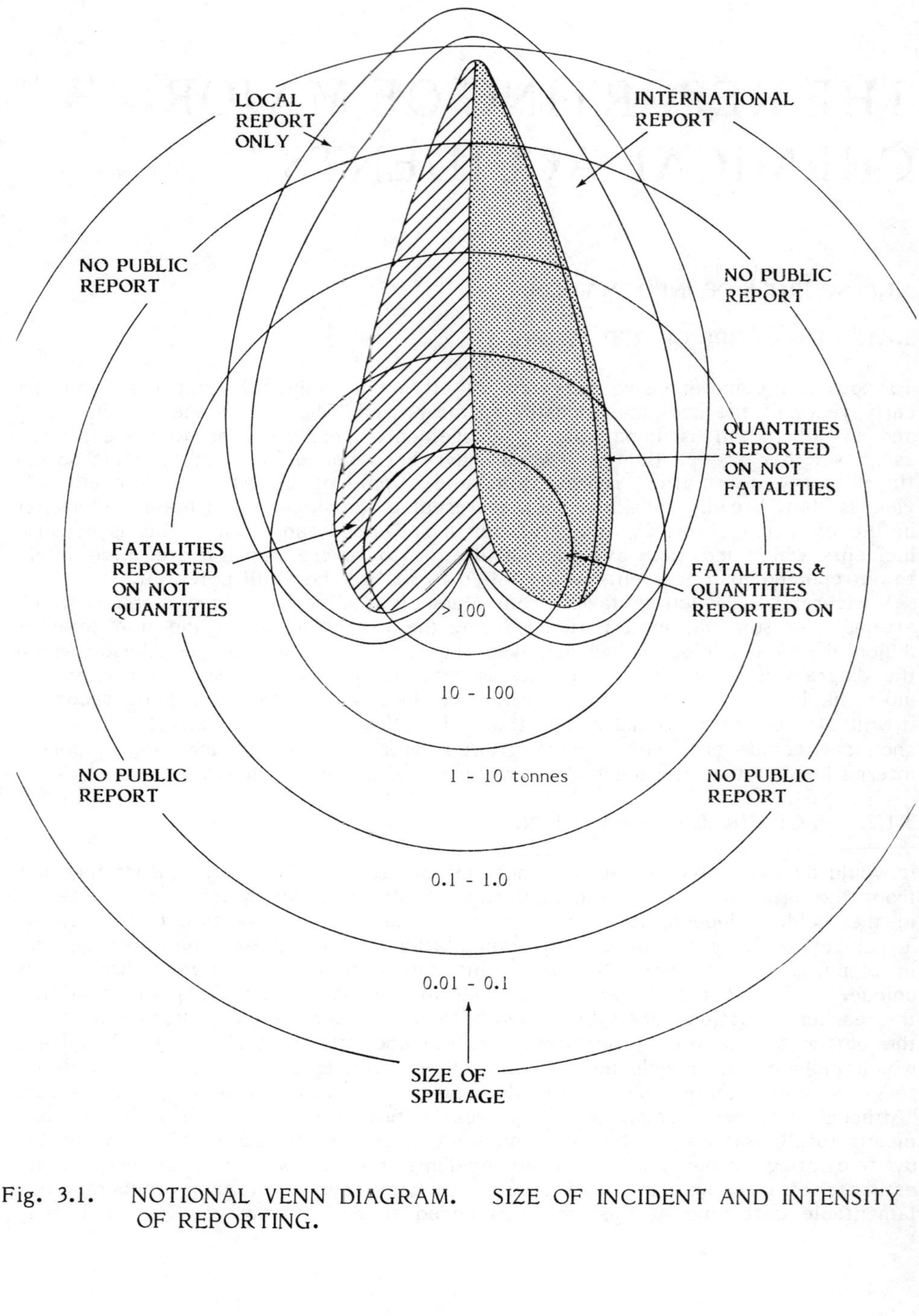

Fig. 3.1. NOTIONAL VENN DIAGRAM. SIZE OF INCIDENT AND INTENSITY OF REPORTING.

be added that Nash claims that he is concerned only with accidents which have killed twenty or more (though he sometimes departs from this principle).

As regards the point made by Griffiths & Fryer about how far the data base is complete, a rough count of Nash's section of "Major Fires and Explosions" shows that he lists approximately 100 examples of explosions of conventional explosives, ammunition or fireworks, whereas Biasutti (1985) lists approximately 900 explosions occurring throughout the world in what he defines as "The Explosives Industry". This definition accounts for why some of Nash's incidents are not listed by Biasutti; this was because they occurred, for example, in ammunition dumps. This consideration makes the discrepancy even wider. However, much depends upon what is meant by 'major'. National Fire Protection Association (1957) claims that between 1900 and 1956, there have been reports of 1,085 dust explosions (excluding mine or military explosions) in the United States and Canada alone. The reference in fact lists 1,120 such occurrences by taking account of a few earlier details of killed, injured and financial losses and gives brief accounts of incidents. In addition there are more detailed accounts of 75 "important" explosions in the reference. Of the 1,000 plus events only 4 led to fatalities of 20 or more and two-thirds entailed no fatalities at all (729 out of 1,120). Of the 4 involving 20 or more fatalities 3 are recorded by Nash though not very accurately. The author could not discover in Nash a record of any other explosion in the United States which is clearly a dust explosion. The compilations of explosions are examined in Chapter 10.

A point which is general to compilations is that they perform a valuable service by giving an investigator a starting point by drawing attention to the date, time and place of incidents and, as a minimum, brief details of the event which took place. A source of general information on major accidents is Whitaker (series). This Almanac, established in 1868, has a section which lists for each year, and briefly, UK major accidents and a few occurring abroad.

3.1.3. SPECIALISED COMPILATIONS.

The specialised compilations of significance to this book are those concerned with cataclysmic fires, explosions and toxic releases and these will be drawn upon below. Suffice it to say that Biasutti (1985) is by far the most comprehensive for the explosion of dense, and National Fire Protection Association (1957) is the best for dust, explosions up till 1956. For cataclysmic fires and unconfined vapour cloud explosions, Gugan (1979) is the most comprehensive compilation published to date. For toxic releases, Advisory Committee on Major Hazards (1979) (which is based upon Marshall (1977)), is the most complete record. However, as Griffiths & Fryer have pointed out: "this lamentable catalogue of misfortune is added to day by day".

3.1.4. THE MASS MEDIA.

That the reporting of disasters by the mass media of press, television and radio, is often inaccurate is a truism. However, the reports of the mass media are valuable as a starting point and often provide the foundation for the compilations discussed above. It is useful, however, to recognise the nature of the bias of the mass media which may be divided into local (i.e. covering mainly one city in the UK) or national (i.e. published daily in London and distributed nationwide). Firstly, the news coverage of an event is likely to be at its most intensive where the media are the media for the locality in which the event occurs. In which case they are likely to feature the event for days, weeks or even months after it has dropped out of the national news. Such local reporting is likely to be strong on "human interest" and weak on technical understanding, as local papers and broadcasting studios are unlikely to be able to draw on the services of a full-time science correspondent. Local broadcasting, however, often calls into the studio someone from a local university or college, but because of lack of first hand contact by the person concerned, or of time to assimilate the facts, this is not always as helpful as it might be.

In general, distance attenuates reporting. This is understandable, as newspapers with pretensions to report both national and international news have much to report and this puts heavy pressure on space. National boundaries, however, override distance as a factor. Most of France, Northern Italy, West Germany and Switzerland, part of Denmark and the whole of Holland are closer to London than the North of Scotland. But no one would pretend that these countries get the news coverage from the UK National Press that Scotland does. But from the point of view of an investigator of MCH the locale is of no significance.

However the further back in time one wishes to go the more newspaper reports become important as the chief source of information. Newspapers have produced special supplements or broadsheets on disasters such as, for example, Sunday Star (1979), which deals with the Mississauga, Toronto, Canada incident of November 1979.

3.1.5. BOOKS ON DISASTERS.

There have been two recent books which deal with disasters. Waltham (1978) is a very detailed and analytical work on disasters involving earth movement. It is, however, concerned mainly with topics which fall outside the theme of this book. Whittow (1980) discusses natural disasters, but includes in his introduction a discussion of the relationship between natural and man-made disasters with estimates of relative frequency and of consequences. There have also been a number of books such as those by Kennett (1975), Canning (1976) and Anon (1975), which report a number of disasters of the various kinds, both natural and man-made, without adding much to what is already known about the few major chemical disasters they deal with.

Few books appear to have been written on specific major chemical accidents. Apart from accounts of the Texas City disaster in April 1947 in which there was a major ammonium nitrate explosion, Nash lists no specific books on major chemical accidents in his bibliography which lists some 150 works on fire, explosions and mine disasters. The Badische Aniline und Soda Fabrik have produced two publications on notable disasters. BASF (1921) is an issue of their house magazine which is wholly devoted to an account of the Oppau explosion of 21/9/21. BASF (1948) is a publication relating to their explosion of 28/7/48. Marshall (1979a) is the present author's book on the Flixborough Disaster.

3.1.6. OFFICIAL SOURCES.

A very significant source of information arises from official reports which are discussed in detail in the next sub-section.

3.2. REPORTS OF PUBLIC INQUIRIES.

3.2.1. INQUESTS.

Any death which is not obviously certifiable as being from natural causes must be reported in England and Wales to the Coroner and, in Scotland, to the Procurator-Fiscal. A Coroner may hold an inquest with or without a jury. Similar procedures apply in Eire, in the USA and in some Commonwealth countries. However, inquests are concerned principally with establishing whether the death falls within certain categories such as from natural causes, accidental death, suicide or murder. Coroners can, and do, make recommendations to prevent accidental deaths. So, in one sense, all deaths due to the realisation of MCH, both in the UK and in certain other countries, are subject to a public inquiry by a Coroner at which lawyers and other representatives such as an official representative of a trade union may ask questions.

However, multiple fatality incidents associated with MCH are likely to be inquired into by way of other forms of public inquiry to be described in paragraph 3.2.2. below. Where such inquiries are set up the Coroner will open the inquest to take evidence of identification and then adjourn it till the findings of the inquiry be made known. The Coroner (or the jury), though not absolutely bound to do so, will almost certainly base the verdict on the findings of such an inquiry. For this reason the proceedings of coroners' inquests contribute little to our knowledge of MCH.

3.2.2. REPORTS OF PUBLIC INQUIRIES.

In the United Kingdom, under certain circumstances, inquiries may be ordered into accidents. Section 14 of the Health and Safety at Work 1974 empowers the HSC, with the consent of the Secretary of State, to hold an inquiry, which may be held in public, into any "accident, occurrence, situation or any other matter . . . for any of the general purposes of this Part (Part 1 of the Act, VM)" and to confer upon a person holding any such inquiry "powers of summoning witnesses to give evidence or to produce documents and power to take evidence on oath . . ." and "cause the report, or so much of it as the Commission thinks fit, to be made public . . .".

The degree of formality of such an inquiry is related to the seriousness of the event. At its most formal, a judge or Queen's Counsel will be appointed as Chairman with assistance from assessors. In such a formal inquiry the Court may be addressed only through barristers. This was the procedure at the Flixborough Inquiry. At a less formal level an inquiry may be conducted by a Factory Inspector or Inspectors. Such an inquiry was conducted into the Appleby-Frodingham Steelworks explosion which was the subject of a report, [Health & Safety Executive (1976)].

In the United States, much valuable information has come to light as the result of inquiries conducted by the National Transportation Safety Board (N.T.S.B). Though the incidents investigated cover all serious accidents involving transportation by road, rail, waterway and pipeline, some of these accidents have involved the release of quantities of flammable or toxic material.

3.2.3. LIMITATIONS OF REPORTS OF PUBLIC INQUIRIES.

In the author's view, the most serious limitation of the reports of public inquiries from the point of view of those investigating MCH is that they tend to devote most of their effort to discovering the causes of disasters and to devote too little effort to discovering the effects of disasters. However, public inquiries are constrained by their terms of reference and by the great expense occasioned by formal inquiries. On the other hand that this is not universally true of official reports is evidenced by Health & Safety Executive (1980) which devotes a great deal of space to the effects, both immediate and long term, of the Seveso disaster. Fortunately in some cases there are subsidiary reports which go a long way to make up for the deficiencies of the major reports. A notable example is Humberside Police (1974) which describes in great detail the aftermath of the Flixborough explosion of 1974 in a way that the Official Report, Flixborough (1975) did not.

3.2.4. REPORTING IN THE FUTURE.

Obviously after a loss of containment of flammable or toxic materials or after an explosion, the first priority is to save life and to control the further spread of fire or toxic materials. It will also be a matter of urgency to recover the bodies of those killed in the incident. However, there are many areas in which it is important that early forensic examination by personnel who can devote their undivided attention to it, can provide information which will shed valuable light on what happens when the potential of a MCH is realised.

To give an obvious example of the data which needs to be collected, it is important to record the positions from which injured persons are recovered and to record the position of bodies before removal. This last may be of considerable significance in identifying victims. It may be noted here that failure to do this created problems in the San Carlos disaster in Spain in 1979 where some 150 bodies were removed to temporary mortuary facilities without taking a record of the position in which they were found. Had this been done, it is claimed, identification would have been easier as the victims in many cases could have been related to their motor cars or other belongings.

Roberts (1980) has discussed the general question and his conclusions are summarised in Table 3.1. Roberts lays stress upon a number of points of which perhaps the most important is the need to avoid excessive commitment to any particular hypothesis. Roberts's paper is, to the best of the present author's knowledge, the only published paper to set out these points.

Two further points may be added to what Roberts has said in his paper:

1. The importance of recording damage to motor cars. These are ubiquitous and may be closer to the scene than houses.
2. The importance, particularly in the case of toxic releases, of recording the position and species of dead animals and birds.

TABLE 3.1. Checklist of Information required from the Investigation of a Major Incident.

General	Short narrative account of the incident. Description of site with plan of plant layout. Description of surrounding district with up-to-date map. Estimated timing of events.
Release Conditions	Size and position of breach. Normal operating conditions in vicinity of breach. Inventory of fluid available for release. Formation of liquid pools by bunds etc.
Dispersion Effects	Wind speed, atmospheric cloud cover, insolation. Size of released cloud as a function of time. Movement of released cloud across terrain.
Fire and Explosion Effects	Extent of flame spread. Thermal radiation effects. Blast effects. Missile projection.
Toxic Effects	Population exposure. Vegetation damage. Absorbed chemicals.

After Roberts (1980).

Chapter 4

THE NATURE OF HAZARDS AND RISKS

4.1. THE PROBLEM OF MEANINGS.

4.1.1. THE NEED FOR DEFINITION.

Until comparatively recently not only was there no generally agreed system of terminology in the field of the analysis of hazard and risk, but it seems true to say that there was no perceived need to have an agreed terminology. A number of specialist words were in use, though two words were used more than others; these were the words **hazard** and **risk.** Some authors used them interchangeably; some still do, this being especially true of the mass media; others have attributed consistent meanings to them but have differed from other authors in the meanings so attributed.

The present author raised the question of the need for an agreed terminology in his paper Marshall (1975). In this he argued that there was a "jargon jungle" and that a failure to agree on meanings was inhibiting clarity of thought on the subject. He commented also that though hazard and risk have had, at least until recently, no clearly accepted meanings, this has not set them apart, so far as everyday speech is concerned, from other words, some of which have clearly defined meanings in science and engineering though not in common speech.

To take an example, the words "stress" and "strain" have clearly defined meanings in science and in engineering. But, in everyday life, they are used interchangeably in such expressions as "the **stress** of business responsibilities" or "the **strain** of business responsibilities. Similar considerations apply to terms such as **energy**, **force** and **momentum** which in common speech or in works of literature do not have precise meanings and are often used interchangeably.

Yet, in science and engineering, precise definitions are quite indispensable. Stress, strain, energy, force and momentum can be expressed as combinations of the physical dimensions of length, time and mass. They can be shown to differ from each other dimensionally and hence to differ qualitatively. As will be demonstrated below, hazard and risk may be defined in such a way as to set them apart and so that they are seen to be qualitatively different.

4.1.2. HISTORICAL DEVELOPMENT OF MEANING.

An examination of the historical development of the meanings of hazard and risk shows that they are both now pejorative words. Yet hazard, for example, originally meant little more than one event among a number and seems to have originated in the name "Castle Hasart" in Syria where the Crusaders learned to play dice. This original meaning still survives in the rules of the game of billiards which still speak of "winning hazards" and "losing hazards".

4.1.3. THE TERMINOLOGY OF INSURANCE COMPANIES.

The word "risk" is extensively used by insurance companies. These companies exist because organisations or individuals fear that lives or property may be lost, and hence that financial damage may ensue, as the result of such occurrences as fire, flood or theft.

Insurance companies estimate the probability that such events may occur and this is done very largely on the basis of past experience. By multiplying together the financial consequences of fire or flood or theft and the frequency of the occurrence of such events, they can estimate how much they are likely to pay out in any year. This sum, multiplied by a factor to take account of the cost of administration and the insurance company's profit, then determines the premiums charged. There are three factors:

1. Physical situations (expressed as a potential for damage translated into financial terms).
2. A probability of realisation.
3. The product of (2) and (3). From these the premiums are calculated.

If insurance companies had had a consistent terminology for these three factors it might have formed a basis for the terminology of the study of hazard and risk. But have insurance companies such a consistent terminology?

It is instructive in this regard to examine the article "Insurance" in Chambers Encyclopaedia [Chambers (1973)]. This article is built up of sections written in the main by four authors, all very eminent practitioners in the field. It is reasonable to expect, therefore, that the terminology used in the article will reflect current practice in insurance companies. What the article reveals is that "risk" as used in insurance is an omnibus word which can mean any of the three concepts discussed above and that it has other meanings besides.

The impression that in the insurance business the word "risk" means a number of quite different things is reinforced by reading the "Handbook of Risk Management" [Kluwer-Harrap (1974)]. This is another book written by people of high standing in the insurance world. In this book the authors admit that risk can mean at least six different things.

> ". . . . the first task set for this handbook is to arrive at a clear definition of what is meant by risk management. This in turn implies that we describe what is meant by "risk" . . . . "
>
> "The word risk in insurance terminology is used to refer to the thing which is insured (e.g. the plant or the company owning it, VM), the insured peril (e.g. an explosion, VM), the expected claims cost for a given policy (the peril quantified in money terms, VM), or as a general term for unwanted or uncertain future events. To compound these problems, economists and statisticians use the word risk to measure the range of possible outcomes which may occur at some future time . . . suppose we refer to the "degree of risk" . . . one might be referring to the expected frequency or size of a loss or some combinations of the two . . . All these interpretations of the word "risk" have their place in risk management but in context the meaning of the word should be clear".

The conclusion must be drawn that there is nothing to be gained in precision of understanding from adopting the terminology of the insurance companies. In fact the contrary is true.

4.2. AN ILLUSTRATIVE EXAMPLE.

4.2.1. THE DIFFICULTIES.

It is best to recognise, at the outset, that the subject of chemical and process plant safety is complex both qualitatively and quantitatively. The appreciation that this means that any figures which are produced are subject to error, and that the error may be as much as one or more orders of magnitude, is fundamental to the understanding of the subject.

It may be useful in analysing the problem to take a general example remote from the chemical and process industries, indeed to take one where the problems are created by nature and not by man. The example chosen is that of a mountain. It will be shown later that the problems of industry correspond to the problems produced by nature.

4.2.2. AN EXAMPLE DESCRIBED.

There are a number of events which arise around a mountain which may kill or injure humans or animals or which may cause damage to property. These include avalanches, rock falls or the bursting of the banks of mountain lakes. Such occurrences are caused by changes in the environment of the mountain as, for example, by the action of such agencies as rain, snow, ice, lightning or wind, or the action of vegetation. Or they may arise on account of slow chemical and physical changes in the rocks of which the mountain is composed or by changes in the earth's crust underlying the mountain. Thus the events may be assumed to have causes and not to occur for reasons which are completely inexplicable.

Study of historical records may lead to the possibility of attributing to each class of events an index of frequency. However, within each class there may be a considerable range of severity and it may be necessary to categorise events as, for example, severe, moderate or slight and to examine the frequency of each of them. What this is doing is to characterise each class or sub-class with an index, the dimensions of which are reciprocal time. This may be loosely referred to as the measure of the "probability" of the event occurring.

However, it should be noted that this index is not a probability in the strict, mathematical, sense. Probability in this sense is the ratio between the numerical value of a sub-set and the numerical value of a set of which the sub-set forms part. It is thus a pure number which lies between zero and one. To take an example, in a single throw of a dice the probability of throwing a 7 is zero, of throwing either 1 or 2 is one third, and that of throwing one of the numbers 1, 2, 3, 4, 5 or 6 is one.

Therefore, in the case being studied, if what is being expressed is the likelihood that if A should occur, B will follow, then this may be expressed as a probability as it will have a value between zero and one.

In the case of the mountain under discussion, an analysis of the records over 100 years may disclose results as in Table 4.1.

TABLE 4.1. Frequency of events.

Event	No. of occurrences.	Frequency (events per annum)
Severe rock falls	23	23×10^{-2}
Moderate rock falls	201	201×10^{-2}
Slight rock falls	930	930×10^{-2}

If the records are then searched for the extent to which human injuries are sustained, a new table may be drawn up which discloses the frequency of injury consequent upon falls of differing severity.

TABLE 4.2.

Event	A No. of occurrences	B No. of falls causing injury	A/B Probability that A will give rise to B.	Frequency of A p.a.	Frequency of B p.a.
Severe rock falls	23	15	0.65	0.23	0.15
Moderate rock falls	201	20	0.10	2.01	0.20
Slight rock falls	930	2	.002	9.30	0.19

There are some problems to be noted immediately. The time span of 100 years may include some years in which the records were less well kept than in recent years and so the records, especially of the smaller falls, may be incomplete. Secondly, the number of people liable to be injured may have risen in recent years because of a growth in tourism. To reduce the time span may give a more accurate picture of present-day conditions but it may be statistically less satisfactory. One method of refining the statistics would be to study not just the one mountain but to combine together the statistics of all the mountains in the range to produce a mean figure from a much larger sample. This would still be subject to the uncertainty principle described above but the larger population of mountains may make possible a shorter time span which may more clearly represent present-day conditions.

4.2.3. PREDICTING THE FUTURE.

Statistics gathered in the way described above may enable the authorities to make predictions as to the future so that, for example, they may be able to determine how many people they will need for mountain rescue teams.

However, such predictions would be based upon the assumption that the future will follow the pattern of the past. The assumptions may have to be modified if there have been obvious changes in circumstances.

4.3. IMPLICATIONS OF THE EXAMPLE.

4.3.1. NOMENCLATURE.

The words "hazard" and "risk" do not appear in the description of the example given above. This was deliberately done as such use might have pre-empted discussion of their applicability. If current trends in attributing meanings be followed, these words may be applied appropriately in the example above.

The word **hazard** could be applied to the potential for avalanches, rock falls and floods. Thus hazard may be regarded as being continually present.

On the other hand, for example, the event of a rock fall is something which occurs only at intervals and may be described as the **realisation** of the hazard. Such realisations occur because of changes in that part of the mountain due, for example, to the action of the elements or because of internal geological changes as have been described above and they may also range in severity from major events to minor events. A realisation may thus fall short or very short of the full potential for a rock fall.

Risk may then be the measure of the frequency of an event. It may be the product of a frequency and a probability as for example the frequency of rock falls which kill people. Here there are two events, the event of the rock fall, and the event of people being present. The final measure, the risk, is a product of the frequency of rock falls (reciprocal time) and of the probability of people being present (a number between 0 and 1) also giving a frequency.

There remains the problem of expressing some measure of the product of the outcome of rock falls (for example number of people killed) and the frequency with which such events occur. It is this which has given rise, and still does give rise, to controversy. This is because some wish to use the word "risk" to describe this as well. But to do this leads to confusion as one word is used for two quite different meanings.

The present author sought to break this deadlock in his paper Marshall (1981) by proposing that "risk" used in the latter sense always be qualified by an adjective such as "individual" or "societal". This proposal has since met with a degree of acceptance.

4.3.2. INDIVIDUAL RISK.

To apply this concept to the mountain, this would be done by examining the number of people who had been killed by rock falls over a given period. Such a figure must then be related to the number of people living in locations where there is a likelihood of the population suffering injury. Such people may be described as being "at risk". The level of risk to be attributed to a given person will depend on a number of factors relating to place or time. The great majority of people change their position with time and are only in their homes for a portion of any day. They leave their homes for part of the day for work, recreation or shopping etc and may be away for days, weeks or months on holiday or for work. There needs thus to be an occupancy factor, a number which lies between zero and one, and which represents the probability of the individual being present in a given location in the event of the realisation of a hazard.

It is also necessary to take account of location. One house may lie in the path of a rock fall whereas another, a few hundred metres away, may be protected by an outcrop of rock. Determining the boundary between being at risk and not being at risk may not be easy. Then again the occupancy factor discussed above is not simply a matter of being at home or not. A shopping trip may take the person into a region of the town with a higher probability of injury or alternatively to a lesser one.

A further factor is that concerned with the severity of injury. This, for purposes of the immediate discussion, may be classified as fatal, disabling, severe but non-disabling, moderate or slight. The risk of suffering any form of injury, i.e. any of the above categories, must always be greater than that of sustaining any particular category, as such a category is a sub-set of the whole category of injury. The ratio between the numbers of each category may, in principle, be determined from records.

To summarise, the attribution of levels of individual risk must take account of the nature of the injury, the proportion of time spent at risk and the location of the individual who is at risk. To give an example, an individual A lives in a small village of 300 inhabitants. The records show that 10 people living in the village have been killed by rock falls and 200 injured in some degree over 50 years and that the population has been fairly constant. A is absent from the village for 40 hours a week to work in a nearby town and is on holiday for 4 weeks in a year. A is away on business trips for 2 weeks each year but spends the rest of the time in the village. The individual risk of fatal injury to A from a rock fall would be given as follows:-

$$\frac{10 \times 46 \times 128}{50 \times 300 \times 52 \times 168} = 4.5 \times 10^{-4}$$

The individual risk to A of sustaining any level of injury from a rock fall would be given as follows:-

$$\frac{(10 + 200) \times 46 \times 128}{300 \times 50 \times 52 \times 168} = 9.5 \times 10^{-3}$$

Such a calculation is valid for the individual A, but individual risk would be much more usefully expressed as an average for the population as a whole by taking a mean figure of occupancy for a typical inhabitant.

It may be noted that one assumption involved above is that the locality is so small that the individual risk is the same for each inhabitant with the same occupation factor. Should this not be true, the area might have to be sub-divided into zones of differing levels of risk.

A further factor to be noted is that the individual risk quoted is for rock falls only. The inhabitants of the village may also be subject to individual risks arising from avalanches or floods. These are capable of addition to give a combined individual risk from all causes connected with the nearby mountain.

4.3.3. SOCIETAL RISK.

Values for individual risk convey no information on the potential for disaster from rock falls. The 10 deaths quoted above could have occurred in 10 rock falls each of which killed 1 person, or they could have occurred in one rock fall which killed 10 persons. For reasons which need not be discussed at this stage it is desirable that, in addition to the calculation of individual risk, there should be an analysis of this aspect, for which the term "societal risk" has gained favour. The hypothetical case given above, in which 10 persons were killed over 50 years, may be further analysed using the data given in the table.

TABLE 4.3. Rock Falls with Associated Fatalities.

Date of fall.	1/8/34	3/10/41	4/12/52	3/1/63	1/2/71
No. killed (N)	2	1	1	4	2

The figures may be further tabulated and analysed.

TABLE 4.4. Frequency and Cumulative Frequency.

N	No. of incidents involving N fatalities	Frequency of incidents involving N fatalities p.a.	No. of incidents involving N or more fatalities	Frequency of incidents involving N or more fatalities
1	2	.04	10	0.20
2	2	.04	8	0.16
3	0	Nil	4	0.08
4	1	.02	4	0.08
5	0	Nil	Nil	Nil

A weakness of using col.2 as a predictor is that it yields zero frequency for falls which kill 3 people. It would be absurd to suggest that the future frequency of falls killing 3 people is zero and therefore col.4, which suggests a frequency .08 for 3 or more fatalities, is a more realistic forecast.

The data may be plotted in the form of a step diagram as below.

FIGURE 4.1. FREQUENCY OF N OR MORE FATALITIES FROM ROCK FALLS.

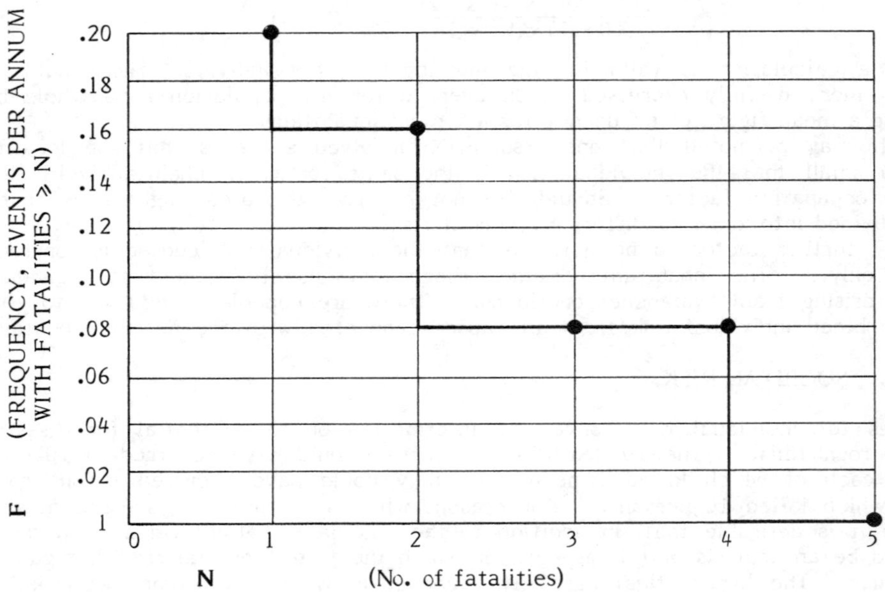

Such a diagram, when there are extensive data with a wide range of values, may be more conveniently plotted on logarithmic paper. It may then be approximated to a curve.

Such plots are known as F v N or F/N diagrams though strictly speaking, in their usual form of a plot of cumulative frequency, they are F v≥N diagrams. The expression F/N diagram when used below will imply, unless otherwise stated, a plot of cumulative frequency. An example of the use of an F/N diagram occurs in Rasmussen (1975) where it was used to compare the disaster potential of nuclear power stations with other hazards such as aircraft accidents.

Unlike individual risk, societal risk is little concerned with precise geographical location. It is a summation of the consequences of the realisations of a particular hazard, though clearly such realisations are likely to be limited to a restricted radius of action. The figure for ≥1 is the same figure for the starting point of the calculation of individual risk. This latter figure is an integral of frequency and time but tells us nothing about the shape of the function which is being integrated.

4.3.4. FURTHER USES FOR F/N DIAGRAMS.

F/N diagrams may be used to display the frequency of realisations of hazard according to magnitude. For example the severity of rock falls may perhaps be expressed in terms of the mass of rock which is estimated to be involved in a realisation. This could generate a diagram of the form of that shown in Fig. 4.2.

FIGURE 4.2. F/N CURVES FOR MAGNITUDE OF ROCK FALLS.

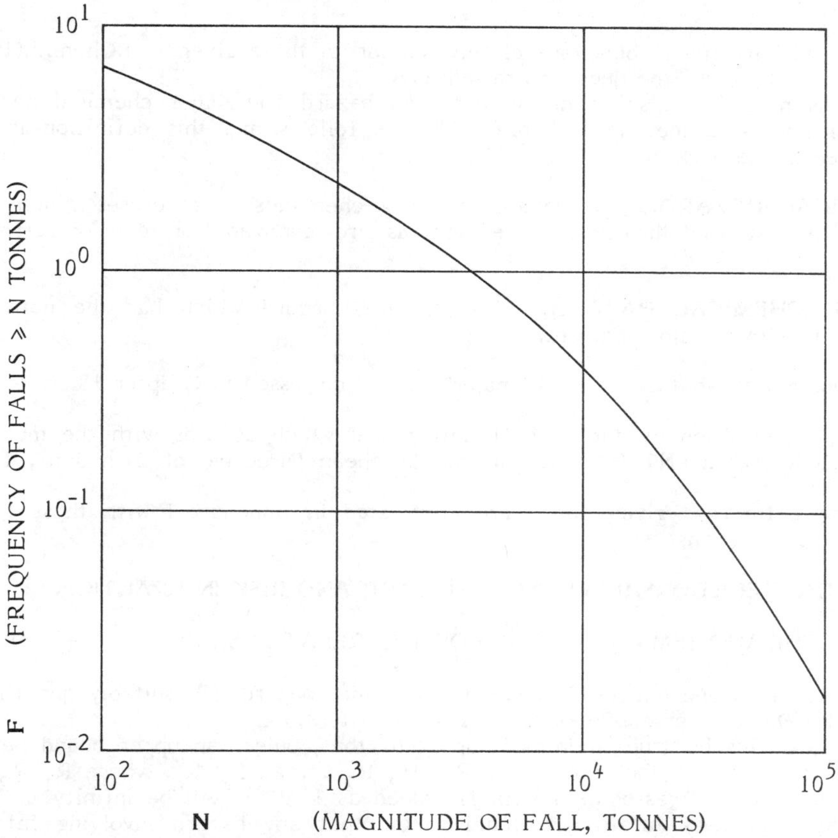

4.4. DEFINITIONS USED IN BOOK.

Taking the sense of the usages in the illustrative example some basic terms may be defined as follows:-

HAZARD. A physical situation with a potential for human injury, damage to property, damage to the environment or some combination of these.

RISK. The likelihood of the event of the realisation of a hazard. It may be expressed as frequency with dimensions of reciprocal time or as a probability that event B will succeed event A. In this latter case, it is a number between zero and one. The product of risk as frequency and risk as probability is a frequency.

INDIVIDUAL RISK. The frequency at which a given individual may be expected to sustain a given level of harm from the realisation of specified hazards.

SOCIETAL RISK. The relationship between frequency and the number of people sustaining a given level of harm from the realisation of specified hazards.

AT RISK. Of an individual, or social group, exposed to a significant level of individual or societal risk in the event of the realisation of a specified hazard or hazards.

The definitions given above are closely similar to those given in I.Chem.E.(1985) which also discusses the questions raised above.

It is necessary also to define chemical hazard and major chemical hazard. The former is defined in I.Chem.E.(1985) as follows and this definition is the one used in the book.

CHEMICAL HAZARD. A hazard involving chemicals or processes which may realise its potential through agencies such as fire, explosion, or toxic or corrosive effects.

MAJOR CHEMICAL HAZARD. A chemical hazard which has the potential to give rise to a major accident.

The question of what constitutes "major" will be discussed in Chapter 18.

This is a definition by the present author and which accords with the meaning of **MAJOR ACCIDENT HAZARD** in the European Directive of 24th June, 1982.

The definitions given above are contrasted in Appendix I with those given in other publications.

4.5. THE RELATIONSHIP BETWEEN HAZARD AND RISK IN PRACTICE.

4.5.1. THE MATHEMATICAL FORM OF THE RELATIONSHIP.

A simple view, and one which some take, is that hazard (H), suitably quantified, and risk (R) are inversely related, i.e. HR = constant.

But, if this be true, then it is apparent that, unless an upper bound be set to the value of H_1, the sum of $H_1 R_1$, $H_2 R_2$ . . . . $H_n R_n$, where R_1 R_2 R_n is the level of risk associated with H valued as 1, 2, n, will be infinity. Thus if H be valued as fatality it would follow that any hazard involving fatality would kill the entire population of the globe in an instant. It follows, therefore, that either H has an upper bound or else that the function H_n R_2 has a finite sum and thus cannot be inverse, or some combination of the two.

The plotting of actual data on H and R and consideration of real circumstances suggests that both of these are true. The notion that a chemical explosion in Sunderland may kill someone in Southampton is in total conflict with common sense. Equally, actual plots suggest that the slope of the line which connects H and R for high levels of H is steeper than would be expected from the relation HR = constant.

4.5.2. HAZARD AND RISK FROM FIRES.

Figure 4.3. demonstrates the relationship between H and R for fires in the United Kingdom in the form of an F/N diagram where F is plotted against N (note that this is not the same as plotting F against $\geqslant$N). The source is UK Fire Statistics (series) and the data from which the diagram is drawn are set up in Table 4.5.

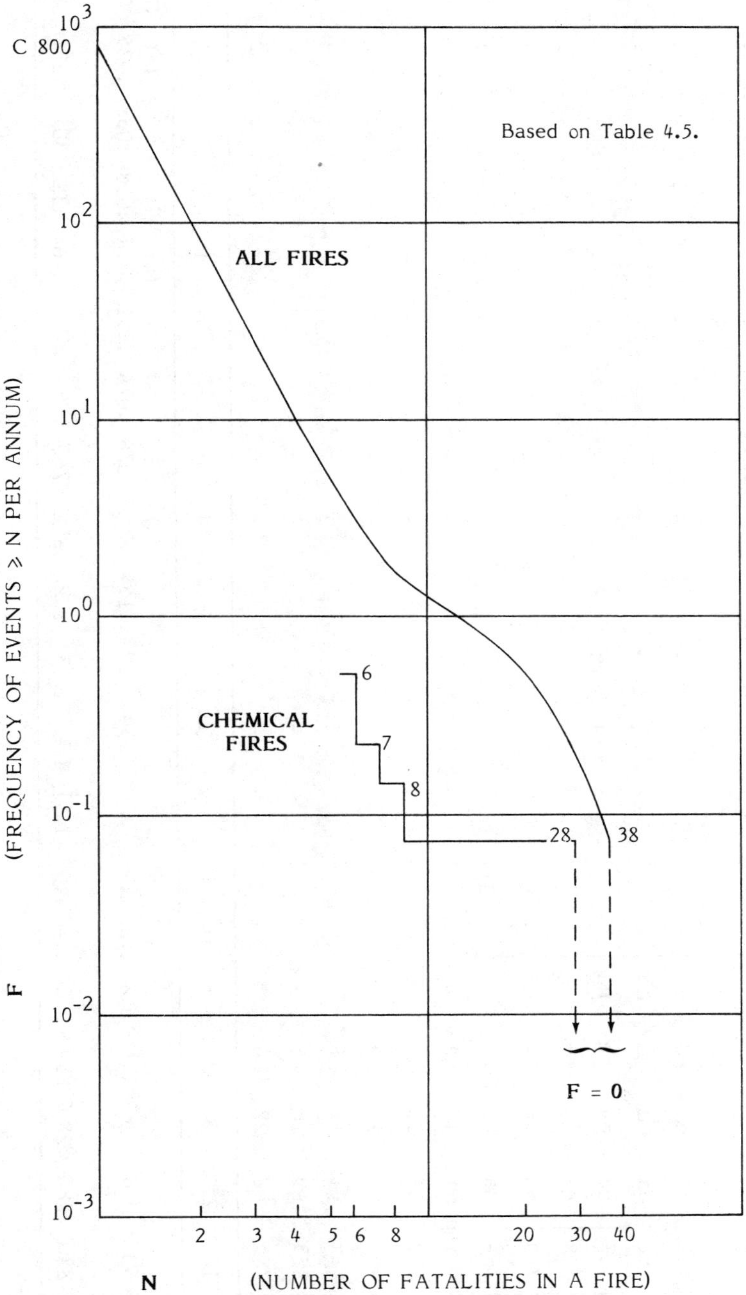

Fig.4.3. F/N PLOT OF UK FIRE STATISTICS 1968-80.

TABLE 4.5. NUMBER OF FIRES IN WHICH A GIVEN NUMBER OF FATALITIES OCCURRED IN UNITED KINGDOM 1968 - 1980.

Year	Total Fires with Fatal-ities.	Number of fatalities per fire																				
		1	2	3	4	5	6	7	8	9	10	11	12	14	15	18	21	22	24	28	30	37
1968	670	na	na	na	na	na	1	1	-	-	-	-	-	-	-	-	-	1	1	-	-	-
1969	716	na	na	na	na	na	3	-	-	-	-	1	-	-	-	-	-	-	-	-	-	-
1970	707	na	na	na	na	na	2	1	-	-	-	-	-	-	-	-	-	-	-	-	-	-
1971	666	567	na	na	na	na	2	1	-	1	-	-	-	1	1	-	1	-	-	-	-	-
1972	911	798	na	na	na	na	-	1	-	-	-	-	-	-	-	-	-	-	-	-	1	-
1973	856	750	na	na	na	na	2	1	-	-	1	-	-	-	-	1	-	-	-	-	-	-
1974	756	764	na	na	na	na	1	1	1	-	-	-	-	-	-	1	-	-	-	1	-	-
1975	na	708	na	na	na	na	2	1	-	-	-	-	-	-	-	-	-	-	-	-	-	-
1976	759	682	36	28	11	1	-	-	1	-	-	-	-	-	-	-	-	-	-	-	-	-
1977	609	544	38	14	4	5	1	2	-	-	-	1	-	-	-	-	-	-	-	-	-	-
1978	709	629	61	11	6	-	-	-	-	1	-	-	1	-	-	-	-	-	-	-	-	-
1979	963	873	69	10	8	-	1	1	-	-	1	-	-	-	-	-	-	-	-	-	-	-
1980	863	785	46	22	4	3	1	-	-	-	1	-	-	-	-	-	-	-	-	-	-	1
Totals		7101	250	85	33	9	16	10	2	2	3	2	1	1	1	2	1	1	1	1	1	1
Frequency, Events per Year		710	50	17	6.6	1.8	1.23	.76	.154	.154	.23	.154	(4) .038	.076	.076	(4) .051	(4) .038	(4) .038	(4) .038	(4) .025	(4) .015	(4) .025
Frequency ⩾n fatalities.		788.6	78.6	28.6	11.66	5.06	3.46	2.39	1.63	1.48	1.23	1.00	.846	.760	.690	.615	.46	.38	.301	.231	.154	.076

Notes (1) Spaces marked "na" denote that statistics are not available. Columns for 1-5 deaths show frequencies derived from those years in which statistics are available.
(2) Statistics are incomplete for some years because of industrial action.
(3) Mean values are based on number of years for which statistics are available.
(4) Mean value based upon mean of a group of values of n.

It will be noted from this table that in the 13 years it covers there had been at least one fire corresponding to each discrete value of n up to n = 12. No fire in the period killed 13, 16, 17, 19 or 20 persons and above 20 gaps are even more frequent. There are no fires in which 38 or more fatalities occurred. To produce meaningful values at the upper end of the table for a value of n in which fatalities occurred and where the frequency associated with, say, n-1 or n+1 was zero, values of n have been grouped together and the frequency calculated has been that associated with n, divided by the number of integers in the group. Thus if there is a gap of one integer next to n the frequency associated with n is divided by 2. If there is a gap on either side it is divided by 3.

The plot is a step diagram not a curve but a curve can be fitted, especially for the higher values of n, from which a modest degree of extrapolation may be obtained.

It might be commented also that as n grows larger the degree of discrimination between n and n+1 becomes finer and finer. When n=1, n+1 is twice as large as n, but when n=30, n+1 is only about 3% larger than n. This is another reason why nil values for frequency occur more and more often as n gets larger and larger.

Figure 4.3. also plots F v ≥N from the same data. This latter plot thus displays the societal risk from fire for the United Kingdom.

Figure 4.3. also plots the societal risk from chemical fires using statistics from Chapter 8.

4.5.3. THE WORK OF RASMUSSEN.

The relationship between hazard and risk was explored in Rasmussen (1975) in an exercise which was intended to compare the societal risks arising from US nuclear power with other societal risks such as those from air crashes or chlorine leaks. Rasmussen's plot is shown in Fig. 4.4. in which F (frequency) is plotted against ≥n fatalities.

The present author has superimposed the F v ≥n values for UK fires drawn from Fig. 4.3. [When compared with US fire statistics as revealed by Fig. 4.4. it would seem, noting that the population of the UK is only one fifth of that of the US, that either levels of fire safety are much better in the US or that Rasmussen's statistics on fire fatalities are in error and are too low].

4.6. A TAXONOMY OF MAJOR CHEMICAL HAZARDS.

4.6.1. THE NATURE OF TAXONOMIES.

The work "taxonomy", which is derived from the Greek "taxis" (= "arrangement"), is defined by the Shorter Oxford English Dictionary as: 'Classification, especially in relation to its general laws or principles; that department of science, or of a particular science or subject, which consists in, or relates to, classification'.

Taxonomies have played a very important role in science. This is especially true of the life sciences which could not have developed without the development of a classification of living beings. In chemistry the periodic table has played an indispensable role in unifying the subject. The history of science shows that the construction of taxonomies can provide new insights, for example, the construction of the periodic table laid the foundations for a new approach to chemistry.

4.6.2. A TAXONOMY OF HAZARDS.

The author believes that the discipline of constructing a taxonomy of hazards

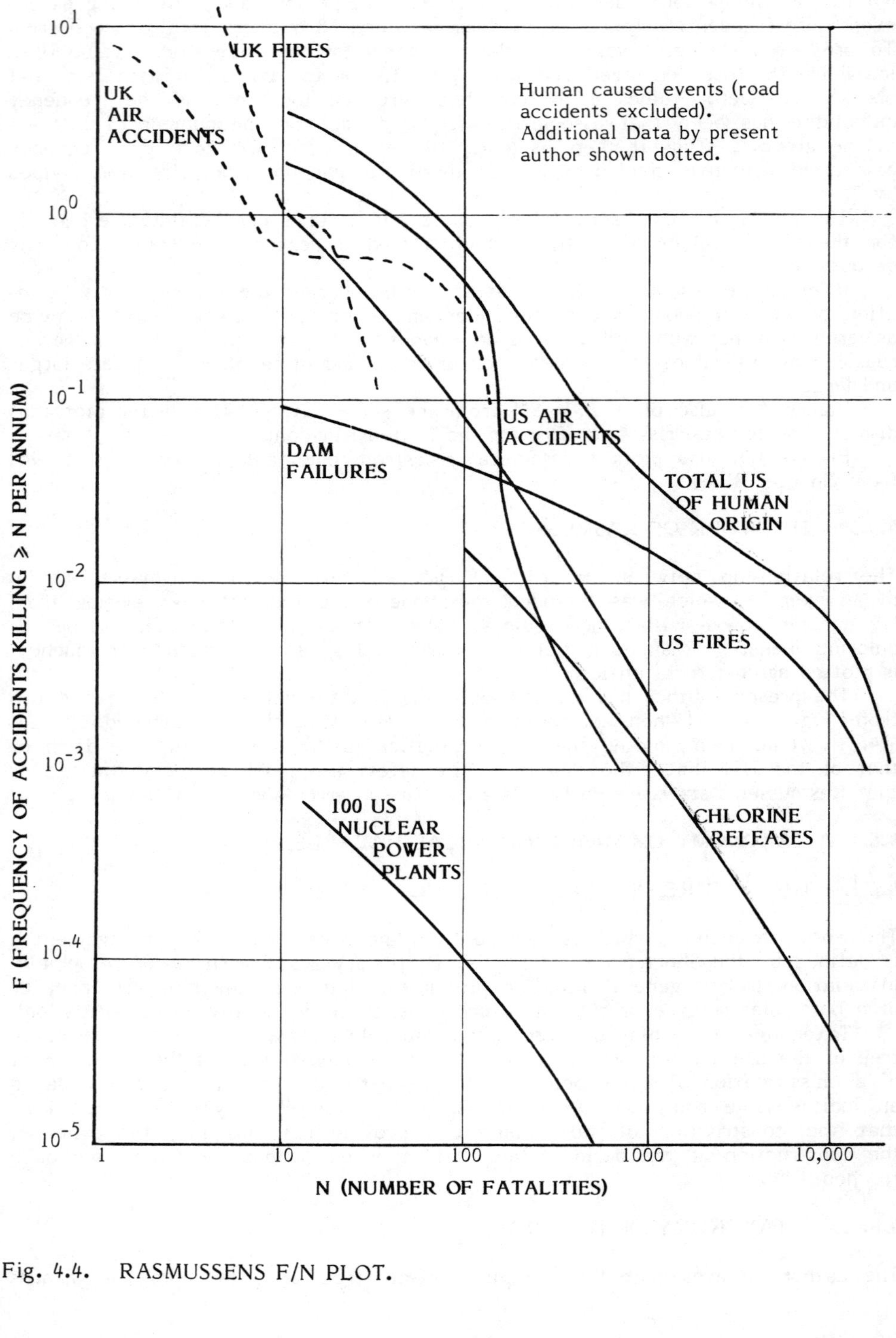

Fig. 4.4. RASMUSSENS F/N PLOT.

may also lead to new insights into their characterisation, quantification and control. He believes that also one of the most serious problems which has hampered rational discussion of the problems of hazards has arisen through the non-analytical approach of many of the protagonists in the field. Attention has already been called to the difficulties raised by use of "risk" as an omnibus word and it may be argued that harm can equally arise through a failure to recognise that hazards can exist in many forms and manifest their potential in many different ways. The author believes that it is important therefore to put major chemical hazards into perspective as one group of hazards among many such groups. Though hazards have been analysed in a number of works such as those of Lowrance (1976), Rowe (1977), Council for Science & Society (1977), Royal Society (1981) and Griffiths (1981), none of these attempted any form of systematic classification of hazards. Generally speaking, the technique adopted by the above was to study exemplars but without attempting to fit the exemplars into a structured analysis of the nature of hazards.

That the author has not encountered a previous attempt to formulate a taxonomy does not, of course, prove that none has been made. However it may be said that the taxonomy below owes nothing to any work previously directed towards this end.

The discussion will be confined to discussion of hazards to humans and will not include hazards to the environment.

4.6.3. THE TAXONOMY - A MULTI-DIMENSIONAL PROBLEM.

Even a very cursory overview of hazards discloses that it is impossible to place hazards along a single continuum as can be done, for example, in classifying electromagnetic radiation. It soon emerges there are many ways of looking at the problem and that there are numerous continua along which hazards can be placed. Some hazards can, if realised, produce high temperatures, others produce high pressures or high concentrations of harmful substances. Some are capable of killing only single individuals, others of killing hundreds of people. Some hazards kill in seconds, others need years to kill exposed individuals.

Some hazards arise from nature, others from the activities of man. To make matters even more complex, it is possible that any given hazard may appear in more than one category. The discussion cannot encompass all known hazards but it is intended that it shall call attention to the principal types of hazards which are significant for the subject discussed in the book.

4.6.4. CLASSIFICATION BY EFFECT ON THE HUMAN BODY.

4.6.4.1. The general effect of environmental change.

Any marked departure from the set of conditions or norms to which man has become conditioned by a long process of biological evolution will lead to bodily injury or to ill health. The chief environmental factors which are of possible significance to the subject under discussion are:-

(a) Temperature.
(b) Ambient pressure.
(c) Localised pressure on the body.
(d) Oxygen concentration.
(e) Concentrations of toxic or corrosive substances.
(f) Concentration of pathogenic organisms.
(g) Level of non-ionising radiation.
(h) Level of ionising radiation.
(i) Level of electric potential.

(j) Level of sound.

(a) Effects which raise or lower the temperature of the human body, either externally or internally, may lead to injury or death. Such effects can include radiation, convection or conduction to or from the skin, the breathing in of excessively cold or hot air or the ingestion of excessively cold or hot liquids or solids.
(b) Sudden changes in ambient pressure which arise from blast waves may lead to injury or death. The effects of blast waves will be discussed at the appropriate point in the book.
(c) Excessive localised pressure leads to mechanical injury. Mechanical injuries include cuts, lacerations, crushing, bruising or breaking of bones. At their most severe, where these are injuries to vital organs such as the brain, the heart or the lungs, or where there is serious loss of blood, such mechanical injuries can speedily cause death. Alternatively, they may cause permanent disablement.
(d) A lowering of oxygen concentration in the atmosphere can lead to injury or death. The cessation of oxygen supply can occur through events such as drowning or by being buried by solid materials. On the other hand excess oxygen is also hazardous. At concentrations above about 25% oxygen fire hazards become far greater as sparks can set clothing or even flesh on fire.
(e) That the presence of certain substances in the environment can lead to disease or death is a truism. The nature of these effects for a number of substances is examined in detail later in the book.
(f) Again it is a truism that exposure to certain levels of pathogenic organisms is harmful as it leads to infectious disease.
(g) At all wavelengths of electromagnetic radiation there is a limiting intensity above which health or even life is endangered.
(h) The human system is apparently unharmed by the level of background radiation which arises from natural radioactivity and cosmic rays together with a small component arising from man-made activities (nuclear power, medical X-rays etc). Increasing levels may give rise to chronic fatal illness; high levels may lead to rapid onset of death.
(i) The human system is hardly sensitive to changes in potential of the order of tens of volts. However, a difference in potential across the body of the order of hundreds of volts, whether DC or AC, may well lead to fatal results.
(j) This is a hazard which is only likely to manifest itself in chronic effects of a non-fatal character.

4.6.4.2. Classification by the time which hazards take to realise their potential.

The medical profession has long used the terms "acute" and "chronic" to describe diseases according to whether they are of rapid onset with short duration, i.e. acute, or of slow onset with long duration, i.e. chronic. The medical profession has never quantified the meaning of short or long. In medical usage, acute and chronic do not in any sense describe the severity of disease. This usage will be applied to hazards in the discussion below.

Whilst it is relatively easy to recognise that the terms represent opposite ends of a spectrum there are considerable difficulties in the way of drawing a line which divides one from another. So far as the present analysis is concerned, acute may be used as relating to hazards which effectively can realise their potential within an hour, and chronic as relating to hazards which take a month or more to realise their potential. Hazards which lie in between can be regarded as lying between acute and chronic.

Table 4.6. illustrates the use of the terms acute and chronic. The times taken are given in rank order with a rough quantification of the time taken to produce death or injury.

4.6.4.3. Classification by the numbers of potential fatalities.

Even very small hazards may, in special circumstances, give rise to fatality. For example a small bubble of air may, if it gets into the circulation system of the body, prove fatal, and people have drowned in a few inches of water. On the other hand sometimes major explosions, fires or toxic releases have occurred without leading to fatality.

This has created many difficulties in public discussion of hazards in that some have argued in terms of the "worst possible case" as the measure of hazard. Such arguments lead, in the author's view, to absurdity. People do not order their everyday lives on the basis of assessing the worst possible consequences of any action they take and such considerations should not be applied to the social domain.

However, it is possible to see that certain hazards can readily give rise to major accidents. On the other hand many hazards, if they give rise to fatality, seldom kill more than one person. Examples of one-fatality hazards are those which are associated with most workshop machine tools which seldom kill more than the operator. Tables 4.7. and 4.8., which are derived from Nash (1976), illustrate this, as does Table 4.9.

4.6.4.4. Classification as personal or as group hazards.

An extension of the classification in paragraph 4.6.4.3. is to consider that there may be a polarisation between personal accidents on the one hand and group accidents on the other and that though they lie on a quantitative continuum there may be considerable qualitative differences between them as set out in Table 4.9. These differences may have implications for the control of major as opposed to minor hazards. Certain of the differences tabulated are discussed in detail in appropriate places below.

4.6.4.5. Classification by level of risk.

Hazards differ enormously in their associated risk. It might be said that the relationship between hazard and risk is "inverse", using the term in its non-mathematical, everyday, sense. Thus high hazard, in the sense of potential to kill large numbers, is often associated with low risk, and vice versa. The subject is discussed more extensively below.

4.6.4.6. Classification as specific or non-specific.

Certain hazards are highly specific in the way they produce fatality. For example if certain poisons enter the system they will produce a well defined syndrome and will attack a specific organ or organs. A notable example is chlorine which attacks the lungs and respiratory passages and kills by oedema of the lungs in which the victim drowns in lung fluids. It is true that in actual incidents of toxic releases people have been killed by other means in the course of trying to escape, but this has occurred only very infrequently.

Explosions are much less specific in their mode of action. People have been blown to pieces in the explosion of dense explosives but this does not happen with gaseous explosions. Deaths following upon explosions may be due to being struck by missiles, or being hurled against structures, or being buried in rubble or being burned or being asphyxiated.

TABLE 4.6. TIME SCALE OF HAZARDOUS EVENTS.

Time Scale	Nature of Events
10^{-2} seconds 1 - 30 seconds	Explosion of dense explosives Fire Balls
2 - 3 minutes 10 - 60 minutes	Asphyxia, drowning Gassing by chlorine
Days	Onset of chloracne from dioxin
Months to years	Duration of chloracne Exposure to lead or other metallic poisons Silicosis Asbestosis

TABLE 4.7. WORST MAN-MADE DISASTERS IN 20TH CENTURY.

Type	No. killed	Location
Air Transport	346	Near Paris
Explosion and Fire	2,000	Lanchow, China
LPG	550 plus	Mexico City
Railway	600	Mexico
Toxic Release	2,500 plus	Bhopal, India
Mining	3,000	Manchuria
Marine	6,000	Manchuria

TABLE 4.8. RELATIVE SCALE OF FATALITIES.

	Everyday accidents (home, workplace, road)	Major Accidents	Man-made Disasters (excluding war)	Natural Disasters
Range of fatalities.	1 to 9	10 upwards (1)	20 (2) to 5,000	20 (2) to 10,000,000
Range of order of magnitude.	1	-	3	6

(1) Author's definition. (2) Nash (1976) definition.

TABLE 4.9. CHARACTERISTICS OF PERSONAL ACCIDENTS AND GROUP ACCIDENTS.

Personal Accidents	Group Accidents
Change in environment small or non-existent.	Change in environment necessarily major.
Prevention usually possible by tactical$^{\phi}$ measures.	Prevention requires a strategic$^{\phi}$ approach.
Victim often contributes to accident.	Victims do not usually contribute.
Ergonometrics important study.	Ergonometrics less important.
Control over individual important to accident prevention.	Control over process vital to accident prevention.
Damage slight.	Damage major.
Largely ignored by public and mass media.	Seized upon by public and mass media.
Can be caused by a single mistake.	Often require a series of mistakes.
Characteristic of the labour-intensive manufacturing industries.	Characteristic of the capital-intensive process industries.
Ratio Non-Fatal : Fatal = 10^2 to 10^3 : 1	Ratio Non-Fatal : Fatal = 10^0 to 10^1 : 1

$^{\phi}$The significance of tactical and strategic management of hazard is discussed in Chapter 20.

4.6.4.7. Voluntary and involuntary hazards.

Some writers on the subject have lain stress upon the differences between hazards, the presence of which is voluntarily accepted, whereas others are imposed, i.e. they are involuntary. Thus, working as a coal miner is regarded by some as a "voluntary" acceptance of hazard whereas having a nuclear power station sited close to one's home implies an "involuntary" exposure to hazard.

This subject is an important component of Chapter 17 and will not be discussed further here.

4.6.4.8. Occupational and non-occupational hazards.

These distinguish between the hazards of the workplace (which now has a wide meaning) and other hazards. The United Kingdom Health and Safety at Work Act (HASAWA 1974) includes provisions for safe-guarding those not at work from the activities of those at work. Thus, certain hazards are recognised as occupational though the persons affected by them are not at work. Examples are:

(a) being injured at home by a factory explosion;
(b) being injured in the street by a brick falling from a building site.

In the UK, however, being injured in the street by a public service vehicle is not construed legally as being injured by the activities of a person at work. Such accidents are governed by traffic legislation.

Only a small fraction of accidental deaths in the UK arise from occupational accidents as is shown in Table 4.10. And, of these, only a small fraction are due to chemical hazards.

TABLE 4.10. PARTIAL STATISTICS FOR FATAL ACCIDENTS IN U K

Nature of accident	No. of fatalities	Source
Road accidents	6,357	Central Statistical Office (1980)
Accidents in the home and in residential institutions.	5,619	Royal Society for the Prevention of Accidents (1978)
Occupational accidents	610	Health & Safety Executive (1977)

Detailed discussion of the relationship between occupational fatalities in factories and those caused by the activities conducted in factories but occurring outside the factory fence, forms an important part of the subject of this book and is dealt with in the appropriate section.

4.6.4.9. Static and mobile hazards.

In the broader sense most railway accidents, road traffic accidents and air accidents are the realisation of mobile hazards. But discussion of mobile hazards will be mainly devoted, in this book, to the discussion of mobile chemical hazards. It might reasonably be concluded that such hazards, if realised on public roads, are likely to kill more of the public than equivalent hazards realised inside factories.

4.6.4.10. Concentrated and disseminated hazards.

There would seem to be few, if any, hazards to which the entire population of the globe is equally subject. The only example which immediately arises is that of the background radiation arising from cosmic rays and this is a hazard which is generally regarded as negligible.

However there are obviously considerable differences in the level of individual risk posed by, for example, major chemical hazards according to their degree of proximity.

A central problem which will be discussed later is that the benefits of a technological civilisation are usually widely disseminated among the population at large whereas the hazards can sometimes be highly concentrated.

4.6.4.11. General conclusions.

Occupational hazard has many factors and it is difficult, if not sometimes impossible, to discuss some of these factors in isolation from the others. It is necessary to ask questions such as what injury is to be feared, how large is the threat to society, what is the risk of realisation, etc, etc.

To illustrate some of the relationships, Figure 4.5., which is based upon that given in Marshall (1977a), expresses the relationship between the time scale of man-made hazards and the number of persons affected. Thus the differences between personal accidents, group accidents and group occupational diseases are brought out. It is shown that, for example, the medical component in prevention is of growing significance as the time scale becomes longer. Figure 4.5. at the same time illustrates the hazard/risk relationship. Personal accidents are shown to be low in hazard (in the sense of the number likely to be affected in any one incident) but high in risk. Group accidents exhibit the contrary tendency. Personal diseases, i.e. diseases which affect only small numbers out of an exposed population, may also be low hazard/high risk whereas group diseases are certainly high hazard but may be regarded as low risk because of the long time of exposure required.

Figure 4.6. illustrates the number and kinds of hazard to which an individual may be exposed in the course of a working week. This exposes the fact that, even for those living in its neighbourhood, a major chemical hazard is only one hazard out of a large number. The figure, moreover, does not disclose the actual values of individual risk; this subject will be discussed in Chapter 17 which is concerned with the control of major chemical hazards.

4.6.5. THE CHARACTERISTICS OF MAJOR CHEMICAL HAZARDS.

4.6.5.1. Major chemical hazards as a special category of occupational hazards.

The section has been concerned with classifying occupational hazards in general and it remains to extract from the characteristics of the general category of occupational hazards those which are specific to major chemical hazards.

4.6.5.2. The general effect of environmental change.

Major chemical hazards may injure by heat or by cold, or by excessive ambient or local pressure. This may cause oxygen depletion (asphyxia) or injury as the result of enhanced combustion arising from excessive oxygen concentration. They may give rise to toxic effects by skin contact or inhalation. Bodily injury may ensue from contact with corrosive substances. Injuries from non-ionising radiation in the major chemical hazards context are likely to arise solely from heat radiation, i.e. from fires.

The other categories listed in para 4.6.4.1, though occupational hazards, do not fall within the subject area of this book.

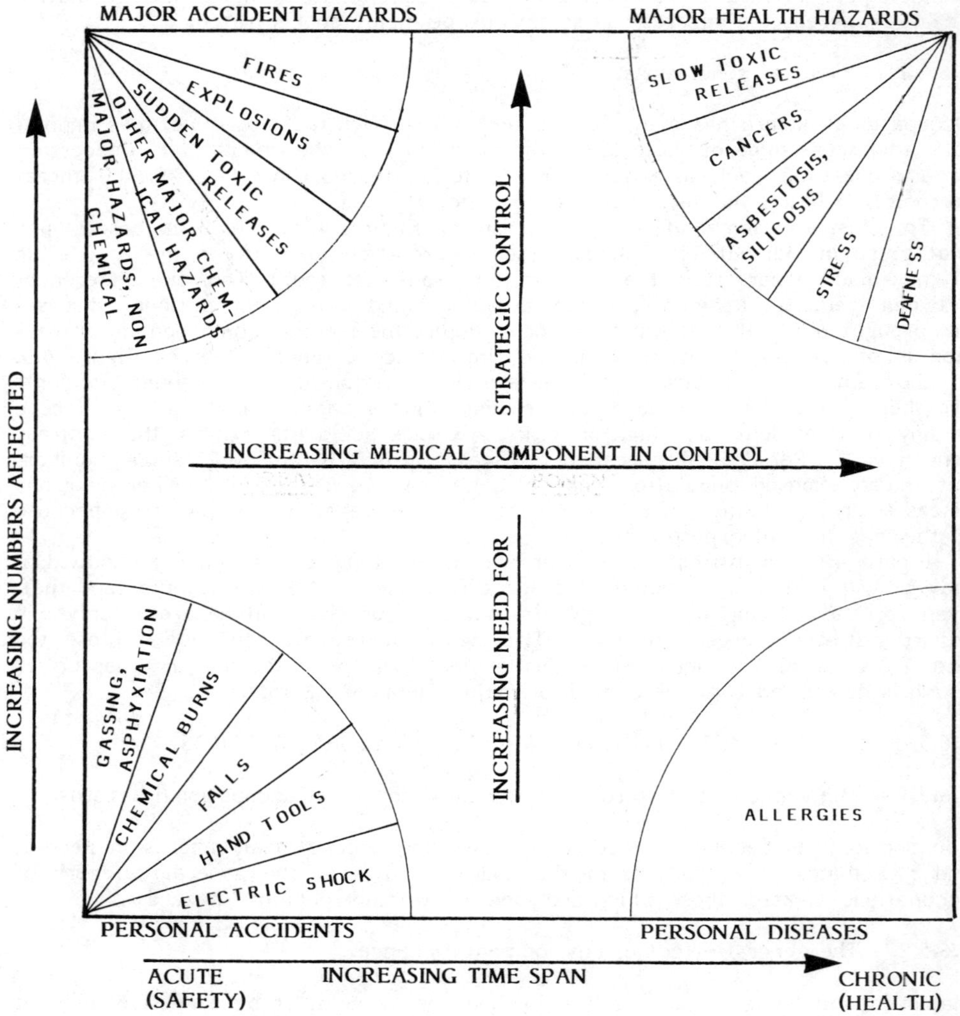

Fig. 4.5. A TAXONOMY OF HAZARDS.

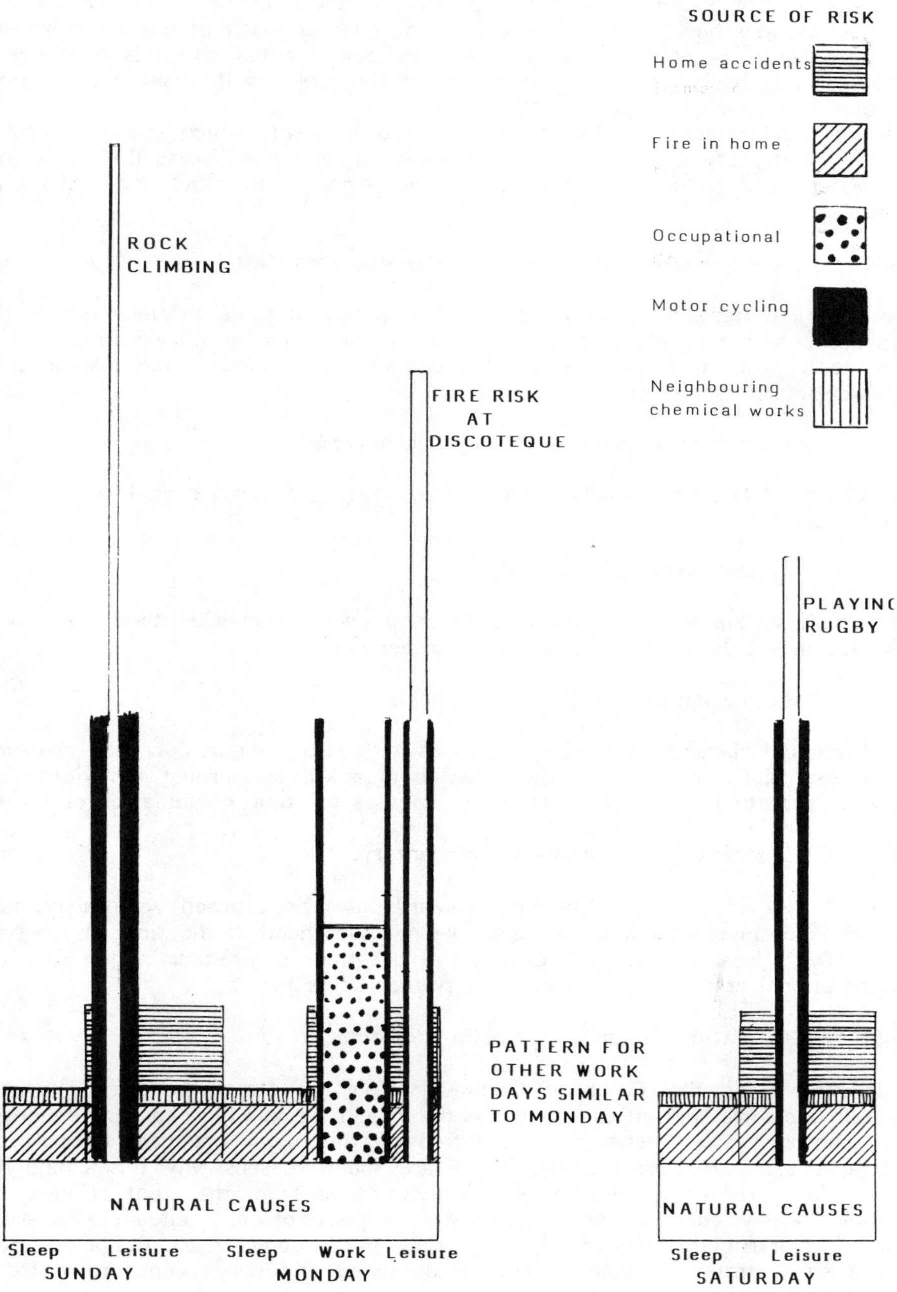

Fig. 4.6. EXPOSURE TO HAZARDS OVER A WEEK.

4.6.5.3. Classification by the time which hazards take to realise their potential.

The realisation of major hazards is typically an acute event. This is clearly true of explosions but the duration of fires may be as much as days. However it is the early stage of chemical fire, the first few minutes, which is most likely to give rise to bodily injury especially when the fire results from the ignition of a cloud of flammable vapour.

In certain instances, of which the Seveso incident, which is discussed in detail in Chapter 15, is an example, an acute event may give rise to the deposition of persistent and harmful substances, chronic exposure to which may give rise to injury.

4.6.5.4. Classification by the numbers of potential fatalities.

Major Chemical Hazards by definition have the potential to cause major accidents. The present author's definition of a major accident hazard is one which, in a typical realisation, will give rise to ten or more fatalities. The justification for this definition is discussed in Chapter 18.

4.6.5.5. Classification as personal or as group hazards.

Major Chemical Hazards clearly must be classified as hazards capable of causing group accidents.

4.6.5.6. Classification by level of risk.

Major Chemical Hazards are low risk hazards; even world-wide their frequency of realisation is only of the order of 20 or so per annum.

4.6.5.7. Classification as specific or non-specific.

Major Chemical Hazards may be specific, as with toxic releases, or non-specific, as with explosions or fires. These latter may kill by burning the victim or by oxygen depletion or by exposure to toxic gases or some combination of these.

4.6.5.8. Classification as voluntary or involuntary.

Persons at risk from Major Chemical Hazards may be exposed voluntarily, i.e. be at work, or involuntarily by being in the neighbourhood at the time of realisation. This classification also covers the question of whether major hazards are occupational or non-occupational. They may be both.

4.6.5.9. Classification as static or mobile accidents.

The legislation of the European Community on the Major Accident Hazards of certain industrial activities and the national legislation which has implemented it, is concerned solely with the regulation of the major chemical hazards of industrial sites. However, though there are sound reasons why this should be the case for purposes of legislation, the distinction, from the point of view of discussing the physico/chemical and engineering aspects of major chemical hazards, is somewhat artificial. Though the book is primarily concerned with the hazards of fixed sites, transport aspects will be discussed and conclusions drawn from case histories of transport accidents.

4.6.5.10. Classification as concentrated or disseminated.

Major Chemical Hazards are primarily concentrated hazards. The most severe effects are likely to be encountered within hundreds of metres and serious injury is not likely to ensue more than ten kilometres from the point of realisation.

4.6.5.11. An extended definition of major chemical hazards.

Based upon the above, Major Chemical Hazards may be characterised as hazards either in a static or mobile situation which, through an acute event, may in a typical realisation give rise to ten or more fatalities through the agencies of fire, explosion or toxic release. Theoretically at least, and provided that a sufficient quantity of hazardous substance be present, the agencies may also include intense cold, deprivation of oxygen, enhanced combustion due to high oxygen concentration and the dissemination of substances corrosive to tissue.

Major Chemical Hazards, though they clearly represent high levels of hazard, are nevertheless characterised by a low level of risk. Fires and explosions are non-specific realisations in that fatalities may result from a number of mechanisms. Other realisations will generally lead to a specific cause of fatality.

The realisation of Major Chemical Hazards leads to group accidents which may involve both those working at an installation or members of the general public.

Chapter 5

THE HANDLING AND STORAGE OF LIQUIDS AND GASES

5.1. THE USE OF PHASE DIAGRAMS.

The handling and storage of materials may conveniently be considered from the standpoint of the phase diagram for the substance. For simplicity of treatment it may be assumed, unless otherwise stated, that a single substance is being considered and that it changes phase without chemical decomposition.

Such a diagram, in an idealised form, is shown in Fig. 5.1.

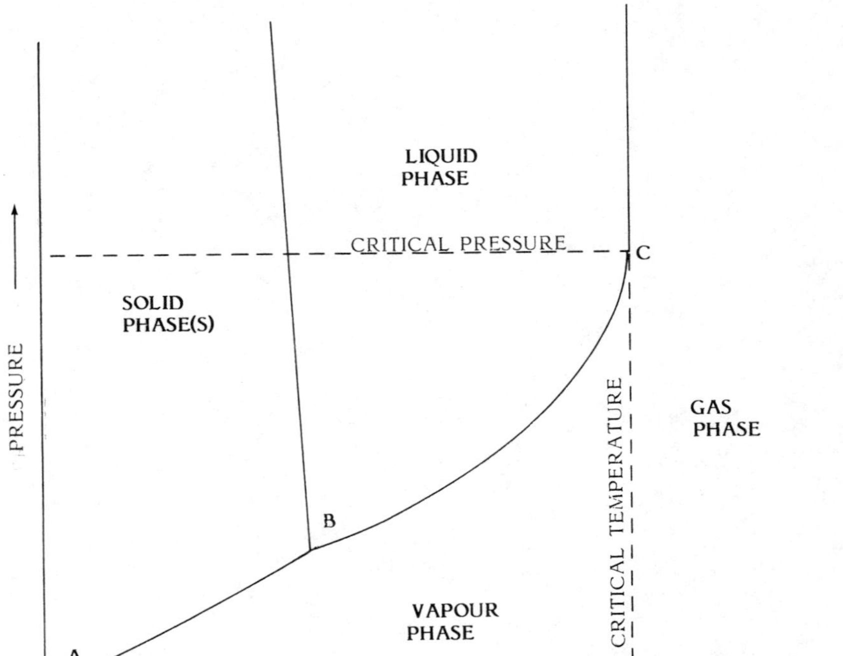

Fig. 5.1. A PHASE DIAGRAM FOR A HYPOTHETICAL SUBSTANCE.

The line A → B represents the vapour pressure/temperature relationship for the solid phase. The line B → C represents the vapour pressure/temperature relationship for the liquid phase. The point C corresponds to the "critical temperature". Above this temperature the substance cannot exist in the liquid phase. The pressure corresponding to C is the "critical pressure", i.e. the pressure which, at the critical temperature, is capable of just liquefying the gas phase. At point C the physical properties of both liquid and gas phases are identical, i.e. the density, specific enthalpy, viscosity etc have the same values. The latent heat of vaporisation/condensation is zero.

The gas phase has a sub phase, the "vapour" phase which is that phase where the temperature lies below the critical temperature and thus lies in the regime in which the gas phase can be converted to liquid by the application of pressure alone. At temperatures above the critical temperature there is no special name given to that portion of the gas phase. However, in the nineteenth century, it was believed that such gases as oxygen and nitrogen, unlike, for example, carbon dioxide, could not be liquefied by pressure alone and were given the name "permanent gases". We now recognise that this was because the ambient temperature was very much higher than the critical pressure of the substances concerned. Today all such gases have been successfully liquefied by processes which involve a stage or stages of chilling to reduce their temperature below the critical temperature.

Table 5.1. sets out the critical properties of some gases and vapours.

TABLE 5.1. PROPERTIES OF SOME GASES AND VAPOURS.

Substance	Atmospheric Pressure Boiling Point °C	Critical Temperature °C	Critical Pressure Bars
Hydrogen	- 252	- 240	12.8
Nitrogen	- 196	- 147	34
Oxygen	- 183	- 118	50.5
Methane*	- 164	- 82	46.5
Ethylene (Ethene)	- 103	9.5	50.2
Ethane	- 88.6	32.2	48.3
Propylene (propene)	- 47	92.0	45.5
Propane	- 42	96.8	42.1
Chlorine	- 34.5	144	77
Ammonia	- 33	132	113
Butane	- 0.5	152.3	37
Cyclohexane	80.7	280.0	40.1
Water	100	374	218

*Principal Constituent of Natural Gas.

Unfortunately the terms "vapour" and "gas" are not used today in a consistent manner. Liquefied petroleum **gases** (LPGs) are actually liquefied petroleum **vapours.** The expression liquefied natural **gas** (LNG) is however correct. On the other hand a **vapour** cloud explosion may involve what is strictly speaking, a flammable **gas,** examples being hydrogen or methane.

In this book existing usages in this area, even where inconsistent, will be followed as they are well established and do not seem likely to give rise to ambiguity.

5.2. IS THE PHASE DIAGRAM GENERALLY APPLICABLE ?

For some substances the temperature at which they chemically decompose is less than the temperature corresponding to point B. Such substances only exhibit a solid and vapour phase. For others decomposition may occur at a temperature between those corresponding to B and C. All these substances therefore exhibit only a partial phase diagram. Others, of which water, sulphur and phosphorus are examples, exhibit more than one solid phase. Hydrogen and helium have anomalous properties at low temperatures. However none of these exceptions are of significance to the present discussion.

5.3. THE INFLUENCE OF DENSITY ON THE EASE OF HANDLING AND STORING VOLATILE SUBSTANCES.

Density, or its inverse, specific volume, is a factor which profoundly affects the economies of the handling of volatile substances. Gases and vapours are of such low density that it is very expensive to store them in the gaseous state.

There are two basic techniques of storage for gases, (a) constant pressure, variable geometry and (b) variable pressure, constant geometry. If gases are stored at atmospheric pressure, though the container may be a relatively lightly built structure, it has to be of large dimensions. Two types of containers have been used for storing town's gas, the wet, telescopic type, and the dry, piston type. These have had maximum sizes in the range 250,000 to 300,000 m^3 [Guiness Book of Records (1983)]. Gases are also stored today in underground caverns. Alternatively, gas may be stored under pressure in suitable tanks. These latter are of course much more expensive to build per unit volume than atmospheric pressure storages, though they are much smaller for a given mass of gas. Storage tanks for North Sea Gas in the UK are to be found which store it at a pressure of 30 bars and which have volumes of ca 1,000 cu metres.

Table 5.2. gives values for specific volume for a number of gases and vapours. The table shows that these liquids at their atmospheric pressure boiling point occupy only about 1/250 to 1/450 of the volume of an equal mass of their saturated vapour in equilibrium with them, and about 1/3 of the same mass of vapour/liquid at its critical temperature.

Fig. 5.2. shows the variation in specific volume of propane with temperature. The use of a logarithmic scale for specific volume does not fully bring out the wide range in its value.

5.4. THE SIGNIFICANCE OF THE AMBIENT TEMPERATURE.

5.4.1. GENERAL SIGNIFICANCE.

What primarily determines the nature of the technology of handling and storage for any given substance, is the relationship of the band of the ambient temperature to its phase diagram.

The extent of this band of ambient temperature varies from one part of the globe to another. In the UK it is seldom outside the range -10 to 30°C.

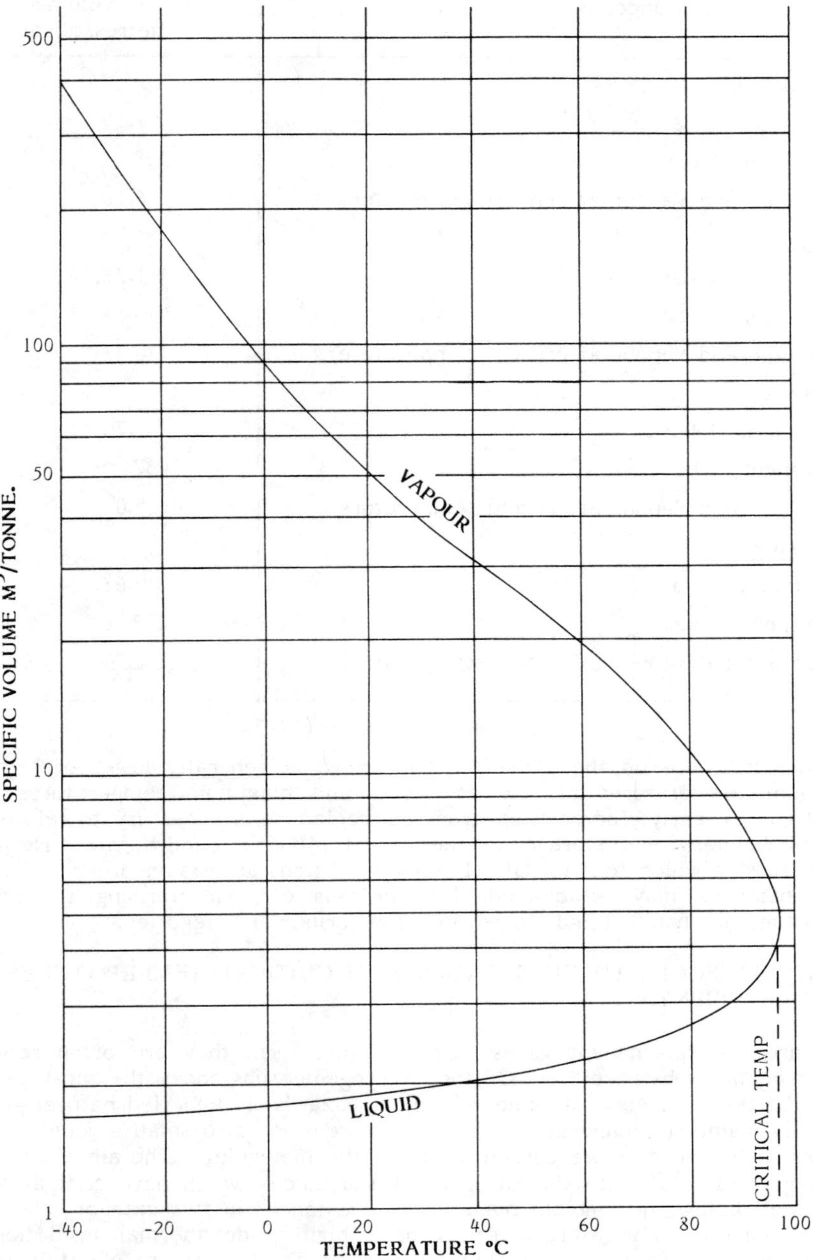

Fig. 5.2. VARIATION WITH TEMPERATURE OF THE SPECIFIC VOLUME OF PROPANE.

TABLE 5.2. SPECIFIC VOLUME OF A NUMBER OF GASES AND VAPOURS.

Substance	Specific Volume Cu metres/tonne
Methane	
Liquid, 1 bar	2.357
Vapour, 1 bar	558
Liquid and Vapour at critical temperature	6.16
Ethylene	
Liquid, 1 bar	1.78
Vapour, 1 bar	478
Liquid and Vapour at critical temperature	4.37
Propane	
Liquid, 1 bar	1.70
Vapour, 1 bar	437
Liquid and Vapour at critical temperature	5.0
Chlorine	
Liquid, 1 bar	0.62
Vapour, 1 bar	272
Liquid and Vapour at critical temperature	1.74

Derived from Perry (1973).

In other locations on the globe the band may be generally lower or higher with minimum temperatures as low as -40°C and maximum temperatures as high as 50°C. Clearly, for each geographical region, it is necessary to relate conditions of handling and storage to the actual climatic conditions. However, in the discussion which follows, UK climatic conditions are assumed.

Substances may be divided, for purposes of characterising the modes by which they are handled and stored, into four principal categories.

5.4.2. CATEGORY (1). SUBSTANCES WITH CRITICAL TEMPERATURES BELOW AMBIENT.

Such are the "permanent gases" and in liquid form they are often referred to as "cryogenic substances". Of these cryogenic substances the most important, from the point of view of major chemical hazards, is liquefied natural gas (LNG) which is mainly methane but which usually contains also small amounts of hydrocarbons with two or more carbon atoms in the molecule. The air gases, nitrogen or oxygen, also fall into the category of substances which have critical temperatures well below the ambient band. For substances in this category the technology of handling and storage relies upon high grade thermal insulation which may entail the use of vacuum jackets. It is not feasible to maintain methane, oxygen or nitrogen in the liquid phase by means of refrigeration, as this could only be done by the use of even colder liquids. The inevitable boil off may be used as it is generated, or it may be reliquefied and returned to the storage, or it may be purged to atmosphere.

The behaviour of cryogenic substances on spillage is governed by the rate of heat input from the surroundings and is discussed more fully in Chapter 7.

5.4.3. CATEGORY (2). SUBSTANCES WITH CRITICAL TEMPERATURES ABOVE AMBIENT AND BOILING POINTS BELOW AMBIENT.

All such substances are liquefiable by pressure alone and, in liquid form, fall into the very important category of liquefied gases. They include LPG, propane, butane, ammonia and chlorine. They are distinguished by their ability to "flash", i.e. if containment be lost they instantly vaporise a fraction of the liquid and cool down to their atmospheric pressure boiling point. Because they can give rise to dangerous vapour clouds, they form a considerable portion of the major chemical hazards problem. Such substances are capable of being stored at ambient temperature by pressure alone, though they may also be refrigerated.

5.4.4. CATEGORY (3). SUBSTANCES WITH CRITICAL PRESSURES ABOVE AMBIENT AND WITH BOILING POINTS ABOVE AMBIENT.

This class includes all substances which are liquids at atmospheric pressure. Those with the lowest boiling points intermingle with the liquefied gases. In cold weather butane falls into the category of a liquid at atmospheric pressure (B.Pt. ca $0^{\circ}C$) and in hot weather ethylene oxide is a liquefied gas (B.Pt. $13.5^{\circ}C$).

All substances which are solid at ambient temperature also fall into this category. But solids are only significant as major hazards if explosive or in a finely divided form in air.

5.4.5. CATEGORY (4). SUBSTANCES AT ELEVATED TEMPERATURES.

Liquids in Category (3) discussed immediately above may behave, under process conditions, as liquefied gases if they are maintained by heat input and pressurised containment at temperatures above their atmospheric pressure boiling point. The most notable example of this is the steam boiler. A further example of this category is cyclohexane (B.Pt. $80^{\circ}C$, T_c $280^{\circ}C$) which was maintained at the Nypro Works, Flixborough at $70\text{-}80^{\circ}$ above its atmospheric pressure boiling point under a pressure of ca 9 bars absolute. Loss of containment of this category 4 substance caused the Flixborough disaster of 1974.

5.4.6. OTHER CATEGORIES.

Other categories of handling problems arise with substances when the ambient temperature straddles two of the above categories as, for example, between sub-critical and super-critical conditions. It may be assumed, for purposes of this discussion, that the technology will be such as to avoid such transitions.

5.5. BEHAVIOUR ON SPILLAGE.

5.5.1. CATEGORY (1) LIQUIDS.

When cryogenic liquids are spilled they are in equilibrium with their vapour at, or near, atmospheric pressure. Accordingly they will boil immediately if heat is transferred to them and the rate of boiling is proportional to the rate of heat input.

The temperature difference between spilled LNG and its surroundings is ca $180^{\circ}C$ and for liquid oxygen it is ca $200^{\circ}C$. However, these very high temperature differences do not guarantee high rates of heat transfer. This is because of the well established transition between two regimes of boiling coefficients, the nucleate and the film boiling regimes which are treated in detail in standard works on heat transfer and for which there is an up-to-date account in Özişik (1985).

Briefly, for modest temperature differences between a boiling liquid and a solid boundary, the heat flux is roughly proportional to ΔT^3. At a certain point this relationship breaks down as nucleate (or nucleata) boiling gives way to film boiling and a vapour blanket separates the boiling liquid from the solid wall. [A domestic example is when water is poured into a hot frying pan. Initially the boiling is film boiling and, if the quantity of water added is small enough, it runs over the surface in small drops which are like tiny hovercraft. As the pan cools, the regime changes to nucleate boiling].

The heat flux thus reaches a maximum as the temperature difference increases and it then declines rapidly to a minimum known as the Leidenfrost point after its discoverer. From then on further increases in ΔT produce increased values of heat flux. The heat flux in this regime is proportional roughly to $\Delta T^{0.75}$.

The regimes are illustrated, for liquid nitrogen, in Figure 5.3.; liquid methane behaves in a similar way.

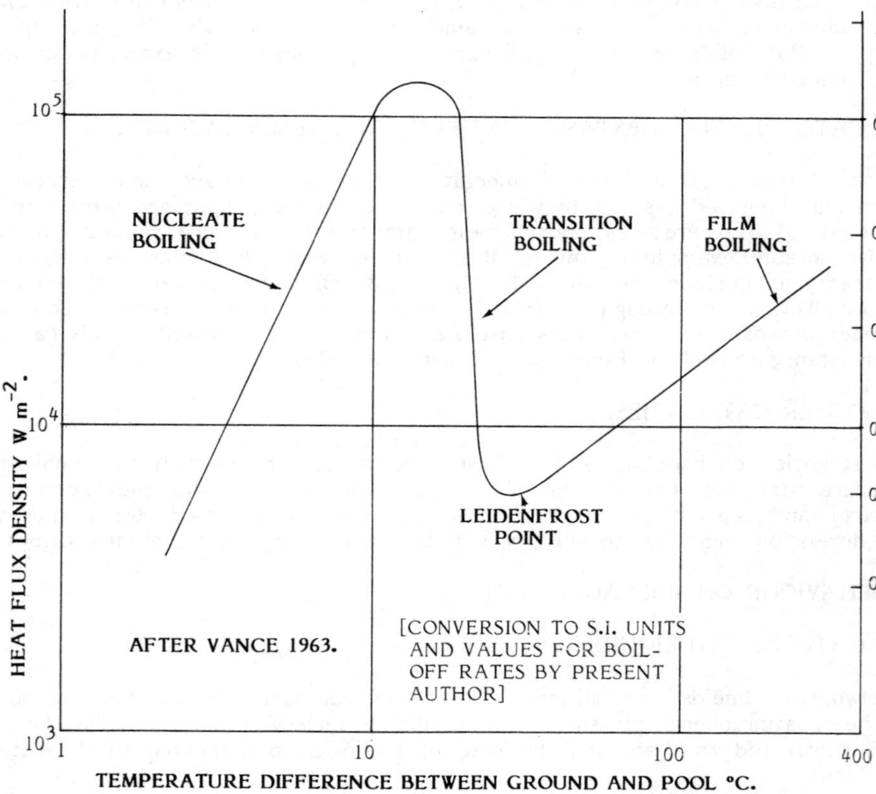

Fig. 5.3. POOL BOILING HEAT TRANSFER, LIQUID NITROGEN.

When liquid nitrogen is spilled on to surroundings which have a temperature of ca 20°C (ΔT=ca 210°C), the value for the boil off rate will be ca 0.10 kg s^{-1} m^{-2}. However the spillage will chill the surroundings and their temperature will fall and with it the heat flux, which will reach a minimum, at the Leidenfrost point, when the surroundings have cooled, in this particular case to circa -170°C. If the surroundings cool below this point the flux will increase to a maximum when they have reached ca -180°C. The actual temperature attained by the surroundings will be determined by their thermal conductivity or by convective effects in the case of liquids. There have been a number of studies of the spillage of methane, for example, on to water and Kneebone & Prew (1974) have reported that, when LNG was spilled on the sea, because of turbulence in the water, icing did not occur. Thus the surroundings in such a case did not fall below 0°C.

With solid surroundings things are likely to be very different and surfaces will cool rapidly. The lowest surface temperatures will occur with materials of low thermal conductivity.

Evidence that equilibrium is only slowly reached was provided by pit storage of LNG at Canvey Island. Here the surrounding ground acted as a heat sink to storage pits each of circa 40,000 cu metres capacity. These are described by the East Anglia Area Factory Inspectorate (1981). The use of these pits has since been discontinued as the area of frozen ground in the vicinity seemed to be increasing even after years of use of the pits.

Some special considerations apply to mixtures of which LNG is an example. In the case of mixtures, fractionation will occur and the substances with the lowest boiling points will vaporise first. With LNG this means that the pool of spilled liquid will become richer in higher hydrocarbons and its temperature will rise. This may lead, for considerations discussed above, to a change in the heat flux. This, it is believed, can give rise to rapid phase transition (RPT) or "flameless explosions" which have occurred when LNG was experimentally spilled on water [Koopman et al (1980)].

The hazards arising from the spillage of cryogenic materials are discussed in Chapter 16.

5.5.2. SPILLAGE OF CATEGORY (2) LIQUIDS.

The spillage of liquids with critical temperatures above ambient and which are maintained as liquids by pressurised containment such as LPG, presents features which differ from those of cryogenic liquids.

5.5.2.1. The phenomenon of flashing.

The chief difference lies in the phenomenon of "flashing". Flashing occurs when a liquid, in equilibrium with its vapour, suffers a loss of pressure. A new equilibrium is then established in which the temperature of the boiling point of the liquid is lower. A significant example is where a liquid is released from containment to the surrounding atmosphere. For example, if a tank of propane were to fail catastrophically, the initial and final conditions could be as below:-

	Initial Conditions	Final Conditions
Temperature °C	26.9°	-42.1
Pressure Bars Absolute	10 bars	1 bar

In moving from the initial conditions to the final conditions, partial vaporisation occurs. If the process is adiabatic (i.e. heat neither enters nor leaves the system) this means that the enthalpy of a unit mass of liquid under the initial conditions is equal to the enthalpy of the fraction of the liquid which is not evaporated plus the enthalpy of the fraction of the original unit mass of liquid which has been vaporised.

This fraction may be calculated from tables or charts of the thermodynamic properties of the substance concerned. In practice there are a variety of ways in which thermodynamic properties are presented in the form of charts. There are charts which plot pressure, temperature, enthalpy, entropy and quality (i.e the fraction of the mass of the material which is liquid and the fraction which is vapour) as variables. They differ in which of the variables form the axes, as for example pressure v enthalpy or enthalpy v entropy. [The charts are often referred to collectively as "Mollier Diagrams" but some authorities restrict the term "Mollier Diagram" to the chart in which enthalpy is the ordinate and entropy is the abscissa]. The charts are usually arranged to display the properties other than those of the axes in the form of parameters.

The most convenient forms of chart for the purpose under discussion are those which plot pressure against enthalpy or temperature against enthalpy. As pointed out above, charts are often multi-purpose, however, and the information sought may be extracted from them even if less conveniently. Perry (1973) provides thermodynamic properties of the substances discussed in this section in the form of tables or diagrams. Figure 5.4. plots enthalpy v pressure for propane. It is based on the data for propane in Table 3.2.5.2. of Perry (1973). Though the present author has converted this data into S.I. units to maintain consistency of presentation in the book, it is not essential to do so for the calculations under discussion as what is significant is the **ratio** between the enthalpy of the liquid and the enthalpy of the vapour at the final state and this is, of course, independent of the units employed.

Figure 5.5. plots the theoretical adiabatic flashing fraction (TAFF) v initial temperature for propane, propylene, ammonia and chlorine. They have been calculated from the following formula which gives the same results as are given by the method discussed above.

$$\text{TAFF}_{T^o} = (H_T - H_x)\, L_x$$

where TAFF_{T^o} = theoretical adiabatic flashing fraction at any temperature T^o

H_T = specific enthalpy of liquid at T^o

H_x = specific enthalpy of liquid at atmospheric pressure boiling point

L_x = specific latent heat at atmospheric pressure boiling point

However, the fraction so deduced is theoretical and, for reasons given below, it does not represent the full extent to which vapour may be released almost immediately on spillage.

The calculation of TAFF assumes:-

(a) That the vapour flashed off is in equilibrium with the liquid fraction remaining. This is not true in practice as the vapour initially released will have a higher temperature than the residual liquid. Calculations by the present author suggest that this effect is only marginal.

(b) That the process is adiabatic. This cannot be true as some heat will enter from the surroundings. However, the flashing process is very rapid and therefore

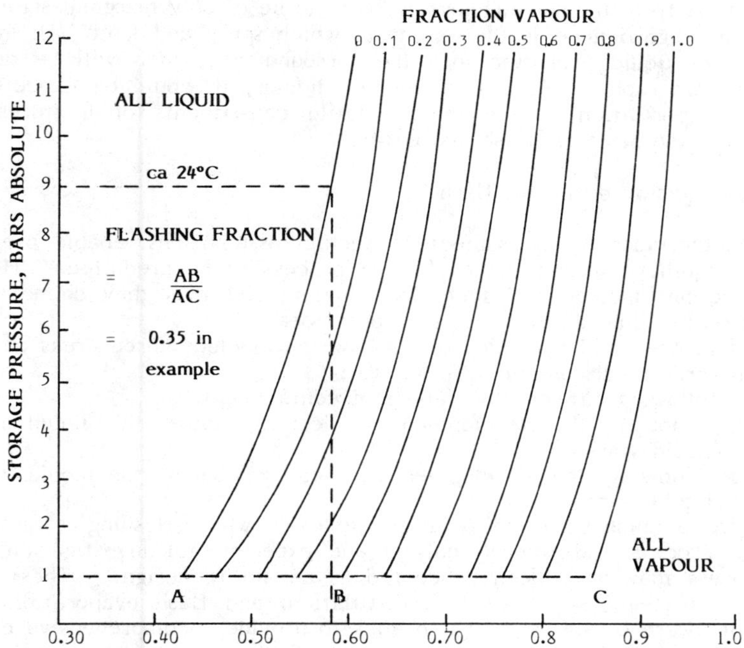

Fig. 5.4. ENTHALPY v PRESSURE, PROPANE.

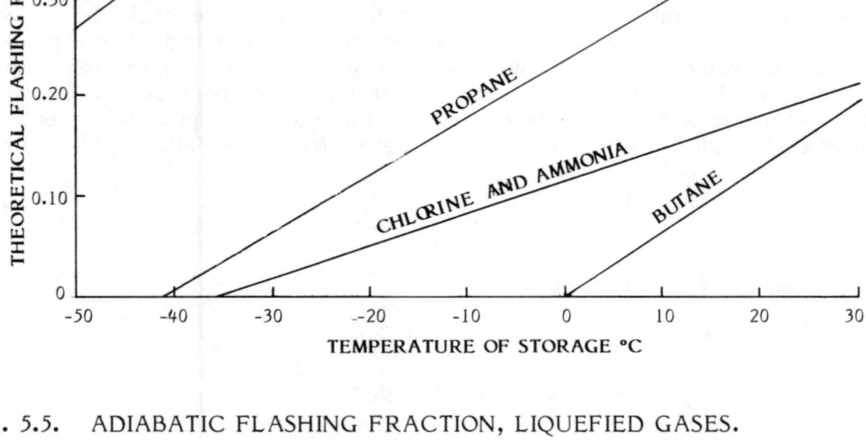

Fig. 5.5. ADIABATIC FLASHING FRACTION, LIQUEFIED GASES.

the flow of heat from the surroundings is likely to be of only marginal significance. Of far greater significance is the extent to which spray and froth will influence the quantity of liquid projected into the surroundings, which will be discussed in the next paragraph. As with cryogenic liquids, it would be expected that differential vaporisation of the more volatile constituents of mixtures would occur. This is the basis of "flash distillation".

5.5.2.2. How quickly do liquids flash ?

The laws of thermodynamics, subject to certain assumptions, enable predictions of the final equilibrium state of a flashing process to be predicted. However, these laws do not include the dimension of time and thus they do not provide information on the fluid dynamics of such processes.

Analysis of the fluid dynamics of flashing may include three areas of significance for the present discussion. These are :-

(a) Flashing following catastrophic failure of containment.
(b) Flashing following the development of leakage above the liquid level in a vapour/liquid system.
(c) Flashing following the development of leakage below the liquid level in a vapour/liquid system.

There are a number of industrial processes in which flashing is an integral part of the process and where analysis and experimental investigation of the flashing process may have been conducted as an aid to design. These include flash boilers in steam systems, flash distillation, and flash evaporation. The author is not aware of any work which aims to provide a comprehensive overview of the phenomenon of flashing liquids, but reference will be made below to relevant work in particular areas.

5.5.2.3. Flashing after catastrophic failure.

Catastrophic failure of a pressure vessel, which is defined by the author as its sudden disintegration into a number of roughly equal sized parts, is, with the exception of such failures in steam boilers, a fairly rare event. Nevertheless there have been a number of such failures which have released flammable or toxic vapours. Some of them are the subject of case histories in this book. It may be useful to attempt to evaluate an approximate time scale for such events bearing in mind that there is little point in trying to apply sophisticated methods of calculation to situations the precise circumstances of which are impossible to predict.

The theoretical minimum time for flashing to take place might be deduced from a simple model in which it is assumed that, at the end of the process, the vapour generated is a cloud, unmixed with air. The time taken for flashing is the time which it would take for the first increment of vapour released to travel at sonic velocity from the surface of the flashing liquid to the edge of the ultimate cloud. This could be calculated from the following formulae:-

$$T_f = \frac{R_c}{C_v}$$

when T_f = time taken to flash

R_c = radius of vapour envelope

C_v = speed of sound in vapour

The radius of a hemisphere is given by:-

$$V = 2/3 \pi r^3$$

therefore $r = \sqrt[3]{V\,0.48} = 0.78 \sqrt[3]{V}$

Strictly speaking the calculation ought to estimate the volume of the cloud and hence its radius by adding together the volume of vapour released and the volume of liquid after flashing and then calculate the radius of the cloud. The radius travelled by the increment would be this radius minus the radius of the liquid before flashing. However for most purposes it is accurate enough to subtract the radius of the initial liquid from the radius of a hemisphere with a volume equal to that of the vapour flashed off.

therefore $T_f = 0.78 \sqrt[3]{\dfrac{V_v - V_L}{C_v}}$

For 100 m^3 of propane at 10 bars.

$$TAFF = 0.38$$

$$E_f = 250$$

$$C_v = 300 \text{ m s}^{-1}$$

therefore $T_f = 0.78 \sqrt[3]{\dfrac{(100 \times 250 \times 0.38) - 100}{300}}$

$$T_f = 0.78 \sqrt[3]{\dfrac{9500 - 100}{300}}$$

$$= 0.055 \text{ secs.}$$

It is of some interest to compare this with the time scale for catastrophic failure. If this be assumed to be due to a crack spreading round the periphery of the hemisphere at the speed of sound in steel the time taken would be $2\pi r/C_s$ secs.

Since r for 100 m^3 hemisphere = 3.63 metres

Then periphery = 22.8 metres

$$C_s = 3{,}200 \text{ m sec}^{-1}$$

$$T = .007 \text{ secs.}$$

The situation described above is artificial if only because hemispheric containers do not exist and such an event would almost certainly create missile projection with serious distortion of the local atmosphere. Nor could the initial cloud be formed without admixture with air. Also the vapour would start at rest and would not be likely to attain sonic velocity even initially. As the pressure fell, sonic velocity would, after falling to a certain critical point, not be attainable even theoretically. Therefore actual time to complete flashing would be greater than that calculated above. Perry (1973), P 5-44, comments, in relation to flashing flow in pipes, "For the flow of flashing mixtures the critical velocity may be much less than the velocity of sound in the vapour phase".

The mechanism of flashing is, in fact, violent. Once the outer surface of a mass of flashing liquid has disengaged its vapour, the disengagement of the layer below can only take place by disintegrating the outer layer. It must be supposed that during the period of flashing, the liquid will turn into a mass of froth. Droplets projected by the violent disintegration may be propelled outside of the theoretical envelope of vapour. At the same time, the momentum in the expanding vapour will carry it into the surrounding atmosphere where it will mix with the air to form a cloud of vapour/air mixture. Kletz (1977) has suggested that flashing is likely to entrain a mass of liquid equal to that vaporised and this view has been accepted by the Advisory Committee on Major Hazards (1979) (see also Addendum).

It may be expected too that the expansion of the vapour, even if subsonic, will compress the air in front of it to produce a shock wave which may, to bystanders, not be distinguishable from a chemical explosion.

Though in the model above it was assumed that the tank was full of liquid, in practice unless the tank had been overfilled and had failed through hydraulic rupture, there would be a vapour phase present. This would expand on rupture. Therefore, for catastrophic failure of tanks of liquefied gases, the resulting vapour cloud would depend for its magnitude on the degree to which a vessel was filled with liquid at the time of rupture. In the model example given above, a tank full of liquid would give rise to 100 times its volume of immediately released vapour. A tank virtually empty of liquid at 10 bars vapour pressure would give rise to only about 10 times its volume.

5.5.2.4. Flashing after puncture above the liquid level.

Where a tank containing flashing liquid is punctured above its liquid level, a small leak could release vapour at the pressure in the tank until all the liquid had evaporated. The surroundings would provide latent heat though the contents would chill to a degree dependent upon the size of the aperture. The rate of emission would be a function of the size of the aperture and of the pressure in the tank. Flow might be critical, depending on pressure and given sonic flow rates. Similar consideration could apply to the severing of a branch connected to the vapour space. Calculation of flow rates would follow standard techniques.

The question of whether entrainment would be significant would depend upon the boil off rate and the height available for disengagement. Efficient Use of Fuel (1964) claims that in flash boilers in which condensate from high pressure heating coils is allowed to flash off, low pressure steam entrainment becomes significant at disengagement velocities in excess of 3 m s^{-1}. Coulson & Richardson (1956) suggest that in distillation columns with wide plate spacings, 2 m s^{-1} is the threshold for entrainment. Thus for disengagement velocities of the order of 2-3 m s^{-1} a puncture in a vessel will give rise only to vapour emission.

5.5.2.5. Flashing after puncture below the liquid level.

Where leakage is below the liquid level, a puncture will permit liquid to emerge. It might be expected that the regime in a hole in a flat plate will be single phase liquid flow and that flashing will occur externally to the leak. Where the leakage is due to the severance of a pipe, flashing in the pipe is likely to give rise to two phase flow. The flow rate will be restricted by the flashing to a lower level than would obtain for single phase liquid flow for the same pressure differential [see Perry (1973) P 5-40 to 5-44]. Nevertheless a puncture below the liquid level is likely to discharge a higher mass flow rate than that of a similar-sized puncture above it (see also Addendum).

5.5.3. SPILLAGE OF CATEGORY (3) LIQUIDS.

The behaviour of these liquids is a function of their volatility which may range from substances approaching their boiling point and hence merging with the category of class (2) liquids. Depending upon the temperature on spillage, liquids of low volatility may exhibit, if hot, characteristics similar to those of cold but more highly volatile liquids.

From the point of view of major chemical hazards, only leakages below the liquid level will be significant. The rate of such leakages will depend upon the pressure on the liquid which may be the static head above the point of rupture or there may be pressurisation of the storage which is likely for the more volatile of the liquids in this category as, for example, acetaldehyde. Without such pressurisation, highly volatile liquids may be difficult to pump.

Behaviour after leakage will depend on the local lie of the land. Generally speaking, bunds will be provided to contain liquid spillages. Sometimes these have been ill-designed. When of volume equal to that of the tank contents, account needs to be taken of the dynamic behaviour of spilled liquids as waves can be generated which can slop over a bund wall. The provision of an internal lip may prevent this from happening (cf. the design of sea walls and promenades). Also, leaks fairly high up a tank may result in a liquid jet which comes to ground outside the wall. A further problem with bunds is the need to dispose of rain water without destroying the integrity of the bund.

Unlike spillages of cryogenic liquids where the surroundings provide latent heat, or of liquefied vapours where there is flashing plus latent heat from the surroundings, category (3) liquids evaporate only because of the influence of the wind. The rate of evaporation depends upon the size of the pool and upon the wind speed. Fig. 5.6. relates wind speed to evaporation rate. The possibilities of pumping spillage of category (3) liquids back into storage is much higher than it is of returning category (1) or category (2) liquids into storage.

5.5.4. SPILLAGE OF CATEGORY (4) LIQUIDS.

5.5.4.1. Resemblance to spillage of Category (2) liquids.

Category (4) liquids, which are liquefied vapours at a temperature above their atmospheric pressure boiling point, resemble, in their behaviour on spillage, category (2) liquids in that flashing will occur. However departure from adiabatic conditions in the case of category (4) liquids arises from loss of heat to the surroundings which, being relatively cool, will condense part of the vapour released.

5.5.4.2. The Flixborough disaster.

A most notable example of a disaster which involved a category (4) liquid was that at Flixborough which is described in Flixborough (1975) and in detail in later passages in this book. A sketch of the incident is shown in Fig. 5.7. Flixborough (1975), the Official Report of the disaster, made no reference to the quantity of cyclohexane vapour discharged beyond referring to it as a "massive vapour cloud". The Report did speak (paragraph 93) of nothing untoward being noted more than 2 minutes before the explosion, which implies that the flashing which occurred took 120 seconds or less. Calculations of discharge rate are given in Cremer & Warner (1974). Figure 5.7. shows that after the collapse of the by-pass, the interconnected Reactors 1, 2, 3 and 4 flashed off vapour

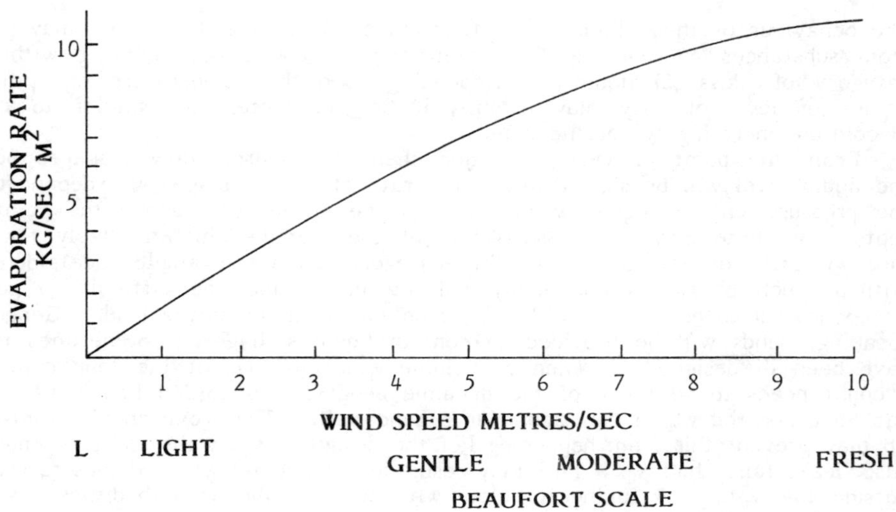

Fig. 5.6. SPILLAGE VERSUS EVAPORATION RATE.

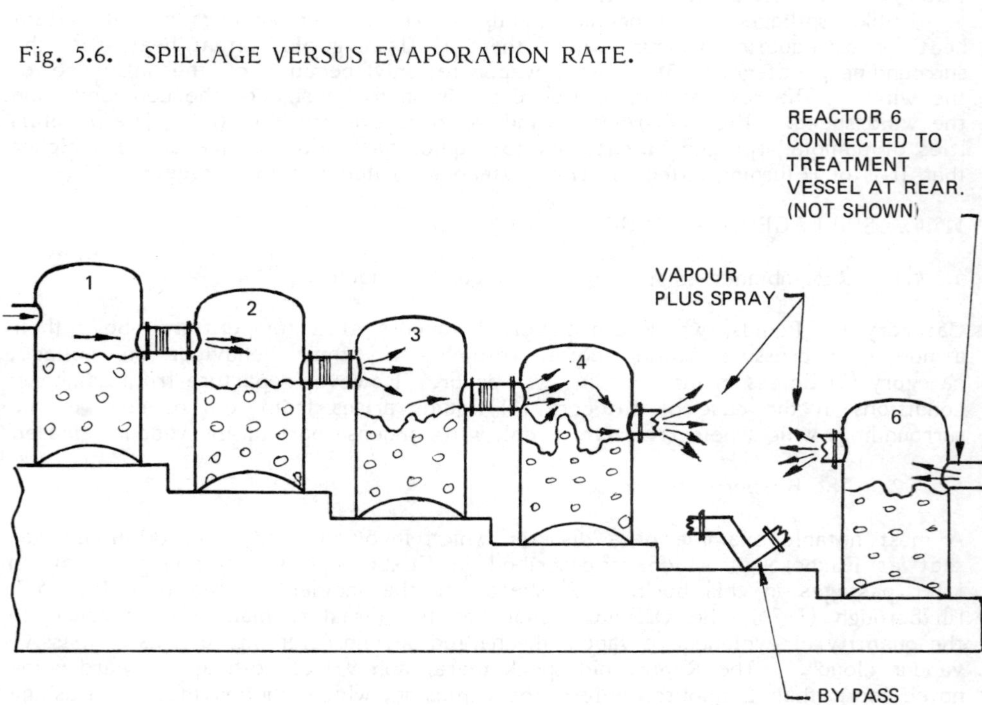

Fig. 5.7. SKETCH OF FLASHING FROM REACTOR AT FLIXBOROUGH.

through a branch which, in fact, was 0.7 m (28 inches) in diameter and that Reactor 6, which was connected also to a treatment vessel, flashed off through a branch which was of the same diameter.

The calculations in Cremer & Warner (1974) assumed that the initial flow through the 0.7 metre diameter branches either from Reactor 4 or 6 would have been governed by the sonic velocity of cyclohexane vapour. The values so calculated for the flow from Reactor 4 and from Reactor 6 were not identical because it was considered that an internal bend in Reactor 6 would have reduced the flow below that which obtained from Reactors 1-4.

It should be noted that the branch on Reactor 4 was throttling the discharge from 4 reactors, whereas in the case of Reactor 6 it was only throttling the flow from 1 reactor plus a treatment vessel. Reactor 6 will be discussed as being more typical. The reference claims that a first order estimate of the flow rates came out as follows:-

		Cumulative quantity of vapour discharged.
up to 5 secs	0.56 tonnes s^{-1}	2.80 tonnes
at 10 secs	0.43 tonnes s^{-1}	5.30 tonnes
at 20 secs	0.25 tonnes s^{-1}	8.70 tonnes
at 22 secs	0.22 tonnes s^{-1} (End of sonic limited regime)	
at 30 secs	0.15 tonnes s^{-1}	10.90 tonnes
at 48 secs	?	12.55 tonnes

It must be emphasised that these calculations are acknowledged in the reference to be very rough.

From a drawing of a Flixborough reactor in the present author's possession, it appears that the internal diameter of the reactor was 3.55 metres (11' 8"). The liquid surface was thus 9.89 (say 10 m^2). The initial flashing rate was 0.56 tonnes s^{-1}. If it be assumed that this all originated from the liquid surface in Reactor 6 and that none came from the liquid surface in the treatment vessel, then the rate of disengagement

$$= \frac{0.56}{10} \text{ tonnes } m^{-2} \; s^{-1}$$

$$= 56 \text{ kg } m^{-2} \; s^{-1}$$

The volumetric disengagement rate, assuming an internal pressure of 8 bars absolute and a molecular weight of 84 for cyclohexane would be

$$\frac{56}{84} \times \frac{22.4}{1} \times \frac{1}{8} \times \frac{400}{273} = 2.73 \; m^3 \; s^{-1} \; m^{-2}$$

i.e. a velocity of 2.73 m s^{-1}. This corresponds with the figure of 3 m s^{-1} as the limit in steam flash tanks quoted above in para 5.5.2.4. However, the figure is higher than would have emerged if it had been assumed that some flashing also occurred in the treatment vessel. Confirmation for this comes from Smith (1982). Smith, who performed a series of calculations for one of the interested parties at Flixborough, concluded that, apart from an initial phase lasting about one second at Flixborough when liquid was ejected from the reactors, the rate of entrainment thereafter was of the order of 1%.

It is clear from the discussion above that flashing rates at Flixborough were very much less than if catastrophic failure of one of the pressure vessels had occurred. The flow of vapour was severely constricted by the 0.7 metre diameter outlet pipes, i.e. 0.38 m^{-2}. The disengagement surfaces were about 40 square metres total in Reactors 1-4 and about 20 square metres in Reactor 6 plus treatment vessel. The ratio of the discharge aperture to the disengagement surface was thus

Reactors 1-4 $\text{Ratio} = \frac{0.38}{40} = \text{ca } .01 \text{ to } 1.$

Reactor 6 + vessel $\text{Ratio} = \frac{0.38}{20} = \text{ca } .02 \text{ to } 1.$

5.5.5. SUMMARY.

The behaviour of liquids on spillage is governed by their phase diagram and by the relationship of the band of ambient temperature to this diagram.

There are considerable economic advantages in handling and storing substances in the liquid state as opposed to the gaseous state and therefore, particularly for large-scale operations, liquefied gas technology is very important. However spillages of liquefied gases have given rise to many serious incidents. The section distinguishes four main categories of liquids:-

(1) Those with a critical temperature below ambient, the so called "permanent gases". These can be only liquefied if they are chilled as well as compressed. In liquid form they are called "cryogenic liquids". They are stored in thermally insulated containers. On spillage the rate of evolution of gas is a function of the rate of input of heat from the surroundings. Heat transfer rates are governed by the complex laws which govern the relationship between heat flux and the temperature difference between a boiling liquid and its surroundings.

(2) Those which are gases at ambient temperature and which have a critical temperature above ambient. Strictly speaking their gas phase is called a vapour. Such gases may be stored at ambient temperature in pressure vessels. On spillage they "flash" i.e. a portion is rapidly vaporised and the remaining liquid falls to the atmospheric pressure boiling point. The process may be very violent and an appreciable fraction of the liquid may be projected as froth or spray. The most serious incidents involving spillage of liquids have arisen from liquefied gases in this category.

(3) Those which are liquid at ambient temperature. Vapour evolution on spillage is far slower than in the two categories above and is determined by wind velocity.

(4) Those which fall into category (3) when at ambient temperature but are maintained at an elevated temperature. When spilled they flash like category (2) liquids.

Chapter 6

CAUSES OF LOSS OF CONTAINMENT OF LIQUIFIED GASES

6.1. INTRODUCTION.

6.1.1. PURPOSE OF CHAPTER.

The purpose of this chapter is to review briefly some common causes of the failure of pressurised systems, including pressure vessels, which contain liquefied gases. The subject has been chosen as being of particular relevance to the major chemical hazards problem because of the significance of liquefied gases as a major component of such hazards. Pressurised systems include pressure vessels, which normally contain the bulk of the inventory of the system, together with pipework, valves, pumps or compressors, instrumentation and other accessories.

Figure 6.1. shows pressure regimes important to the chemical and process industries and it is necessary to state why the chapter does not concern itself with the higher pressure ranges shown in Figure 6.1. which, at first sight, would appear to present the greatest hazard. The answer is that the systems which operate at the highest pressures contain far smaller inventories of flammable or toxic substances than do systems containing liquefied gases. This is partly because of the impracticability of building vessels of several metres in diameter capable of withstanding the necessary pressure. Even though there is a considerable energy content in the compressed gas in such systems, it falls well below the energy content equivalent to the threshold inventories of high explosive envisaged by the authorities who regulate major chemical hazards (this subject is discussed in Chapter 19). This is not to deny, however, that rupture of such vessels may not bring about severe effects, but these would be localised. It is also made clear in Chapter 5 (see Table 5.1.) that the critical pressure of many hydrocarbons is of the order of 40 bars and, for a number of reasons discussed in Chapter 5, they are handled as liquefied gases at, or around 10 bars. This is also true of chlorine and ammonia.

Loss of containment in systems designed to operate in the pressure region 10-20 bars has constituted in the past the cause of a number of serious disasters, including Ludwigshafen, Flixborough and San Carlos.

Particular attention will be paid in the chapter, therefore, to the causes of failure of pressure vessels and pressurised systems operating in the 10-20 bar range.

6.1.2. CHARACTERISTICS OF PRESSURE VESSELS.

Nichols (1969) has defined a pressure vessel as being:

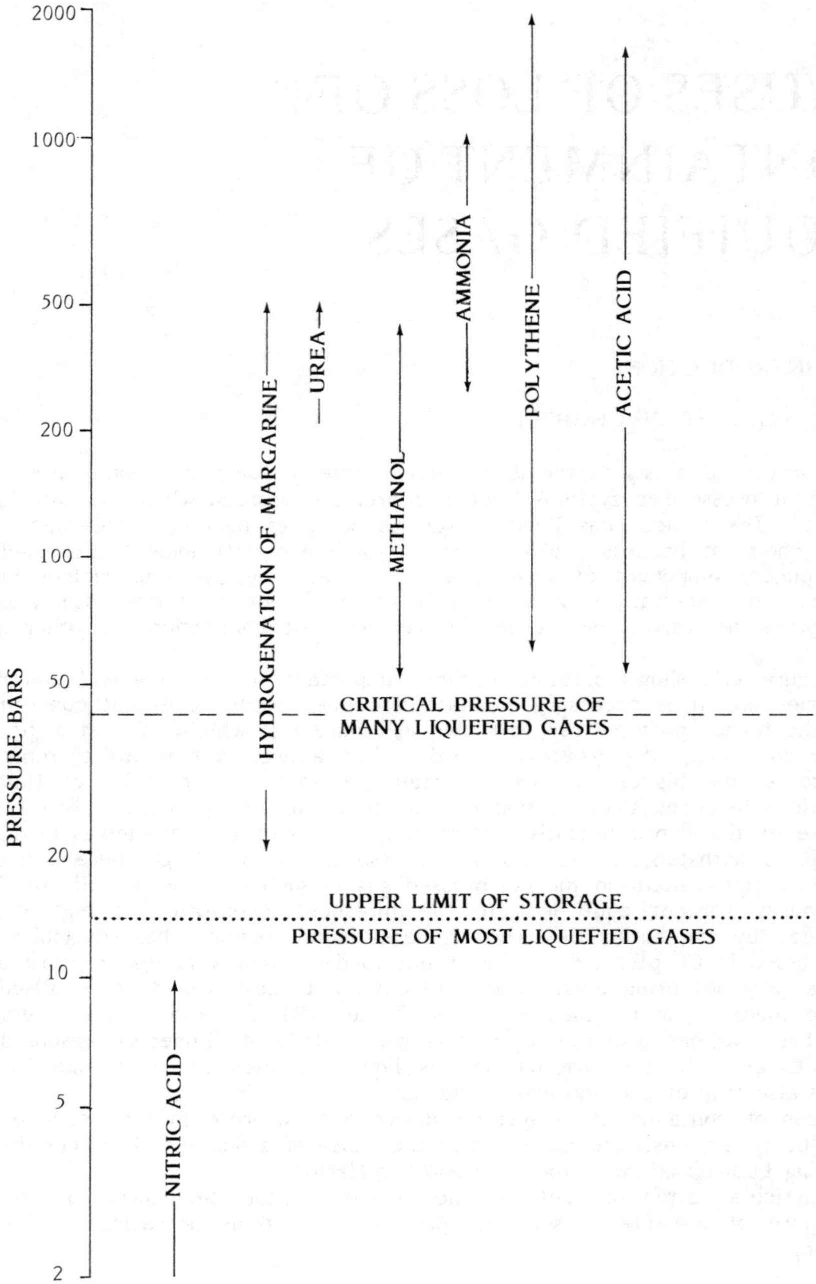

Fig. 6.1. SOME PRESSURE REGIMES IN THE CHEMICAL INDUSTRY.

"A vessel intended to contain a fluid at a pressure greater than atmospheric pressure which, in general terms, is designed and built according to a pressure vessel code such as BS 1500 and BS 1515 [now superseded by BS 5500 (VM)] in this country or ASME Section VIII in the USA and/or is subjected to inspection requirements . . ."

There is a considerable literature on the subject of the design, fabrication, testing, inspection and operation of pressure vessels. This takes the form of reviews, textbooks and papers. There are reviews in Perry (1973) Pages 6-91 to 6-105 incl., and High Pressure Technology Association (1975). There are textbooks such as Harvey (1974), Bednar (1981) and Chuse (1977). One textbook, Thielsch (1965), is of particular relevance to this chapter as it deals with defects and failures. The American Society of Mechanical Engineers (ASME) is a major source of published material which includes the JHPT (series) and the very import-and ASME Code, ASME (series) which has a significance which extends far beyond the USA. British publications include Nichols (1969) (which contains a useful glossary of terms), and the design code BS 5500 (1982). Both the ASME code and BS 5500 are subject to frequent revisions and up-dating.

The limitations of pressure discussed in section 6.1.1. restrict the field of interest to thin walled vessels defined in High Pressure Technology Association (1975) as having a ratio (K), outside diameter to inside diameter, of 1.2:1 or less. These present far less complex problems of design than do thick walled vessels which have a value for K $\geqslant$ 1.2:1.

Figure 6.2. is an outline sketch of a pressure vessel which may be regarded as typical of the vessels under discussion. Though it is, in fact, a transport vessel the dimensions are reasonably typical.

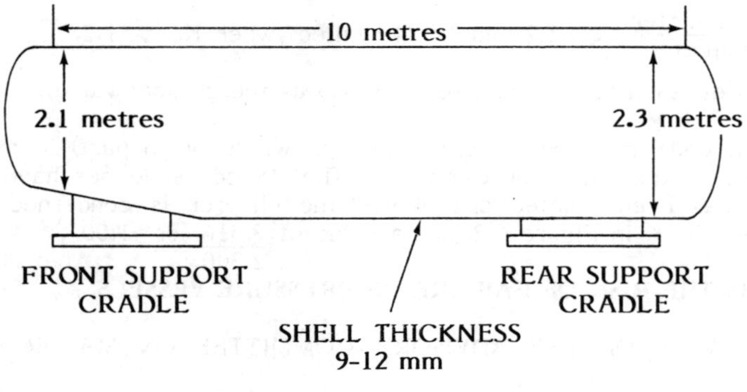

Fig. 6.2. OUTLINE SKETCH OF PRESSURE VESSEL.

6.1.3. FRACTURE MECHANICS, A BRIEF REVIEW.

The fracture of pressure vessels takes place through the medium of the formation and propagation of cracks. It is a matter of common experience, however, that some materials fracture in a brittle fashion. The best example of brittle fracture is that exhibited by glass or cast iron. Other materials fracture in a ductile fashion. An example is the failure of an overheated bean can. In the first case, that of brittle fracture, there are numerous fragments of various sizes. In the other case, that of ductile fracture, there are a relatively few large pieces. (Sometimes there may be only one piece, much distorted from its initial shape and with tears in it).

In some materials, a crack, once it reaches a critical length, will continue to propagate and to branch provided that the material remains stressed above a certain critical value. In others, crack propagation is hindered or relieved by the ductile nature of the material.

The subject was first scientifically investigated by Inglis (1913) and the work was extended in Griffith (1921). The subject is given a general review in Gordon (1968) and in Marshall (1975a).

Put briefly, the theory holds that when an object is stressed, typically, the strain energy is not evenly distributed and that in the neighbourhood of certain features known as "stress raisers" the stress may have a level which is two to three times or even, at the tip of a crack, up to a hundred times that of the mean value. Stress raisers include sharp corners, notches and cracks. The art of designing pressure vessels includes providing not only a wall thickness adequate to withstand the pressure but in avoiding built-in stress raisers or, where this cannot be avoided, by providing an adequate reinforcement of material. The art also includes the selection of suitable materials of construction such that under service conditions (or apprehended departures from them) the material will have adequate strength and toughness to withstand the stresses imposed. Even so, design must take account also of agencies such as creep, fatigue and corrosion which may reduce the strength of the vessel.

6.1.4. CONSTRUCTION CODES.

As stated above, pressure vessels are classified as "thin walled" when

$$K = \frac{\text{outside diameter}}{\text{inside diameter}} \leqslant 1.2 \text{ and "thick walled" when } K \geqslant 1.2.$$

(Some authorities, such as BS 5500, set $K \leqslant 1.3$ as the critical value).

The ASME code is a legal requirement (in whole or in part) in many states of the US and in the whole of Canada. The BS codes do not have this legal status. As has been pointed out above, the chapter is concerned only with thin walled vessels; K in Figure 6.2. has a value of $\frac{2.316}{2.300}$ or 1.007.

6.2. WHAT IS THE RISK OF FAILURE OF PRESSURE VESSELS ?

6.2.1. THE VIEW OF THE ADVISORY COMMITTEE ON MAJOR HAZARDS.

It was the view of the Advisory Committee on Major Hazards (1979) that "As regards pressure vessels themselves, provided that it has been shown that the appropriate degree of strength is present and maintained by correct design, fabrication, testing and inspection, it may be accepted that they will not fail; and under the same strict conditions the same acceptance may be extended to certain short, large diameter, simple pipes" (Para 147). The Report went on to say that "the following assumptions would normally be mandatory:-

(1) That there could be a failure of containment in the lines around each pressure vessel equivalent in area to the cross section of the largest line."

Paragraphs (2) and (3) which followed then specified the behaviour of the contents.

Considered in the light of the discussion which follows, it may be argued that this assessment begs a number of questions in that despite the provisos, there have been a number of incidents which have involved failures of pressure vessels in fixed installations, and a greater number of serious failures in transport vehicles.

However, it is important, before considering the potential modes of failure, to attempt to make some estimate of the frequency with which such events occur. There are no comprehensive statistics on the failure of pressure systems. There are, however, statistical analyses based on partial data which enable an appraisal to be made of the likely rate of failure of pressure vessels themselves.

Such an analysis is that of Smith & Warwick (1981).

6.2.2. THE ANALYSIS OF SMITH AND WARWICK.

This analysis was carried out for the UKAEA primarily in order to make a judgement on the likelihood of failure of pressure vessels in the nuclear industry. The analysis was a joint effort between the Safety and Reliability Directorate (SRD) of the UKAEA and the Associated Offices Technical Committee of the UK insurance industry. It represents the fourth Report of a series which updates and consolidates the data on pressure vessel failure in the UK during the period 1962-1978. Comparisons are made with the USA and West Germany.

The authors have adopted two classes for failure, (a) potential, i.e. discovered by inspection, and (b) catastrophic or disruptive.

A total of 20,000 vessels forms the population studied and the total of vessel years is greater than 3×10^5. These vessels have the following characteristics :-

(a) Built to class 1 requirements of a recognised design code.
(b) Unfired. (This means that failures of vessel parts exposed to flame are not included but that failures of the outer shell of shell boilers are included).
(c) Wall thickness greater than 9.5 mm (3/8").
(d) Welded or forged.
(e) Working pressure in excess of 7 bars.
(f) Age less than 40 years.

The general conclusions are given below.

There were 216 potential failures and 13 catastrophic failures out of 3.1×10^5 vessel years. These yield values of 6.9×10^{-4} potential failures per vessel year and 4.2×10^{-5} catastrophic failures per vessel year. Statistical analysis would suggest 99% confidence limits for the upper bounds of 8.0×10^{-4} and 8.3×10^{-5} for potential and catastrophic failures respectively. Of the 216 potential failures, 94% were attributed to cracking. Table 6.1, taken from the reference, shows the distribution attributed by the authors of the cause of cracking. It may be noted that the authors draw attention to the early years of a vessel's life as being the ones in which failure is most likely to occur, a point which is also taken up on Bush (1975). Figure 6.3. is a notational chart of this relationship which has a shape similar to that of the curve which relates the probability of a person dying in a given period following a given date versus the age of the person at that time. Such a curve is called a "bath tub" curve.

TABLE 6.1. CAUSES OF CRACKING IN PRESSURE VESSELS.

Cause.	No. of Cases	%
Defects prior to service	63	29
Not ascertained	61	28
Fatigue	52	24
Corrosion	30	14
Other causes, including creep	10	5
	216	100

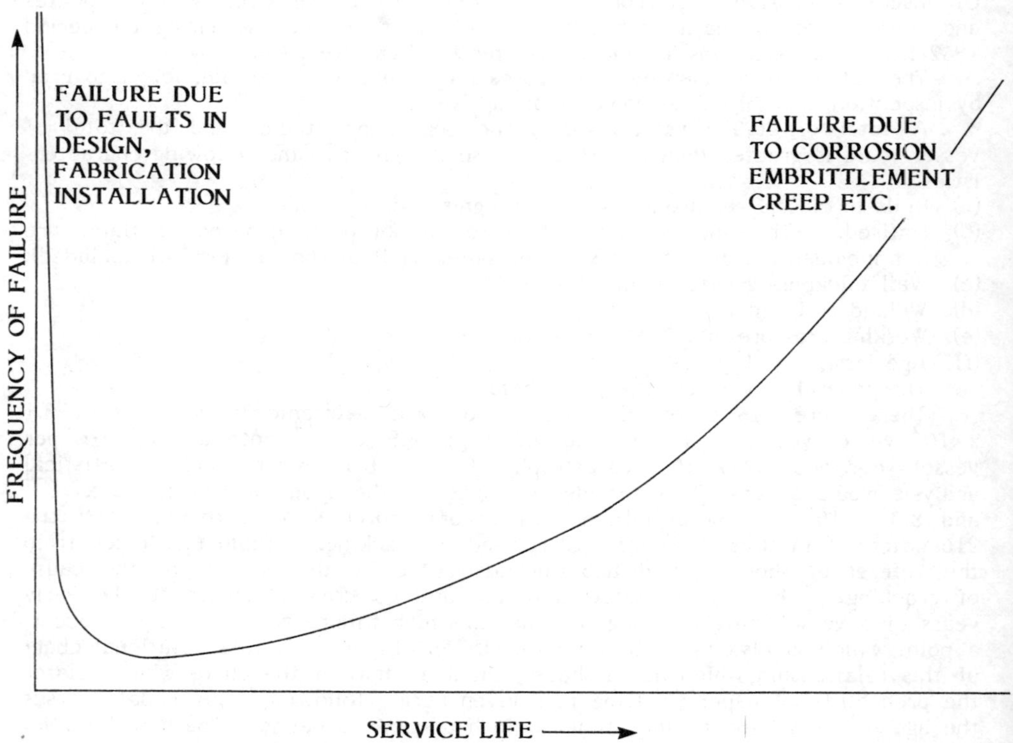

Fig. 6.3. "BATH TUB" CURVE OF PRESSURE VESSEL FAILURE.

6.2.3. THE ANALYSIS OF BUSH.

Bush (1975) is primarily concerned with US data but makes comparisons with data from the UK and in the FDR (West Germany). However, the paper inevitably makes the comparison with an earlier paper by the authors cited in sub-section 6.2.2, i.e. Smith & Warwick (1974).

Bush draws attention to the difficulty of establishing a reliable data base for US statistics comparing the situation unfavourably with that which obtains in the Federal Republic of Germany where there is a system of compulsory data collection and where, in spite of the smaller size of the economy, there are more extensive data. The Germans had data in 1972 for 470,000 vessels and 4.3×10^6 vessel years. US data is primarily for boiler drums in fossil fuelled power station. In the population of power stations there has never been a catastrophic failure of a steam drum built to ASME code section I or VIII.

Bush notes for LPG vessels in the US in the year 1973 that there were:-

(a)	In service	1 accident	0 deaths	3 injuries
(b)	Filling and handling	3 accidents	12 deaths	106 injuries

Bush considers the possibility of extrapolating from the statistically significant number of potential failures to obtain a more statistically reliable catastrophic failure rate but concludes that there are too many difficulties in establishing this ratio. He draws attention to Kellermann (1968) who claims that 75% of all failures during operation, of a population of 3×10^5 pressure vessels, were discovered by inspection.

Bush finally concludes that:-

(a) The 99% confidence upper bound for catastrophic failure of ASME section I steam drums is 1×10^{-5} per vessel year.
(b) For pressure vessels in the FDR the figure is $\leqslant 10^{-5}$ per vessel year.
(c) Welding defects are the principal cause of vessel failure.

It will be noted that the figures in (a) and (b) are an order of magnitude greater than those determined for the UK by Smith & Warwick.

6.3. MODES OF FAILURE.

6.3.1. THE DIVISION OF PROFESSIONAL RESPONSIBILITY.

There are a number of professional activities connected with the manufacture and continued operation of pressure vessels in the chemical and process industries. Some of these activities fall within the area of mechanical engineering, some within the area of chemical engineering and some fall within both. Some also fall within the area of metallurgy and possibly of other professional activities. These are set out in Table 6.2.

Human failure in any of the seventeen inputs listed (and the list is not necessarily exhaustive) may lead to vessel failure.

Vessel failure may be potential, i.e. inspection will show evidence that, unless corrective measures are taken, failure will occur. Alternatively vessel failure may be catastrophic, i.e. there is a total failure without warning. There are, of course, intermediate stages ranging from pinhole leaks to total failure.

It is not intended to attempt to make a comprehensive review of the ten activities listed. Instead there will be a discussion of those aspects which are mainly of interest to chemical engineers from the point of view of their likely influence on vessel failure. Attention is drawn to the implications which arise from the involvement of different people at different stages, for example though design and maintenance are listed as "mechanical engineering", in the general case those responsible for design will not be the same people as those

TABLE 6.2. PROFESSIONAL ACTIVITIES CONNECTED WITH PRESSURE VESSELS.

Activity	Mechanical Engineering	Chemical Engineering	Metallurgy
a Specification		X	
b Materials selection	X	X	X
c Design	X		
d Fabrication	X		
e Testing	X		
f Commissioning	X	X	
g Operation		X	
h Periodical inspection	X		X
i Maintenance	X		X
j Modification	X	X	X

responsible for maintenance. It may be noted, in this connection, that Table 6.2. lists ten activities, whereas, for example, Hutchings (1978), in reviewing service failures in the chemical industry, lists only six. Hutchings does not list "specification" separately from "design" though, as will be discussed below, there are important distinctions between these two functions. Again Hutchings combines testing with commissioning and omits periodical inspection and modification from consideration. The ten categories will be discussed, in order, below.

6.3.2. SPECIFICATION OF PRESSURE VESSELS.

The specification of a pressure vessel constitutes, together with the graphics, and the fabrication history, the description of the vessel. Taken on their own, the graphics constitute only an inadequate description of a vessel.

In the chemical and process industries the specification of a pressure vessel arises from the flow sheet. It is therefore drawn up by the process designer who usually will be a chemical engineer. The design of the vessel, described in a later sub-section, will usually be undertaken by a specialist mechanical engineer.

The specification will include the following items:-

(a) Total internal volume.
(b) Volumetric and gravimetric measures of contents.
(c) Description of contents including anticipated impurities and whether liquid or gaseous.
(d) Working pressure.
(e) Approximate dimensions and geometry.
(f) Attitude, i.e. horizontal or vertical.
(g) Whether for indoor or outdoor service.
(h) Range of temperature under service and fault conditions.
(i) Whether thermally insulated.
(j) Dimensions and position of any piped connections.
(k) Essential accessories including pressure relief system, instrumentation, water sprays, etc.

In many cases a pressure vessel will perform functions other than those of storage; for example, it may constitute a distillation column, boiler, heat exchanger or reaction vessel and the internals will be specified at the same time.

It must be noted that any departure from the specification during subsequent stages will constitute a modification as discussed under sub-section 6.11.

6.3.3. MATERIALS SELECTION.

Though the majority of pressure vessels which operate at above ambient temperature are made of carbon steels of various specification, for low temperatures alloy steels are also used as well as copper, brass and aluminium. Where carbon steels are used for handling liquefied gases the transition temperature, (see below), may be of considerable significance.

6.3.4. TRANSITION TEMPERATURE OF STEELS.

It is essential that every endeavour be made to ensure that brittle fracture will not occur as, unlike ductile failure, such failure gives no warning. However, as is pointed out in Thielsch (1965), steels which are normally considered to be ductile sometimes fail in a brittle mode at low temperatures. According to Thielsch the majority of cases have occurred in ships and bridges; of 3000 welded ships constructed in the US in World War II, 1200 sustained major hull fractures, 250 of which were considered very hazardous to the safety of the vessel, and at least 20 vessels broke into two sections. Some pressure vessels have failed in service by brittle fracture and in a few cases pipes have failed. Such failures are attributed by Thielsch to "notch brittleness", notch brittle materials being those at which fracture can be initiated at notches or weld defects at stresses less than the yield point stress. Thielsch gives nine case histories of brittle failure in the chemical and process industries. The transition temperature is defined by Thielsch as follows:-

"The transition temperature for any steel is the temperature above which the steel behaves in a predominantly ductile manner and below which it behaves in a predominantly brittle manner".

It might be noted that Thielsch points out that the transition temperature for any steel is hard to define precisely and that different test methods give different values. Thielsch's conclusions are embodied in Table 6.3., in which the present author has converted Thielsch's figures to SI units.

The behaviour of materials of construction, both metallic and non-metallic, is described in detail in Wigley (1978) and the Charpy V-notch impact test is described in BS 131 (1972). The Charpy impact test is also described in standard works on the strength of materials, as for example in Clark (1953). The numerical values obtained by Charpy tests are closely similar to those obtained by Izod tests which are also described in the same work.

For purposes of comparison the lowest outdoor ambient temperature in the UK is of the order of $-18^{o}C$. However, a propane leak on a mild steel vessel at ordinary ambient temperatures may produce, locally, temperatures tending towards $-40^{o}C$ which is below the mild steel transition temperature. Figure 6.4. illustrates the temperature brittleness relationship for a structural steel.

It is obvious from the above that mild steel is a material which is quite unsuitable for cryogenic liquids. The tanks of the sea-going "Methane Pioneer", which transports LNG at around -160°, are made of aluminium. These are not, however, pressure vessels in the sense discussed above. However, spillages of LNG on to mild steel supporting members or the mild steel deck, or the hull as the result of rupture of a tank in a methane carrier, could make such steelwork as brittle as glass.

TABLE 6.3. LOW TEMPERATURE LIMITATIONS OF VARIOUS MATERIALS OF CONSTRUCTION.

Low Temperature Limit	Material with ASTM Designation	Comments
- 18°C	Mild steel (A53,A120)	No special requirements
- 23°C	Mild steel (A53, A106, A135)	Charpy impact test ⩾ 20 joules at design temperature
- 45°C	Killed or limited carbon steel (A201, A212, A333)	⩾ 20 joules at design temperature
- 100°C	3 1/2% NI Steel (A203, Gr D,E. A333 Gr 3, A410)	⩾ 20 joules at design temperature
- 195°C	9% NI Steel (A353)	⩾ 20 joules at design temperature
- 250°C	Austenitic Stainless Steel (Types 304, 315 etc.)	Not normally subject to notch brittleness
No limit	Copper, brass, aluminium	Not normally subject to notch brittleness.

A particular point of danger arises in commissioning or recommissioning pressure vessels to store liquefied vapours. The first liquid to enter will flash and this will reduce the temperature of the lowest parts of the vessel to the atmospheric pressure boiling point of the liquid. Matters will be worse if the tank is evacuated to remove air before filling. In this case lower temperatures will be achieved after flashing. Methods of filling, in which the tank is filled first with vapour or pressurised with nitrogen, will avoid such dangers.

6.3.5. MATERIALS OTHER THAN CARBON STEEL.

In certain cases composite walls have been used. The reaction vessels at the Nypro caprolactam works at Flixborough (which operated about 130°C above ambient temperature) were of composite construction being made of 1/2" (12 mm) mild steel plate with an inner lining of 1/8" (3 mm) stainless steel bonded to it [Flixborough (1975)]. The object of this proceeding was to combine the corrosion-resistance of stainless steel, which is expensive, with the strength of the much cheaper mild steel. The failure of reactor 5 at that works will be discussed in Chapters 9 and 13. Some special problems relating to lined vessels are discussed in I.Chem.E. (1978). Aluminium, which is not subject to notch brittleness, has been used for pressure vessels. [It may be noted that aluminium is incompatible with mercury, contact with even small amounts of it such as are used in some instruments can ruin an aluminium pressure vessel completely].

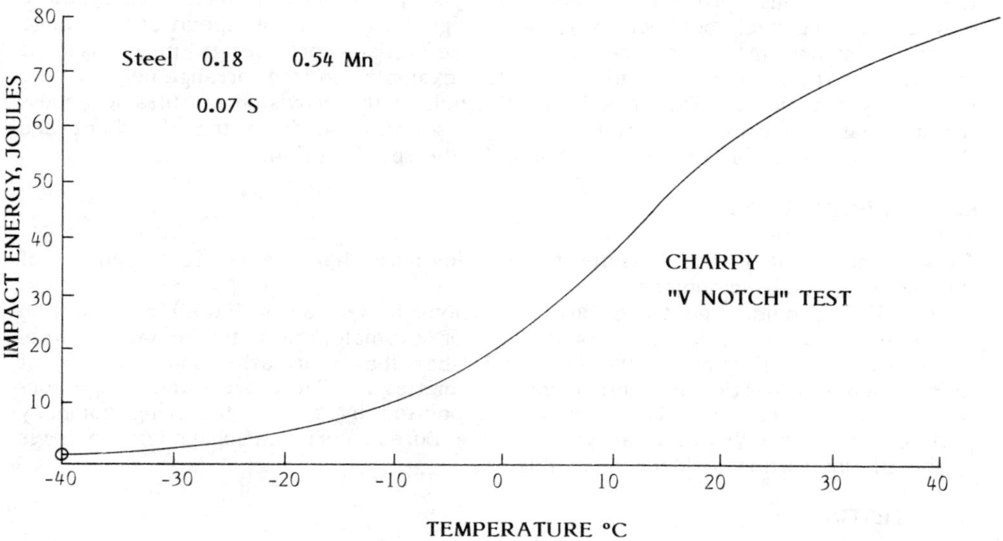

Fig. 6.4. NOTCH BRITTLENESS OF A STRUCTURAL STEEL VERSUS TEMPERATURE.

Alloy or stainless steels have been used extensively for pressure vessels as these steels are capable of withstanding corrosive conditions much better than carbon steels. However, there are problems connected with the use of stainless and other alloy steels and these problems are discussed in I.Chem.E.(1978). One special problem it discusses is that of zinc embrittlement of stressed austenitic stainless steel at high temperatures. The topic is raised in Flixborough (1975) but that it had been recognised for some years before that disaster as evidenced by the bibliography in I.Chem.E.(1978).

6.3.6. SPECIAL PROBLEMS POSED BY PARTICULAR SUBSTANCES.

Chlorine, when handled in steel, must be handled dry. Influx of water, as for example residual water left after hydraulic pressure testing, may produce corrosion in service. Anhydrous ammonia is liable, with certain steels, to produce embrittlement [Chemical Industries Association (1975)].

6.4. DESIGN.

Pressure vessels are designed to meet a specification and any departure from this specification will constitute an input failure which may lead, in some circumstances, to the failure of the vessel. Pressure vessels must be designed to a code which, in the UK, will normally be BS 5500. (Reference has been made above to the ASME code). Failure to comply with the appropriate code may lead to vessel failure.

As indicated previously, it is not possible to give a comprehensive review of all of the area and no attempt will be made to give a detailed review here of the entire field of pressure vessel design. It is necessary, however, to make a general observation. This concerns the inter-relationship between the specification, the materials selection and the design stages. The activities in the first instance must proceed in the order just given but, in many cases, there may be looping back between, say, the design stage and the specification stage as the designer may conclude that a more satisfactory design may ensue if the specification can be modified as, for example, in the arrangement of the connecting pipework. Design will usually include the provision of suitable access for internal inspection, a feature which does not arise from the flowsheet and this, for example, may give cause to modify the specification.

6.5. FABRICATION.

As with design it is not feasible to provide more than the briefest mention of this highly specialised subject.

BS 5500 provides guidance on fabrication in Section 4 (steel) and Section 4a (aluminium) including such topics as, for example, heat treatment of steel to achieve normalisation. BS 5500 prescribes that fabrication shall be subject to continuous inspection by "Inspecting Authorities". These are either inspectors representing insurers or else inspectors appointed by a manufacturing company holding a Pressure Vessel Quality Assurance Bureau certification under the aegis of the UK Institution of Mechanical Engineers.

6.6. TESTING.

BS 5500, in Section 5, lays down conditions for the testing of pressure vessels. This includes non-destructive testing (NDT). NDT is described in Birchon (1975) and includes such techniques as radiographic testing or ultrasonic testing. Radiographic testing is especially effective in detecting volume defects, i.e. cavities in welds. Ultrasonic testing on the other hand is most effective in detecting cracks. Surface flaws may be detected by magnetic particle or dye penetration techniques. Birchon lists 12 techniques in use in 1975.

Some form of pressure testing is obligatory; BS 5500 recommends hydraulic as opposed to pneumatic though, under certain conditions where hydraulic testing is impracticable, pneumatic testing is permitted. Even under the moderate pressure regimes discussed in this chapter there is no question but that hydraulic testing is safer as the amount of strain energy locked up is very much smaller than in the case with pneumatic testing.

The subject of pressure testing is discussed with case histories in Loss Prevention Bulletin (1979), where it is pointed out that at pressures of less than 20 bars (the range with which this chapter is mainly concerned), the strain energy contained in water is less than that of air at the same pressure by a factor of at least 10^3. Values quoted for the ratio W_A/W_W where W_A is the work done by air on rupture and W_W is the work done by water on rupture are:-

TABLE 6.4. RATIO OF WORK DONE ON RUPTURE VERSUS PRESSURE.

Pressure	2.4	5.8	13.6	19	26
W_A/W_W	0.6×10^3	6.4×10^3	4×10^3	2.8×10^3	2.25×10^3
Pressure	67	135	400		
W_A/W_W	10^3	5.5×10^2	2×10^2		

(Pressures are gauge pressures, in bars)

The above does not take account of the strain energy locked up in the vessel walls. Water, it is pointed out in the reference, is still dangerous, particularly at high pressures. Whichever mode is practiced pressure testing may be destructive and produce dangerous missiles. Test temperatures higher than 7°C are recommended to avoid brittle failures.

BS 5500 prescribes a test pressure derived as follows:-

$$P_t = 1.25 \quad P \frac{fa}{ft} \times \frac{t}{t-c} \qquad \text{where}$$

P_t = test pressure

P = design pressure (usually safety valve pressure)

f_a = nominal design stress at test temperature

f_t = nominal design stress at design temperature

t = nominal thickness

c = corrosion allowance

The test pressure must be maintained for 30 minutes. For hydraulic tests water is favoured, presumably on account of its low compressibility but where this creates difficulties, as, for example, where minute traces of water may cause corrosion, other liquids may be used even though their compressibility is higher. The BS warns of the dangers of testing vessels made of austenitic stainless steel with water containing chlorides. The present author has been told that the use of sea water recently for hydraulic testing in the Persian Gulf area led to the failure of a stainless steel vessel through bacterial attack. Vessels subject to hydraulic testing must be so vented, during the initial filling, as to guarantee that no air pockets remain.

6.7. COMMISSIONING.

Commissioning (start-up) procedures for chemical or process plant do not appear to have been made the subject of a monograph. There have been a number of papers scattered in the literature such as Gans (1976) and Whiston (1977). There has been an I.Chem.E.Symposium on the subject, I.Chem.E.(1974), and there is a paper on "Commissioning" in Loss Prevention Bulletin (1977).

The papers, however, have been concerned with the general principles of how to organise a commissioning operation, including the hazards associated with commissioning but the author has not been able to find one which deals specifically with the commissioning of pressurised systems. Though the paper by Whiston is concerned with the commissioning of an ethylene plant it does not deal with the particular problems of pressurised systems.

Commissioning is usually a joint exercise between a contractor who supplies the plant and the client, in which the contractor demonstrates the method of start-up to the client's operating team. This team will typically participate in the actual operation of start-up under the supervision of the contractor's commissioning engineers. These engineers usually remain on site until the acceptance trials are complete.

It is widely accepted that chemical plant has a greater risk of failure during the period of starting up than it has during steady operation and the commissioning period, the period of the first start-up, has an even greater risk than subsequent start-ups. These problems, so-called "teething troubles" are well recognised.

They can be usually minimised by careful attention to precommissioning in which the entire plant is first cleaned out to remove all erector's debris and any rust or dust. This is followed by leak testing of all closed volumes. Additionally, there is the need for mechanical checks of supports, brackets and steelwork.

Under no circumstances should commissioning be attempted unless in accordance with the manufacturer's manuals which, in the present author's view, should highlight the safety aspects in well-defined sections devoted to that purpose, and that purpose only, and which should explain carefully the reasons behind the dos and don'ts. One circumstance which must be particularly noted is that the experience of commissioning sometimes gives rise to a need to conduct modifications. Such modifications are an interference with the original specification and must not be attempted without careful assessment of their consequences (see Section 6.11).

6.8. OPERATION.

6.8.1. CAN PRESSURE VESSELS FAIL IN OPERATION ?

From a simplistic point of view it might appear that a vessel correctly specified, designed, fabricated and tested would not suffer failure in operation.

This, however, begs the question as to whether it is possible to achieve these aims. Where the materials being handled are materials with decades of operating experience to call upon, this is more easily achievable than with novel applications. Even so there are many possibilities of operational mischance.

6.8.2. THE ENTRY OF FOREIGN SUBSTANCES.

One example of operational mischance is the accidental introduction of foreign materials into the system, of which water or air are the two most likely. The admission of water into a storage tank containing methyl isocyanate has been blamed for the Bhopal disaster of 1984. Water may also suddenly reduce the boiling point in distillation systems with consequent sharp pressure rise. Air may form explosive mixtures with the contents of pressure vessels. Other process materials may enter. For example, it is believed that the serious explosion at Zarnesti, Roumania, 1939, was caused by hydrogen getting into a chlorine storage.

6.8.3. CORROSION.

Corrosion is an all-pervading problem, it can occur both internally and externally. It may be enhanced by the admission of air or water to the contents or by impurities in process liquids. The subject of pressure vessel corrosion is comprehensively reviewed in Hutchings (1978).

6.8.4. FLAME IMPINGEMENT.

Though it is practicable to design for low ambient temperatures, it is usually impracticable to design unfired pressure vessels to resist flame impingement through inherent strength. However some fired pressure vessels, such as steam boilers, rely upon maintaining temperatures in the walls of tubes which are close to those of the boiling point of the liquid by the cooling effect of the high heat transfer co-efficients which boiling liquids exhibit.

Where, due to operational mischance, flame impinges upon the walls of pressure vessels under such conditions that on the pressurised side there is gas or vapour, then wall temperature of the order of 700–800°C becomes feasible. As will be seen from Figure 6.5., at such temperatures steel becomes very weak

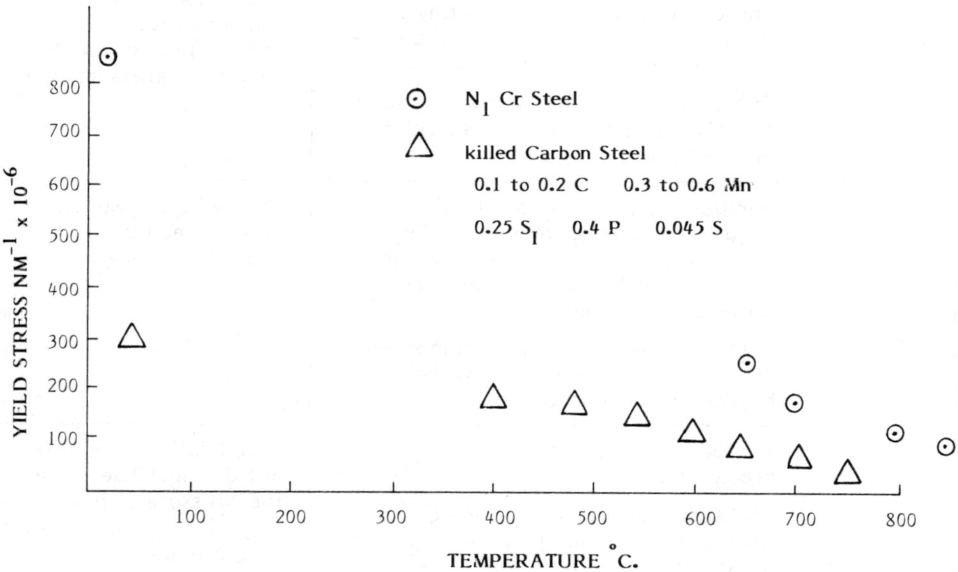

Fig. 6.5. STRENGTH OF STEEL VERSUS TEMPERATURE.

and catastrophic failure such as that giving rise to a "petal" fracture can occur. This phenomenon has been called a BLEVE (Boiling Liquid Expanding Vapour Explosion). It will be further discussed in Chapter 9 in studying the case history of the disaster at Feyzin in 1968. There will also be discussion of the shortcomings of the term BLEVE. The dangers of rupture following flame impingement are dealt with by careful design of the auxiliaries of pressure vessels, such as those intended to avoid fires starting underneath them, and by the provision of drench systems.

6.8.5. HYDRAULIC RUPTURE.

A further possibility of operational mischance is the possibility of hydraulic rupture. The basis of the phenomenon is well established in that it is well known that enormous pressures can be exerted by the thermal expansion of liquids. As pressure vessels in installations today have to be fitted with relief valves, it would seem that such pressures would be dissipated and are not capable of causing vessel failure.

However, transport vehicles may not be so fitted. Table 6.5. outlines the circumstances in which hydraulic rupture may occur. Hydraulic rupture has to be alleged as the cause of at least two major disasters, those of Ludwigshafen, 1948 and of San Carlos, 1978. The first of these was a rail car, the

TABLE 6.5. THE SEQUENCE OF EVENTS CONSEQUENT UPON HEATING A LIQUID STORED IN A TANK WITHOUT PRESSURE RELIEF.

Stage	Description	Pressure Regime
1	The tank is loaded with liquid.	Vapour pressure of the liquid, increasing with temperature. Failure may occur due to vapour pressure alone.
2	The contents expand on heating but the expanded volume is less than the internal volume of the tank.	
3	The expanded contents have a volume exactly equal to the internal volume of the tank.	
4	Further expansion of the liquid occurs which is accommodated by stretching of the tank walls within their elastic limit. The compressibility of the liquid is a factor.	Hydraulic pressure of liquid is added to vapour pressure.
5	Further expansion is accommodated by stretching of the tank walls beyond the elastic limit.	
6	A crack develops and rapidly propagates.	Rapid fall in pressure but during these phases the pressure is higher than the vapour pressure of the liquid.
7	The strain energy in the liquid and tank walls is dissipated; this may be accompanied by missile projection.	
8	The internal energy in the liquid is dissipated. In the case of a liquefied gas this will be a function of the energy which would be required to liquefy the gas. This dissipation will occur through the medium of flashing.	Pressure falling from vapour pressure of liquid towards atmospheric pressure.
9	Where flashing occurs reactive forces may operate, tending to propel the whole or the parts of the tank.	
10	The flashing will produce a vapour cloud with a degree of admixture with air.	Atmospheric pressure.

NOTES. The table is intended to cover the general case of a tank loaded with a liquid which therefore may, or may not, be a liquefied gas. Obviously in the vast majority of cases the sequence stops at stage 2. In some cases of malpractice the sequence may start at stage 3, i.e. where the tank is completely filled with cold liquid at loading. For non-volatile liquids the sequence stops at stage 7.

second a road car, and neither was fitted with a relief valve. The particular circumstances of these cases are discussed in Chapter 9. It is far more likely that hydraulic rupture will occur in pipes than in vessels. This possibility is well known to the chlorine industry for example and Chemical Industries Association (1975a) draws attention to it and the precautions which must be taken to avoid it happening.

6.8.6. RUPTURE OF REACTORS BY RUNAWAY REACTIONS.

6.8.6.1. The nature of runaway reactions.

There are numerous possibilities of mechanical failure of reactors, particularly in those operated above atmospheric pressure, such as by breakage of a sight glass or a branch. However, what is much more serious, though not often on the scale of the realisation of a major chemical hazard, is the possibility of rupture of a batch reactor, which is designed as a pressure vessel, due to a runaway reaction. Such a situation may arise when, for any reason, the temperature begins to rise. Temperature rise leads to increased reaction rate and hence increased output of heat. Unless this can be checked by increased cooling the temperature then rises still further with a further increase in temperature. If once the rate of heat output exceeds that which the cooling system can cope with, the reaction may be, by then, out of control. The situation is not self-correcting, reaction rates are an exponential function of temperature and hence so is the heat flux. On the other hand, rates of cooling are a linear function of temperature difference.

Runaway reactions are of relevance to this discussion only if they give rise to overpressure beyond the design limits of the system. This can arise in two ways. In the first, the overpressure is caused by the presence of volatile constituents whose vapour pressure may rise due to the rise in temperature. The second is by thermal decomposition, of which the most extreme example is that arising in the manufacture of explosives.

The subject has long been of interest to chemists and chemical engineers. The Institution of Chemical Engineers has organised two major symposia on the topic, [I.Chem.E.(1979) and I.Chem.E.(1984)]. The papers presented at these symposia demonstrated, inter alia, that there had been many other symposia, principally in the USA, devoted, in whole or in part, to this topic.

Barton and Nolan (1984) have provided a partial listing of 263 incidents of failure of batch reactors due to runaway reactions. The listing is mainly of UK incidents and is drawn mainly from the period 1962-82. The authors conclude that only very broad lessons can be drawn from the listing because of under-reporting both of the number of incidents and of the causative circumstances. The largest single cause seems to have been a basic lack of knowledge of the chemistry and thermochemistry of the process. In some cases, laboratory preparations had been transferred to full-scale manufacture without proper attention being paid to the factors involved in scaling up.

6.8.6.2. Causes of runaway reactions.

There are many batch processes in which reaction rates in the cold are very slow but when the reactants are heated to, say, 80°C to 120°C, an exothermic reaction takes place. Clearly such vessels then have to be cooled to dissipate the heat of reaction. This may create considerable problems of control. For example, if one of the reactants is being added continuously and the vessel is temporarily over-cooled and the reactant continues to be fed, the increase in concentration of this material in the unreacted form may lead, if the contents be reheated, to heat release rates which are beyond the power of the cooling system to control. Similarly, cessation of agitation or failure of cooling water or both may lead to the reaction getting out of control.

6.9. INSPECTION.

By inspection in this section is meant periodical inspection by a specialist. This may be either at the instigation of an inspecting authority or at the instigation of the operator. The activity is very similar to that of post-fabrication inspection but there are additional phenomena to be looked for which arise due to the passage of time. These include the effects of corrosion and erosion and such metallurgical phenomena as fatigue and creep. Clearly any cracks which appear will require close attention and possibly repair or, in the extreme case, the scrapping of the vessel. As has been shown in section 6.2., potential failures revealed by inspection considerably out-number those of catastrophic failure.

6.10. MAINTENANCE.

As with the maintenance of pressurised systems, which will be discussed below, maintenance must take place against a background of the vessel's standards (specification, material selection, design and fabrication). Maintenance must preserve and not deteriorate the vessel's standards.

6.11. MODIFICATION.

Modification, which can be unintentional or intentional, is a very hazardous activity. Any departure from the vessel specification may, in principle, lead to a failure. Operators must be ceaselessly vigilant to maintain the vessel standards by recognising the implications of any such departure as, for example, by changes in the chemical composition of the materials being handled or by changes in operating conditions.

Where deliberate modifications are intended, the stages of specification, material selection, design, fabrication, inspection and testing must be followed with the same meticulous care as for a new vessel.

6.12. PIPEWORK, PUMPS, ETC.

6.12.1. PURPOSE OF SUB-SECTION.

The purposes of this sub-section are to discuss process units of which one or more pressure vessels form part, but with especial reference to the pipework, valves and equipment for pumping fluids which form part of such a unit. As is demonstrated below the **system** is far more likely to fail than is the **pressure vessel** though the size of a leak from a system failure is likely to be much smaller than from a vessel failure. However, severe escalating effects have followed from pipework failures. At Feyzin, 1968, a sample point leak led to a fire which led to the rupture of a number of pressurised storages (see Chapter 9). At Flixborough in 1974 (see Chapters 9 and 13) a pipework by-pass collapsed, and in 1982 a small bore pipe leak on an ethylene plant led to severe damage to a major plant believed to have been in Canada [High (1982)].

6.12.2. RELATIVE LIKELIHOOD OF FAILURE.

The likelihood of failure may be ranked as in Table 6.6. Failure rates are taken from Green and Bourne (1981). These have been converted to frequency per 10^3 years by the present author.

TABLE 6.6. RANK ORDER OF LIABILITY TO FAILURE OF ITEMS OF EQUIPMENT.

Equipment Item.	Failure Rate $\times 10^{-3}$ years.
Pipe fittings, e.g. bends, tee junctions and small bore instrument connections.	350
Valves	260
Rotating shafts on pumps or agitators.	60
Expansion bellows	40
Pressure hoses	35
Pressure vessels	17
Gaskets	4
Straight piping	2

In general, pressure vessels are large unique, custom-built items, about which confusion over replacements is most unlikely to arise. On the other hand individual items of pressurised systems may be closely resembled by equipment in stock which may be intended for other, less severe, duty. For example, gasket material for joints may be easily confused.

Pumps and compressors are perhaps the most vulnerable items of pressurised systems because they contain moving parts which may revolve at speeds of up to 3,000 rpm, 1,450 rpm being a common speed. Pumps are subject to erosion and cavitation and they and compressors produce vibration which, in its turn, may lead to fatigue failure. Most pumps and compressors have an external motor and have a revolving shaft which must enter the equipment item through a seal and be supported by bearings. Both seals and bearings are liable to failure.

Agitator systems also present difficulties. Though they operate at much lower speeds than pumps they are subject to more severe mechanical stresses. Joints are a further vulnerable item not least because, on occasions, they have to be broken to allow for other items to be replaced. Errors in gaskets by using the wrong material or even, on occasions, omitting to insert one at all, have occurred.

Expansion bellows, wrongly used, were the immediate initiating cause of the Flixborough disaster which is discussed in Chapters 9 and 13. Expansion bellows, correctly used, promote safety by taking up thermal expansion. However they are bound, inherently, to be thinner than the pipework which they serve. They are, therefore, more liable to mechanical damage. If subjected to non-axial loads they may "squirm" as in Plate 11 of Flixborough (1975). (See also Teddington (1974)).

Valves, because they have moving parts, are more vulnerable than pipes and fittings. Except for diaphragm valves they have spindles which must be sealed. Diaphragm valves are, on the other hand, liable to diaphragm failure.

Pipe fittings are much more vulnerable than straight pipe because there is usually a change of direction or of cross-sectional area which may lead to erosion. However small bore pipe, leading, for example, to a pressure gauge, is particularly vulnerable to mechanical damage. High (1982) has reported an incident (believed to have occurred in Canada in the summer of 1982) in which a compressor house was wrecked, as well as a control building and other plant items, following the fracture of a small bore pipe at its junction with a pressure gauge on an ethylene plant.

Though straight piping is, so far as its intrinsic properties are concerned, probably the least vulnerable item it is still subject to failure. It can fail through hydraulic rupture or mechanical damage, say from a moving vehicle, or thermal expansion or contraction.

It has to be remembered that in a pressurised system the numbers of such items as pumps, valves, lengths of pipe etc are present in far greater numbers than the pressure vessels. The likelihood of failure somewhere in the system is, therefore, many orders of magnitude greater than that of the pressure vessels themselves.

6.12.3. DESIGN AND MODIFICATION OF PRESSURISED SYSTEMS.

Pressurised systems need the same care in specification, material selection, design, fabrication, inspection and testing as do the pressure vessels which they serve. Especially, they need to be properly supported and provision made to protect from collisions and to avoid thermal stresses or shocks.

As previously stated, the Flixborough disaster arose because of a modification which led to the failure of an expansion bellows system in which there were two expansion bellows at either end of an improvised by-pass. This was condemned in Flixborough (1975) which said:-

"The disaster was caused by the introduction into a well designed and constructed plant of a modification which destroyed its integrity".

It may be noted that the following faults occurred at Flixborough:-

(a) There was no formal specification; the modification was drawn in chalk in the workshop.
(b) No stress calculations were carried out.
(c) The system was tested at design pressure and not at the higher pressure called for by the British Standard then current.
(d) The test was a pneumatic test which violated the British Standard then current.
(e) Proper supports were not provided; the system was held up on scaffolding poles.
(f) In the course of the assembly of the by-pass the manufacturer's instructions on the safe use of bellows were ignored.

It is true, however, that the material was correctly selected and that no fault was to be found with the fabrication of the by-pass piece.

6.12.4. MAINTENANCE OF PRESSURISED SYSTEMS.

Reference has been made above to the possibility of confusion over replacement parts. The present author knows of an incident in a chlorine plant in which a titanium flange was fitted in error. The flange caught fire ! The incident draws attention to the necessity for the most careful system of spares control in connection with the maintenance of pressurised systems. This task may be facilitated by the use of computerised stock control.

Serious incidents, sometimes with fatal results, have attended errors made in the course of maintenance. Examples are those in which the system has been opened up before the pressure has been vented off, or the wrong flange has been broken or a valve has been opened inadvertently. Kletz (1985) draws attention to the dangers which attach to illogical numbering systems such as where a line of pumps has them labelled 1, 2, 5, 3, 4 in that order. There is a general discussion of the dangers associated with maintenance in Health & Safety Executive (1985). This reference claims that a quarter of all fatal accidents reported to it arise from maintenance.

The activity of maintenace in pressurised systems is one which must be controlled by a "permit to work" system. This system, which is more fully discussed in Chapter 20, is a highly formalised system in which maintenance is under the control of a single authority which lays down in writing the procedures to be followed before maintenance can commence or continue.

The key to the successful maintenance of pressurised systems thus lies in a number of measures. In the first place it must be recognised as an activity which demands a high level of engineering skill at all levels. The activity must be highly formalised and controlled by written permits and by measures such as a well-organised system of stock control for spare parts.

6.13. EXTERNAL FACTORS.

6.13.1. INTRODUCTION.

There are a number of modes of failure which fall outside the considerations normally encountered in the examination of flow sheets. Some of these are discussed below but the topics discussed are not, in any sense, put forward by the author as an exhaustive list. In any given location, attention must be paid to the local geographical factors.

6.13.2. STRUCTURAL COLLAPSE.

6.13.2.1. General.

Even the simplest installation has some structural elements, as there have to be foundations and supports for vessels and pipework. In the most elaborate cases, complex supporting structures in steel and/or reinforced concrete may be provided. Supporting structures may collapse for a variety of reasons and, in consequence, ruptures of vessels or pressurised systems may occur.

6.13.2.2. Fire.

A common cause of structural distortion, or even of outright collapse, is fire. Structural steelwork is readily weakened by fire unless encased in a suitable thermal insulation which gives time for the fire to be brought under control before the steelwork distorts. A case in point are the legs which support Horton spheres which are usually today encased in thermally insulating material.

6.13.2.3. Corrosion.

Structures may collapse because of corrosion. This is especially liable to occur near or at ground level. Bolts which hold the structure to the foundations may be corroded due to seepage under badly designed sole plates. Corrosion damage may not show up until the stucture is subjected to unusual stress for causes such as those described below.

6.13.2.4. Faulty design.

This can occur in a great variety of ways. One fault would be a failure to provide adequate bracing.

6.13.3. MECHANICAL CAUSES.

Mechanical impact can occur through collisions with vehicles or their loads. A cause of growing importance is the mal-operation of mobile cranes. This is illustrated in Kletz (1985) Fig. 3.2., which is reproduced as Figure 6.6. Structural steelwork not designed to act as a lifting beam, if so used may lead to local distortion or even to collapse.

Fig. 6.6. COLLAPSE OF A MOBILE CRANE (PHOTO).

6.13.4. WIND LOADING.

There have been many examples of failure of structures because of errors in estimating the effects of wind loading. These include the collapse of the Tay Bridge, Scotland (1879), the collapse of the Tacoma Narrows, USA, suspension bridge (1940), and three cooling towers at the Ferrybridge, England, power station (1965). The author is not aware of any incident in which a pressure vessel or pressurised system has suffered serious damage on-shore from wind loading but the possibility of such damage cannot be excluded.

Techniques for estimating wind loads for simple structures have existed since the eighteenth century [Smeaton (1759)]. The techniques of estimating wind loadings for structures such as chemical plant are necessarily more complex and fall outside of the subject matter of this book. They lie in the domain of mechanical and civil engineering. A standard work on the subject is Sachs (1972). However, there can be little doubt that the problems of the design of off-shore oil platforms which have similarities to chemical and process plant structures and where wind loadings are extreme, will have done much to refine practice in this area. An example of a work in this field is Dawson (1983).

6.13.5. EARTH MOVEMENT.

However well designed as a self-supporting structure capable of withstanding wind loading, there remains a residue risk of failure of chemical plant due to earth movement. The subject is discussed in Waltham (1978). This gives a technical review, with an extensive bibliography, of the problems of earthquakes, volcanic eruptions, landslips and subsidence. Though, in principle, all of these may affect pressurised systems, the discussion below will concentrate on earthquakes as being the most likely cause of failure of pressurised systems. There are other phenomena of earth movement which are not discussed by Waltham which may have significance. These are "frost heave" and the effects of conducting heat in "permafrost". This is ground which is frozen permanently in the absence of human activity and which provides apparently good foundations. However, if heat is conducted into it from process plant, it can turn into mud as has occurred in the USSR.

Earthquakes have been intensively studied as the cost in human lives and in physical damage in some parts of the world arising from earthquakes has been enormous. As might be expected, the greatest part of the research conducted into earthquakes has concentrated on the seismic design of buildings, bridges, roads, and other civil engineering structures. Building codes for earthquakes were evolved in Japan in 1923 and the US in 1925.

In recent years there have been extensive studies of the likely effects of earthquakes on nuclear power stations, especially in the US which has a large number of nuclear installations and also a number of areas of high seismic activity. Alderson (1985), which has an extensive bibliography, reviews the seismic design criteria for the UK. This reference includes case histories of earthquakes. In one of these, at the Paloma plant in Kern County, California, in 1952, a vapour cloud explosion was followed by a severe conflagration. This was caused by the collapse of two butane storages. A number of process plant structures exhibited damage to foundation bolts. However, examination of the other nine case histories in the reference shows little evidence of damage to process plant or public utilities. Alderson comments that "Severe earthquakes and modern complex single stream plants have not, as yet, interacted". Alderson goes on to discuss the types of damage which would occur to a hypothetical UK Major Hazard installation as the result of an earthquake with a risk of 10^{-4}.

Muir Wood (1985) has concluded that earthquakes pose a significant threat to the safety of petrochemical installations. Industrial activities which have not been seismically designed are located at a number of sites in Europe which are subject to earthquakes. He considers that some fairly recent geological changes (which carry with them an enhanced earthquake hazard) are the very ones which provide the planners with what seems to be ideal sites for petrochemical operations.

Earth tremors arising from man-made activities as, for example, mining subsidence, may exhibit some features of earthquakes.

The lesson from the above seems to be that all major chemical hazards sites should be appraised by geologists and that the structures should be appropriately designed to resist earthquakes.

Chapter 7

THE FORMATION AND DISPERSAL OF VAPOUR CLOUDS

7.1. THE HAZARDS ASSOCIATED WITH VAPOUR CLOUDS.

As has been discussed in Chapter 5, vapour clouds as a component of major chemical hazards arise principally from flashing liquids or from the spillage of cryogenic liquids. Of the two, the clouds which arise from flashing liquids are much the more serious problem.

Vapour clouds may give rise to three sorts of hazards, cataclysmic fires, vapour cloud explosions, or toxic effects, and a number, of which ammonia is an example, are both flammable and toxic. Flammability and explosibility are closely related and it is difficult to predict, if a flammable cloud be ignited, whether it will result in a fire or an explosion.

Flammable vapour clouds are flammable only within certain limits of composition which vary according to the substance involved. Figure 7.1. shows flammability limits for a number of substances which are significant major chemical hazards. With the exception of hydrogen and methane, the gases and vapours shown have a lower flammability limit between 1.5% and 3.0% by volume in air, the value being roughly inversely related to the molecular weight of the gas. It will be noted that olefines have a wider range than the corresponding paraffins. The explosive range is somewhat narrower than the flammable range shown in Figure 7.1. Thus for flammables the hazard is mostly related to concentrations in excess of $1.5 - 3.0 \times 10^4$ parts per million by volume.

Matters are very different with toxic gases where there is a wealth of evidence that the lethal concentration may be less than 10^2 parts per million by volume.

On these simple grounds it is apparent that, other things being equal, toxic gas clouds are likely to be dangerous at much greater distances from the point of release than are flammable clouds.

7.2. THE FORMATION OF VAPOUR CLOUDS.

There are two extreme manifestations of vapour cloud phenomena. The first is that of a continuous release such as could occur where the material is released at a steady rate over a period of, say, an hour or more. The second is the near instantaneous release which could occur in the event of the catastrophic failure of a vessel containing a substance capable of flashing. This manifestation is sometimes termed in the literature as a "puff". The first manifestation is analogous with a continuous chemical process in that such properties of the system variables as the dimension of the cloud and the material concentrations, remain relatively constant with time. The second manifestation is analogous

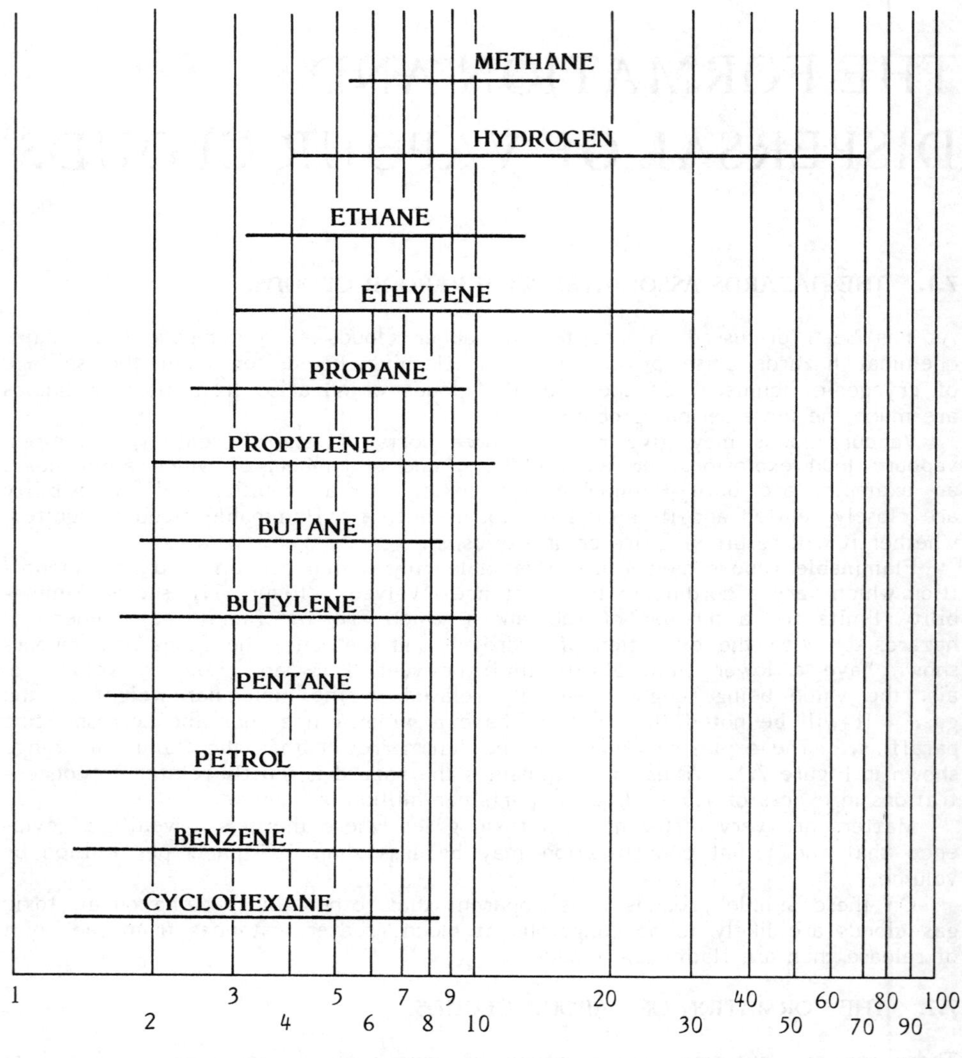

Fig. 7.1. FLAMMABILITY LIMITS FOR SOME GASES AND VAPOURS.

with a batch process in that these properties vary continuously with time. The theoretical treatment of the first manifestation is simpler than the treatment of the second. The second, the instantaneous release, is described below in some detail. The first manifestation is a problem which is fairly readily capable of solution once the necessary techniques have been developed to solve the second.

The two manifestations described above, and the infinite number of intermediate or other possibilities, some of which have been treated in Chapter 6, are described in the literature as "source terms". They represent the basic assumptions upon which the necessary algorithms for predicting vapour cloud behaviour are founded. Such algorithms may form the basis of computer programmes such as "DENZ" described in Fryer and Kaiser (1979).

7.3. THE DISPERSION OF VAPOUR CLOUDS.

There are two principal stages in the dispersion of a vapour cloud formed as the result of a catastrophic loss of containment of a liquefied gas. The first of these stages is one in which the cloud slumps under gravity. [This assumes that the vapour cloud is heavier than air; the validity of this assumption is discussed below]. During this slumping phase, the duration of which is of the order of a minute, the influence of the ambient weather conditions is not significant.

The second phase is the dispersal of the slumped cloud by admixture with air. This phase may have a duration of tens to hundreds of minutes and is influenced by and, at considerable dilution, dominated by the ambient weather conditions.

7.4. TRIALS WITH HEAVY GASES.

Trials of releases of heavy gases have been conducted at a number of sites. Some of these are tabulated in Table 7.1. and will be discussed in greater detail later in the chapter.

McQuaid (1979) has reviewed progress in the field up till 1979. Experiments involving the ignition of flammable gases have also been conducted at a number of sites. They are not listed above but instead will be discussed in Chapter 8, Cataclysmic Fires.

7.5. THE MODELLING OF THE RELEASE OF HEAVY GASES.

Considerable attention has recently been directed to the modelling of heavy gas releases concurrently with the trials which have been conducted at the Chemical Defence Establishment, Porton Down, England, and those which have taken place at Thorney Island, West Sussex, England, (see Fig. 7.2.). The proposed Thorney Island experiments were described in McQuaid (1982).

There is a considerable financial incentive to produce a reliable model as full-scale trials are extremely expensive. Even the Thorney Island experiments, which are expected to cost in excess of £1 million (1982), do not represent the full scale of industrial release. Thus the ability to transfer experiments to a comparatively inexpensive wind tunnel, backed by a model which has been verified by full-scale testing, would represent a major achievement. However, there are serious problems in the way of achieving such a model. The question of models and the problems of scaling are comprehensively reviewed in McQuaid (1979, 1982, and 1982a) and in Havens (1982). Field and other trials will be further discussed below.

TABLE 7.1. SOME TRIALS WITH HEAVY GASES.

Date	Place	Sponsored by	Agent	Continuous or Instantaneous	Quantity Used	Initial Density of Cloud	References
April 59	At sea between Algeria and Canvey Island	North Thames Gas Board	L N G	Continuous, $3m^3$/min	$20m^3$ Liquid	Not known	Kneebone, A. & Prew, L.R. 1974
July 73	At sea, 100 km west of St. Nazaire, France	Shell	L N G	Continuous, $3m^3$-$20m^3$ per minute	$627m^3$ Liquid (in total)	Not known	Kneebone, A.& Prew, L.R. 1974
Autumn 73	30 km west of Rotterdam	Ministry of Social Affairs, Netherlands	Freon 12* without smoke	Instantaneous, 2 trials	1000 kg Liquid	1.25 times air	Uldern, A.P. van 1974, M S A 1975
1978	Chemical Defence Establishment, Porton Down	H.S.E.	Freon 12 plus air with smoke	Instantaneous, 42 trials	$43m^3$ Freon/Air Mixture per trial	1.03 to 4.20 times air	Picknett, R.G. 1978, McQuaid, J. 1979
1981 to date	Warren Spring Laboratory, Stevenage, UK Wind Tunnel	H.S.E.					Hall, D.G. 1982
1982 to 1984	National Maritime Institute, Thorney Island, West Sussex, England	Consortium of 30 public and private corporations	Freon 12 plus air with smoke	Instantaneous, 5 trials (minimum programme)	$2000m^3$ Freon/Air Mixture per trial	2.00 times air	McQuaid, J. 1982 & McQuaid, J. 1982a

* Dichlorodifluoromethane, $C\,Cl_2\,F_2$

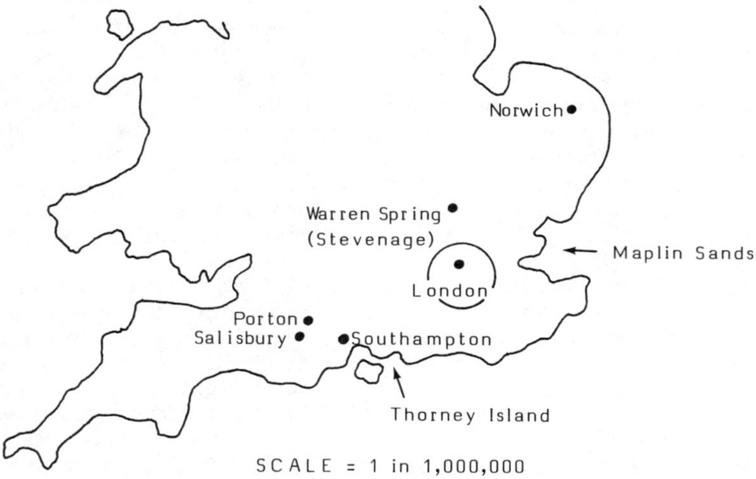

Fig. 7.2. LOCATION OF FIELD AND OTHER TRIALS IN THE UK.

7.6. THEORETICAL APPROACHES TO HEAVY GAS DISPERSION.

7.6.1. INTRODUCTION.

The discussion below divides the phenomenon of the dispersal of a cloud of gas which is more dense than air into two phases. The first is the gravity slumping phase which is then followed by the mixing and dispersion phase. There is no sharp boundary between the two phases, in fact mixing occurs all the time, even during the initial formation of the cloud. However, during the gravity slumping phase, mixing may be a secondary effect.

Because of the way the subject has developed historically, it is convenient to discuss the influence of the weather first, even though it plays only a minor role during the gravity slumping phase which precedes mixing and dispersion.

7.6.2. DISPERSION AND METEOROLOGY.

The understanding of the mechanism of gas dispersion requires an understanding of at least a part of the science of meteorology, that part which is known as "micrometeorology" or "boundary layer meteorology" which is concerned with the behaviour of air masses at heights less than 100 m above the surface of the ground.

It is not possible to give more than an outline of the relevant factors here; the subject as a whole is treated in standard textbooks. Of these, the most authoritative and relevant is Pasquill & Smith (1983). An earlier work which is still useful is Sutton (1953). In this book Sutton discusses the subject of micrometeorology or the physical processes in the lower layer of the atmosphere. The same author summarises the subject in Sutton (1974). Two works which treat the climate more locally are Lamb (1964), which deals with the northern hemisphere but especially the climate of the British Isles, and Chandler (1965) which treats of the climate of London.

7.6.3. STABILITY AND INSTABILITY.

The rate at which a cloud of gas of a hazardous concentration may be reduced to a concentration so low as to be harmless depends greatly upon a property of the atmosphere known as its "stability". It is well known that the pressure of the atmosphere falls as the distance from the surface increases. This has, as a concomitant, a general tendency for the temperature to fall as the distance from the surface increases. This fall in temperature is known as the "lapse rate" which is the rate at which temperature falls with height under adiabatic conditions. As such it is known as the "dry adiabatic lapse rate" (D.A.L.R.) and is denoted below by Γ. It has a value in low altitudes which is almost exactly equal to $1^{o}C/100$ m.

Though Γ has this value on the average there may be marked variations from it in actual conditions. In some circumstances the values of the lapse rate may be greater than Γ (superadiabatic) or lie between Γ and zero (subadiabatic) and, in certain cases the lapse rate may be "inverted", i.e. the temperature increases with height and thus the lapse rate has a negative value. These departures are most marked close to the ground. In general, lapse rates $\geqslant \Gamma$ are classed as "unstable" and lapse rates $\leqslant \Gamma$ are classed as "stable", negative lapse rates being the most stable form. In general, in clear weather, there are high lapse rates during the day and negative lapse rates (inversions) during the night. In an English summer, near the ground at the hottest time of the day, lapse rates exceeding 600 Γ have been observed [Sutton (1974)]. In general, in clear weather, instability exists for most of the day in summer and the reverse is true in winter. Cloudy or overcast weather considerably damps down both instability and stability. The subject is treated in detail in Pasquill & Smith (1983) and Sutton (1953).

In general, the rate of dispersion of gas clouds is highest for high values of lapse rate and lowest for negative rates.

7.6.4. THE PASQUILL CATEGORIES.

Pasquill (1961) investigated the dispersion of windborne material and this work has since been extended by that author [Pasquill & Smith (1983)] and by Bryant (1964). These authors are frequently quoted in the literature on dispersion but it is now generally accepted that some of their conclusions are inapplicable to releases of gases or vapours whose density is markedly greater than air. However, Pasquill's categories of weather remain central to the present approach to the problem and they are displayed in Table 7.2.

7.6.5. THE BRYANT APPROACH.

The work of Bryant was directed in the main towards the prediction of the spread of windborne radioactive contamination, principally from vents and chimney stacks, though her calculation methods included some for ground level releases. This work was initiated by the U K A E A following the incident at Windscale, Cumbria, UK on 10th October, 1957, in which a quantity of radioactive material was discharged from a chimney. This incident gave rise to much adverse publicity. Bryant's method has the following characteristics:-

(a) It is founded upon the previously cited work of Sutton and Pasquill, and for long releases, upon Beattie (1963).
(b) It is applicable to short releases (i.e. up to several minutes' duration or to long releases (i.e. up to six hours' duration) or a continuous release (indefinite duration).

TABLE 7.2. PASQUILL'S STABILITY CATEGORIES.

Pasquill Category.	Typical Wind Speed $m\,sec^{-1}$	Description	Lapse Rate	Frequency of occurrence in UK.
A	1	Very sunny and cloudless.	$>>> \Gamma$	.017
B	2	Sunny and warm.	$>> \Gamma$	.084
C	5	Partial cloud during day.	$> \Gamma$	.168
D	5	Overcast day or night.	Γ	.410
E	3	Partial cloud during night.	$< \Gamma$	.118
F	2	Clear night.	Negative	.203

Bryant adds a Category G which is "very calm with clear sky in winter, inversion persisting all day". This would represent the highest level of stability.

(c) It assumes that the concentration profile measured in both cross wind and vertical directions has a Gaussian distribution.
(d) The dispersed material has neutral buoyancy.

The reference gives a table which shows the frequency of occurrence of Pasquill's stability categories for different places in England and Wales and in Scotland. However, Bryant's approach is today regarded as being inapplicable to releases where the buoyancy of the cloud differs markedly from air, a consideration which applies to virtually all problems associated with liquefied gases.

7.6.6. THE WORK OF VAN ULDEN.

Van Ulden was the first to call attention to the inapplicability of the Bryant model to situations of negative buoyancy. In the reference cited he analysed the results of work discussed below which was conducted by the Netherlands Ministry of Social Affairs [N M S A (1975)]. This work was conducted in part at least to test out his theories. In his paper, van Ulden (1974), he set out first a basic equation for the velocity of the wave front of a vapour cloud slumping under gravity.

$$u_v = c \sqrt[2]{g \frac{\rho 1 - \rho 2}{\rho 1} h_w}$$

where the symbols have the following meanings:-

Symbol	Description	Dimension
u_v	velocity of vapour front	LT^{-1}
g	acceleration of gravity	LT^{-2}
$\rho 1$	density of denser phase	ML^{-3}
$\rho 2$	density of lighter phase	ML^{-3}
h_w	height of wave front	L
c	constant	dimensionless.

In the case of water from a dam burst he argued that

$$\rho 1 \; >> \; \rho 2 \quad \text{and hence} \quad \frac{\rho 1 - \rho 2}{\rho 1}$$

He drew attention to the experimental results which showed that for low values of $\frac{\rho 1 - \rho 2}{\rho 1}$ the advancing front has a distinct head or roller but suggested that, nevertheless, the assumption of a vertical front and a horizontal top agreed well with the experiment. Van Ulden's view was that the constant c, both on theoretical grounds and from experimental results, could be taken as equal to unity.

Thus van Ulden's equation would reduce to:-

$$u_v = \sqrt[2]{g \left[\frac{\rho 1 - \rho 2}{\rho 2}\right] h_w}$$

Van Ulden also derived the equation

$$u_v(r) = c_r \sqrt[2]{g \left[\frac{\rho o - \rho a}{\rho o}\right] \frac{V_o}{\pi}}$$

where the symbols have the following meaning:-

Symbol	Meaning	Dimensions
$u_v(r)$	Velocity at radius r	LT^{-1}
c	constant	dimension-less
r	radius	L
g	acceleration of gravity	LT^{-2}
ρo	density of gas cloud at zero time	ML^{-3}
ρa	density of air	ML^{-3}
π	constant	dimension-less
V_o	velocity at zero time	LT^{-1}

Van Ulden claimed that vertical mixing was negligible where $u_v >> 2u$ and where u is the "friction velocity" described in Monji and Businger (1972). Thus transition from gravity slumping to turbulent mixing occurs when $u_v \simeq 2u$ and when $\rho r \simeq \rho a$. Van Ulden gives a value of $0.25 \; m^{-1}$ for u at a time of 80 secs from initial release under the conditions of the Netherlands experiments. Van Ulden's conclusions from the experiments were as follows:-

(1) The Gaussian model cannot be used for heavy gases without a proper correction for density spread.
(2) Strong winds and insolation reduce the effects of density spread.
(3) Density spread may occur against the wind and hence an upwind position is not necessarily safe.
(4) For releases during the night, very considerable density spread can occur.

7.6.7. THE McQUAID APPROACH.

This approach is described by McQuaid in a series of papers, McQuaid (1979,

1981, 1982, 1982a and 1982b). It is the approach now adopted by the UK Health and Safety Executive.

In McQuaid (1979), he reviewed both the theoretical approaches at the time of writing and the preliminary results of the Porton Down trials of 1978. A principal conclusion of this paper confirmed the view of van Ulden that models based upon neutral buoyancy are inapplicable to heavy gas spills. However, McQuaid took the opposite view to van Ulden who had claimed that "As a result of density spread higher concentrations over larger areas should be expected" [van Ulden (1974)]. McQuaid's view was that "The pendulum has now swung the other way, in that recently developed models . . . predict a shorter hazardous distance than applies for a neutrally-buoyant material emitted at the same rate". In short, McQuaid considers that the Pamela Bryant approach when applied to heavy gases gives pessimistic answers and therefore that neutral buoyancy models exaggerate the range over which concentrations of vapours are harmful.

The reference also reviewed the recommendations made by the Advisory Committee on Major Hazards for research in the field and outlined a proposed scheme of experiments which would meet these recommendations. In McQuaid (1982b), the author discusses "Box Models" of vapour clouds. He draws attention to the ambiguities in the terminology. The term "box", in his view, can apply to the physical shape of the cloud, which may be envisaged as a cylinder within the boundaries of which the concentration of the vapour air mixture is assumed to be uniform. It can also relate to the concept of a "black box", a concept rather widely used to denote a system in which the internal workings are unknown or disregarded but where knowledge exists of the relationship between input and output. An every-day example is that of a motor vehicle which can be satisfactorily driven by persons ignorant of the principles of the operation of the engine, transmission, steering or brakes. A more sophisticated example is that of the home computer or the hand-held calculator where only a tiny minority of the users are acquainted with micro-chip technology.

It may be noted that the model described in Fryer and Kaiser (1979) which is used as the basis of their computer programme DENZ assumes a Gaussian concentration distribution within the box boundary. McQuaid used this programme to predict the "lethal dosage envelopes" presented in Barrell (1981), acknowledging [in paragraph 4.4. of McQuaid (1981)], that the DENZ model is one with a non-uniform concentration profile.

The question as to whether the lethal dosage envelope approach is a valid predictor of fatalities will be discussed below in Chapter 18.

7.6.8. OTHER APPROACHES TO THE MODELLING OF HEAVY GAS DISPERSAL.

The brief review above of the work of van Ulden and McQuaid may not do justice to the work of many other investigators, many of whose conclusions are incorporated in McQuaid's approach or have been used by Fryer and Kaiser in constructing the DENZ programme.

The recently published Dense Gas Dispersion issue of the Journal of Hazardous Materials [Britter and Griffiths (1982)] gives a comprehensive overview of the work of investigators such as Havens and Fay. Haven's work on classifying models, [Havens (1982)], will be discussed below. An example of Fay's work is Fay (1980).

7.6.9. REVIEWS OF PROPOSED MODELS.

Two important reviews in the area were published in 1982. One is that of Blackmore et al (1982), and the other is that of Havens (1982). There have been, in addition, a number of reviews of the closeness of fit between mathematical models and the results of field trials; these will be referred to in Section 7.7.

Blackmore et al review 15 models and the paper cited is claimed to represent an abbreviated version of a survey conducted by a working group of the UK Chemical Industries Association. In common with other investigators Blackmore accepts as a basis the concept that heavy vapour systems cannot be adequately described by models suitable for neutrally buoyant or positively buoyant clouds. Blackmore draws attention to the practical importance of models which can predict the spread of clouds generated by the spillage of up to 25,000 m^3 (ca 12,000 tonnes) of LNG from sea-going tankers. Blackmore distinguishes between two basic types of model.

(a) K-type, or "eddy diffusivity" models.
(b) Slab models (these are the same as "box" or "top hat" models).

Both basic types are claimed to oversimplify reality.

The K-type models integrate numerical equations of mass, momentum or energy conservation and these can be integrated in two or three dimensions. Mass transfer occurs by eddy diffusion and is proportional to the concentration difference.

Slab models assume that mass transfer takes place across the density interface of a cloud which has an assumed geometry (usually cylindrical with vertical sides and a flat top). Internal mixing is assumed to be so rapid as to produce a uniform concentration within the cloud. Air entrainment rates depend upon turbulence levels, density differences and cloud speeds. For both basic types it is assumed that a point is reached where there is a transition (abrupt or gradual) to nearly neutral buoyancy and that the model must incorporate criteria for determining when this point is reached.

In addition to the division between the two models described above there is a further sub-division between continuous sources such as are provided by steady-state pools or small ruptures and instantaneous releases such as those due to catastrophic failures sometimes known as "puffs". As has already been discussed in Chapter 5, many real situations have characteristics of both of these.

Blackmore (1982) tabulates five K-type models and ten slab-type models with their characteristics, giving in successive tables an overview of capabilities, references to comparison with experimental data, lists of experimental discharges, details characterising the models including transition to neutral buoyancy criteria and a list of obsolete models and their replacements. There is, in addition, a table of models of jet release for elevated sources.

The paper lists the assumptions common to all or most of the models.

Havens (1982) approaches the problem mainly from the standpoint of the analysis of the mathematics with particular reference to the form and structure of the equations and to the values of the parameters adopted by the different modellers. Havens recognises that some of the models, particularly the K theory models, require long runs on powerful computers and draws attention to Fay (1981) which is a detailed comparison of top hat models.

7.7. FIELD AND OTHER TRIALS OF HEAVY GAS DISPERSION.

7.7.1. OVERVIEW OF FIELD TRIALS.

In an overview of field trials by Puttock et al (1982), attention is called to the wide discrepancies to be observed when different models are used. For example, in Havens (1978), the distance travelled to reach 5% CH_4 concentration from a 25,000 m^3 spillage of pure LNG was predicted on one model to be 0.75 miles (1.3 km) and on another model to be 25-50 miles (44-88 km). These were both calculated for 5 mph (8 km/hr) and Pasquill D conditions. This discrepancy represents a factor of between ca 35 to 1 and ca 70 to 1.

The paper points out that laboratory scale wind tunnel and experiments using liquids of differing densities in tanks are very important, and may be even more important in the future, if it can be shown that they can model large-scale releases successfully. But this cannot be done without large-scale trials as it is not possible to extrapolate, with confidence, from the laboratory to full scale at the present time.

Field trials conducted up to the present have been of three basic types.

(a) In which the vapour cloud is generated from liquefied gas by rapid evaporation and without ignition.

(b) As (a) but with subsequent ignition.

(c) In which the vapour cloud is released at atmospheric temperature and pressure as a mixture of a heavy inert gas with air, usually incorporating a smoke as a marker.

Most of the field trials under (a) and (b) to date have involved LNG (95% methane) because of the desire to produce rapid results in an area where there is considerable public concern. Some such experiments have been on a 10 tonne plus scale.

However, as is pointed out in Puttock (1982), spillages of LNG introduce complications as there are considerable thermal effects, and the gas itself, at atmospheric temperature, is lighter than air. Freon/air mixtures and, to a lesser extent, liquid propane, do not have these disadvantages. Freon/air mixture experiments, though they enable gravity slumping and dispersion to be studied in isolation from the thermal and other effects noted in connection with LNG have, up to the present, been conducted on a scale well below that of many notable examples of spillage associated with serious consequences. For example, there was a release of ca 30 tonnes in the disaster at Ludwigshafen 1948, and of ca 20 tonnes in the disaster at San Carlos 1978.

20 tonnes of propane, density 1.45, would have a volume of 10,000 m^3 at NTP which compares with the 2,000 m^3 projected volumes for the Thorney Island trials (see Table 7.1. and discussion below). The Thorney Island trials thus were planned to be roughly one-fifth of the scale of commercially important spillages.

7.7.2. SPILLAGES INTO BUNDS.

Experiments involving spillages into bunds are reviewed in Puttock (1982). Some which involved liquid oxygen are described in Lapin and Foster (1967). Others involved 100 - 1,500 kg of LNG and were conducted by, or for, Gaz de France and the American Gas Association.

7.7.3. LIQUEFIED GASES SPILLED ON WATER.

Puttock lists 13 series of tests involving spillages on water. The US Bureau of mines conducted a series of trials in which 20 - 250 kg quantities of LNG were spilled onto ponds. These are described in Burgess et al (1970) and Burgess et al (1972).

Reference has been made previously to the paper Kneebone and Prew (1974), which reported on the trials in which LNG was jettisoned by the 75,000 m^3 Shell LNG carrier "Gadila" at a scale of up to 100 tonnes of LNG. This paper drew attention to the phenomenon of the "flameless explosion".

Kneebone and Prew also reported on the April 1959 tests in which 20 m^3 of LNG were discharged at about 160 m^3 per hour rate from the ship "Methane Pioneer". It may perhaps be noted, that though Puttock characterises the composition of LNG as being more than 95% methane, this was not true of the "Gadila" trials where the LNG was only 87% methane, C_2 hydrocarbons being present at 9% concentration and C_3 at 2.75% concentration. This markes a further complication in that, as compared with field trials involving Freon, LNG is not a pure chemical substance. The principal conclusions drawn by Kneebone and Prew were:-

(a) There was no detectable gas concentration on the ship even when stationary and with a wind speed of only 3.8 knots (1.95 m sec^{-1}).
(b) Using a 51 mm (2") nozzle there were no LNG pools on the surface.
(c) Using a 100 mm (4") nozzle there were a few patches of LNG on the surface and the vapour cloud was 60-70 metres away from the ship.
(d) The vapour cloud dispersed 5-10 minutes and 15-20 minutes after the discharge was completed from the 51 mm and 100 mm nozzles respectively.
(e) Static charges developed were too low to constitute an ignition hazard.

7.7.4. LIQUEFIED GASES SPILLED ON LAND.

The Netherlands tests, which were reported on in van Ulden (1974), NMSA (1975), and in Buschmann (1975), used 1000 kg quantities of Freon 12 which were rapidly evaporated by being discharged into hot water to give an initial mixture with a density about 1.25 times air. The atmospheric conditions were stable and the ground wind speed was about 3 m sec^{-1}. The height of cloud initially was 5 m but it slumped to 0.20 metre as its lowest point recovering to 10 metre at 1 kilometre distance from the point of release. The trials gave a considerable degree of confirmation to van Ulden's theories.

The Naval Weapons Centre, China Lake, California 1978 trials (the Avocet series) have been reported in Koopman et al (1980) and the 1980 trials (the Burro series) have been reported for Burro 2 to Burro 9 in Koopman et al (1982). The China Lake trials in essence were spillages of 10 to 18 tonnes of LNG onto a 58 m diameter lake to provide evaporation at spill rates from 5-8 tonnes per minute under wind speeds of 1.8-9.1 m sec^{-1} and unstable to slightly stable atmospheric conditions.

A notable feature of these trials were the occurrence of rapid phase transit (RPT) explosions in two of the spills. (These are "flameless explosions" of the kind referred to above). Overpressures of up to 5×10^3 ρa (0.70 psi) arising from these were recorded at 30 m from the spill point. This maximum level was calculated as equivalent to 6.3 kg TNT. Koopman concludes that it cannot be ruled out that RPT explosions may be sufficiently energetic to detonate the ethane rich gas present at this stage of the evaporation of the spill. It was not clear why RPTs occurred in only two of the spills.

Ermak et al (1982) have used the results of the Burro series to check the accuracy of a number of dispersion models. They found that a modified Gaussian plume, the GD model, gave a cloud which was narrower and higher than the actual by a factor of 1.5 to 3.0. This seems to contradict the view expressed by McQuaid and quoted above. A slab model on the other hand gave a good prediction of cloud width and distance to the lower flammability limit. It also predicted well the time to disperse to below the lower flammability limit. It tended, however, to overestimate the vertical concentration gradient, particularly under conditions of high wind speed.

The FEM 3 model, which is a fully three-dimensional conservation equation model, due to Chan et al (1981), gave generally better agreement with experiment than the slab model. FEM 3 also predicted correctly a bifurcated cloud structure for one of the series which was conducted at low wind speed but not at the low elevation at which the phenomenon was actually observed. Table 7.3. below, taken from Ermak (1982), compares the experimentally determined distance of the lower flammability limit of the clouds with that predicted by the three models studied. The experimental limit is accurate only to -40 to +20 metres. It is of importance also in demonstrating the considerable distance that such clouds can travel and still be capable of ignition. The distance correlates inversely with wind speed.

TABLE 7.3. MAXIMUM DOWN WIND EXTENT OF LFL IN BURRO SERIES, CHINA LAKE, 1980.

Test No.	Quantity M^3	Spilled Tonnes	Wind Speed $m\ sec^{-1}$	Experiment	G D	SLAB	FEM 3
				Distance of LFL in metres			
3	34	16	5.4 ± 1.8	255	126	215	190
7	39.4	18.5	8.4 ± 1.2	200	150	264	210
8	28.4	13.4	1.8 ± 0.3	420	661	418	630
9	24.2	11.4	5.7 ± 0.7	325	235	315	330

The wind speed is from Table 1 of Koopman (1982).

7.7.5. MAPLIN SANDS TESTS.

There is a preliminary report on these tests in Puttock (1982). They were conducted by Shell Research Ltd at a remote site near Foulness Island on the north bank of the Thames Estuary using LNG and refrigerated liquid propane. In total there were 34 spills of which 11 were purposely ignited (The ignition effects are discussed below in Chapter 8). The facilities permitted up to 20 m^3 (ca 10 tonnes) to be discharged at a time. The fullest-scale spillages were from a submersible barge onto water contained by a low dike which gave a pond of 300 m diameter. This arrangement, in which trials were conducted only with off-shore winds, gave a test ground which was almost flat for 10 km at low tide. Two tests were conducted with the sinking barge using LNG and propane. The remainder were continuous with discharge rates varying from 1 to 5.6 m^3 (area 0.5 to 2.7 tonnes) per minute. The wind speeds varied from 1.2 to 10.0 $m\ sec^{-1}$. Table 7.4., which is derived from Puttock (1982), gives some basic data. The paper compares the LFLs for the 6 spills in Table 7.4. with those predicted by the slab model HEGADAS which is described in Colenbrander (1980). It is found that this model overestimates LFL distances.

TABLE 7.4. SELECTED DATA FROM MAPLIN SANDS TRIALS.

Test No.	Substance	Discharge conditions.	Wind Speed $m\ sec^{-1}$	Measured LFL* Distance
46	Propane	2.8 $m^3\ min^{-1}$	7.9	245 ± 35
47	"	3.9 " "	5.2	340 ± 80
54	"	2.3 " "	3.6	400 ± 100
56	LNL	2.5 " "	4.8	110 ± 30
39	"	4.5 " "	4.5	130 ± 20
15	"	2.7 " "	3.9	150 ± 30

*LFL is 5% $\frac{Vol}{Vol}$ for LNG and 2.1% for LPG.

Though not referred to in Puttock (1982) it was disclosed by Gee (1981) that RPT explosions were experienced in the instantaneous LNG spillages at Maplin Sands. Gee reported that RPT explosions had been noted in the 1968 and 1969 trials conducted by the US Bureau of Mines and that these explosions are discussed in Enger and Hartman (1972).

Further information on the Maplin spills is contained in the reference, Shell Research (1981).

7.7.6. THE PORTON DOWN TRIALS.

These trials, which were conducted in 1976, are reviewed briefly in Puttock (1982) and more comprehensively in Picknett (1978) and in McQuaid (1979). The salient features of these trials have been tabulated in Table 7.2. above.

The Porton Field trials, conducted at the Chemical Defence Establishment, Salisbury Plain, involved the release of 40 m^3 (approximately 0.2 tonne) of Freon 12 which is a material of low toxicity. The gas was marked by having orange smoke mixed with it before discharge. The gas was discharged by filling a plastic tent and then collapsing the walls, which took 0.5 second.

The releases, which were the first trials in which substantial quantities of a dense gas were released under isothermal conditions were monitored by photography, including a fish eye lens, from a height of 15 m. There were 10 continuous concentration monitors on site as well as a number of bag samplers and diffusion samplers. There was also a set of instruments for recording the meteorological conditions, such as temperature, relative humidity, wind velocity and its variation with height, and wind direction. The Pasquill categories could be evaluated from these and other data. Some 42 trials were conducted but the number for which the recording systems worked fully was smaller than this. It is not appropriate to give more than a short account of the results of the Porton trials, as any conclusions drawn from them may well be superseded by the results of the larger-scale 1982 Thorney Island trials.

The behaviour of the cloud generally accorded with the behaviour of the dense clouds discussed above and confirmed the basic concepts of van Ulden. However, though many models assume that it takes the form initially of a right circular cylinder, this seems only approximately true and a better description might be that of a torus with a raised annular front, and in which a vortex motion is set up. After the cloud begins to move with the wind it temporarily assumes a horseshoe or bifurcated shape. This shape was predicted by England et al (1978) using the eddy diffusion SIGMET model, a model which is analysed in detail in Havens (1982a). It was noted in Ermak et al (1982) that bifurcation occurred in the Burro trials. Eventually the cloud assumes the shape of an oblique ellipsoidal cylinder leaning with the wind and with the major axis aligned with the wind.

Picknett (1978) gives a number of equations for calculating heavy gas dispersion based upon a statistical analysis of the Porton trial in which constants are empirically determined. These are set out in Table 7.5.

TABLE 7.5. DIMENSIONLESS CONSTANTS IN DISPERSION CALCULATIONS.

Constant	Description	Best Fit Value
α	Edge Mixing	0.82
β	Top Mixing	0.15
γ	Gravity Spread	0.94
λ	Wind Shear	0.57

Van Ulden proposed a value for α of 1.0 though he defined it somewhat differently. Van Ulden assumed the value of β to be zero but it appears in this he was mistaken and that edge mixing is in fact significant.

Picknett suggests that the relative density at which the transition takes place from gravity slumping to normal atmospheric disturbance varies from 1.01 in calm conditions to 1.10 in a 6 m sec^{-1} wind.

7.7.7. THE THORNEY ISLAND TRIALS.

The Thorney Island heavy gas dispersal trials (HGDT) were conducted between 1982 and 1984. Finance (ca £1.7 million) was provided by 38 organisations together with the Health and Safety Executive, the work being contracted out to the National Maritime Institute (now British Maritime Technology Ltd). In total there were 29 trials conducted in the HGDT programme. At the time of writing only a preliminary analysis of the vast amount of data arising from the trials has been carried out. McQuaid (1985) has edited the proceedings of a symposium held at Sheffield, England, at which 23 papers were presented. These papers covered such topics as the design of the experiments, methods of instrumentation and recording, comparison with models and overviews of the results obtained.

Barrell and McQuaid (1985) have reviewed the Thorney Island trials in the context of the overall programme of the HSE which is designed to provide the information which is necessary for the prediction of the behaviour of the dispersal of heavy gas. The reference sees three areas of application to the interests of the HSE. These are:-

(a) To predict the consequences of releases.

(b) To assess the dependability of safety cases provided under the 1984 CIMAH Regulations (these are discussed in Chapter 19 of the book).

(c) To analyse past incidents in order to relate them to postulated incidents.

After reviewing the Porton Trials the reference reviews, in broad terms, the results of the Thorney Island trials. Particular attention is drawn to the considerable degree of validation which the Thorney Island trials have provided for physical models, especially wind tunnel models, though their limitations are also recognised.

The view of the potential user of the results of the Thorney Island trials is given in Harris (1985). Harris tabulates potential releases from transportation or from fixed installation and points out that Thorney Island trials corresponded to ca 5 tonnes of hazardous material, whereas releases of up to 100 tonnes are feasible for on-shore fixed installations. Practical considerations limited any continuous releases to the very low level of 1-2 kg s^{-1}. The reference draws attention to a newly established system which aims to collect weather data and to analyse them by criteria other than those of Pasquill.

Thorney Island, Harris notes, has provided useful data on the effect of obstructions. Harris concludes that the trials were successful, but that the results may take many years to be assimilated.

7.8. WIND TUNNEL TESTS.

7.8.1. THE ADVANTAGES OF WIND TUNNEL MODELLING.

Wind tunnel simulation of the dispersion of buoyant and neutrally buoyant plumes has a long and successful history and has been shown to give a close correspondence with practice, as is discussed in Turner (1973) and Hall (1982a).

However, it is not obvious on first principles that the modelling of dense gas releases in wind tunnels can also give accurate prediction of full-scale releases. Wind tunnels cannot accurately model full-scale systems, for reasons discussed below. Therefore they cannot with entire confidence, at the present time, take the place of full-scale trials. However, if it proves possible to validate them against the full scale then they might be substituted for costly full-scale trials. This could arise if some of the factors which cannot be accurately modelled prove not to be significant. To this end the HSE has commissioned both full-scale trials and wind tunnel experimentation with a view to examining the degree of confidence with which, in the future, it may be possible to predict the course of events with full-scale spillages on the basis of wind tunnel experiments alone.

It should, however, be noted at the outset that, up to the present, it has not proved possible to reproduce, in a wind tunnel, conditions corresponding to markedly unstable Pasquill categories. However, such conditions are believed to have little influence during the gravity slumping phase and they represent, in any case, the least dangerous set of conditions.

7.8.2. WIND TUNNEL TRIALS AT WARREN SPRING.

The wind tunnel experiments conducted in the Air Pollution Division, Warren Spring Laboratories, Stevenage, UK, which aimed to simulate heavy gas dispersion, have been described in a series of papers, Hall et al (1974), Hall (1979), Hall (1979a), Hall (1982) and Hall et al (1982a). As will be discussed below, the latter reference, which is concerned with reconciling wind tunnel experiments with the Porton trials, is the most useful to summarise.

Hall (1982a) is the most useful single reference because, in addition to discussing the Porton trials, it is the most up-to-date of the papers cited and it also contains a review of wind tunnel modelling in general.

7.8.3. BASIS OF WIND TUNNEL MODELLING.

Wind tunnel modelling is based upon the application of boundary layer theory, a subject which is comprehensively treated in standard works such as Schlichting (1968). It is also an example of "model studies", the theoretical basis of which is discussed in a number of works such as Johnstone and Thring (1957) and Langhaar (1951). It involves two sorts of modelling:-

(a) Physically similar modelling, which is modelling in which a prototype and a model differ, so far as the features being studied are concerned, only in scale; scale being the number by which some physical property such as length, mass or viscosity has to be divided to produce correspondence between the two.

An Ordnance Survey map is a notable example of a physically similar model; the popular edition being one in which the linear scale of the map is 1/50,000 of that of the territory being mapped. It should, however, be noted that the model is flat whereas the territory is slightly curved, being part of the earth's surface, and that the map is two-dimensional whereas the territory is in three dimensions. It is relevant also that though the linear dimensions are in the ratio 1/50,000 the area dimensions are in the ratio $1/50{,}000^2$ demonstrating that it is impossible to model simultaneously, to the same scale, all the characteristics of a prototype.

(b) Physically dissimilar modelling in which the model is, for example, a mathematic equation (though it could have other forms, such as an electric circuit). The ultimate object of wind tunnel modelling is to produce a physically dissimilar model, a set of mathematical equations, which will enable the physical behaviour of the prototype to be predicted; these mathematical equations being generated on the basis of experimental work on the physically similar model.

7.8.4. THE ROLE OF DIMENSIONLESS GROUPS.

The use of dimensionless groups in the solution of experimental data has been extensively discussed as for example in Johnstone and Thring (1957) and Langhaar (1951). It is summarised in Perry (1973).

In wind tunnel experiments the following variables are of significance:-

Size of gas release	$= L_1 \times L_2 \times L_3$
Density of gas released (ρg)	$= ML^{-3}$
Density of air (ρa)	$= ML^{-3}$
Kinematic viscosity of air (γ)	$= ML^{-1}T^{-1}/ML^{-3}$
Wind velocity (u)	$= LT^{-1}$
Time scale	$= T$
Acceleration of gravity (g)	$= LT^{-2}$

Dimensional analysis shows that the following dimensionless groups are of significance:-

$$\frac{uL}{\gamma} = \text{Reynolds Number (Re)}$$

$$\frac{u}{\sqrt{gL}} = \text{Froude Number (Fr)}$$

$$\text{or} \quad \frac{u^2}{gL} = \text{Fr}$$

$$\frac{\Delta\rho}{\rho a} = \frac{\rho g - \rho a}{\rho a} = \text{Density Difference Number}$$

Reynolds Numbers are very difficult to scale in wind tunnels because it is almost impossible to operate the wind tunnel successfully at low values of Re. However, it is considered that this is not important provided that the flow is fully turbulent, i.e. that it satisfies the criterion Re $\geqslant$ 40,000.

Similarly there are problems with matching the Froude Number because it implies that the wind speed should reduce as $\sqrt{L}$ and thus a prototype windspeed of 2 m sec^{-1} would become 0.14 m sec^{-1} in a model with a scale of 1/200. This tunnel windspeed is inconveniently low.

It has been found feasible to match in both model and prototype the product of the density difference number with the inverse of the Froude Number. This product, the densimetric Froude Number, is otherwise known as the Richardson Number, Ri. [see Schlichting (1968) and Richardson (1920)].

$$R_I = \frac{\Delta\rho}{\rho a} \times \frac{Lg}{u^2} \quad \text{or} \quad g\frac{\Delta\rho}{\rho a} \times \frac{L}{u^2}$$

This means that a wind tunnel model can be run at higher windspeeds than if the Froude Number is the sole scaling parameter. However, it is only in the outer regions of the cloud where $\frac{\Delta\rho}{\rho a}$ is close to unity that the model strictly applies. This is known as the "Boussinesq approximation" [see Turner (1973)].

For the still air case where u = 0 the characteristic time scale is used. Richardson's number now becomes:-

$$R_I = \frac{\Delta\rho}{\rho a} g \frac{T^2}{L}$$

7.8.5. DESCRIPTION OF THE WARREN SPRING WIND TUNNEL.

The No. 1 wind tunnel at the Warren Spring Laboratories, Stevenage, was used for the trials. It had the dimensions of length = 22 m, height = 1.5 m and width = 4.3 m. When the Porton trials were modelled the scale was 1/25 and a model of the Porton collapsing tent source was made to that scale.

The dense gas used was the refrigerant and fire extinguishing agent BCF, otherwise known as bromochlorodifluoromethane ($CClBrF_2$) or Freon 12 B1. The density of the gas relative to air is 5.74 but it was mixed with air in appropriate proportions to give values of $\frac{\Delta\rho}{\rho a}$ of 0.3 to 2.56. It was coloured with smoke to aid photography.

Cap screws in the wind tunnel, set at a 50 mm triangular pitch, were used to give a value for Z_o (the aerodynamic roughness height) of 2 mm in the tunnel compared with 100 mm at Porton. In the modelling of one trial an upward slope of 1 in 13 was built into the tunnel.

7.8.6. COMPARISON OF WIND TUNNEL AND PORTON TRIALS.

Hall [Hall (1982a)] summarised the results which arose from six trials selected to simulate selected Porton trials.

(a) The gas clouds simulated in the model were of the same scaled size and shape as the prototypes.

(b) Model cloud spread rates and travel times were in good agreement with the prototypes.

(c) Comparisons of concentration sometimes showed good agreement but also sometimes poor agreement.

7.9. GAS DISPERSION IN PRACTICE.

7.9.1. LIMITATIONS OF MATHEMATICAL MODELS AND OF WIND TUNNEL TRIALS.

Attention has been drawn above to certain factors which, prior to the Thorney Island trials, have limited the attempts which have been made to predict the course of heavy gas spillages. These are:-

(a) Mathematic models which take account of the effects of the slope of the ground have so far not been devised.

(b) Wind tunnel trials are valid only for neutral stability conditions.

(c) Wind tunnel models are not, as yet, validated for the prediction of full-scale results because of inherent doubts over the modelling process.

To these may be added that no attempt has been made to model mathematically the effects of buildings and other obstructions though some work has been done in this area in field trials and wind tunnels.

Barrell and McQuaid (1985) have concluded that preliminary analysis of the Thorney Island trials show good agreement with wind tunnel experiments with regard to shape, size, rate of spreading and down wind travel distances. There is, however, a much poorer correlation between field and wind tunnel for levels of concentrations of the heavy gas. The correlation is highest for still air releases and is low for low Richardson numbers, i.e. where the negative buoyancy is least. Unfortunately it has been found that reproducibility of concentrations in the wind tunnel between one run and another is poor and can vary, in some circumstances, by an order of magnitude.

7.9.2. GAS DISPERSION IN ACTUAL INCIDENTS.

7.9.2.1. Gas dispersion in the First World War.

The First World War provided numerous examples of deliberately contrived releases of heavy gases. These incidents far outweighed, in both the numbers of releases and in the quantities released, the numbers and quantities of heavy gases released for purposes of study in the recent decade.

It would appear that the experience afforded by these releases has been largely forgotten or ignored even though the releases were carefully planned, and their effects studied, by German scientists as eminent as Professor Fritz Haber [see Goran (1967)]. The present author called attention to this in Marshall (1977b). Recently, Withers (1985), in assessing the toxicity of chlorine, has drawn upon First World War experiences.

It must be emphasised that the First World War heavy gas releases had source terms which differed markedly from those currently under consideration. The April 15th, 1915, Ypres release, which was personally supervised by Professor Fritz Haber, is a case in point. The release took the form of spraying liquid chlorine from cylinders each of which was equipped with an auxiliary compressed air supply. 168 tonnes were discharged over a roughly straight 7 kilometre front from cylinders spaced in pairs about 2 metres apart. The evidence, such as it is, suggests that no one was gassed in Ypres, the Belgian town which was about 5 km from the front where the gas was released [Spence (1976)]. Ypres is, however, situated on a low ridge.

The release thus had a geometry very different from the "point" sources of Porton and Thorney Island. Moreover, it took place over terrain which was criss-crossed by trenches and heavily cratered by shell holes, i.e. with a high roughness factor and with a capacity for trapping heavy gas.

Goran (1967) refers to field experiments to determine the proper humidity and wind conditions. "Weather of the wrong kind was capable of making poison gas useless". He quotes Haber as saying "When reed grass is merely moved by breezes, one can pack up one's gas equipment and go home".

There are a number of extant photographs of First World War gas clouds as for example those in Hammerton (1916) and Wren (1972). Two of these are aerial photographs. The subject of gas attacks will be further discussed in Chapter 15.

7.9.2.2. Gas clouds in chemical factories and arising from transport incidents.

There have been a number of gas clouds arising from loss of containment in chemical and process installations in which it has been possible to map out the extent of the cloud. Such cloud profiles have often departed markedly from a circular plan and they have sometimes been highly unsymmetrical as a result of the influence of buildings.

These cloud outlines will be discussed in detail in the consideration of such cases studied as those of Flixborough and Beek (Chapter 13). In addition, there have been notable examples of clouds formed as the result of transport accidents. The cloud outlines, where established, will be discussed in the case studies of San Carlos (Chapter 9), Decatur (Chapter 13) and Houston (Chapter 15).

Chapter 8

CATACLYSMIC FIRES

8.1. AIMS OF THE CHAPTER.

The aims of the chapter are:-

(1) to review the problem of cataclysmic fires as agencies for the realisation of major chemical hazards.

(2) to set the experience of major chemical fires against the background of major fires in the UK.

(3) to examine the particular characteristics of major conflagrations and fire balls, illustrating these by case studies.

(4) to examine briefly whether the fire storm phenomenon is relevant to the study of major chemical hazards.

8.2. DEFINITIONS.

A cataclysmic fire may be defined, for the purposes of this discussion, as one which is distinguished by its intensity and/or its speed of development from industrial fires as are ordinarily encountered. Such fires include major conflagrations and fire balls. These are discussed in detail below and they are separately defined at the appropriate point.

It may be noted that none of these terms are defined in BS 4422 (1969). However, BS 4547 (1972), gives a classification of fires as follows:-

Class A. Fires involving solid materials, usually of an organic nature, in which combustion takes place with the formation of glowing embers.

Class B. Fires involving liquids and liquefiable solids.

Class C. Fires involving gases.

Class D. Fires involving metals.

The discussion will be concerned, in the main, with Classes B and C. Class D fires will not be discussed at all.

8.3. FATALITIES IN FIRES IN THE UNITED KINGDOM.

This question has been discussed previously in the context of the relationship between hazard and risk in Chapter 4.

Based upon the statistics for 1968-80 as displayed in Figure 4.3., it would appear that the probability of a fire killing ten or more people in the UK is of the order of 1.25, i.e. such an event on the average occurs about five times in four years. Since the total death toll from fires in the UK is 900 - 1000 per annum, it is apparent that fires on the scale under discussion do not make a very considerable contribution to the total, even if they figure very prominently in discussion in the mass media. Figure 4.3. shows the cumulative frequency of chemical fires for fires causing 6 or more fatalities (see also Table 8.1).

It is necessary to be aware of the possibility of double counting between fire and explosion. The data under discussion include the Flixborough disaster which may appear in other compilations as an explosion.

TABLE 8.1. CHEMICAL FIRES LOSSES AS A PERCENTAGE OF ALL FIRE LOSSES IN LARGE (1) FIRES.

Year	1976	1977	1978	1979	1980	Totals
Cost of chemical fires (2)	7.545	6.724	6.724	4.894	10.5	
Cost in 1982 pounds	15.54	11.96	11.02	6.84	12.70	58.06
Cost of all fires	143	168	181.6	252	314	
Cost in 1982 pounds	294	299	297	352	425	1667
Mean cost per chemical fire	0.167	0.233	.181	.128	.238	
Mean cost in 1982 pounds	0.344	0.410	0.296	.180	.287	
Mean cost of all fires	0.133	0.1467	.181	.130	.242	
Mean cost in 1982 pounds	0.274	0.260	0.296	.182	.292	

(1) Large fires are ≥ £25,000 in year quoted, not in 1982 pounds.

(2) All figures are in £million.

A study which compares UK fatalities and fire losses, the latter as a % of Gross National Product, with those of 13 other countries is given in Fire Prevention (1982). The statistics in this paper are given (in part) in Table 8.2. The reference does not discuss the reasons for the considerable spread of the index (mean fatalities per million of population).

TABLE 8.2. INTERNATIONAL COMPARISON OF FIRE FATALITIES AND FIRE LOSSES 1978-1981.

Country.	Mean fatalities per annum.	Mean fatalities per million of population.	Losses as percentage of GNP.
South Africa	882	32	0.19
Canada	884	31	0.29
United States	7630	29	0.23
Norway	77	19	0.40
Great Britain	992	18	0.18
Sweden	140	16	0.26
Australia	157	11	0.24
Austria	73	7	0.24
France	283	6	0.27
Netherlands	81	6	0.21
Switzerland	68	3	0.08

Notes (1) GNP = Gross National Product

(2) Countries with incomplete data, Fire Prevention (1982) table,omitted.

(3) Mean fatalities per million quoted for South Africa does not accord with usually accepted population figure of 22 million.

Inspection of the figures suggest that there is little evidence of correlation between financial loss and fatality. The present author has determined the correlation coefficient and found it to be only +0.27.

It may be noted that only one fire involving 20 or more fatalities in the UK occurred in a chemical plant in the period under review. This was the previously mentioned Flixborough disaster.

8.4. THE COST OF FIRE DAMAGE IN GREAT BRITAIN.

The damage discussed is for Great Britain (England, Wales and Scotland). Northern Ireland is excluded because of the distortion produced by politically motivated arson.

A problem with analysing financial damage is that brought about by inflation. The present author has converted losses for years before 1982 into 1982 prices using an index provided by Webster [Webster (1982)]. This information is presented graphically in Figure 8.1. which shows the factor by which losses expressed in pounds sterling in earlier years must be multiplied to make them comparable with 1982 losses. The curve has been extended to 1985 by the present author

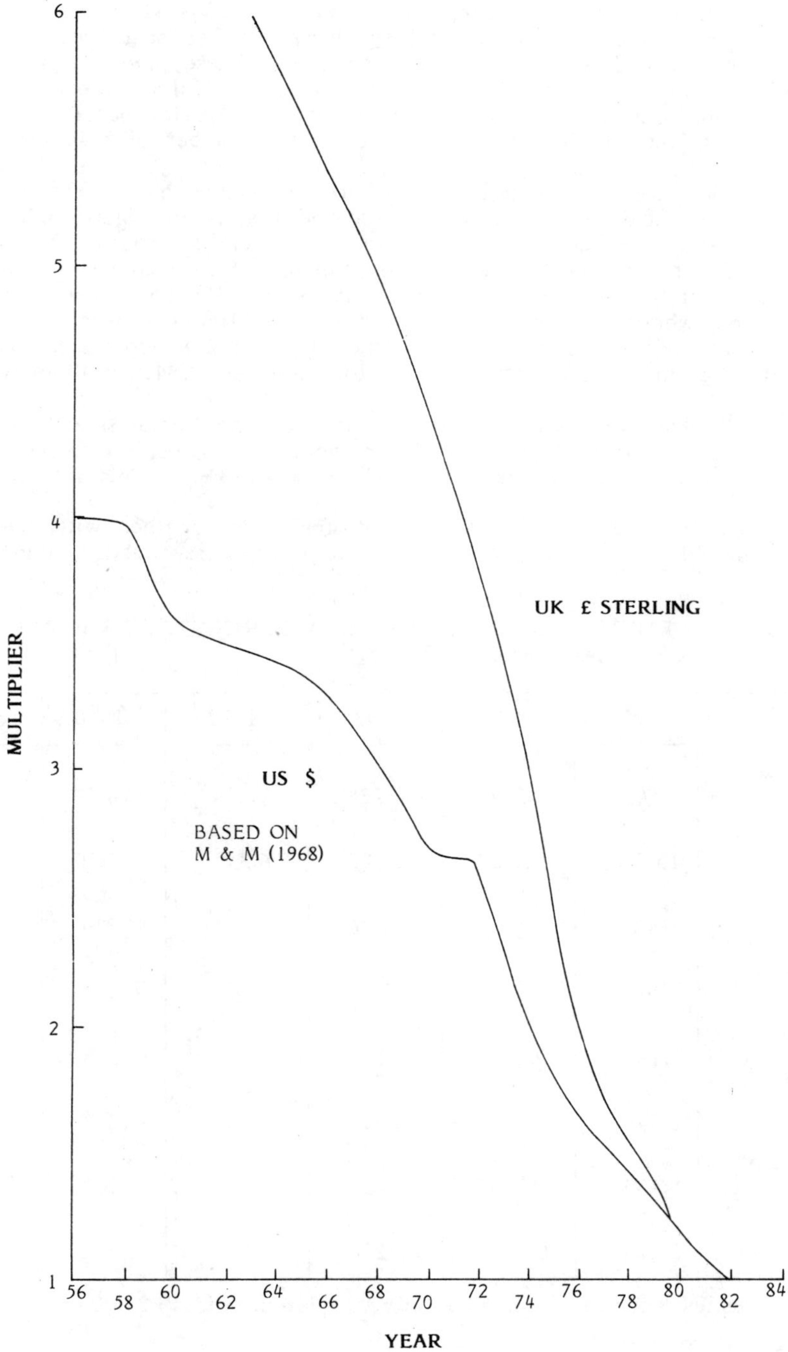

Fig. 8.1. MULTIPLIER TO CONVERT LOSSES TO CURRENT VALUES.

based on the official rate of inflation. Figure 8.2. is a curve relating cumulative frequency of all fires entailing financial losses ⩾n pounds sterling. This is for the UK for the years 1976 to 1980 (inclusive). The basic data is taken from FPA Series(Nos. 123, 128, 136, 141 and 149). These give tables showing numbers of fires in loss ranges £25,000 to £99,999 etc. Unfortunately no allowance is made in these tables for inflation and thus the data appear to suggest that, at the top end of the range particularly, the number of high cost fires is increasing.

After making certain assumptions, the present author has corrected these data for inflation and presents the consolidated results in Figure 8.2. which is for all fires in the United Kingdom. After correction for inflation it appeared that there had not been much change in the number of fires in the various loss ranges. In plotting the graph, class intervals £x to $\sqrt{10}$x and £ $\sqrt{10}$x to £10x have been used, where appropriate, to give equal logarithmic intervals. Figure 8.2. also shows chemical fire losses for the years 1968-80 abstracted by the present author from the FP data referred to, for losses ⩾£1.5 million in 1982 values.

Figure 8.2. suggests that though the ratio of the cumulative frequency, all fires to chemical fires, is circa 10:1 for the largest fires, for losses ⩾circa £50,000 (1982 levels) the ratio is circa 30:1 or about 3%. This is confirmed below.

The values for the chemical and allied industries together with the coal and petroleum industries as a percentage of total fire losses, is given in Table 8.3.

TABLE 8.3. CHEMICAL FIRES LOSSES AS A PERCENTAGE OF ALL FIRE LOSSES IN LARGE (1) FIRES.

Year	1976	1977	1978	1979	1980	Totals
Cost of chemical fires(2)	7.545	6.724	6.724	4.894	10.5	
Cost in 1982 pounds	15.54	11.96	11.02	6.84	12.70	58.06
Cost of all fires	143	168	181.6	252	314	
Cost in 1982 pounds	294	299	297	352	425	1667
Mean cost per chemical fire	0.167	0.233	.181	.128	.238	
Mean cost in 1982 pounds	0.344	0.410	0.296	.180	.287	
Mean cost of all fires	0.133	0.1467	.181	.130	.242	
Mean cost in 1982 pounds	0.274	0.260	0.296	.182	.292	

(1) Large fires are defined as ⩾£25,000 in year quoted, not as £25,000 in 1982.

(2) All figures are in £million.

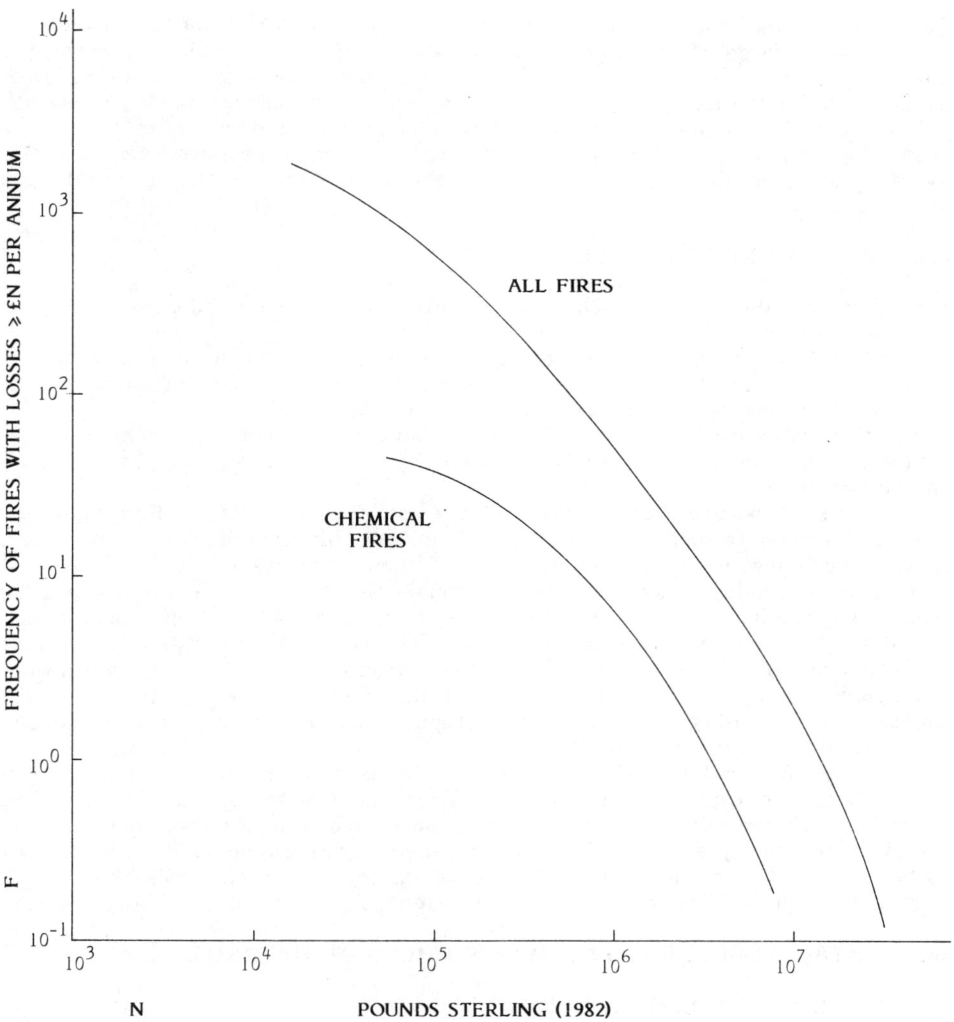

Fig. 8.2. NUMBER OF UK FIRES WITH LOSSES ≥ £n PLUS DATA ON CHEMICAL FIRES.

$$\text{Chemical fire loss as \% of total loss over period} = \frac{58}{1667} \times 100$$

$$= 3.50\%$$

$$\text{Mean cost of large chemical fire} = \frac{58.06}{\text{total No.}} = \frac{58.06}{218} = £0.266 \times 10^6$$

$$\text{Mean cost of all large fires} = \frac{1667}{5738} = £0.290 \times 10^6$$

Too much should not be read into the apparent close agreement between the mean cost of large chemical fires and of large fires in general. The way the data is presented makes this measure of agreement inevitable. Large fires, as defined, constituted only about 0.2% of the total number of fires recorded in 1976 and 1% of the number of fires in occupied buildings in that year. In presenting this very small proportion of the total, the more meaningful statistic, average loss over all chemical fires, is unobtainable from the data at the disposal of the author.

8.5. WORLD FIRE STATISTICS.

According to Microinfo (1982), a new initiative has been adopted to collect fire statistics on, initially, a European basis.

The reference claims that though fire is costing many countries 1% of Gross Domestic Product, few nations keep reliable fire cost statistics. This gravely handicaps efforts to cut costs. The largest items of fire costs are direct fire losses and the cost of protecting buildings against fire. Other significant items are the cost of fire brigades and of fire insurance, indirect fire losses and human losses.

To encourage better national fire statistics, a World Fire Statistics Centre has been formed in Geneva and is helping the United Nations in a pilot scheme involving European countries. Although estimates for national fire costs can be readily calculated, few national statistics for fire costs are being produced annually. If all twelve countries in the eventual study produced annual statistics for each of seven items included in national fire costs, a European collection of eighty-four valuable fire statistics would be available for worthwhile practicable use. At present only around thirty of these eighty-four statistics appear to be published. In fact, no European country appears to be producing complete annual fire cost statistics.

At present the World Fire Statistics Centre is co-operating in a pilot scheme organised by the United Nations and the Economic Commission for Europe under which the United Kingdom, Denmark, France and Hungary provided national statistics for the years 1978/79 covering seven vital elements of national fire costs. The Centre is also organising a top-level fire statistical seminar in Geneva in March 1983 which is to be attended by international fire experts.

8.6. CATACLYSMIC FIRES AND MAJOR CHEMICAL HAZARDS.

8.6.1. THE NEED FOR A TAXONOMY.

As will be demonstrated there is now a considerable literature concerned with the problem of major chemical fires which are defined, for the purposes of the present discussion, as major fires involving substances used as feed stocks, intermediates or final products in the chemical and process industries.

However, though over recent years a number of glossaries of terms have appeared in the literature, as, for example, in Stull (1977), Gugan (1979), Advisory

Committee on Major Hazards (1979), Canvey (1981), and in British Standards, the present author has not been able to find any satisfactory systematic classification of the different phenomena of fire arising from the realisation of Major Chemical Hazards. After discussing the nature of the phenomena in the sub-section below, an attempt is made to construct such a taxonomy at least in relation to fires involving liquids, gases and vapours.

8.6.2. THE NATURE OF CHEMICAL FIRES.

Except for certain reactions of solids which are loosely described as "fires" and which will receive brief reference later in the chapter, the fires to be discussed will be assumed to consist of gas or vapour phase oxidations.

The burning of a candle is an illustrative example. The common candle is made of a mixture of paraffin wax (or approximate formula $C_{25}H_{52}$) mixed with tallow. Paraffin wax is combustible rather than flammable, i.e. it burns, but with difficulty, as its vapour pressure, when cold, is negligible. The presence of the wick, made from cotton, is an essential feature without which a candle does not burn. When a candle is undergoing steady burning the heat of the flame melts the wax (MP ca 50°C) which mounts the wick by capillary action. In the wick it vaporises and/or decomposes (cracks) to shorter chain paraffins and it is the vapour and its decomposition products which burns in the flame. During combustion the long chain paraffins are further broken down and some free carbon is liberated which gives rise to smoke.

This process exemplifies many of the features of chemical fires of which the chief are that the flame may provide latent heat of fusion, of vaporisation and heat of decomposition. (In the case of liquids, it is latent heat of vaporisation which is needed, in the case of gases and vapours no latent heat is required at all). Thus combustion is least vigorous with solids of low vapour pressure and most vigorous with flammable gases or vapours.

There is a spectrum between these extremes which includes flammable liquids whose vapour pressure at ordinary temperatures is less than the lower flammable limit (LFL), and highly flammable liquids whose flash point (FP) is below ambient temperature. These characteristics will be tabulated below in a taxonomy of major chemical fire hazards.

8.6.3. A TAXONOMY OF MAJOR CHEMICAL FIRE HAZARDS.

The phenomena which accompany the ignition of a spillage or loss of containment of flammable liquids are dependent upon the quantity of vapour above the spillage and not upon the total quantity spilled. The phenomena are also dependent upon the degree of admixture of the flammable vapour with air.

It seems convenient to divide liquids liable to give rise to fires into six classes as below:-

CLASS 1 LIQUIDS.

These are liquids with negligible vapour pressure at ambient temperature. Such liquids require considerable heat input, sufficient to raise their temperature by several hundred degrees centigrade before they enter into the combustion process. Though incapable of self-sustaining combustion they may burn as part of a larger fire. They are best described as "difficultly combustible" rather than "flammable" and do not constitute a major chemical hazard. Lubricating oil is an example.

CLASS 2 LIQUIDS.

These are liquids of high flash point which require a fair degree of heat input to produce a vapour pressure equal to the lower flammable limit (LFL). Such liquids may be described as "flammable" but they do not constitute a major chemical hazard. p-Xylene, F.Pt. = 39°C, is an example.

CLASS 3 LIQUIDS.
These are liquids which have a flash point equal to or lower than a nominal ambient temperature (32°C or 90°C as defined in SI 72). In principle such liquids have, at least immediately adjacent to their surface, a mixture of vapour and air of a concentration above the LFL. At some distance above the surface the concentration will fall below the LFL. However, there are wide variations in vapour concentration according to chemical species even for substances which are classed as highly flammable and which are liquids at ordinary temperature. For example, the FP of octane is 13°C, whereas the FP of diethylether is -49°C.

The classification of the behaviour of flammable substances is made more difficult because ambient temperatures can vary, even in the UK, by at least 30°C. It may be convenient, however, to classify highly flammable liquids into two classes, Class 3 Liquids, which at a given ambient temperature, have a vapour pressure between the LFL and their upper flammable limit (UFL), and,

CLASS 4 LIQUIDS, which at a given ambient temperature, have a vapour pressure above their UFL and less than atmospheric pressure. Thus at 20°C, octane would be a class 3 substance with a vapour pressure of 10 mm Hg (1.3% by volume in air, LFL = ca 1%), whereas diethylether at 20°C has a vapour pressure of about 450 mm of Hg (60% by volume in air, UFL = ca 28%) and is class 4.

CLASS 5 LIQUIDS.
These are flammable liquids which have a vapour pressure of ca 1 bar absolute under conditions of storage. Refrigerated or cryogenic flammable gases such as LNG fall into this category. Though on spillage the layer adjacent to the surface is too rich to burn, dispersion will occur releasing an appreciable fraction of the spillage as a flammable mixture of vapour and air.

CLASS 6 LIQUIDS.
Class 6 liquids are the liquefied flammable gases which, on loss of containment release an appreciable fraction as vapour. This vapour, with entrained liquid droplets may, in certain cases, represent a total transformation of the liquid into vapour plus aerosol. This subject has been extensively discussed in previous chapters. Liquefied propane gas is an example of a Class 5 liquid.

The behaviour of liquids if spilled and ignited is closely dependent upon the properties discussed above. It is also dependent upon the configuration of the surroundings, upon wind speed, and upon the time which elapses between spillage and ignition.

There may be six responses to the presence of an ignition source.

(a) Class 1 liquids. These will not ignite from a nearby ignition source and may not ignite even as a result of flame impingement.
(b) Class 2 liquids. These will not ignite from a nearby ignition source but will ignite from actual flame impingement and may give rise to a self-sustaining pool fire.
(c) Class 3 liquids. These will ignite from a nearby ignition source and will readily give rise to a self-sustaining pool fire. The distance between the point of spillage and an ignition source which is capable of igniting the cloud, is dependent not only on the properties of the cloud but also on the wind speed. This distance is much greater down wind than cross wind or up wind. In some circumstances, where there is a steady wind and the wind speed is greater than the flame speed, the cloud may fail to burn

back to the point of spillage. An ignition source will give rise to a flame which will rapidly traverse and ignite the entire volume of vapour/air mixture which lies above the LFL. This may be described as a "flash fire" or a "cloud fire". There will be no element of volume which is too rich to burn.

(d) Class 4 liquids. These will ignite from a nearby ignition source producing a flash fire and will give rise to a self-sustaining pool fire. The vapour cloud will have an element of volume which is too rich to burn en masse but which will burn at its envelope by diffusion.

(e) Class 5 liquids. On spillage they are in equilibrium with their vapour at one bar absolute pressure. Heat input from the surroundings causes boiling which will increase the vapour cloud volume. Ignition may occur from ignition sources relatively remote from the point of spillage and the resultant flash fire will ignite the envelope of the over-rich mixture. A fire ball, (to be defined later), may possibly occur with a very large spillage especially if there is a long delay between spillage and ignition. Such spillages will give rise to a pool fire.

(f) Class 6 liquids. These are capable of ignition from a relatively remote ignition source giving rise to a flash fire and may give rise also to a pool fire where the flashing fraction is small (say about 0.10 as may be the case for example with butane). There will be an appreciable volume of the cloud which is too rich to burn en masse and which will burn on its envelope by diffusion. Where the spillage is of the order of tonnes the flash fire may escalate into a fire ball. In certain cases there may be a further escalation into a vapour cloud explosion, as will be discussed in Chapter 12.

The difficulty of bringing to bear fire-fighting resources increases from class to class. Fires of classes 1 to 3 may be controlled by foam, but classes 4 and 5 pose difficulties of a serious if not impossible nature. Table 8.4. is a taxonomy of the behaviour with respect to ignition sources of the six classes of liquid discussed above together with gas or vapour releases. In addition, Figure 8.3. summarises the table in graphic form.

The table is based upon the concept of a notional "partition fraction" which, under conditions of spillage at an ambient temperature of 20°C, represents the approximate fraction of the whole spillage which is available as vapour. Though highly approximate the table discloses that the fractions corresponding to the various classes differ by orders of magnitude. In the case of classes 1 to 5, the assumption is that there is a two-metre layer of vapour lying above the liquid which is assumed to be 0.2 metres deep. The table, but not the figure, also includes compressed flammable gases.

Releases of flammable gases or vapours where these are heavier than air, would be expected to behave like class 5 liquids but without any possibility of a pool fire. Pool fires, conflagrations and fire balls are discussed below.

8.7. POOL FIRES.

8.7.1. DEFINITION.

Canvey (1981) defines a pool fire as follows:-

"A pool of flammable liquid burning with a stationary diffusion flame".

8.7.2. LITERATURE SOURCES.

There is now a considerable literature on the subject of pool fires. This was reviewed in Raj (1977) in relation to LNG pool fires and more recently, and including also LPG pool fires, in Mizner and Eyre (1982). There are, in addition,

TABLE 8.4. TAXONOMY OF FIRE HAZARDS FROM LIQUIDS AND GASES IN MULTI-TONNE SPILLAGE.

Author's Classification.	Example	Flash Point	Vapour Pressure at 20°C (Bars Ab)	Partition Fraction (see text)	Probability of Ignition at Metres = x				Probability of Occurrence of		
					Zero	1	10	100	Flash Fire	Pool Fire	Fireball
1 Combustible.	Lube Oil	-	ca .001	Negligible	Nil	Nil	Nil	Nil	Nil	Only in general conflagration	Nil
2 Flammable	p-Xylene	40°C	.008	.0005	Mod	Nil	Nil	Nil	Nil	High	Nil
3 Highly Flammable (a)	Octane	13°C	.013	.0015	High	High	Low	Low	Mod	High	Nil
4 Highly Flammable (b)	Diethyl Ether	-45°C	0.58	.024	High	High	High	Mod	High	High	Low
5 Cryogenic or Refrigerated	LNG	Below -160°C	1 bar (-160°C)	0.04	High	High	High	High	High	High	Low
6 Liquefied Flammable Gas	Propylene	-107°C	10	0.40	High	High	High	High	High	Sometimes no liquid residue	High
Compressed flammable gas	Ethylene			1.00	High	High	High	High	High	Nil	High

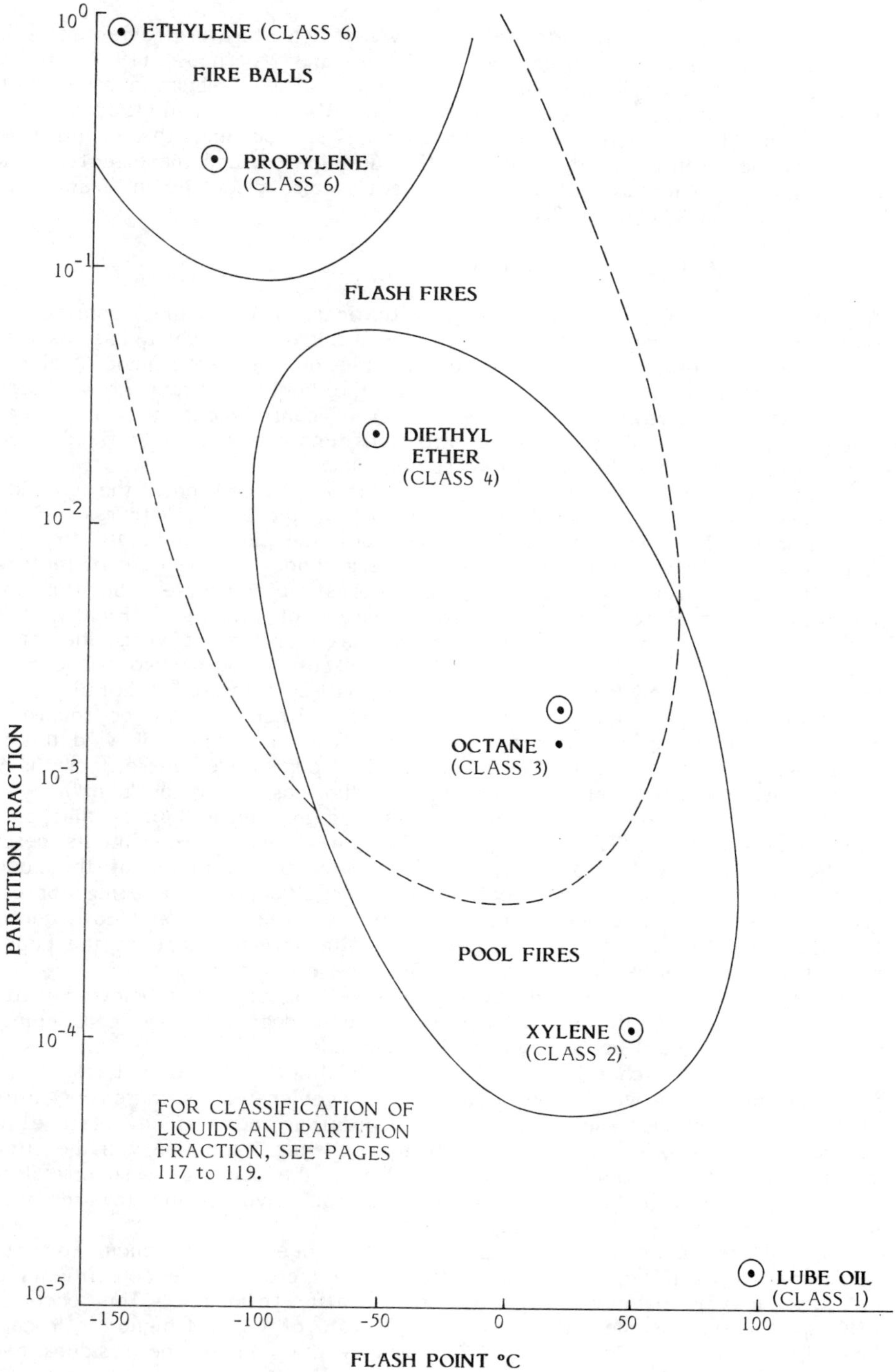

Fig. 8.3. NOTIONAL VENN DIAGRAM FOR FIRE BALLS, FLASH FIRES AND POOL FIRES.

the papers Moorhouse and Pritchard (1982) which is a literature review of radiation hazards from large pool fires and fire balls, and Moorhouse (1982) on scaling factors derived from large-scale experiments. The combustion aspects of the Maplin Sands trials are the subject of the paper Blackmore et al (1982a).

Craven (1976) analyses the circumstances surrounding the occurrences of spillages, the intensity of thermal radiation anticipated, and the effects of radiation on human subjects. This latter subject is also discussed in Crawley (1982) and in Lihou and Maund (1982).

8.7.3. THE NATURE OF POOL FIRES.

A pool fire may occur in a number of situations. At its most confined it can take place in a storage tank, as, for example, in a tank which has lost its top as the result of either internal or external explosion. In tanks made of aluminium the walls may melt down to the level of the liquid and thus grow shorter as the liquid burns away. At the next level of confinement, the fire may take place in a bund. Both of these situations imply a clearly defined boundary and a shape which may be circular or rectangular.

Other pool fires occur after a liquid is discharged onto the ground, the shape and depth of the pool being determined by the local contours. Factories and airports have considerable flat areas but the liquid is likely to run into storm drains where it may burn possibly underground. In the case of motorways, the storm drains usually discharge into the nearest water course. Such unconfined spillages may therefore carry fire for hundreds of metres. Finally there is discharge directly onto water courses, rivers, lakes or the sea where the opportunity for spread is virtually unlimited. In the discussion below two of these situations, fire in bunds and fire on water surfaces, will be discussed in detail.

Pool fires, especially in bunds which are either circular or square, have a shape which is approximated by a cylinder. In the absence of wind this would be a right cylinder, but in the more usual circumstance where there is wind, the cylinder is oblique (referred to by some authors as "tilted" or "bent").

A notable feature of pool fires, illustrated in Figure 8.4., is that of down wind "spill-over" or "flame drag". This spill-over, in experiments described in Mizner and Eyre (1982), was from 25-50% of the diameter of the bund (20 metres) and resulted in the effective diameter of a pool fire being appreciably greater than that of the bund which contained it. The figure of 50% is confirmed in Robertson (1976). It may be noted that the up-wind edge of the pool fires was not observed to be blown down-wind in these experiments.

The magnitude of the radiative effects will be discussed below, but it may be noted here that all the flames were to some degree smoky, LNG being less smoky than LPG and LPG less smoky than kerosine.

Pool fires may change in character with time. There is likely to be an induction period in which the burning rate accelerates as increasing intensity of heat radiation increases the rate of evaporation and a steady state at which balance is achieved. With a relatively pure chemical the fire will die out from steady state as the fuel is exhausted. Where the pool has a sloping floor, as for example in a hollow in the ground, the pool will shrink towards the end with a corresponding fall off in intensity.

Where the flammable material is not homogeneous in chemical composition, there may be a falling off in intensity as the more volatile constituents distil out first leaving behind a relatively non-volatile residue. This occurs with LNG which contains, depending on origin, 5-15% of C_2 and higher hydrocarbons. LNG fires also become more smoky towards the end as the residues become richer in higher hydrocarbons.

Probably the most extreme example of the phenomenon is that of the burning of crude oil which contains hydrocarbons from CH_4 to C_{25} and higher. There

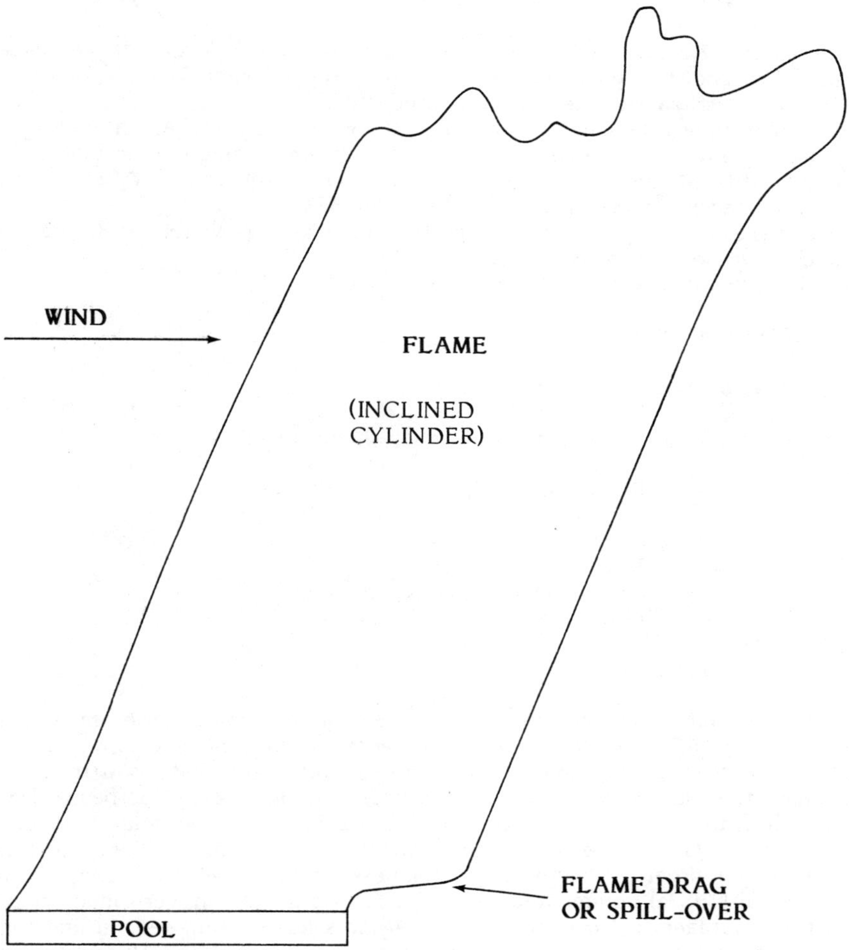

Fig. 8.4. SCHEMATIC DIAGRAM OF POOL FIRE.

was a major pool fire of this sort involving some 40,000 tonnes of crude oil on the sea in Bantry Bay, Eire, starting in the early hours of 8 January, 1979. This resulted in the loss of life of 50 persons and severe damage to the jetty. This disaster is fully reported in Whiddy Island (1980).

8.7.4. TRIALS.

Two of the organisations which have conducted large-scale trials in the UK are Shell and British Gas.

The Shell experiments were conducted at Maplin Sands, Essex. The nature of these trials have been discussed previously in connection with the development of gas dispersion models. The combustion aspects of the trials are discussed in Blackmore et al (1982a). Further work by Shell at an unstated inland location is described in Mizner and Eyre (1982). These trials used a 20 metre diameter, 0.30 metre wall height, bund which was precooled with liquid nitrogen before admitting either LNG or refrigerated LPG.

The British Gas trials were carried out at their Midlands Research Station and are described in Moorhouse (1982). The work was carried out with a pool which measured 12.2 metres by 15.4 metres and 0.5 metre deep. It was capable of sub-division and the pools studied lay in a range from 37 m^2 to 186 m^2. LNG was the test fluid and the bund was precooled.

Test programmes by the American Gas Association (AGA) in which LNG was discharged into ca 2 metre, 6 metre and 25 metre diameter, circular earthen bunds which had not been precooled is touched on in Moorhouse (1982). These trials are reported in American Gas Association (1973).

There were trials involving the ignition of LNG on water at China Lake, USA, which have been reported in Little (1979).

The above trials are summarised in Table 8.5.

8.7.5. THEORETICAL ANALYSIS.

8.7.5.1. Areas of Theoretical Analysis.

There are a number of areas of analysis which include:-

Flame appearance
Flame geometry
Flame speed
Rate of burn up of fuel
Change of composition of fuel
Radiative flux

8.7.5.2. Flame appearance.

This relates to luminosity or smokiness. As an example, those investigators who comment on this area of analysis characterise the flame of LNG as being luminous but with a smoky top. The presence of black soot is attributed to oxygen starvation in Mizner and Eyre (1982). However, Moorhouse (1982), comments that the area of a LNG flame is easier to define than that of the flame of, say, LPG because it is comparatively soot-free. Mizner and Eyre (1982) assess the order of increasing smokiness as methane, LPG and kerosine. Blackmore et al (1982a) commented that LPG flames at Maplin were much smokier than those of methane. None of the investigators have attempted, in any fundamental way, to relate smokiness to such factors as the number of carbon atoms in the molecule or to molecular structure. Nor do they seek to relate it to established data on the thermochemistry of flames.

The presence of smoke has a considerable influence on the radiative flux with which it appears to correlate negatively.

8.7.5.3. Flame geometry.

There is a large measure of agreement that an oblique cylinder is the best representation of flame shape. The ratio, height to length of base, is discussed in Moorhouse and Pritchard (1982), in Moorhouse (1982), and in Mizner and Eyre (1982), who note the difficulties experienced in determining the shape of flames on account of the presence of smoke. The ratio L/D where L is measured along the axis of the flame and D is the pool diameter may be taken, as a rough approximation, as having values in the range 1.75 to 2.50. Moorhouse and Pritchard (1982) recommend the basic equation as:-

$$\cos \theta = 0.75 \times U_D^{-0.49}$$

or $$\cos \theta = 0.75 \frac{1}{\sqrt{U_D}}$$ as a reasonable approximation

TABLE 8.5. POOL FIRES, SOME EXPERIMENTAL TRIALS.

Year conducted or reported on	Location and/ or Organisation	Land or Water	Materials Tested	Dimensions of Pool	Quantity Used	Factors investigated	References
1973 (Reported)	American Gas Association	Land	LNG	2m, 6m, 25m circular earth bunds, not pre-cooled		Flame geometry burning rate	AGA (1973) Moorhouse (1982)
1978 (Conducted)	China Lake, USA	Water	LNG			Radiative flux and others	Little (1979)
Summer 1980 (Conducted)	Maplin Sands, Essex, UK Shell	Water	LNG, Refrigerated LPG, continuous and instantaneous	300m dia. shallow pond	$20m^3$ (max) ca 9 tonnes.	Flame geometry Flame speed Radiation flux Overpressure	Blackmore et al (1982a). Preliminary Report
1982 (Reported)	Shell Research	Land	LNG, Refrigerated LPG, Kerosine	200m dia. precooled		Flame geometry Mass burning rate and composition of pool Radiation flux	Mizner and Eyre (1982)
1982 (Reported)	British Gas Midland Research Station	Land	LNG	12.2m x 15.4m. Capable of subdivision to give pools of $37m^2$ to $185m^2$ precooled		Flame geometry Radiation flux	Moorhouse (1982)

θ = angle of deviation from vertical

U_D = dimensionless wind velocity

$$U_D = \frac{Uw}{\left(\frac{m}{\rho v} \; g \; D \right)^{0.333}}$$

where m = mass burn up rate kg/m^2s

ρ = density of vapour kg/m^3

g = acceleration of gravity m/s^2

D = pool diameter m

Uw = velocity of wind m/s

This equation is attributed to Attalah and Raj in American Gas Association (1974).

Flame drag has been previously discussed; there is little theoretical analysis of the phenomenon available.

8.7.5.4. Flame speed.

Blackmore et al comment that at Maplin flame velocities in the cloud were about 12 m/sec with some transient speeds up to 28 m/sec.

8.7.5.5. Rate of burn up.

Factors which may influence rate of burn up are chemical species, pool size and wind speed. Moorhouse and Pritchard (1982) conclude that chemical species has a considerable influence on the mass burning rate (kg/sec metre^2) being highest with LNG and lowest for kerosine and that there is an inverse correlation between mass burning rate and boiling point. Values for LNG lie in the range .05 to .08 for small pools whereas methanol has a value around .02. Larger pools show somewhat higher burning rates.

8.7.5.6. Change of composition of fuel.

LNG, LPG and the majority of liquid fuels are not pure substances. Though they may be distilled or otherwise treated to remove substances which may be harmful to the combustion process there is little point in securing sharp fractions of hydrocarbon in material for which the sole purpose is to be burned.

It has already been noted that LNG may contain substantial amounts of ethane and higher hydrocarbons. When evaporation occurs the residue becomes enriched in C_2 and C_3 hydrocarbons. Mizner and Eyre (1982) present data on LNG pool fires which showed that in an LNG which initially contained 89%CH_4 and 11%C_2H_6 as molar concentrations, after six minutes of burning there was about 50% of each.

8.7.5.7. Radiative flux.

The subject of radiative flux is of such importance that it will be dealt with in a separate section below, where it will be treated along with the question of the effects of thermal radiation.

8.7.6. POOL FIRES AS MAJOR HAZARDS.

Pool fires clearly can vary in size from small pools less than a metre in diameter and a few millimetres in depth which cannot be classed as major hazards, up to very large pool fires or groups of pool fires. These large-scale events are treated in the next section.

8.8. CHEMICAL CONFLAGRATIONS.

8.8.1. DEFINITION.

The Shorter Oxford English Dictionary defines "conflagration" as a "great and destructive fire". Following this it is possible to define a chemical conflagration as a "great and destructive fire involving the feed stocks, intermediates or final products of the chemical and process industries".

In the case histories discussed below, one major chemical conflagration was that a Flixborough in 1974 which involved a number of pool fires over an area of 60,000 m^2 and the other was an extremely large pool fire at Cleveland, Ohio, in 1944 which inflicted severe damage over an area of about 120,000 m^2.

8.8.2. THEORETICAL TREATMENTS.

There is a growing interest in models for assessing the probability of fire spread and these are reviewed in Ramachandran (1982). Ramachandran distinguishes two basic techniques for assessing fire "risk", the "Point Scheme" and the "Logic Tree" approach.

The point scheme, as practised by some insurance companies, was originally proposed in Gretener (1968) and took the form of evaluating a function X as equal to the product of the inherent hazard of a building and the tendency for a fire to start (Y) divided by the product of the fire precautions, the special fire precautions and the fire resistance of the building (Z). Other point schemes are those of the Dow and Mond indices which are discussed later.

The logic tree or fault tree or decision tree approach is now well established in many areas of loss prevention. Such a system analyses formally the various combinations of circumstances which may lead to an undesired event involving "or" events, i.e. those where any of a number of possible events may occur and lead to the undesired event, and "and" events in which there must be two or more events occurring together before there is a lead forward. Such trees, if they have the probability of each event built into them, are a means of systematically evaluating the probability of the final, undesired, event.

Methods which Ramachandran discusses have been devised to model fire spread. These include deterministic modelling and stochastic modelling. The former uses a theoretical approach based upon knowledge of rates of heat transfer and of flame propagation taken along with the geometry of the surroundings. The latter is based primarily upon the study of past events and therefore has built into it the factors of probability which are exemplified in real circumstances. The two models are in essence the "theoretical approach" and the "historical approach" which are discussed in Advisory Committee on Major Hazards (1979) and below in Chapter 18.

8.8.3. CONSIDERATIONS OF WORKS LAYOUT.

That fire in one part of a site may spread to another is today a well recognised factor in plant layout. The subject is discussed in what may be, even today, the only monograph on the subject, Mecklenberg's "Plant Layout", [Mecklenberg (1973)]. The safety aspects of plant layout have, however, been the subject of considerable study since that book was published and some of the data in it are now superseded.

BS 5908 (BS 1980), which is concerned with fire precautions in chemical plant, makes reference to the need for systematic layout. It does not prescribe distances between storages for flammable materials for example, but it draws attention to publications which make recommendations in this area.

Health & Safety Executive (1976a) gives safety distances for LPG storages and auxiliary equipment. The booklet however, in the present author's view, does not give adequate attention to the problems of fire balls or unconfined vapour cloud explosions which may arise from loss of containment of LPG. It is understood that it is likely soon to be superseded.

8.9. FIRE BALLS.

8.9.1. THE NATURE OF FIRE BALLS.

In June 1977, at the Eurochem Conference on "Chemical Engineering in a Hostile World", in circumstances in which there was little general recognition of the dangers of fire balls in the UK, at the time, the present author had this to say:-

> "That fire can spread from one site to another has long been recognised and the Fire Authorities have developed, on the basis of past experience, the concept of 'safe distances'. It might seem that this is all that is to be said on the subject but unfortunately this is far from being the case.
>
> This is because there is a third possibility with a cloud of vapour, the possibility that it may ignite and burn without exploding. It has been assumed by some that ignition leading to burning (deflagration) is necessarily the lesser of the two evils but it is by no means certain that this is the case.
>
> A cloud of vapour mixed with air but perhaps too rich to explode will start to burn around its outer envelope and will 'lift off' to form a fire ball. Such fire balls are dangerous in the extreme. When formed of hydrocarbons they are luminous and they radiate heat which can cause fatal burns to bystanders and which can ignite wood and paper. As they rise to form a mushroom cloud the stalk consists of a violent upward convection current which can suck up debris, ignite it and scatter burning brands over a wide area.
>
> There have been attempts to predict the size and duration of such fire balls. Formulae, attributed by NASA have been adapted by the present author as below:-
>
> $$D = 55 \times \sqrt[3]{M} \text{ (approximately)}$$
>
> where D = diameter of fire ball (in metres)
>
> M = mass of hydrocarbon of assumed formula C_nH_{2n} (in tonnes)
>
> and $$T = 3.8 \times \sqrt[3]{M} \text{ (approximately)}$$
>
> where T = duration in seconds
>
> M is as above
>
> It is not only the energy release which is important but the rate at which it is released, i.e. its power. If the energy released is a linear function

of M and the time taken to release it is a function of $\sqrt[3]{M}$ then

$$\text{Power} = f\, M^{2/3}$$

Taking a release of 50 tonnes of hydrocarbon

D = ca 200 metres

T = ca 14 seconds

P = ca 170,000 megawatts

The size of the cloud and the rate of burn up are difficult to imagine. Perhaps it might help to say that the cloud would have a volume equal to about twenty times the volume of St. Paul's Cathedral and the burn up rate, ca 3.5 tonnes per second, compares with 0.1 tonnes per second which is the rate of burn up in the boiler of a 1000 megawatt power station, or the rate of burn up of a million motor cars.

Such a phenomenon is completely outside the range of conventional fires and can clearly carry fire far beyond the normal safe distance." [Marshall (1977b)].

It was just over a year later, on 7th November 1978, that a fire ball at San Carlos, Spain, occurred at a seaside holiday camp giving rise to an eventual death toll of 215. This death toll was higher than the death toll in the worst VCE disaster, that of Ludwigshafen in July 1948 which killed 207 people.

Advisory Committee on Major Hazards (1979), quoting Marshall (1977b), called for further study of the phenomenon.

8.9.2. REFERENCE SOURCES.

The literature in the field of the study of fire balls is reviewed in Moorhouse and Pritchard (1982).

The study of fire balls seems to have been initiated in High (1968). This study was concerned with fire balls which had arisen from dumped rocket fuel and which related diameter and duration to propellant mass. High was concerned, inter alia, with the prediction of the properties of the fire balls which might arise from the Saturn V moon rocket then in course of development. This rocket, if it proved necessary to dump its fuel, would discharge a mass of fuel which would be more than an order of magnitude greater in mass than had previously been experienced. Hence the need to devise a correlation for such properties as fire ball radius and thermal pulse. The work was not directed to the prediction of fire balls from discharges of flammable gases into air. Few such events had been recorded prior to 1968; it may be noted that all the incidents in Table 8.6. on which there is sufficient data to attempt a correlation, have occurred subsequent to High (1968).

High's formulae were used in Brasie (1976) as a basis for calculating safety distances in chemical plant in circumstances in which High's mass of propellant was uncritically equated with mass of flammable vapour.

Hasegawa and Sato (1978) discusses the circumstances which determine whether a given release on ignition will be predominantly a pool fire or whether it will be a fire ball.

Roberts (1982) is a study of the circumstances and effects of fire balls which is extended in Roberts (1982a).

Crawley (1982) discusses fire balls in a general review of the effects of the ignition of fuel spillages. Lihou and Maund (1982) present data derived

TABLE 8.6. SOME FIREBALL INCIDENTS.

Index Number	Incident	Date	Material	Quantity Spilled (Tonnes)	Fireball Radius (Metres)
1	Aberdeen, Scotland.	17/1/74	Butane	2	35
2	Lynchburg, Virginia.	9/3/72	Propane	9	60
3	St. Paul, Minnesota.	11/1/74	LPG	10	50
4	Eagle Pass, Texas.	29/4/79	LPG	25	170
5	Kingman, Virginia.	5/7/73	Propane	45	150
6	Goldanna, Virginia.	28/12/77	LPG	70	160
7	Crescent City, Illinois.	21/6/70	Propane	75	75-100
8	Belt, Montana.	26/11/76	LPG	80	150
9	Lewisville, Arkansas.	29/8/78	VCM	110	155
10	Houston, Texas.	19/10/71	VCM	165	150
11	Donnellson, Iowa.	4/8/78	LPG	435	305

from laboratory studies of fire balls. Some sources present evidence of a negative character, that trials with LNG and refrigerated propane did not lead to fire balls. Such references include Blackmore et al (1982a), and Mizner and Eyre (1982).

8.9.3. CONDITIONS WHICH GIVE RISE TO FIRE BALLS.

8.9.3.1. How appropriate is experience with rocket propellants?

It is necessary first to look at the fire ball phenomenon studied in High (1968) to examine whether it is an appropriate model for the understanding of fire balls arising from hydrocarbons ignited in air.

There are, in fact, major differences between the two situations. Rocket propellant for the main engines of high thrust rockets is a mixture of an oxidising agent, liquid oxygen (Lox) with a reducing agent such as kerosine (RP1) which reaches a temperature in the engine of ca 3,400°C. [Shreve and Brink (1977)]. The choice of fuel is determined by a number of factors of which probably the most important is to maximise specific impulse which is expressed as "seconds" having the units (lbs of thrust/lbs of fuel burned per second). [The pioneers of rocketry cancelled out pounds in numerator and denominator disregarding the difference between lbs force and lbs mass to give the spurious residual dimension of "seconds"]. Specific impulse is largely determined by the ratio $\sqrt{T/M}$ where T is the absolute temperature, M is the mean molecular weight of the exhaust gases. This implies that a low molecular weight in the exhaust maximises thrust. Since hydrogen has the lowest molecular weight, it is advantageous to use less than stoichiometric oxygen to leave free hydrogen in the exhaust. In practice, and in approximate terms, the O_2 is equivalent to the carbon in the fuel. Thus a propellant consisting of $C_n H_{2n}$ with oxygen would be 0.30 hydrocarbon : 0.70 oxygen. This is almost exactly the ratio quoted in Chambers (1973), Vol. XI, which quotes the North American Aviation Inc. Papers such as Brasie (1976), which equate mass of fuel with mass of propellant in the equations in High (1968), thereby underestimate fire ball diameters.

However, there are other factors. It was stated above that the engine temperature for Lox/kerosine is ca 3,400°C. This compares with theoretical flame temperatures of 1900-1950°C for hydrocarbons in air. Furthermore rocket propellant is a mixture of liquid reactants and it may be expected that the reaction will be much more rapid than occurs when hydrocarbon vapours have to mix with air.

Taking account of all these factors, it would seem that data derived from rocket propellants should be applied to hydrocarbon air mixtures with great caution.

8.9.3.2. What are the limiting conditions?

Hasegawa and Sato (1978) consider that liquefied flammable gases which have a TAFF of 0.35 or above are capable of giving rise to fire balls and that they do not give rise to pool fires. Roberts (1982) comments that this implies that the ratio, entrainment : TAFF, is greater than 1.

This approach accords with "notional partition fraction" which was set out in Table 8.4. above and which is presented also in graphical form in Figure 8.3.

The figure 0.35 for TAFF proposed in Hasegawa and Sato (1978) cannot be an exact one as the circumstance of rupture, i.e. whether it be catastrophic or not, must play a considerable role. Where the aperture is above the liquid level the results may be markedly different from a situation where the aperture is at the bottom of a vessel.

8.9.3.3. The geometry of fire balls.

High (1968) describes the history of fire ball formation from rocket propellant as follows:-

> "In fire balls associated with propellant explosions, as the pressure of the detonation products decreases to ambient pressure, the density of the gas is considerably below ambient density and the resultant buoyant force causes the gases to begin to rise. The bulk of the propellant is then entrained in the fire ball and rapidly burned. A hemispherical shape is maintained during most of this period until the fire ball growth rate is exceeded by the buoyancy and a spherical shape develops. After the spherical shape is completely formed, the fire ball lifts from the ground, and surrounding air is drawn into the fire ball by a combination of convective forces and a vortex motion to continue mass addition. A stem is created with propellant spilled on the ground to produce a mushroom-type cloud. The hot fire ball continues to change into an oblate spheroid and finally a toroid. Combustion between the fuel-rich gases and entrained air continues until a stoichiometric mixture is produced, after which the entrained air dilutes and cools the gases. The radiation losses also contribute to a rapid cooling process. The dynamic behaviour of the fire ball can

be described as a spherical "thermal" composed of the hot fire ball gases on top and entrained air below. The development and rise of this thermal are controlled by thermal radiation, addition and combustion of propellant, and air entrainment."

The paper assumes a knowledge of the circumstances which surround the spillages and passes little comment on them. It speaks of "propellant explosions", of "missile failures" and of "nominal abort conditions".

It seems reasonable to assume that the circumstances that High (1968) refers to are the enforced extremely rapid dumping of propellant followed by ignition on the ground. This dumping (presumably in the form of cryogenic fuel/oxidiser mixture) the paper quantifies for Saturn V as a total of 5.5×10^6 lbs (2495 tonnes) at a rate of 700,000 lbs (317 tonnes) per second. USIS (1968) claims that the Saturn V rocket after launching, consumed, in its first stage, a total of 2375 tonnes of propellant with a burn-up rate of 18.26 tonnes per second (3.65 tonnes per second per engine). It would be expected that propellant would be capable of being dumped as a liquid via the engines much more rapidly than its combustion products could be ejected as a jet streak, hence the discrepancy between dumping rate and burn-up rate.

As commented above, rocket fuel dumping has important differences from hydrocarbon spillage. High (1968) speaks of "propellant explosions" and "detonation products", the meaning of which is not clear to the present author and which does not seem profitable to explore further.

Instead it is proposed to assume, based upon the present author's arguments about partition fraction, the views expressed in Hasegawa and Sato (1978), and the historical experience of fire balls, that fire balls will only occur in the chemical and process industries as the result of catastrophic loss of containment of liquefied flammable gases such as LPG, propane, propylene or vinyl chloride monomer. This being so, the history of a fire ball will have as its precursor the history of vapour cloud formation and dispersal arising from such an event. A fire ball, per se, commences its history at the time that such a vapour cloud encounters an ignition source.

It is necessary here to refer once more to the source term which is dependent upon the mode of loss of containment. This may lie between two extremes; at one extreme lies a fracture which causes a stream of hydrocarbon to emerge in a highly directional form, at the other extreme lies a catastrophic failure in which a vessel disintegrates into a number of pieces. The former case, particularly if the emergent jet be parallel to the ground, would give rise to an elongated cloud; the latter case would give rise to a more symmetrical cloud approximating to an oblate hemispheroid. In either case, as has been previously noted, and based upon the results of many trials, gravity slumping is rapid. So rapid is this process that the assumption made in Roberts (1982a) that the initial fire ball is hemispherical may be open to doubt though the assumption contained in it, that the initial fire ball is not spherical, seems indisputable.

The source term will be examined in more detail in two case histories in the next chapter. One is that of the Feyzin, France 4/1/66, disaster where there was a directional jet, the other is the previously noted San Carlos, Spain disaster of 11/7/78 where the vapour cloud may have been less directional.

Approximate estimates of the theoretical radius of a fire ball, spherical in shape, are relatively easy to make. Assuming stoichiometric composition the radius R may be obtained from an equation which has the form:-

$$R = KM^{1/3}$$

where M = Mass of fire ball

and K = Constant

The value of K does not differ much as between paraffins and olefines, but may differ significantly for hydrogen. In the discussion below R, the radius, is used instead of D, the diameter. This is because R is a measure of the distance from the centre of a spillage at which certain effects will occur; this corresponds to the practice in analysing the effects of explosives. Tonnes are used as the units of M as it is only the tonnage range which is important from the point of view of major chemical hazards.

K may be evaluated by calculating the volume of a stoichiometric cloud containing M tonnes of material and then calculating the volume of the combusted cloud using the well known Charles' Law. In fact there is little difference between such a volume as calculated and the volume which the original uncombusted cloud would have if multiplied by the factor arising from Charles' Law. If it be assumed that the cloud reaches the theoretical flame temperature, the expansion factor deduced from Charles' Law is about 8 : 1. This implies a doubling of the radius. The author's paper, Marshall (1977b), suggested for a theoretical spherical fire ball

$$R = 27.5\ M^{1/3}$$

Roberts (1982a) suggested

$$R = 29\ M^{1/3}$$

(converted to M in tonnes by the present author)

These equations do not settle the question of the radius of flame engulfment of a shallow vapour cloud. For example, if a cloud at the moment of ignition has dimensions R (radius) and H (height) such that $R \gg H$, will the burned cloud have dimensions ca 2R and ca 2H ? This seems unlikely and certainly aerial photographs of the ignited Maplin clouds do not bear it out. What seems more likely, in such a case, is that expansion takes place mainly in the vertical direction as it does with pool fires.

Roberts (1982a) draws attention to initial unignited cloud expansion effects immediately following the rupture of the container. This question will be further discussed in the San Carlos case history. Attention is drawn in Table 8.7. to the relative radii and volumes for propane, which assumes storage at 10 bars absolute.

TABLE 8.7. EXPANSION FACTORS FOR A PROPANE RELEASE.

	Initial Liquid	Vapour unmixed with air (0.35 TAFF)	Stoichiometric cloud (unburned)*	Stoichiometric cloud (burned)
Relative volume	1	80	9142	73,000
Relative radius of equivalent sphere	1	4.3	21.0	42.0

*Assuming that all entrained liquid has evaporated.

The effects of the initial vapour expansion are thus only felt close in. However, the concept of an initial cloud of pure vapour is somewhat artificial.

High (1968) attempted to correlate the diameters of actual propellant fire balls with case histories and arrived at the following equation, based upon a least squares regression:-

$$D = 9.82\ WP^{0.320}$$

where D = maximum diameter (ft)

WP = weight of propellant (lbs)

Transformed into metres and tonnes by the present author, this becomes

$$R = 17.6\ WP^{0.320}$$

Transformation of the equation in High (1968) to fuel rather than propellant has difficulties as the correlation is based on four different fuel/oxidant combinations. However, kerosine (RP1)/Lox provides the majority of the points and the correlation seems to fit these points, taken on their own, quite well. The ratio of 2.24 Lox : 1 hydrocarbon previously quoted may then be used to convert the High (1968) equation as follows:-

$$R = 27.5 \times M^{1/3}$$

It was this transformation which appeared in Marshall (1977b). It should be noted that this equation does not represent a theoretical approach, as does the equation in Roberts (1982a).

Table 8.6. (P.130) and Figure 8.5. are based on data from a variety of spillages including those of propane, butane, LPG and VCM (vinyl chloride monomer, which has a vapour pressure in the LPG range). All were transport accidents and, with one exception, they occurred in the USA. The Iowa incident was a pipeline. (That there are no incidents listed as occurring on site proves nothing; most transport incidents in the US have to be investigated by the US National Transportation Safety Board, which is the source of most of the information, whereas no such requirement exists for installations). A regression analysis of the 11 data in Table 8.6. yields

$$R = 27.5\ M^{0.375}$$

This agrees reasonably well with Marshall (1977b) and Roberts (1982a) as Figure 8.5. shows. This figure includes also the correlations in Hardee et al (1977) and in Hasegawa and Sato (1978). All in all, there seems to be a broad measure of agreement as to what is the correlation between radius and mass and there is a fair measure of agreement between theory and practice. However, actual incidents show a considerable scatter in their data. Radii which appear to be larger than theory predicts may be reported in this way either because of subjective reporting or because they markedly depart from a spherical shape when close to the ground.

Prediction of fire ball radius does not of itself determine safety distances from installations. This is because

(a) Injury or death may occur well outside the predicted radius because of thermal radiation. This topic will be discussed below.

(b) As with all vapour clouds, wind can cause a flammable cloud to drift and the greater the delay to ignition the greater will be the displacement from the point of release.

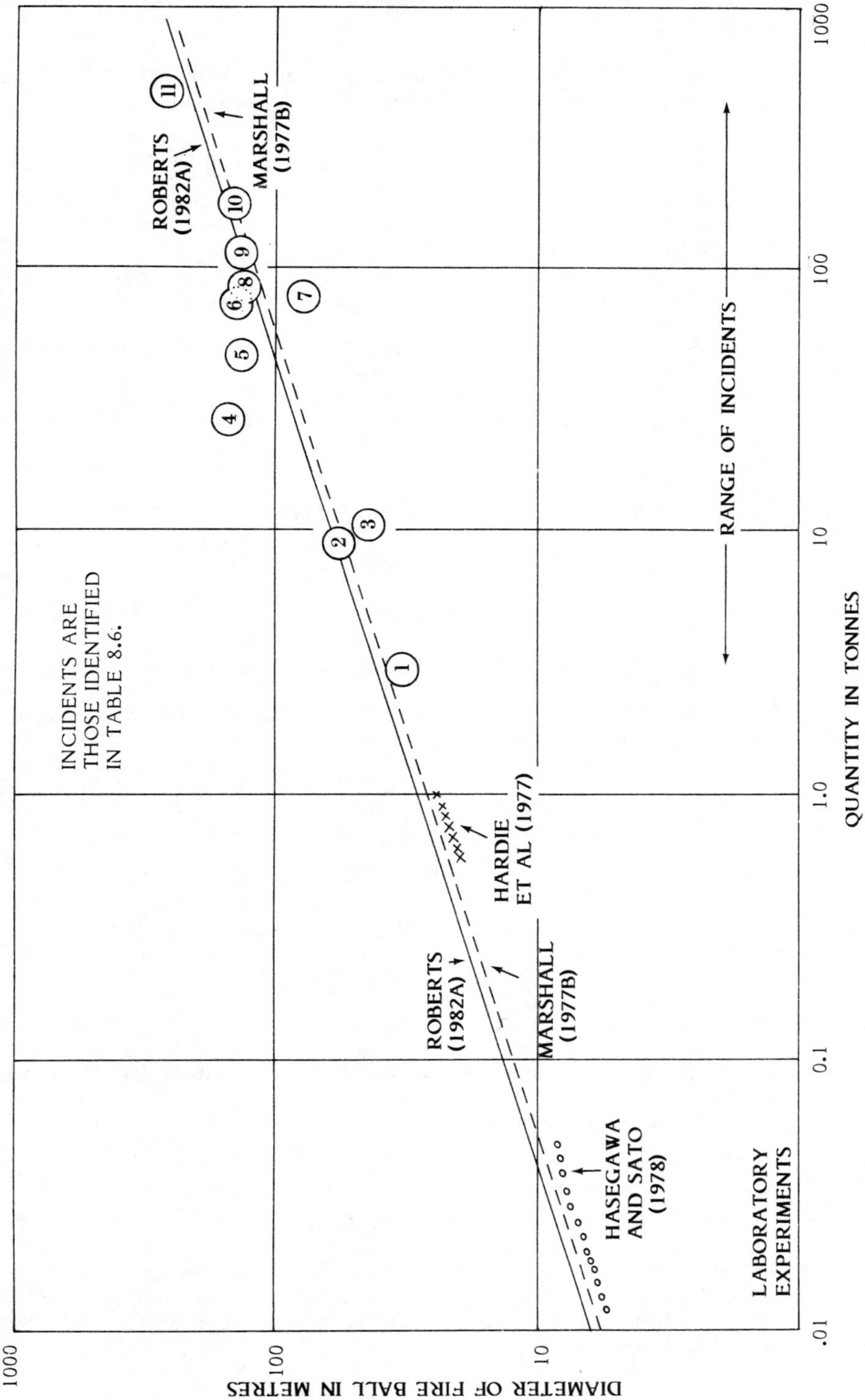

Fig. 8.5. RADIUS OF FIRE BALL VERSUS MASS OF FUEL.

8.9.3.4. Fire ball duration.

The history of a fire ball involves four main stages commencing from time zero (T_o), this being the time at which loss of containment commences.

Time to ignition	$T_o \rightarrow T_i$
Ignition to lift off	$T_i \rightarrow T_{LO}$
Lift off to full formation	$T_{LO} \rightarrow T_{FF}$
Full formation to radiation ceasing	$T_{LO} \rightarrow T_{RC}$

High (1968) correlated $T_i \rightarrow T_{RC}$ with an equation, which the present author has transformed into metric units as follows:-

$$T = 2.72\ M_P^{0.320}$$

or $$T = 3.96\ M_F^{0.320}$$

or $$T = 3.8 \times M_F^{1/3}$$ (Marshall 1977b)

where T = duration in seconds

M_P = mass of propellant

M_F = mass of hydrocarbon fuel in tonnes

Roberts (1982a) puts forward equations for the time stages (the symbols used are those of the present author). For propane he proposes:-

$$T_o \rightarrow T_{LFL} = \frac{35\ M_F^{1/3}}{\alpha}$$

where T_{LFL} = time to lower flammable limit

M_F = mass of fuel in kilograms

α = momentum of release per unit mass

Roberts assumes a value of α = 220 m/s

(It may be noted that this equation is not dimensionally consistent and therefore that 35 is not a number).

For propane, this further reduces to:-

$$T_o \rightarrow T_{LFL} = 1.6 \times M_F^{1/3}$$

where M_F = mass fuel in tonnes

Roberts (1982a) also concludes from a study of the paper, Hardee and Lee (1975), that for releases of up to 300 tonnes the momentum driven phase is the dominant one in the period in which ignition to produce a fire ball is feasible. On the basis of Roberts (1982a):-

$$T_o \rightarrow T_g = T_o \rightarrow T_1 \text{ (Max)}$$

Roberts (1982a) gives, where M_F is tonnes

$$T_1 \rightarrow T_{LO} = 3.16\ M_F^{\ 1/6}$$

and

$$T \rightarrow T_{RC} = 4.5\ M_F^{\ 1/3}$$

The transition point at which T_{RC} (total duration) exceeds T_{LO} is 0.12 tonnes. It follows from this that lift off for propane is not possible for releases of $\leqslant$ 0.120 tonnes. The phenomenon, in the present author's view, is significant only for tonnage releases. Thus Roberts (1982a) gives duration $T_o \rightarrow T_{RC}$ which are about 1.20 times the predictions of High (1968) as transformed in Marshall (1977b) and based on M_F instead of M_P. It would be reasonable to expect propellant fires to be more intense than hydrocarbon/air fires.

When the work of other investigators is reviewed wide discrepancies appear in this area. These are summarised in Moorhouse and Pritchard (1982) and the correlations fall into two classes. In one, based on small-scale tests, $t = \propto M_F^{\ 1/6}$ and in the others, based on rocket propellant, $t = \propto M^{1/3}$. The former are all based on small-scale trials where $T_o \rightarrow T_{LO}$ is greater than $T_o \rightarrow T_{RC}$ and thus they fall into the $M^{1/6}$ regime. It does not seem valid to proceed to extrapolate these correlations into the tonnage area. Lihou and Maund (1982), which reports recent experimental work on fire balls by the authors, also advances the opinion that it is not reasonable to hope to produce a correlation of the form

$$T = AM^a$$

in which A is constant, and which fits the data over a range of 1 - 10^6 kilograms.

8.9.4. THE SIGNIFICANCE OF THE WORD "BLEVE".

The acronym "BLEVE", according to Stinton (1981), was coined in the 1960s by an American, Wilbur Walls. The word was used by the UK Safety and Reliability Directorate in a report on Canvey Island, UK. This Report defined it, in Canvey (1981), as follows:-

" **BLEVE** (**B**oiling **L**iquid **E**xpanding **V**apour **E**xplosion).

> This event may occur when a pressurised vessel is engulfed in fire. Heat from the vessel boils the liquid and raises the tank pressure. Although a relief valve may prevent the tank exceeding its design pressure for normal conditions the tank metal above the liquid level will become overheated and weaken. Eventually the tank ruptures and the contents are explosively dispersed. If the material is flammable, instantaneous ignition occurs producing a fire ball. Although there is some blast effect from the tank rupture the main hazards are thermal (from liquefied flammable gases) and from flying vessel fragments."

The definition given above is defective in that it only assumes, and does not specifically state, that the contents of the vessel consist of a liquefied gas. Yet similar effects may arise from a cylinder of compressed gas though here no flashing will occur. The engulfing of drums of volatile liquids may also lead to rupture and, at the time of rupture, the contents may constitute a liquefied gas.

The scenario given above has four stages:-

(a) Flame engulfment of a pressurised vessel.
(b) Flame impingement weakens an unwetted portion of the walls leading to failure.
(c) The consequent pressure rupture releases a liquefied gas which flashes.
(d) The vapour from the flashing liquid ignites to produce a fire ball.

The acronym BLEVE can only describe stage (c). There is a long history of incidents involving stages (a) to (c) which arises from the explosion of steam boilers, especially fire tube boilers. In fire tube boilers, either through operational mischance or through operator negligence, a fall in water level below that of the top of the fire tube would produce an unwetted area of wall and lead to an incident involving stages (a) to (c). However, such an implosion could arise due to other causes such as the building up of hard water deposits on the water side of the fire tube, or to metallurgical failures. In such a case stage (b) would not apply.

Such boiler explosions have given rise to severe, if localised, effects. Though the present author has been unable to find a monograph on boiler explosions in general, there is a monograph on locomotive boiler explosions in the UK, that of Hewison (1983). This makes it clear that considerable local overpressure effects occur and that locomotives may be displaced by many metres by the reactive forces generated.

The same is true of pressure ruptures of containers of any liquefied gas. Though vapour cloud explosions are capable of generating significant overpressure effects over many kilometres, the maximum overpressures developed do not seem likely to exceed 2 bars absolute. Pressure rupture, at a point immediately adjacent to the vessel, will have a peak overpressure close to that of the pressure of storage but this is likely to attenuate to atmospheric pressure within a few multiples of the radius of the vessel. Propane at ten bars absolute pressure will give rise, theoretically, to a cloud of about 80 times its original volume at atmospheric pressure, or about 4.5 times its original radius if evenly distributed.

In summary, the expression BLEVE is an acronym of uncertain meaning, the use of which should be avoided. This advice accords with that given in I.Chem.E. (1984).

8.10. FIRE STORMS.

8.10.1. FIRE STORMS AND THE CHEMICAL AND PROCESS INDUSTRIES.

The phenomenon of the "fire storm", which is described below, has raised the question as to whether a conflagration in the chemical and process industries, on a sufficiently large scale, could give rise to a similar phenomenon.

There seems to be no clear answer to this question except to say that it cannot be ruled out but that only the very largest spillages seem likely to be able to give rise to the fire storm phenomenon and then under conditions in which the initial low ambient wind speed is low.

8.10.2. FIRE STORMS IN THE SECOND WORLD WAR.

During the Second World War it is apparent that in a number of air raids on Germany, and in one on Japan, casualties were inflicted on a scale far greater, in relation to the quantity of bombs dropped, than had been experienced in any air raid on the UK or in most raids on Germany. In these air raids fires occurred on a scale previously unknown and it has been generally agreed that these fires were accompanied by hurricane force winds directed towards the centre of the conflagration. It is this latter circumstance which gave rise to the German expression "feuerstorm" or "firestorm" which is now in general use.

These air raids involved roughly 50/50 proportions of high explosive and incendiary material. The raid on Hamburg, 17-28th July 1943, is described in Middlebrook (1984). In this raid, ca 1000 tonnes each of high explosive and incendiary bomb were dropped over the whole city. The fire storm area, on which about 80% of the bomb load fell, covered an area variously estimated as between 10 and 22 sq kilometres. The dead totalled some 40,000 and 16,000 apartment buildings caught fire.

Similar fire storms were noted, though on a less severe scale, at Darmstadt and Kassel, and in raids on other cities. The exceptionally horrendous raid on Dresden on February 13th 1945 has been chronicled in Irving (1980) and in McKee (1982). There is a brief account of the fire storm at Tokyo of March 9th 1945 in Daniels (1982). This reference claims that some 1600 tonnes of incendiary bombs started fires which killed more than 80,000 people and totally burned out about 40 sq kilometres of the city.

A fire storm followed the atomic attack at Hiroshima in August 1945, and there was also an incipient fire storm at Nagasaki [Glasstone & Dolan (1980)].

8.10.3. ANALYSIS OF FIRE STORMS.

There is a considerable literature on fire storms, some of it restricted in circulation, but it does not seem appropriate to review it here. Carrier et al (1985) is a recent paper which conducts a detailed mathematical analysis of the phenomenon and which, inter alia, reviews some of the literature. It sets out what the authors consider to be the principal parameters of fire storms; some of these appear in Glasstone & Dolan (1980).

Carrier et al see many resemblances between fire storms and certain meteorological phenomena and describe a fire storm as a "heat cyclone" or "mesolow". They see it as rotating and producing winds of 20 m/s to, possibly, a maximum of 50 m/s (70-180 kilometres/hour, 40-100 mph), and as producing a convective column perhaps 10 km in height. Such storms are initiated by numerous small fires in a heavily fuel laden urban environment which merge into a single fire. The storm may take half an hour to set in, reach a peak in two hours and die down in 6 to 9 hours. Typically 12 km^2 will be reduced to ashes. Calculations are presented for Hamburg in which an area for the storm of about 12 km^2 was assumed and the burn-up rate was assumed to be consistent with about 160 kg/m^2 of combustible material burning completely in 6 hours. This material was assigned a heat of combustion of 1.86 x 10 J/kg. Thus the mean heat release was 137 kW/m^2. Ground level temperatures, the authors claim, may reach 1000 K. The authors postulate that a minimum area of the order of 1.25 km^2 is needed to sustain a fire storm. Paradoxically, absence of initial wind seems to favour fire storm formation.

They, and other authors including Glasstone and Dolan, see a fire storm as an antithesis of a conflagration so far as spread of fire is concerned. A fire storm restricts fire spread whereas a conflagration will spread fire in the direction of the ambient wind.

8.10.4. COULD A FIRE STORM ARISE FROM THE REALISATION OF A MAJOR CHEMICAL HAZARD ?

Consideration of the parameters arrived at above seems to suggest that such an event is very unlikely under peace time conditions even for a very large site. But the possibility of numerous simultaneous ignitions cannot be discounted as this could arise from a vapour cloud explosion. Such explosions are discussed in Chapter 9, and one preceded the conflagration at Flixborough discussed as a case history in Chapter 9. However, the merging of these fires seems to be very unlikely in any site laid out in accordance with standard UK practice;

this may not apply however in some parts of the world.

It is instructive to examine briefly the conflagration at Flixborough (a detailed case history follows in Chapter 9). According to Flixborough (1975), there were some 1,650 tonnes of volatile liquids on the site, together with an unspecified quantity of combustible caprolactam. The conflagration occurred in part of the site and it measured, roughly, 200 metres by 300 metres [see sketch plan in Chapter 9], i.e. about 60,000 m^2. If all the volatiles were in this area and an arbitrary allowance of 1000 tonnes be made for caprolactam, this amounted to a total of, say, 2,650 tonnes of flammable material. Dividing by the area this gives 27.5 kg/m^2, which is only about a fifth of the minimum loading put forward by Carrier et al (1985). The area of the conflagration was only about one-twentieth of that set as a minimum area by Carrier et al. Thus by both criteria the Flixborough conflagration fell far short of the circumstances associated with a fire storm. In fact no such storm occurred.

It may be instructive too to examine the possibility of a fire storm occurring, under present-day UK conditions in surrounding housing. Current population densities are ca 4,000 per km^2 and, at 3 persons per household, this gives 1,333 households per km^2. The present author has obtained estimates from furniture removers and builders for average weight of combustibles per household. These suggest that furnishings account for 3 tonnes per household and that 5 tonnes of wood are needed to build the average house, making 8 tonnes per household in all. Allowance needs to be made for uninhabited buildings such as shops, schools, churches, petrol filling stations, etc, so that the quantity of flammables, in round figures, may be 10 tonnes per household. Multiplying this by 1,333 gives 13.33 kg/m^2 which is about half that required to give rise to a fire storm. It does not seem likely that the inhabitants of cities like Hamburg were more lavishly provided with furniture or housing than the present-day UK (a difference is that in Hamburg there were stocks of solid fuel but these must have been only a fraction of 1 tonne per person). It would seem to follow that the buildings in these German cities were far more closely packed than is the case in the UK today. Advisory Committee on Major Hazards (1984) concluded that fire storms in housing areas could only be feasible as an accompaniment to a massive spillage of liquid oxygen.

An incident in which a fire storm has been reported is that of Quatar in 1978. Details of this incident are lacking because of pending litigation but reports suggest that, after flashing, there was a pool of liquid propane some 500 metres in diameter. Such sketchy details as exist for this incident are given in Chapter 9.

8.11. ANAEROBIC FIRES.

8.11.1. INTRODUCTION.

Since anaerobic means "without air" (some dictionaries say "without oxygen"), it would seem to be a contradiction in terms to speak of an "anaerobic" fire. However there are some substances which, if raised to an appropriate temperature, undergo chemical decomposition to give rise to incandescent gas barely distinguishable from a flame.

8.11.2. EXAMPLES OF ANAEROBIC FIRES.

8.11.2.1. Monopropellant mixtures.

The oldest monopropellant is gunpowder (or black powder). This is an intimate mixture of grains of an oxidising agent, nitre, with a reducing agent in the form of grains of sulphur and charcoal. It will "burn", under unconfined conditions, with great rapidity as, for example, to form a fuse. In the barrel of a gun it acts as a "low" explosive or propellant. The process may be visualised as one in which, at an elevated temperature, the nitre decomposes to yield oxygen which reacts with the sulphur and carbon. This latter process is highly exothermic and accelerates the decomposition of the nitre giving rise to a violent exotherm.

Notable examples of propellants are solid rocket fuels which are sometimes described as "monopropellants". It was such a "flame" from one of the Challenger booster rockets which, it is being alleged at the time of writing, initiated the failure of the Challenger Space Shuttle in January 1986. The powerful monopropellants which have been developed for space exploration are described in a number of standard reference works such as the McGraw-Hill Encyclopedia (1982). They are also mixtures involving substances such as nitrated cellulose gelatinised with nitroglycerine. The manufacture and handling of such propellants is too specialised for further description here.

8.11.2.2. Nitrated cellulose.

Nitrated cellulose (miscalled "nitrocellulose"; it is a nitrate of cellulose, not a nitro compound) was discovered in 1838 and further developed by Schoenbein in 1845. The properties of nitrated cellulose are much dependent on the degree of nitration. It is theoretically possible to nitrate the cellulose unit, which may be thought of as $[-C_6H_7O_2(OH)_3-]_n$, in three stages each substituting a nitrate ester group for a hydroxyl group. In practice, nitrocelluloses of commercial importance lie in the range between the dinitrate and the trinitrate. They are mixtures of esters, not pure compounds. Read (1942) gives three classes with their nitrogen content expressed as a percentage.

Collodion cotton	(10.7 - 11.7)	Plastics, lacquers.
Collodion cotton	(11.5 - 12.7)	Smokeless powders.
Guncotton	(13.0 - 13.5)	High explosives.

Only the latter is listed in the CIMAH Regulations [Health & Safety Executive (1985a)] where cellulose nitrate of $\geqslant 12.6\%$ nitrogen is classed as an explosive substance.

However, all varieties of cellulose nitrate are capable of "burning". This process may be thought of as an "internal" redox reaction in a molecule in which there are oxygen atoms which are capable of reacting with the carbon and hydrogen of the cellulose unit. There is a short account of cellulose nitrate fires in Chapter 9. Since no oxygen is required, the burning takes place with great rapidity.

8.11.2.3. Ammonium nitrate.

Ammonium nitrate is an anomalous substance. Though it can be detonated to explode with great violence, the pure substance is not classified in the UK as an "explosive". To classify it in this way would create serious problems because of its widespread use as a fertiliser. Mixed with fuel oil to form a slurry, it is widely used for blasting. It is then called ANFO (**a**mmonium **n**itrate **f**uel **o**il). As such it has been used by terrorists in Northern Ireland. It has also been extensively used as a mixture with TNT to give the high explosive "Amatol".

Clancey (1963) has reviewed the properties of ammonium nitrate in relation to fire. He records that fires involving up to 2000 tonnes have occurred without explosion, and he concludes that fires involving ammonium nitrate are much more likely to escalate to explosions if organic matter is present. This can take the form of paper bags or sacks used for packaging.

It would appear that a fire in which ammonium nitrate is involved could be maintained without oxygen being supplied and that, under confined conditions, this may escalate to an explosion. The case history of the Texas City disaster in which this appears to have happened is given in Chapter 11.

8.11.3. SPILLAGES OF LIQUID OXYGEN.

There has been apprehension in the UK for a decade or more that a large-scale oxygen spillage could give rise to a severe conflagration. In a circular from the Department of the Environment [D O E (1972)], local planning authorities were advised to consult the Factory Inspectorate before permitting housing in the neighbourhood of a storage containing as little as 135 tonnes of liquid oxygen. The effect of enhanced oxygen levels is that substances not normally combustible will burn, and that the ignition temperatures of ordinary combustibles will be lowered. For example, clothing becomes highly combustible at ca 25% oxygen. Flame temperatures are increased when oxygen-rich air is utilised in place of air.

It seems just barely conceivable that the apprehensions expressed in Advisory Committee on Major Hazards (1984), that a fire storm could be caused by a massive spillage of oxygen, could come to pass. There could be multiple fires and possibly the much greater rate of burn-up could compensate for what would otherwise be too low a loading of combustible material.

It has to be said, however, that no spillage from an installation anywhere in the world appears to have led to injury to members of the public through enhanced combustion.

8.12. THERMAL RADIATION AND ITS EFFECT ON PEOPLE.

8.12.1. INTRODUCTION.

The human species, if suitably clothed, is capable of living within a relatively wide range of ambient temperature. At the minimum, in the Antarctic, the ambient temperature may approach -90°C and at the highest, in a number of parts of the world, it may approach 58°C. However, the human body itself can only function within a relatively narrow temperature range and therefore for most climates some form of clothing is needed.

Though the human body can be subjected to intolerable heat or cold by contact with hot or cold materials, whether solid, liquid or gaseous, which conduct heat into or out of the human system, this section will be concerned only with the harmful effects of heat, not of cold, and then principally with the effects of fire and its consequences rather than with scalding, i.e. contact with hot liquids where the mechanisms of conduction or convection are governing.

The prediction of the quantity of thermal energy received by the human body from a radiating source is, in principle, soluble by application of the laws of physics. The prediction of the effects of radiation on the human body, though governed overall by the laws of physical science, also entails an understanding of human anatomy, physiology and pathology. Only the barest outlines of these subjects, and only insofar as they are relevant to this aspect of Major Chemical Hazards, can be discussed here.

The subject of "burning" by contact with corrosive chemicals such as mineral acids and alkalis will also be excluded from consideration here.

8.12.2. THE NATURE OF INJURY TO PEOPLE.

Injury by burning is described in Mason (1978). The total number of deaths from thermal injury (including scalds) in the UK in 1975 was 492 with a downward tendency in mortality of about 20% over the previous decade. Mason (1978) notes that the figure is only half the number of the deaths which occur in fires and that this is attributable to the proportion who die in fires through asphyxiation or inhalation of carbon monoxide and other poisonous fumes. For every death from thermal injury there are twenty detained in hospital for an average of 5-6 weeks and 200 who seek treatment in hospital casualty departments. Roughly half of those treated are under the age of 10. Two-thirds of male adults injured are burned at work.

Mason (1978) claims that high explosive bombs are unlikely to create burns. A terrorist bomb of 13.5 kg produces a fire ball of about 9 metres diameter, but this dissipates immediately in the open air. However, in confined spaces, burns may ensue. The injuries from the military agent napalm, which is petrol thickened with an aluminium soap made from coconut oil, naphthenic acid and oleic acid, can be very severe. Gases at temperatures above 1000°C are produced and in confined spaces there is a likelihood of death from asphyxia or carbon monoxide poisoning. Asphyxiation may result from inhalation of the hot gases which will burn the air passages or destroy the alveoli (the air sacs in the lungs).

There is clearly a threshold below which thermal injury does not occur. Bath water usually has a temperature in the range of 36-42°C. However, 10 seconds' contact with water at 60°C will produce partial skin thickness loss and 10 seconds at 70°C will cause full skin thickness loss.

In temperate climates for most of the year solar radiation has little effect on the skin. However, on a few days a year, sunshine is capable of burning the skin. In essence there is an upper limit to skin temperature which, if exceeded, causes injury; pain sets in at about 44°C skin temperature. The prediction of the intensity of radiation required to produce such a temperature is complicated by the existence of cooling mechanisms in the skin such as sweating and blood flow. The reaction of the skin depends not only on the intensity of radiation but upon its duration; this topic is further considered in sub-section 8.12.4.

8.12.3. THE PHYSICAL LAWS OF RADIATION.

The physical laws which govern radiation are established with an accuracy which greatly exceeds the accuracy with which the physiological laws to be discussed later are known.

Thermal radiation forms part of the electromagnetic spectrum as set out in Figure 8.6. and the emission of, or the absorption of, thermal radiation is a manifestation of the exchange of energy and can be quantified as such. The laws of propagation of thermal radiation are similar to those of light.

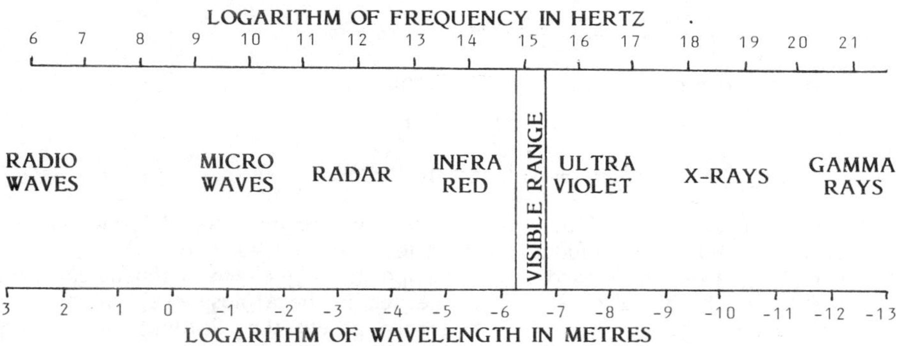

Fig. 8.6. SPECTRUM OF ELECTROMAGNETIC RADIATION.

For example, bodies which reflect light reflect thermal radiation. However, there are considerable differences in the extent to which bodies which are transparent to light are also transparent to thermal radiation and vice versa. The wellknown "greenhouse effect" is an example of this.

The intensity of radiation, denoted here by the symbol I_R, is quantified as power per unit area of the emitter. The units used are the SI units, joules sec^{-1} $metres^{-2}$. The intensity of thermal radiation from a body is a function of its absolute temperature raised to the fourth power (T_A^4) and its emissivity ε is the fraction radiated as compared with what would be radiated by a "black body" or ideal radiator. A mirror is the opposite of a black body and has an emissivity approaching zero.

The equation which governs energy exchange by thermal radiation is the well established Stefan-Boltzmann equation:-

$$I_R = \sigma[\varepsilon Te^4 - \varepsilon Ta^4]$$

where ε_e, ε_a = emissivities of emitter and absorber

T_E, T_A = temperature (absolute) of emitter and absorber

σ = Stefan's constant

= 5.67×10^{-8} joules sec^{-1} $metres^{-2}$ K^{-1}

It will be apparent that where T_E is much greater than T_A, the importance of the T_A term becomes less and less significant. For the case where T_E = 1365 K and T_A = 273 K, $\frac{T_E}{T_A} = 5$.

$$\text{Thus } \frac{T_E^4}{T_A} = 5^4 = \frac{625}{1}$$

Under these circumstances the value of T_A^4 is swamped by the uncertainty attached to the evaluation of T_E^4 and T_A^4. The T_A^4 term will be neglected in the discussion below.

The law which governs the wavelength (or its reciprocal, the frequency) is Wien's displacement law which holds that the wavelength of the most strongly emitted radiation is inversely proportional to the absolute temperature of the body:-

$$\lambda = \frac{b}{T}$$

where b = Wien's constant

= 2.9×10^{-3}m K

λ = wavelength in metres.

From the point of view of the subject of this chapter, Wien's law is not of great application but it is included for completeness of treatment.

In travelling through a medium, radiation is subject to attenuation. The medium with which this discussion is concerned is the atmosphere, which consists of the monatomic gases, argon and the rare gases, the diatomic gases oxygen and nitrogen, and the triatomic gases carbon dioxide and water vapour, with aerosols, such as mist (mainly water droplets) and dust. At the levels of temperature under discussion neither monatomic nor diatomic gases are significant attenuators of thermal radiation. Of the triatomic gases the carbon dioxide content of the atmosphere is fairly constant at about .03% by volume, but the water vapour content is very variable having as its upper bound the saturation vapour pressure of water at the ambient temperature. This is given in Table 8.8. below.

TABLE 8.8. VOLUME FRACTION OF WATER VAPOUR AT SATURATION IN ATMOSPHERE.

Temperature ^{o}C	-10	0	10	20	30
Volume fraction %	0.28	0.60	1.20	2.3	4.2

It is apparent from this that it would be difficult to arrive at a typical figure for the attenuation due to water vapour.

As will be discussed below, some authorities consider that attenuation due to passage through the atmosphere plays a significant role in the assessment of the thermal effects of fire balls. However, there is little published material in this area which is, or so one might suppose, more susceptible to theoretical analysis and laboratory experiment than many of the other subjects relevant to this analysis. Reference will be made below to the data which do exist in this area.

8.12.4. THE PHYSIOLOGICAL ASPECTS OF THERMAL RADIATION.

8.12.4.1. Anatomy.

The skin may be divided into two layers. The outer layer is known as the "epidermis" which is itself divisible into layers. Sweat pores penetrate the epidermis but it does not contain blood vessels or nerve endings.

The lower layer, or dermis, which is much thicker, contains blood vessels, nerve endings, the sebaceous glands and the hair roots. The epidermis and dermis are about 1-2mm thick in total.

Below these two layers is the subcutaneous tissue which houses the sweat glands and larger blood vessels. The blood vessels enable local heating or cooling to be adapted to, but within limits.

The degree to which the skin is damaged by thermal radiation may be expressed in terms of the depth of injury. Mild radiation conditions will damage the epidermis only, with a depth of 1mm, more intense radiation may damage both epidermis and dermis, and the most intense radiation will also affect the subcutaneous layer. These three levels broadly correspond with the well established categories of 1st, 2nd and 3rd degree burns.

The possibilities for survival for the victims of burns are dependent upon a number of factors, the degree of burning, the extent of the burned area expressed as a fraction of the total skin area, the age and the previous medical history of the victim. Broadly speaking the chances of survival diminish as the percentage of burned area and the age of the victim increases. The pathology and treatment of burns lies outside the scope of this book; there is a brief description of these aspects in Mason (1978).

8.12.4.2. Thermal dosage.

There is clearly a threshold below which even indefinite exposure to heat is without injurious effect. With bath water most adults can withstand temperature up to 40-42°C which accords with the pain threshold skin temperature of 44°C claimed in Hymes (1982). Above the threshold the degree of injury is a function both of the thermal energy received per unit area of skin and the time of exposure. Eisenberg et al (1975) puts forward (energy/unit area) 4/3T as the equal injury energy input function. Expressing energy intensity in SI units of W/m^2 and T seconds, according to Eisenberg et al, injury occurs when $[E/TL^2]^{4/3} \geq$ ca 9×10^4. The problem of determining a safety distance is to evaluate a distance such that $[E/TL^2]$ 1.33T does not exceed a certain maximum value.

8.12.5. THE ESTIMATION OF SAFETY DISTANCES FOR FIRE BALLS.

8.12.5.1. Two approaches.

One approach to the estimation of safety distances for fire balls is the "pragmatic" approach. This approach studies actual incidents to determine from the positions occupied by those killed, those injured with varying degrees of severity, and those who escaped, the approximate boundaries between the various classes of injury. This approach is the one which is taken in the field of explosives in Robinson (1944) and in Healy (1965). These references, particularly the latter, attempt to derive safety distances for explosives on the basis of the analysis of actual incidents and in the light of Hopkinson's Scaling Law. Healy (1965) is today the basis of the UK Regulations governing safety distances from stores of explosives and ammunition. Data for these studies, which will be further discussed in Chapter 10, are fairly abundant.

It may be noted that Crawley (1982) draws attention to survival at distances much below those that theory would suggest as fatal. The reference quotes the cases of survival of firemen at distances as little as 30 metres from disrupting tank cars. Eisenberg et al (1975) claims that there is a fireman's "rule of thumb". "Firemen have succumbed to radiation incurred when they were as far away as 75 metres from large fire balls (i.e. arising from US "jumbo" rail tankers)". Figures of this nature appear to be an order of magnitude lower than some theoretically derived figures but unfortunately they are too vague to use as a basis for calculating safety distances.

It has to be stated, however, that data in the area of fire balls are not abundant. Stinton (1978) records that at San Carlos no attempt was made, even when moving those already dead, to record their positions before removal; a process which in Stinton's view also made the difficult task of identifying virtually impossible.

However, in Chapter 9 there will be a discussion which attempts to draw such conclusions as are possible arising from a study of the San Carlos and other disasters.

A second approach is a "theoretical" approach which aims to calculate, either for a particular size of fire ball, or for the generality of fire balls, the intensity of radiation at any particular distance and to relate this to the known physiological effects of such radiation.

8.12.6. THE HISTORICAL APPROACH TO FIRE BALL SAFETY DISTANCES.

8.12.6.1. Analysis of San Carlos.

The general circumstances of this accident are described in Chapter 9. As has been noted above, data on the locations of the people who were killed or injured are sparse. This is partly because, in the case of the dead, they were removed to the mortuary without recording their positions. Secondly, there does not appear to have been any attempt on the part of the Spanish authorities to conduct a systematic analysis of the position of eye-witnesses, whether injured or not. (It should be said that there was also a failure to record the position of the dead in Flixborough (1975).

Thirdly, though there may still be information available from survivors they are now widely scattered and beyond approach. Certain deductions can be made, however, and the results, however crude, are put forward as the best that can be concluded from the incident.

Scilly (1978) gives the area of the site as about 400 metres long by 80 metres deep, a total of 32,000 sq metres. The total number of persons present on the site is given by Stinton (1978), as about 500. The number instantly killed is estimated at about 95 (5 were also killed in a "disco" off site), delayed deaths are given as 115, and hospitalised at 67. Assuming a uniform distribution of population the respective areas are thus:-

(a) Instantly killed $\frac{95}{500} \times \frac{32,000}{1} = 6,080\ m^2$

(b) Instantly killed plus delayed deaths $\frac{210}{500} \times \frac{32,000}{1} = 13,440\ m^2$

(c) Instantly killed plus delayed deaths plus hospitalised survivors $\frac{277}{500} \times \frac{32,000}{1} = 17,728\ m^2$

If it be assumed that the areas affected be semicircles then the respective radii are as follows:-

a = 62 metres
b = 92 metres
c = 105 metres

The fire ball radius, calculated according to Marshall (1977b), and assuming that M was 23 tonnes, would be:-

$$R = 27.5\ M^{1/3} = 78 \text{ metres}$$

There is, however, some measure of uncertainty as to centre of the fire ball.

The conclusion to be drawn from this is that the instantly killed were those engulfed in the fire ball, and that serious injury occurred up to 1.5 fire ball radii. Most factors favoured high mortality. This included paucity of shelter, scanty clothing, and pre-existing warm skin arising from exposure to the sun. However, a factor which may have assisted survival is the nearness of the sea into which a number of people appear to have fled.

8.12.6.2. Good Hope, Louisiana Marine Collision.

This collision, in which butane was released, occurred on 30 August 1979 and is described in National Transportation Safety Board (1980). In the accident, in which 12 were killed and 25 injured, a freighter collided with a butane barge moored at a jetty, striking it amidships and puncturing severely at least one of the tanks. The escaping gas gave rise to a fire ball with flames "hundreds of feet" (100 metres?) high which lasted "less than a minute" or about 30 seconds according to Safety at Sea (1980). There was evidence of overpressure both on the freighter and on the jetty.

National Transportation Safety Board (1980) gives an unusually detailed account including a map with coordinates showing the position of those who died as they were at the time of the fire ball. Unfortunately, the published map is a relatively small-scale map 1:24,000 (ca 21/2" to 1 mile). Figure 8.7. is based on this map and reconstructs the position of the vessels involved as described in the narrative of National Transportation Safety Board (1980).

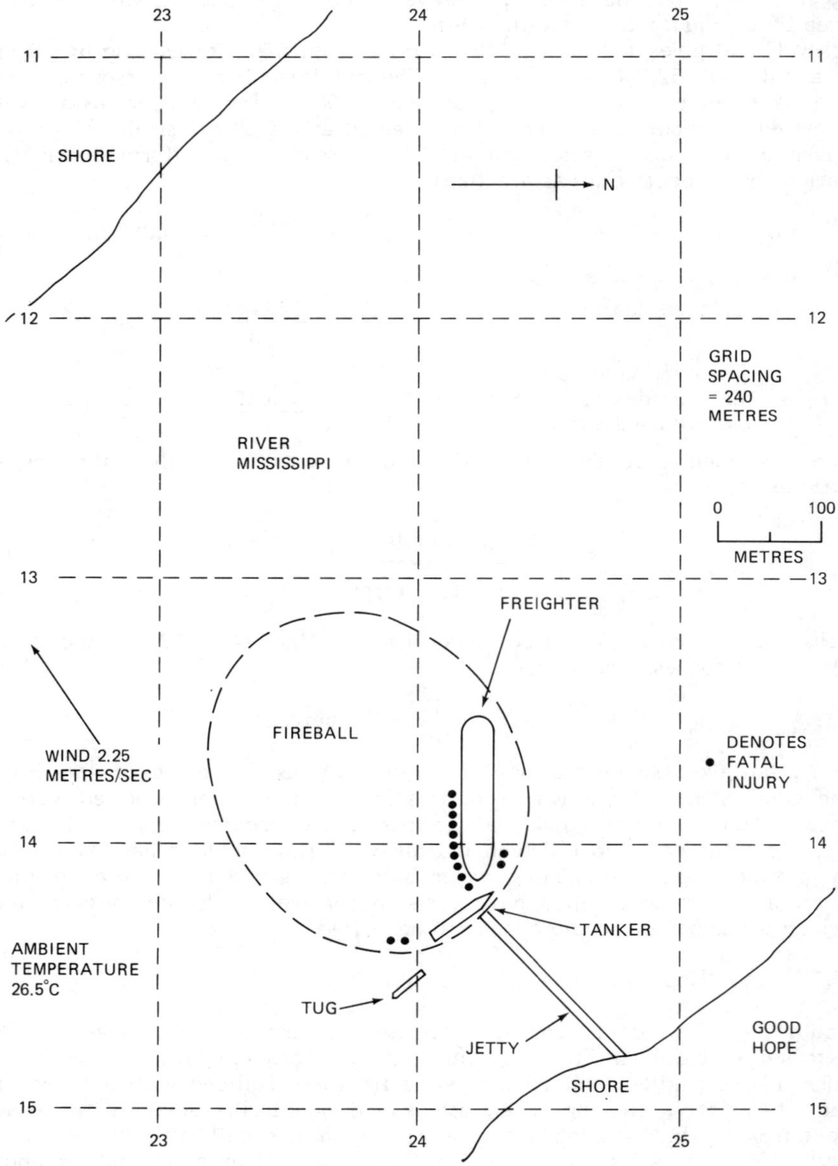

Fig. 8.7. SKETCH PLAN OF FIRE BALL AT GOOD HOPE, LA.

The quantity of butane released is not directly stated. The barge carried 6 tanks in two groups, one fore and one aft, consisting of three tanks side by side. The entire cargo, it is claimed, was 283,000 gallons (US) or about 630 tonnes of LPG, stated to be butane. The fate of the tanks is stated at various places in the Report to have been as follows:-

Tank	
1) 2)	Floated free of barge and drifted downstream. They continued burning for 3 hours after the collision.
3	Stated to be ruptured. Found to be empty afterwards, it sank with the fore part of the barge.
4) 5) 6)	After portion with these tanks drifted downstream on fire, 4 and 5 were cracked, 6 was blowing off its safety valve, eventually found to be intact and to contain 53,000 gallons (US) of butane (118 tonnes).

It seems evident from this that only tank 3 suffered a catastrophic rupture. At the stated ambient temperature of 80°F (26°C) the pressure would be (as a maximum) about 3 bars. How much of the ca 100 tonnes in the tank would have flash vaporised or boiled off in contact with the river water is difficult to say but it would appear that burning was over in less than one minute.

Therefore there are some grounds for assuming that the entire contents, 100 tonnes, may have taken part in forming the fire ball. According to Marshall (1977b), if M = 100

$$R = 27.5\ M^{1/3}$$

$$= 127 \text{ metres}$$

The maps in National Transportation Safety Board suggest a fire ball radius of about 120-150 metres.

At the time of the fire ball there were 39 persons in the vicinity, i.e. within 200 metres of fire ball centre. The locations of these persons and the injuries they sustained are shown in Table 8.9.

TABLE 8.9. LOCATION AND INJURIES SUSTAINED AT GOOD HOPE, LOUISIANA.

Vessel		Nos		Injuries sustained	
Freighter	On bridge	5)	32	Fatal (2 drowned)	10
	On deck	7)		$\geqslant$ 14 days in hospital	6
	In engine room	5)		$\leqslant 14 \geqslant 1$ days in hospital	10
	Unstated	15)		$\leqslant$ 1 day in hospital	6
Tug	On deck	2)	3	Fatal (1 drowned)	2
	Below	1)		$\leqslant 14 \geqslant 1$ days in hospital	1
Butane Tanker	Unstated	1		$\leqslant 14 \geqslant 1$ days in hospital	1
Jetty		1		$\leqslant 14 \geqslant 1$ days in hospital	1
		37			37

Summing up, the injuries were as follows:-

		Burn victims.
Fatal	12 (3 drowned)	9
Severe burns	$\geqslant$ 14 days in hospital	6
Moderate burns	$\leqslant$ 14 $\geqslant$ 1 days in hospital	13
Mild	$\leqslant$ 1 day in hospital	6
		34

Consideration should be given to the numbers who may have been below decks on the freighter (20?) and tugboat (1) who may have escaped the fire ball. Nevertheless everyone was within about 1.5 fire ball diameters and two out of three survived. Only 9 persons died of burns or 24%.

8.12.7. THE THEORETICAL APPROACH TO FIRE BALL SAFETY DISTANCES.

8.12.7.1. Outline of theoretical approach.

The method seeks to determine from a given mass of fuel in a fireball the distance from the centre of the fireball at which certain physiological effects are likely to occur and to determine from this data the "safety distance" to be employed. Such data may also be of value in quantifying the hazard which arises from the storage of flammable substances.

The steps include the following:-

(a) Calculation of theoretical and practical energy releases from a given mass of fuel.
(b) Estimation of the fraction of this energy which will be released as thermal radiation.
(c) Estimation of the duration of the thermal pulse.
(d) Estimation of the intensity of the radiation at various distances assuming zero attenuation.
(e) Estimation of the degree of attenuation by the atmosphere.
(f) Evaluation of the physiological consequences of various values of the "dose" of thermal energy.
(g) Estimation of the "dose" at various radii.
(h) Deciding what is the maximum dose which society will be prepared to tolerate in the event of a fire ball arising from the ignition of flammable vapour.
(i) Relating this level to a radius for a given level of apprehended release.

8.12.7.2. Conclusions of other investigators.

The comprehensive review of the literature in Moorhouse and Pritchard (1982) offers little information on the calculation of safety distances but calls attention to the importance of atmospheric attenuation as a factor in the calculation. Crawley (1982) suggests that the evacuation distance be calculated on the basis that the skin can withstand a flux of 2.1×10^4 Wm^{-2} for 2 seconds without pain and that the safe evacuation distance is therefore given as:-

$$R_{EV} = R_{FB} \left(\frac{I \text{ at fireball surface}}{2.1 \times 10^4} \right)^{0.5}$$

where R_{EV} = evacuation radius
R_{FB} = fire ball radius
I = intensity of radiation

this gives $R_{EV} = 3.1\ R_{FB}$, and, for a surface temperature of 1473 K, this gives

$$R_{EV} = 3.6 R_{FB}$$

Roberts (1982a), for a 100 tonne hemispherical fire ball of R = 171 metres and 21 secs duration (assuming zero attenuation) gives the following distances:-

TABLE 8.10. THRESHOLD DISTANCES AFTER ROBERTS.

	$F_R = 0.3$	$F_R = 0.4$
50% Lethality	395m ($2.30R_{FB}$)	455m ($2.66R_{FB}$)
1% Lethality	520m ($3.04R_{FB}$)	600m ($3.51R_{FB}$)
Blister threshold	925m ($5.50R_{FB}$)	1070m ($6.25R_{FB}$)

where F_R = fraction of heat release which is radiated by fire ball.

Williamson and Mann (1981), which assumes an adiabatic flame temperature for the surface of a multitonne fire ball and with no atmospheric attenuation, gives distances for third degree and second degree burns as are shown in Figure 8.8. This figure may be expressed by the following relationships.

Probable distance to third degree burns = $80M^{0.42}$ metres

Maximum distance to third degree burns = $130M^{0.42}$ "

Probable distance to second degree burns = $150M^{0.42}$ "

Maximum distance to second degree burns = $240M^{0.42}$ "

These distances for a 1000 tonnes spillage correspond to the distances at which burn injuries occurred at Hiroshima for a 12,500 TNT equivalent bomb.

8.12.7.3. Calculation of heat release.

When speaking of an "M" tonne fireball (M_{FB}) this is not necessarily the same thing as the fire ball which arises from a spillage of M tonnes of spilled LPG (M_S). The factors which make M_{FB} tend to be equal to M_S are:-

(1) a high value of TAFF.
(2) circumstances which promote a rapid evaporation of any residual unflashed LPG, i.e. spillage in summer, or onto water, or a catastrophic failure which showers any unflashed LPG over a wide area.

It is evident from previous discussion that a spillage of M_S tonnes can give a vapour cloud which ranges in mass of flammable material from M_S to zero.

The key factors which tend to produce fire balls are high ambient temperatures and spillage on water rather than on land. Therefore regard must be paid to the value of the ratio M_{FB}/M_S in assessing the likelihood and magnitude of fire balls.

The calculation of the theoretical energy release is relatively easy in that there are abundant data on heats of combustion. Useful sources, albeit in British units, are those of Tables 9-18 and 9-20 of Perry (1973), which gives, inter alia, heats of combustion and adiabatic flame temperature of substances in gaseous form to give gaseous products. Some of these data are tabulated in Table 8.11. in which the heats of combustion are converted to megajoules

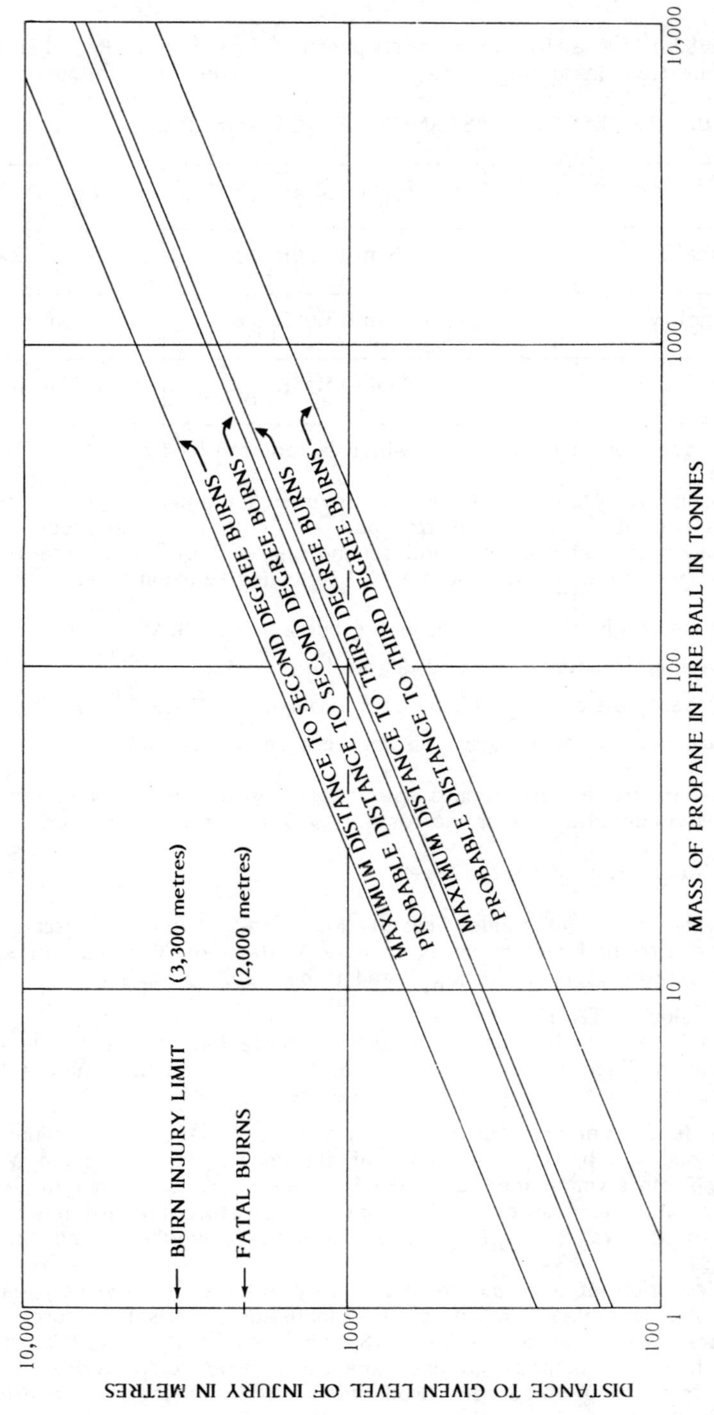

Fig. 8.8. DISTANCE TO A GIVEN LEVEL OF INJURY.

per kg or gigajoules per tonne.

TABLE 8.11. HEATS OF COMBUSTION (GAS TO GAS) AND ADIABATIC FLAME TEMPERATURES.

	BTU /lb	MJ/kg or GJ/tonne	Adiabatic Flame Temp K
Paraffins			
Methane	21,502	50.7	2236
Ethane	20,416	48.2	2244
Propane	19,929	47.0	2250
Butane	19,655	46.4	2255
Pentane	19,499	46.01	2244
Unsaturated aliphatics			
Ethylene	20,276	47.8	2375
Propylene	19,683	46.45	2250
Butalene	19,475	45.90	2319
Butadiene	19,152	45.20	2375
Acetylene	20,776	49.00	2598
Cyclic Compounds			
Cyclohexane	18,846	44.50	2250
Benzene	17,446	41.17	2305
Toluene	17,601	41.53	2344
Miscellaneous			
Dimethyl Ether	-	28.69[(1)]	2270
Ethylene Oxide	-	31.60[(1)]	2411
Vinyl Chloride-Monomer (VCM)		19.10[(2)]	

(1) From Weast (1981).
(2) From KOE (1964).

It is apparent from the table that the substances which most commonly have given rise to fire balls have heats of combustion ranging from about 45 to about 48 megajoules per kg (or gigajoules per tonne) and adiabatic flame temperature of about 2250-2350 K. For purposes of approximation, a heat of combustion of 47 MJ/kg and an adiabatic flame temperature of 2300 K are not unreasonable assumptions.

One example of a substance which has been known to give rise to fire balls but which has a much lower value of heat of combustion than paraffins or olefines is vinyl chloride monomer (VCM). However, fire balls may also arise from tonnage releases of any flammable gas that is boiling at a temperature above its atmospheric pressure boiling point and allowance may have to be made for values of heats of combustion which differ notably from those suggested above.

The discussion up to now has dealt with the factors of the relationship between the mass spilled and the mass in the cloud and on the energy release per unit mass. However, there is a further factor and that is the question of the efficiency of combustion. No quantitative figures are quoted in this area but a number of writers, including Crawley (1982), have commented on the presence of smoke, i.e. unburned carbon in fire balls. Calculation of theoretical diameters assume stoichiometric conditions but these do not favour complete combustion which requires excess air. The quantity of fuel reacted in unit time per unit volume is no greater than is the case with pool fires as is shown below.

The power, i.e. the heat release in unit time, in a typical fire ball is given as follows:-

$$P_{FB} = \frac{47 \times 10^9 \text{ joules} \times M}{3.8 \times M^{1/3}}$$

or $$P_{FB} = \frac{47 \times 10^9}{3.8} \times M^{2/3} = 12.3 \times 10^9 \times M^{2/3}$$

where P_{FB} = power in watts

where M = mass in tonnes

for a 20 tonne fire ball

$$P = 12.3 \times 10^9 \times 7.36$$

$$= 90 \text{ gigawatts}$$

The power density, i.e. power per unit volume, or, if it be assumed that heat release is a linear function of mass reacted, the burn-up rate per unit volume, may be calculated as follows:-

$$P_D = \frac{\text{power}}{\text{volume of fire ball}} = \frac{12.3 \times 10^9 \times M^{2/3}}{\frac{4\pi}{3} \times (27.5 \times M)^3}$$

$$= \frac{12.3 \times 10^9\ M^{2/3}}{8.7 \times 104\ M}$$

$$= 1.42 \times 105 \times M^{-1/3} \text{watts per m}^3$$

The burn-up rate of a fire ball is thus inversely proportional to $M^{1/3}$. For a 20 tonne fireball the power density would be:-

$$\frac{1.42 \times 10^5}{2.71} = 5.2 \times 10^4 \text{ watts/metre}^3$$

A corresponding figure for pool fires may be deduced using the information in Mizner and Eyre (1982). For LPG the burn-up rate = ca 0.10 kg m^{-2} s^{-1}.

For a 10 m radius pool the area is 314 m^2, ignoring the flame drag. The height of the flame would be about 3.4 pool radii. This gives a flame volume of ca 314 x 34 m^3 or ca 10^4 m^3. Thus power density for a pool fire is given as:-

$$P_D = \frac{0.10 \times 314 \times 47 \text{ MW}}{104}$$

$$= 10^{-5} \times -14 \times 47 \text{ MW per m}^3$$

$$= 10^{-5} \times 314 \times 47 \times 10^6 \text{ per m}^3$$

$$= 1.4 \times 10^5 \text{ watts/metre}^3$$

If flame drag be taken into account the value is lower, say 10^5 watts/metre^3. Thus fire balls have a power density comparable with those of pool fires. Hence the quantity reacted per unit volume in unit time in a fire ball is likely to be of the same order as that of a pool fire. On general grounds estimations of the fraction of the heat actually released per mass of fuel as compared with that theoretical available in a pool fire should be indicative of the value of corresponding fraction in a fire ball. Unfortunately, the present author could find no record that this has been investigated. However, all investigators of pool fires have noticed unburned carbon (soot) in the flame and it seems very probable that some of the carbon burns to carbon monoxide only.

The following are sources of inaccuracy in estimating the energy release in a fire ball:-

(a) Uncertainty as to the value $\frac{M_{FB}}{M_S}$

(b) Uncertainty as to the efficiency of combustion

The assumption that $M_{FB}/M_S = 1$ and that combustion is complete are assumptions which are pessimistic in character. The theoretical intensity of the radiation depends also on D_{FB}, the fire ball duration, and hence the accuracy with which it may be calculated is no greater than the accuracy of the equation. However, the effects on the surroundings are characterised at least roughly by the energy pulse, i.e. the product of intensity and duration and, thus, to a degree, the effect of inaccuracy in estimating duration is self-cancelling so far as this end product is concerned.

8.12.7.4. Calculation of the fraction of the energy which is radiated.

A number of investigators, including Moorhouse and Pritchard (1982) and Roberts (1982a), have drawn attention to the convenience which ensues from an approach which assumes that the fire ball may be treated as a point source which radiates a given fraction of the theoretical energy release, the remainder being sensible heat remaining in the products of combustion. In principle, knowledge of the value of this fraction would obviate the need to calculate the radius of the fire ball, and the emissive power of its surface. Unfortunately, there is no means of calculating this fraction directly and recourse has to be had to checking certain assumed values for it against what is known of the surface temperatures of fire balls. No field trials have produced reliable evidence of surface temperatures, but there have been records on film made by journalists at accidental releases in the USA.

Using the equations given above, the deduced mean emissive power (intensity at the fire ball surface) may be derived as follows:-

$$I_S = \frac{P_{FB} \times F_R}{4 \pi R_{FB}^2}$$

where I_S = Intensity at surface, watts metre^{-2}

P_{FB} = Power of fire ball

F_R = Fraction radiated

R_{FB} = Radius of fire ball, metres

$$= \frac{P_{FB} \times F_R}{4\pi(27.5\ M^{1/3})}$$

$$= \frac{47 \times 10^9 \times M \times F_R}{(3.8 \times M^{1/3})\ 4\pi(27.5 \times M^{1/3})^2}$$

$$= \frac{47 \times 10^9 \times F_R \times M}{3.8 \times 4\pi \times 27.5^2 \times M^{1/3} \times M^{2/3}}$$

$$= 1.3 \times 10^6\ F_R$$

It is useful to check the values which arise from different assumptions as to the value of F_R against the theoretical surface temperature which would give rise to these values of intensity. In the calculations E is assumed to have the value of 1.0 and thus what is being studied is the effective surface temperatures. It may be noted that for F_R = 1 all the heat would be radiated and this would require that the flame temperature would have to be the adiabatic flame temperature.,

$$P_{E1} = \sigma T^4 = 1.3 \times 106 \times F_R$$

$$= 5.67 \times 10^{-8}\ Wm^{-2} \times T^4$$

$$T = \frac{13 \times 10^5\ F_R}{5.67 \times 10^{-8}}^{0.25}$$

$$= 2.29 \times 10^{13}\ F_R^{\ 0.25} \quad (2.29 \times 10^{13}\ F_R)^{0.25}$$

$$T = 2.18 \times 10^3 \times F_R^{\ 0.25} \text{ in K}$$

TABLE 8.12. FIRE BALL SURFACE TEMPERATURES v F_R

F_R	0.10	.15	0.20	.25	0.30	.35	0.40
Temp K	1225	1356	1457	1541	1613	1675	1733
Temp °C	952	1083	1184	1268	1340	1403	1460

Crawley (1982) describes the history of a fire ball arising from an 80 tonne spillage as deduced from a film. The temperatures deduced were as follows:-

Time (seconds)	0-1	1-2	2-12	12-17
Temp °C	1300	900-1300 (1100-1200)	1100-1200	?

These results seem consistent with a fraction of about 0.25-0.30. It may be noted that the radius of the fire ball described in Crawley (1982) would be that calculated using the equation D_{FB} = 3.8M. It would have a duration of 16-17 seconds which agrees well with the "frame by frame" count referred to in Crawley. The picture which thus emerges is that F_R may be assumed to have a value of 0.30 corresponding with a mean power intensity of ca 400,000 watts metres^{-2} at the fire ball surface with a mean flame temperature of 1613 K or 1340°C. Figure 8.9. shows the relationship between P_{E1} and flame temperature. Moorhouse and Pritchard (1982) quote values of 170,000 watts metres^{-2} as being derived by Rasbash et al (1956) for pool fires. Lower values are given for liquid fuel fires elsewhere in Moorhouse and Pritchard ranging from 36,000 to 130,000 watts metres^{-2}. Roberts (1982 and 1982a) draw attention to the dependencies of the fraction F on initial vapour pressure, high vapour pressure before rupture leads to more intense combustion and hotter fire balls. Roberts (1982a) claims that values of F of up to 0.40 and values of P_{E1} of up to 250,000 watts metres^{-2} have been recorded for pool fires.

For fire ball models to be realistic they must conform to the law of the conservation of energy. Thus the variables, radius, duration, fraction radiated and effective surface temperature for a given mass of fuel or, more significantly, a given energy release cannot be regarded as independent. This means that the product of fire ball surface, radiation intensity and duration must equal the product of the energy release and the fraction radiated. The assumptions which have been examined above, and which conform as a group to the law of the conservation of energy, may perhaps be accepted as a working assumption.

Probably the most important area of uncertainty applies to the value of F_R which is difficult to calculate from first principles. As has been suggested above, heat release per unit volume in pool fires approximates to the same index for fire balls and so estimation on F_R for large pool fires may shed light on F_R in fire balls. Moreover, the assumed value of 0.30 seems to fit reasonably well with estimates of surface temperature and duration from actual incidents.

Taken together the assumptions tabulated appear to give results which are neither pessimistic nor optimistic for a fire ball of known fuel mass. The assumptions are brought together in Table 8.13.

8.12.7.5. Calculation of thermal pulse received.

The discussion above has assumed that fire balls are spherical. It is obvious, however, that considerable departures from a spherical shape do occur and that the fire ball changes shape during its life. However, it seems likely that fire balls start as hemispheres (possibly oblate), become spherical and finally mushroom shaped, as claimed by High (1968) and which has been largely unchallenged.

It must be agreed that neither hemispheres nor mushroom-shaped fire balls will produce radiant intensities outside the ball which correspond exactly with those to be anticipated from spherical clouds. There is, however, nothing in the literature which enables a quantitative assessment to be made. The present author, therefore, follows other investigators in treating the fire ball, for purposes of analysis, as a sphere. Scilly (1982) has analysed shape factors and concludes that the assumption that a fire ball may be treated as a sphere is valid. Thus it follows that the energy received (E_R) over the duration of a fire ball of mass M by a unit surface at 90° to the direction of incident radiation and at a target distance R_T may be expressed as follows:-

$$E_R = \frac{E_T \, F_R \, M}{\pi \, 4 \, R_T^{\,2}}$$

$$= \frac{47 \times 10^9 \times 0.30}{4\,\pi} \quad \frac{M}{R_T \; 2}$$

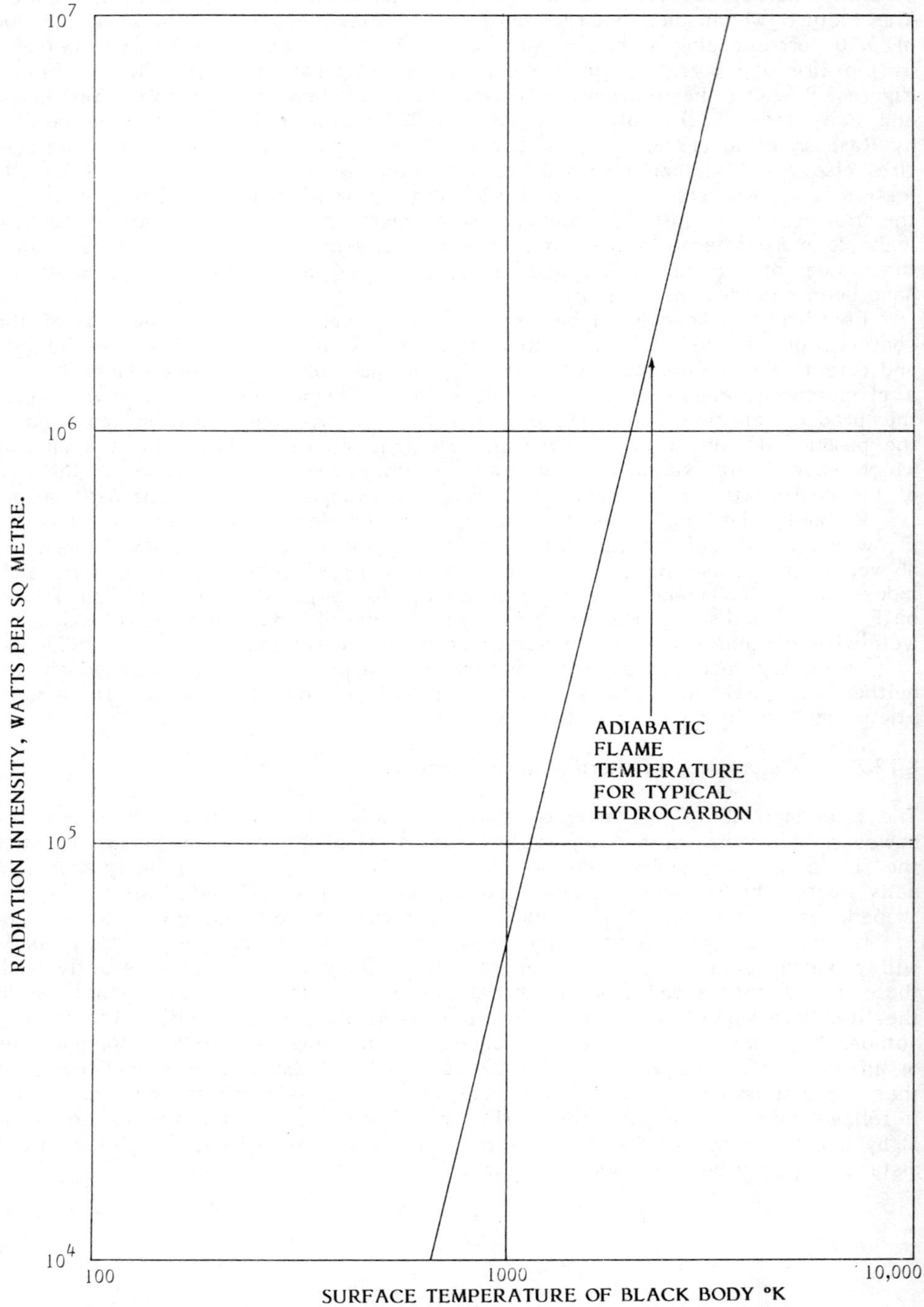

Fig. 8.9. RADIATION INTENSITY OF BLACK BODY.

TABLE 8.13. ASSUMPTIONS RELATED TO HEAT RELEASE AND INTENSITY.

Index Letter	Property Assessed	Value Assigned	Basis of Assessment	Notes
(a)	Mass of fuel in fire ball as a fraction of mass spilled	$M = M_s \times TAFF \times 2$ (Winter) $M = M_s \times TAFF \times 3$ (Summer)	Practical experience	Assumes catastrophic failure
(b)	Heat release	$E_T = 47 \times 10^9 \times M$ joules	Established values	Applies only to paraffins or olefines
(c)	Fraction of heat release which is radiated	$F_R = 0.30$	Based on b, d, e, f, and g	Convenient basis for assessing radiation intensities at targets
(d)	Radius of fire ball (Max)	$R_{FB} = 27.5\ M^{1/3}$ metres	Based on High (1968) corrected for hydro-carbon fuel	
(e)	Duration	$D_{FB} = 3.8\ M^{1/3}$ seconds	Based on High (1968) as above	
(f)	Theoretical radiation at fire ball surface.	$I_{FB} = \varepsilon \sigma T^4$	Well established radiation law	
(g)	Effective fire ball surface temperature	$T_e = \left[\dfrac{E_T F_R}{\sigma D_{FB} 4 \pi R_{FB}^2}\right]^{1/4}$ (°K)	Based on b, c, d, e, f, and observation	Black body assumed. Is 1613 K (1340°C) when $F_R = 0.30$
(h)	Mean intensity at surface	$= \sigma T_s^4$	Based on b, c, d, e, f	Value is 4×10^5 watts metre^{2} ε assumed as unity
(i)	Power of fire ball (radiative)	$P_{FBR} = \dfrac{E_T}{D_{FB}} \times F_R$	Based on b, e	Value is $1.24 \times 10^{10}\ M^{2/3}$ watts
(j)	Power density	$P_D = \dfrac{P_{FB}}{V_{FB}}$	Based on b, d, e	Value is $1.42 \times 10^5 \times M^{-1/3}$ watts per metre^3
(k)	Intensity at distance R	$I_T = 2.94 \times 10^8\ M\ R_T^{-2}$	Based on b, c, e, i	X = 1 for RH = 0.2 = 0.96 for RH = 0.5 = 0.92 for RH = 1.00
(l)	Attenuation	$\tau = X - 0.12 \log_{10} R_T$	Based on Raj (1979)	

$$= 1.125 \times 10^9 \frac{M}{R_T 2}$$

$$\text{or } R_T = \left[\frac{1.125 \times 10^9 M}{E_R}\right]^{0.5}$$

$$= 3.35 \times 10^4 E_R^{-0.5} M^{0.5}$$

Alternatively, the energy intensity at a target I_T may be expressed as follows:-

$$I_T = \frac{P}{4 \pi R_T 2}$$

$$= \frac{1.2 \times 10^{10}}{4 \pi R_T 2} m^{2/3}$$

$$= \text{ca } \frac{10^9}{4} \frac{M^{2/3}}{R_T 2} \text{ watts metre}^{-2}$$

$$R_T = \left[\frac{10^9 m^{2/3}}{I_T}\right]^{0.5} \text{ metres}$$

$$= 3.16 \times 10^4 I_T^{0.5} M^{1/3} \text{ metres}$$

Of the two expressions above I_T is capable of being expressed as a number of fire ball radii and is independent of M whereas E_R is not capable of being expressed in this way.

Though the equations above enable incident intensity and total thermal pulse to be assessed with a fair degree of accuracy, it must be emphasised that the radiation falls in practice on three-dimensional bodies. As a consequence the intensity and the pulse expressed per unit surface area of the body as a whole is less by a factor of at least two. If a stationary lamina be exposed with one face at 90° to the axis of incident radiation (assumed to be in parallel beams) then half its surface facing the incident radiation would receive thermal radiation whereas the rear would not. Thus the intensity and pulse, calculated over the entire surface would be half that calculated by the equations above. Were the lamina to rotate about an axis at 90° to the incident radiation such that it rotated through 180° or through some angle equal to x times 180° (where x is an integer) during the duration of the pulse it would receive on any part of its surface much less than half the pulse because it would interpose its entire surface only twice. If a sphere similarly rotated the incident radiation would be intercepted by a projected area = πr^2. However, the radiation would impinge upon the entire surface which is given by $4 \pi r^2$. Thus the mean intensity and pulse for a sphere would be one quarter of that which would be deduced for a stationary circular lamina. For a cylinder, long in relation to its diameter, with its axis at right angles to the incident radiation, the factor would be $1/\pi$.

It should be noted that the aspect of a solid body is very important; a long cylinder with its axis parallel to the incident radiation would show a factor less than $1/\pi$. Depending too on shape and rate of rotation is the ratio between the maximum intensity and pulse per unit area. This has its analogies in the behaviour of the earth where the incident sunlight, averaged through the year, received at the poles, is a tiny fraction of that received at the equator.

8.12.7.6. Attenuation.

Reference has been made above to the effects of attenuation. This is a complex question. It involves the problem that the atmosphere can vary from place to place or from one day to another or even from minute to minute in its ability to attenuate thermal radiation. At the one extreme lies high altitude and clear frosty weather, at the other lies a tropical monsoon. A further complication is that differing wavelengths of radiation may suffer different degrees of attenuation.

The estimation of effective fire ball temperature is important here. At higher temperatures there is a shift towards the visible spectrum and ultraviolet radiation becomes an increasingly significant component. These smaller wavelengths are more highly attenuated. Thus the attenuation anticipated would increase if a higher fire ball effective surface temperature be assumed.

The subject of attenuation is discussed in Glasstone and Dolan (1980). Using the arguments in that work, attenuation by a uniform atmosphere would modify the equation,

$$E_R = \frac{E_T F_R M}{4 \pi R_T^2}$$

to $$E_{AR} = \frac{E_T F_R M}{4 \pi R_T^2} e^{-KR_T}$$

where E_{AR} = Pulse of radiation after attenuation

e = Base of natural logarithms (2.718)

K = Absorption coefficient over whole spectrum of wavelengths.

However, Glasstone and Dolan (1980) point out that scattering plays a role also and that this is a formula which may be adapted to the following:-

$$E_{AR} = \frac{E_T F_R M}{4 \pi R_T^2} \times \tau$$

τ = Transmittance, i.e.

= fraction of radiation which is transmitted

τ may be greater than 1 in some circumstances, but the values given for τ apply to nuclear explosions and may not be applicable to the fire ball situation. Moorhouse and Pritchard (1982) also draw this conclusion. This latter reference provides values for τ for various values of relative humidity (RH) for a black body source at 1150 K taken from Raj. The graph so presented shows that up to 100 metres from the source (characteristics otherwise unstated), Glasstone's values for transmissivity are higher than those of Raj, and below about 100 metres they are much less. [At 300 metres transmittance is 0.70 at 0.20 RH, and 0.60 at 1.00 RH]. The graph is fitted by the equations:-

$$\tau = 1 - 0.12 \log_{10} R_T \quad (RH = .20)$$

$$\tau = 0.96 - 0.12 \log_{10} R_T \quad (RH = .50)$$

$$\tau = 0.92 - 0.12 \log_{10} R_T \quad (RH = 1.00)$$

where R_T = distance from source to target. Whether RH is the relevant factor or whether it is the partial pressure is not made clear by this work.

Roberts (1982a), which quotes Eyre as a private communication, selects 0.75 as the value for τ at 50 metres and 0.61 at 500 metres which corresponds to the Moorhouse and Pritchard (1982) graph for RH = 0.50 and therefore it goes back to Raj et al (1979). It would appear, therefore, that there is little hard information apart from the latter reference which suggests that even in dry climates attenuation is significant and that transmittance should be assessed, as a maximum, as equal to $1-0.12 \log_{10} R_T$.

8.12.7.7. Evaluation of relationship between radiation intensity and duration and injury.

The knowledge that there is a threshold level of radiation intensity below which injury is zero, no matter how long the duration, shows that the equation

$$\text{dosage} = f(I_T \times D_{FB})$$

cannot be strictly true. This is a conclusion which is analogous with that for dosage levels of toxic agents where it has long been known that the product $C_T D_e$, where C is concentration of toxic material and D_e is duration of exposure, may correspond to different effects for differing values of C_T and below a certain value of C_T no harm can be detected from a lifetime's occupational exposure (say 10^5 hours). This is the basis of threshold limit values for the control of toxic atmospheres.

However, in the matter under discussion, it is not necessary to find an expression which is true for values of D_{FB} which exceed, say, 60 seconds. A form for $f\ I_T$ which gives a product $f\ I_T \times D_{FB}$ which is valid for $D_{FB} \geqslant 1 \leqslant 60$ seconds is therefore all that is sought.

Such a function (or series of functions) is given in Eisenberg et al (1975). The data in this reference are derived from a study of the effects of nuclear weapons and combine the data of White (1971) with Glasstone (1962). The investigators concluded that probability of fatality is best expressed by the product $I_T^{4/3}\ D_{FB}$ giving the probit equation:-

$$\text{Probit} = -14.9 + 2.56 \log_e (D_{FB}\ I_T^{4/3} \times 10^{-4})$$

where I_T is in joules metres^{-2} sec^{-1}.

The dosages for differing proportions of fatalities, when D_{FB} is in seconds and I_T is J m^{-2} sec^{-1}, are given in Table 8.14.

TABLE 8.14. PROBABILITY OF FATAL BURNS versus DOSAGE.

Fraction Fatally Burned	Dosage Δ_I $D_{FB}\ I_T^{4/3}$
.01	1.0×10^7
.50	2.30×10^7
.99	6.5×10^7

For first degree burns the investigators put forward:-

$$D_{FB}\ I_T^{1.15} = 5.5 \times 10^5$$

i.e. with a lower value of the exponent of I_T

It should be noted that dosage does not have recognisable physical dimensions. It is denoted below as Δ_I, or dosage index, and treated as the number which arises from the product $D_{FB} I_T$ where a has a value recommended by Eisenberg et al (1975). It is not clear whether Eisenberg et al assume a "square wave" impulse. Such an assumption may be slightly optimistic as a square wave will give a lower value for $I^{4/3}$ than will the summation of $I^{4/3}$ for the elements of any other wave form which has the same mean value of I over a given time interval.

8.12.7.8. Specimen calculation for 20 tonne fire ball.

$$\text{Power} = \frac{E_T}{D_{FB}} = \frac{47 \times 10^9 \times 20}{3.8 \times 20^{1/3}}$$

$$= \frac{47 \times 10^9 \times 20^{2/3}}{3.8}$$

$$= 91 \times 10^9 \text{ watts}$$

$$\text{Fraction radiated} = 0.30$$

$$\text{Radiated Power} = 27.3 \times 10^9 \text{ watts}$$

Values of I_T at differing radii R_T are given by

$$I_T = \frac{27.3 \times 10^9}{4 R_T^2}$$

$$= \frac{2.17 \times 10^9}{R_T^2} \quad \text{for a 20 tonne fire ball.}$$

The values of I_T are set out in Table 8.15. The values for attenuation in the table are based upon Raj et al (1979) for 50% humidity and a further set of figures is calculated for I_T after attenuation (I_{TA}) and for I_{TA} D_{FB} and for $I_{TA}^{1.15}$ D_{FB}, this latter to give values for the dosage appropriate to non-fatal injury. The values in the table for dosages corrected for attenuation are displayed also in Figure 8.10. This figure also includes values for a 100 tonne fire ball which are not set out in the table but are calculated in a similar way. The figure also shows values for a rotating figure which may be assumed to approximate to a human victim turning away from the source of radiation.

Injury threshold		= 330 metres
Fatality threshold		= 210 metres
Fatality probability	= 0.50	= 170 metres
Fatality certain		= 110 metres

The injury threshold of 330 metres for a 20 tonne fire ball should be contrasted with the previous Figure 8.8. which was based on Williamson and Mann (1981). This latter reference predicts that, at a radius of 330 metres, for a 20 tonne fire ball, third degree burns would be sustained.

8.12.7.9. Theory and practice compared.

Earlier discussion based upon San Carlos, Spain and Good Hope, Louisiana, suggested that the radii for given levels of injury or of death from fire balls may be a good deal less than theory would suggest.

TABLE 8.15. DOSAGE VERSUS DISTANCE FOR A 20 TONNE FIRE BALL.

R_T (metres)	74.6 (1)	100	141	200	282	400	565	800	1131	1600	2262
I_T (watt metres^{-2}) $2.17 \times 10^9 \times R_T^{-2}$	3.9×10^5	2.17×10^5	1.09×10^5	5.42×10^4	2.72×10^4	1.35×10^4	$6 \times 80 \times 10^3$	3.37×10^3	1.696×10^3	8.47×10^2	4.27×10^2
$I_T^{1.333} \times D_{FB}$ (2)	2.92×10^8	1.34×10^8	5.36×10^7	2.11×10^7	8.39×10^6	3.30×10^6	1.32×10^6	5.19×10^5			
Attenuation Factor(3) $= 0.96 - 0.12 \log_{10} R_T$	0.735	0.720	0.70	0.683	0.66	0.647	0.63	0.61	0.59	0.575	0.555
I_{TA} Intensity after Attenuation	2.79×10^5	1.56×10^5	7.6×10^4	3.70×10^4	1.81×10^4	8.70×10^3	4.28×10^3	2.05×10^3	1.00×10^3	4.87×10^2	2.37×10^2
$I_{TA}^{1.33} D_{FB}$	1.86×10^8	8.61×10^7	3.3×10^7	1.26×10^7	4.87×10^6	1.837×10^6					
$I_{TA}^{1.15} D_{FB}$					8.11×10^5	3.49×10^5	1.54×10^5	6.6×10^4	2.90×10^4	1.26×10^4	5.54×10^3

(1) Radius of fire ball.

(2) $D_{FB} = 3.8 \times 20^{1/3} = 10.3$ secs.

(3) After Raj. Relative humidity = 0.50.

There is additional data in Eisenberg et al (1975), on case studies of fire ball incidents. Reference has previously been made to Crawley who claimed that Eisenberg had quoted a "fireman's rule of thumb that for jumbo tankers (assumedly of capacity of 500 tonnes) that firemen had succumbed to radiation when they were as far away as 75 metres from large fire balls", but Crawley over-estimates the capacity of US tank cars. These contained about 50 tonnes of flammable liquid in the case studied by Eisenberg namely the vinyl chloride monomer incident of Houston, Texas 19/10/71 and Crawley's figure may be therefore a misprint. The quotation is otherwise accurate. The fire ball originated in the catastrophic failure of a vinyl chloride monomer car of some 45 tonnes. It killed one fireman and injured non-fatally many others. Most of these injured firemen seem to have been little more than 30 metres from the ruptured car.

A further case studied by Eisenberg et al is that of a propylene tanker on the New Jersey Turnpike 21/9/72. The tanker contained some 15 tonnes of propylene and persons were burned but not fatally at the following distances from the tanker

1	at	30 metres
1	at	60 metres
3	at	80 metres
1	at	100 metres
1	at	130 metres
1	at	180 metres

In theory 6 or 7 should have died and all but 1 should have had a more than 50/50 chance of dying.

At Lynchburg, Virginia 9/3/72, a fire ball which may have contained about 10 tonnes of propane (say 60 metres radius, though eyewitnesses claim it was 120 metres in radius) killed the driver at 90 metres and severely burned 3 witnesses at 140 metres. A further fatality occurred at 130 metres and there were two persons severely but not fatally injured at that distance. This latter incident seems to accord more with theory.

Nevertheless, there seems a reasonable body of evidence from actual cases which is at variance with the effects predicted by the theoretical approach. It is the view of the present author however, that the calculations which lead to the prediction of the intensity of radiation at a given distance from a given mass of material in a fire ball are reasonably accurate. Instead, it seems that the explanation or explanations lie elsewhere. In the first place there may be considerable errors in the prediction of the fraction of the material spilled which enters into the fire ball. The second is that human beings are not stationary flat surfaces orientated at 90° to the direction of radiation. One problem with the Eisenberg approach is that it is based upon experience with nuclear weapons; these data, at least so far as human subjects are concerned, was derived from thermal pulses of a duration which was relatively short compared with that of fire balls. Glasstone and Dolan (1980) give the following expression for the duration of the thermal pulse of a nuclear weapon:

$$T_{max} = .0417 \ W^{0.44} \text{ seconds}$$

where W = yield in kilotonnes

and T_{max} = time to maximum value of thermal intensity

Glasstone and Dolan further suggest that $10\ T_{max}$ is a rough estimate of the duration of the thermal pulse. The yield at Hiroshima was about 12.5×10^3 tonnes

(TNT equivalent) and at Nagasaki it was about 20×10^3 tonnes.

Thus: Pulse length at Hiroshima $= .0417 \times 12.5^{0.44} \times 10$

$= 1.44$ seconds

Pulse length at Nagasaki $= .0417 \times 20^{0.44} \times 10$

$= 1.56$ seconds

Turning through 360° a cylinder would have a mean intensity over its surface equal to $1/\pi$ times that of a plane surface at 90° to the radiant beam.

Figure 8.10 shows the effect of this reduction on the dosage. The thresholds are compared below.

TABLE 8.16. DISTANCES TO A GIVEN EFFECT FOR A 20 TONNE FIRE BALL.

Effect	Stationary (metres)	360° Rotation (metres)
Injury Threshold	320	180
Fatality Threshold	210	125
0.50 Probability of fatality	170	90
0.99 Probability of fatality	110	65*

* Less than fire ball radius.

Rotation through 360° is unlikely for a human subject exposed to a fire ball but the distances quoted for such a rotation seem more in keeping with the effects of actual exposures. The calculations moreover do not take account of the reduction in exposure consequent upon running away.

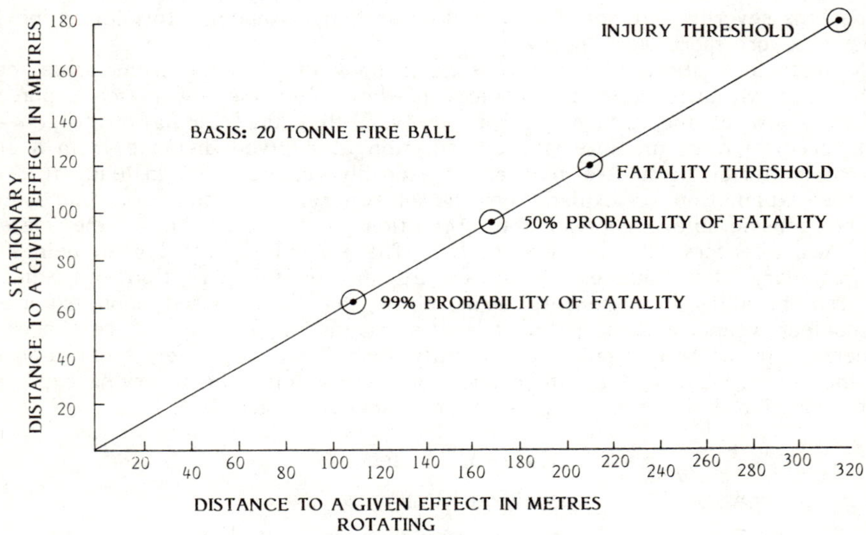

Fig. 8.10. COMPARISON OF DISTANCE TO A GIVEN LEVEL OF INJURY FOR A STATIONARY BODY AND A ROTATING BODY.

Thus for Japanese data the pulse duration approximates to 1.50 secs which is close to the value quoted by Eisenberg et al. (It may be noted that the intensity/time relationship for the thermal impulse nuclear weapons departs markedly from a square wave).

However the point which emerges from this is that under Japanese conditions there was not any possibility of evasion. Japanese victims experienced "profile" burns; Glasstone and Dolan quote the case of a man writing at a window who suffered severe injuries to the hands only. In short, the injuries were confined to those areas of the skin which were facing the explosion centre at the instant that the nuclear weapon exploded.

For thermal pulses from fire balls which would last for 10-20 seconds, which are likely in the case of fire balls from hydrocarbons, the victims may turn round completely and thus suffer burns of lower intensity but distributed over a larger area of the body. Obviously also the victims may attain shelter or run distances of 50-150 metres in that time depending upon physical condition. Work done on megatonne weapons was conducted upon anaesthetized pigs; it is not stated in Glasstone and Dolan whether these were stationary though this may be presumed.

The effect of clothing is important. Clothing may catch fire and give rise to flame burns. The kind of clothing worn will influence the extent of bodily injury. Firemen are probably the best protected, apart from individuals wearing specially fire-proofed clothing. Firemen's uniforms have evolved over a century or more to provide the maximum of protection consistent with freedom of movement. Such uniforms are dark which, of itself, is consistent with high absorption. Nevertheless a fireman in a woollen serge uniform and wearing a helmet with protection for the neck, with back turned to a fire ball, will have a far better chance of survival than a man or woman in ordinary clothes. Additionally the training of firemen is such as to make panic less likely.

In conclusion it would appear that theoretical models are pessimistic as they do not treat human subjects as three-dimensional and with a capacity to evade the effects of thermal radiation.

There is therefore a clear need for further research:-

(a) To study the historical record

(b) To seek to reconcile theory with practice.

Chapter 9

CASE HISTORIES OF CATACLYSMIC FIRES

9.1. INTRODUCTION.

The case histories presented are intended to provide examples of chemical fires to illustrate major conflagrations, fire balls and anaerobic fires.

9.2. CLEVELAND, OHIO, 20th OCTOBER 1944, MAJOR CONFLAGRATION FOLLOWING A SPILLAGE OF LIQUEFIED NATURAL GAS.

9.2.1. INTRODUCTION.

There were two spillages of LNG in a gas works. Initially there was a spillage of about 1900 tonnes, followed 20 minutes later by a further spillage of about 1000 tonnes. The LNG quickly ignited and some of it ran into storm sewers which were later damaged by internal explosions. As well as destroying the gas works, the fire destroyed or severely damaged 10 business premises and 80 houses which were situated up to 400 m from the point of spillage. The death toll established by the Cleveland Coroner was 128 and the number injured was estimated to be between 200 and 400. Property damage was \$6.8 x 10^6 (1944 values)(ca £20 x 10^6 in 1983 values). Figures 9.1. and 9.2. are sketch plans of the area.

9.2.2. DISCREPANCIES IN THE ACCOUNTS OF THE INCIDENT.

The accounts of the incident are marked by many discrepancies, even including disagreement over the date (see Table 9.1.). Of these, Elliott et al (1946), the principal source for this case history and referred to below as 'the Report', is the most authoritative. This was the enquiry set up by the US Bureau of Mines, conducted shortly after the event, with full access to the site and with the opportunity to conduct a thorough forensic examination including metallurgical examination of the debris.

Nash (1976) is a popular account although it is inaccurate on most points. The account in Davis (1979) is much more accurate than Nash and gives a useful general picture.

9.2.3. THE PLANT, THE PROCESS AND THE STORAGE CONTAINERS.

The plant was the East Ohio Gas Company, Cleveland, Ohio, USA, and the disaster was initiated at about 14.40 hours on 20th October 1944. Figures 9.1. and 9.2. show sketch plans of the plant and its environs. It was the first of its

Fig. 9.1. LOCATION OF GAS WORKS IN RELATION TO CENTRE OF CLEVELAND.

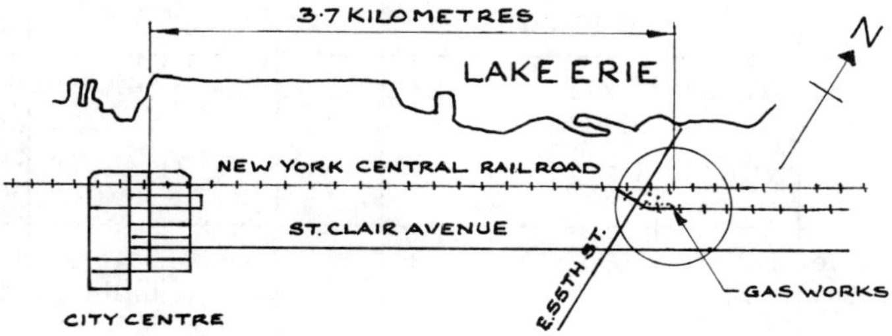

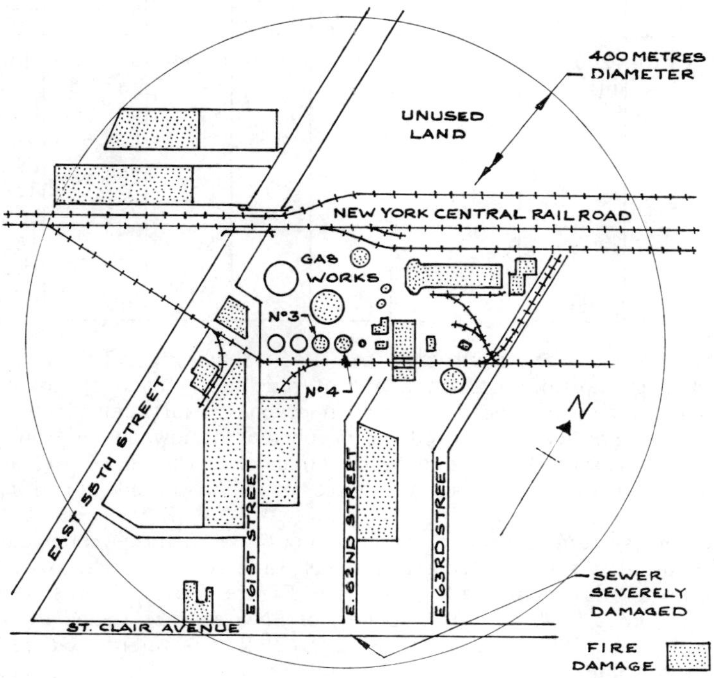

Fig. 9.2. PLAN OF GAS WORKS AND NEIGHBOURING STREETS.

TABLE 9.1. DISCREPANT ACCOUNTS OF THE CLEVELAND DISASTER.

Source	Attributed Date	Quantity Spilled	Quantity Spilled converted to tonnes of liquid	No. killed	No. Missing	No. Injured	Property Damage (1944 values)	General description of Event
Elliot (1946)	20th October, 1944	150,000 plus 83,000 cu ft.	1,900 plus 1,000	128	-	200-400	$6.8 x 10^6$	Fire with explosions in buildings and sewers
Strehlow (1972)	15th November, 1944	1.2 x 10^6 US gallons	2035	136	77	-	Several million dollars	Initial vapour cloud explosion triggered a holocaust
Nash (1976)	20th October or 21st October, 1944	40 x 10^6 cu ft	3080 (total spillage)	112	104	-	50 blocks $15 x 10^6	East Ohio Gas Co. exploded. 3 Thermos jug shaped NG tanks. 1000 homeless.
Davis (1979)	20th October, 1944	4200 m^3 initially	1900	130	-	Hundreds	30 acres burned out $6.8 x 10^6 200 cars destroyed	Huge blast in yard. Formed explosive mixture in sewers. Temperature reached 1650°C. 35 fire brigades involved.

kind in the world and was constructed for 'Peak Shaving'. This was a project, aimed at reducing capital costs on a set of pipelines, each approximately 250 kilometres in length, by making do with a line capacity sufficient for only average demand. Briefly, gas was extracted during periods of low demand and liquefied, stored and then regassified to meet peak demands. The liquefying and storage plant, it was claimed, cost about one-third of what an additional pipe would have cost.

This storage system consisted of four containers, three spherical and one of a rather unusual design and referred to as Tank No. 4. This was a vertical tank (known as a 'torosegmental' tank), see Figure 9.3. The spherical tanks could each hold about 2350 m^3 (1050 tonnes) of LNG at -156°C and the cylindrical tank about 4250 m^3 (1900 tonnes) of LNG. (For equivalent volume of gas in m^3 multiply the volume of the liquid by 600). The tanks were not adequately bunded as the walls round them were close to the tanks and thus they could only accommodate small leaks. An overflow tank was provided which was the pit of a former wet gas holder. The spherical tanks were thermally insulated by cork segments or granulated cork in the annular space which measured approximately 1 m. The cylindrical tank was insulated by rock wool of similar thickness.

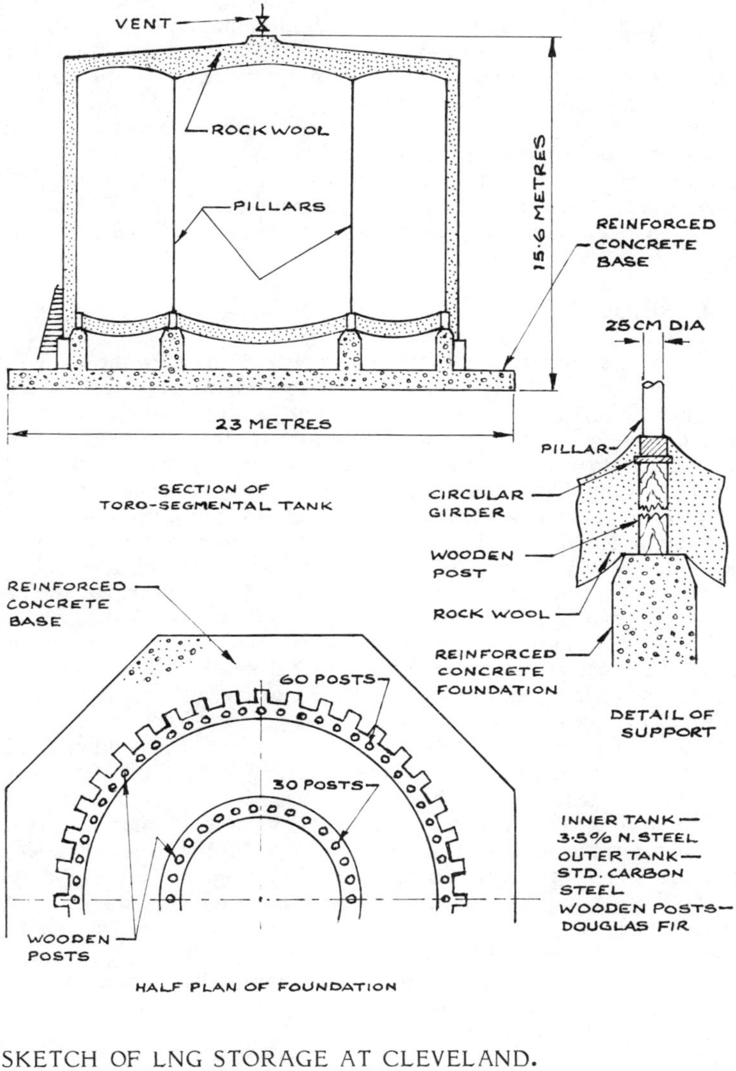

Fig. 9.3. SKETCH OF LNG STORAGE AT CLEVELAND.

The atmosphere in the annular spaces was maintained in a dry condition by circulating about 100 m^3 per day of natural gas through them. The annular spaces were fitted with pressure relief systems.

The inner tanks were constructed of low carbon 3.5% Ni steel, of a thickness which the Report does not state, and which was considered to possess a safe Charpy impact test value at temperatures below -45°C. The outer shells were made of open hearth mild steel. The three spherical tanks of 17.4 m radius were the first to be installed. The cylindrical tank (21 m diameter, 13 m high) was built later and the reason for the change in design was attributed to the greater confidence the manufacturers had in stress calculations on a cylindrical tank as opposed to a spherical tank.

LNG was first introduced into the cylindrical tank during commissioning and a crack developed in the bottom of the tank. A piece measuring about 0.40 m by 0.65 m had to be cut out and replaced. It was after this that the tanks were bunded to a height of about 2 m to take care of small leaks. The bund round the cylindrical tank was fitted with an annular skirt leaving a gap of only about 0.60 m between it and the tank, and similar measures were taken with the spherical tank. These modifications were not approved of by the manufacturers of the tank, as this action restricted natural air circulation and thus made the bottom of the tanks colder than the design specified. The operators therefore admitted a certain amount of open steam to compensate for this lower temperature. A drainage system was provided from the bunds to a vented sump improvised from a disused gas holder pit of about 45 m in diameter, referred to above.

9.2.4. THE INCIDENT.

At the time of the disaster, all the tanks were full and the liquefaction plant was being closed down. At about 14.40 some witnesses felt an earth tremor or were conscious of a low rumble. Then several observers saw streams of gas or liquid issuing from the south/south-east face of the cylindrical tank. A stream of LNG flowed with the lie of the ground in the direction E.S.E., engulfing a number of buildings and flowing down a neighbouring street where some of it ran into the storm sewers. Clouds of vapour were given off and these were blown by a N.N.E. wind in roughly the same direction as the liquid flow. Fire broke out shortly afterwards. There seems strong evidence that there were a number of confined vapour cloud explosions, both in the yard of the gas works where two of them occurred under the skirts of the spherical storage tanks, and others in houses or commercial buildings as a result of leakage into basements. Explosions also occurred in the sewers and road surfaces were severely cracked or buckled. A sewer explosion left a crater 8 m deep, 10 m wide and 20 m long at a spot some 350 m from Tank No. 4. This crater engulfed a fire appliance. Twenty minutes after the collapse of Tank No. 4, Tank No. 3 collapsed and disintegrated spilling a further 1000 tonnes of LNG which immediately ignited. The flames were claimed to reach 900 m into the air. (This argues for a pool fire of about 300 m in diameter - present author). There was considerable fire damage to property to the west of the plant which the Report considers was initiated by heat radiation. This property lay within 200 m to 400 m of Tank No. 4; the wind was not blowing in that direction.

The main body of the fire was under control within 24 hours and within 40 hours employees were able to commence rehabilitation work. Numbers 1 and 2 tanks survived the blaze and were eventually purged of their contents to atmosphere.

9.2.5. ANALYSIS OF CAUSES AND CIRCUMSTANCES.

The Report ruled out the liquefaction plant as a cause of the disaster, seeing no evidence to substantiate it.

Attention was, therefore, focussed on Tank No. 4 with special attention paid to the materials of construction. The manufacturers were fully aware of the phenomenon of low temperature embrittlement and provided the investigators with a list of materials in rank order of increasing 'notch brittleness'. The 3.5% Ni steel adopted was the least suitable, one reason claimed was that it was the cheapest and another was that the choice was dictated by wartime shortages, a claim not made by Elliott et al (1946). The Report quotes the Chief Engineer of the firm which supplied Tank No. 4 as saying that the alloy steel was to all intents and purposes brittle at -16°C despite a satisfactory Charpy impact value. He indicated that when a sheet of nickel alloy steel was at a low temperature, a sledge (hammer) might be driven through it. Thielsch (1965) gives the opinion "though the origin of the failure was not definitely established, the plate material had not been normalised as had been specified". However, in the present author's view, the suppliers of the tank, in choosing this grade of steel, were sailing very close to the wind. Moreover they had had the experience of cracking on contact with LNG. (This crack was skilfully repaired and the Report did not consider that the crack had anything to do with the subsequent failure).

It is apparent, however, that any initiating event which could start a crack could lead to catastrophic failure in a tank so brittle. The Report concluded that the most likely cause was the transmission of a ground shock directly to the floor of the tank either from a passing locomotive or from a steam hammer. Witnesses attested to such ground shocks in the past. The liquid level chart for the cylindrical tank, which survived intact, showed that the level of LNG in the tank dropped like a stone after holding steady for at least five hours. The outer shell of mild steel would have been embrittled at once after failure by contact with LNG and there could have been a number of reasons why it would fail quickly. No attempt was made by the investigators to try to reconstruct the tank from the debris; some parts of it were washed 50 m or more by the flowing LNG. The collapse of neighbouring spherical Tank No. 3 is attributed in the Report to the bending of its support legs through flame engulfment. The tank, whose contents alone weighed 1000 tonnes, would then have ruptured due to mechanical impact with the wreckage of Tank No. 4.

The effect of the discharge of LNG into the sewers and the subsequent explosions is treated in the Report and by other authors entirely as a process in which a flammable gas mixture was produced which then found an ignition source. But there is no need to postulate this as the discharge of any cryogenic fluid into a sewage system would rapidly pressurise it, blowing water seals on gullies and blowing off inspection covers as occurred at Cleveland. Moreover the discharge of LNG of composition 85% methane could give rise to 'flameless explosions' when the higher boiling constituents changed from nucleate to film boiling. Sewers are not built to withstand internal pressures of this order. The general problem of the spread of fire or other effects following the discharge of cryogenic fluids into sewers deserves further study. There seems little doubt that at Cleveland the sewers played a role in spreading the fire but whether this had anything to do with the fires to the west of the site is unproven.

Though a number of reference sources speak of 'blast' or 'explosion', the Report found little evidence for this on the site. The Report accepts that there were confined explosions in buildings both on or off site and in the sewers (however these may have been caused). Using explosion in its widest sense to include pressure ruptures it might be said that both Tanks Nos. 4 and 3 'exploded'. But there is no evidence for a VCE (Vapour Cloud Explosion). The photographs in the Report clearly show many brick walls still standing within 50 m to 100 m of Tank No. 4 and furthermore about 50% of the windows in the compressor house were intact according to photographs. The point that there was not a VCE as at Flixborough is made in Warner (1976). "Careful reading of the Report (i.e. the Report under discussion above - present author) however indicates that the gas cloud only caught fire but did not explode . . . . "

9.2.6. GENERAL CONCLUSIONS.

The severe death toll (128) was primarily due to the massive quantity of material spilled, circa 3000 tonnes of LNG in total. In comparison, the San Carlos disaster of 7 November 1978, which killed 215 people, involved the spillage of only about 20 tonnes of liquid propylene.

The salient reason for these discrepancies is that LNG is, tonne for tonne, far less dangerous than liquefied flammable gases stored under pressure. This is because LNG needs the input of latent heat to make it vaporise whereas on release liquefied flammable gases flash off a substantial fraction of their mass more or less instantaneously. (If stored under refrigeration they might be expected to behave similarly to LNG). Having said this there seems little doubt that there were a number of factors which made this disaster far more damaging than it might otherwise have been. Taken in order from the general to the particular they are:-

(a) The location of the plant in the middle of a densely populated area containing a mixture of housing and business premises was totally unsuitable. This peak shaving plant should have been located outside the built-up area.
(b) The layout with the four storages separated by barely 3 m each was a recipe for rapid fire spread. So was the failure to protect the supports of the spherical tanks from flame impingement.
(c) The presence of numerous ignition sources, which stems from (a).
(d) The absence of a suitable bunding system to impound the spillage and prevent it from entering the storm drains.
(e) The use of a material of construction for the tank which was acknowledged by the designers to be brittle at operating temperature.

Further to (e) some have argued that the choice of material was forced on the designers by the unavailability of suitable alloy steels which had been diverted to the atomic bomb programme. But this is nowhere argued in the Report. Moreover the Report was not published until after the war was over and could hardly by that time, have been constrained by security considerations from referring to this point. More germane is the fact that the three spherical tanks were also made from 3.5% Nickel steel and that they were designed, erected and commissioned a year before the US entered the war.

9.3. FEYZIN, FRANCE, 4th JANUARY 1966. A BLEVE.

9.3.1. THE INITIATING EVENT.

On the morning of this disaster, at about 6.40 hours, an operator attempted, as a daily routine, to run off the water which had settled out of the propane stored in a spherical tank of 1200 m^3 capacity at the Feyzin Refinery near Lyons, France. The sphere is believed to have been three-quarters full at the time and thus to have contained about 450 tonnes of propane.

The ambient temperature was $0^{o}C$ and it seems possible that the run off system illustrated in Figure 9.4. had become frozen or blocked with propane hydrate. The operator opened both cocks full and liquid suddenly gushed out. The one and only cock key dropped off the lower valve onto the ground and could not be retrieved. A vapour cloud formed in a situation where there was little wind and the cloud drifted in all directions. Ignition occurred, it is believed, as the result of a car passing on a nearby service road at circa 7.15 hours, i.e. after some 35 minutes after the escape commenced.

The pressure in the sphere was initially 7 bars but this rapidly rose as the result of the fire which was now burning underneath and the safety valves at the top blew, giving rise to a vertical flame 30-40 metres high.

At 8.45 (2.05 hours after the incident started and 1.30 hours after the fire started) the sphere ruptured and there was a fire ball which killed 17 people, 11 of them firemen, and injured some 80 people, 40 of them seriously. The flames injured people up to 300 metres' distance from the sphere. In the general conflagration which followed four further spheres ruptured in a similar way and a number of petrol and crude oil tanks caught fire. The conflagration took 48 hours to bring under control.

9.3.2. REFERENCE SOURCES.

A general account is given in Anon (1966). The incident features in the following compilations:- Strehlow (1972), Commercial Union (1976), Davenport (1977), Slater (1978), and Gugan (1979). Unfortunately, these compilations add little to knowledge of the incident as, where the source is acknowledged, it is Anon (1966). There are discrepancies in the death toll, some giving 16, some 17 and some 18, but more importantly there are differing views as to the nature of the occurrence. Slater describes it as a vapour cloud explosion but the others do not concur.

There is, in addition, Leonetti and Brats (1966), which is an official French Report which includes a sketch of the run-off system and a ground plan of the storage area which have been reproduced in Figures 9.4. and 9.5.

The present author also has a private communication, Backmann (1974), which gives the views of a senior safety adviser to the French chemical concern, Rhone Poulenc, on the causes and circumstances of the disaster. This communication includes a photograph taken after the disaster (Fig. 9.6).

9.3.3. SOME DETAILS OF THE INCIDENT.

It is impossible to tell from the evidence given in the references above what bunding, if any, was provided though its absence may not have had any marked influence on the course of the disaster. Though it appears that the tank originally held ca 400 tonnes, it is impossible to determine how much propane remained in the tank at the moment of rupture. Liquid propane had been discharging for 2 hours and the safety valves (aperture unstated) had been blowing for one and a half hours. Perhaps an estimated 200 tonnes would be of the right order.

The shell is stated in Anon (1966) to have been 14 metres diameter (45.5ft) and to have a shell thickness of 45 mm and to have been designed to withstand an internal pressure of 80 bar. The volume, based on a radius of 7 m would be 1436 m^3 so the stated capacity of 1200 m^3 would refer to the rated liquid capacity with an ullage of 20%. The volume of the metal walls the present author has calculated, based on the dimensions above, to be ca 27 m^3 (or ca 220 tonnes). Anon (1966) claims that a piece of the sphere weighing 15 tonnes was projected for 300 metres. (This would measure, laid flat, about 6 metres by 6 metres) and would represent about 1/14th of the shell. Backmann (1974) speaks of pieces of the sphere having fallen vertically to the ground and having become folded up. Some of the fragments were only 19 mm thick instead of 42 mm originally and it is estimated that these portions of the sphere had reached a temperature of 1000°C. Backmann claims that a piece weighing 100 tonnes (i.e. just under half the weight of the vessel in toto) had been projected for 300 metres. Anon (1966) says:-

> "Now a sea of fire spread out on all sides and it was from this that the fire-fighters came running enveloped in flames. Even those within a distance of 300 metres were burned."

Backmann (1974) speaks of a ball of fire, "s'eleva dans le ciel jusqu'a 700/800m de hauteur"; this seems to refer to the height it reached rather than its diameter. Thus there is not much information on the relationship between fire ball diameter and mass of propane involved. If people were burned up to 300 metres, this argues for a radius smaller than 300 for the actual fire ball. This thus might therefore have measured 150-200 metres in radius. The quantity of propane remaining in the vessel is also uncertain; it may have been in the range 100-300 tonnes. Figure 9.7. shows that the lines "best fit for incidents", Marshall (1977b) and Roberts (1982a), all pass through this range of possibilities.

Detailed descriptions of the subsequent explosions involving the other spheres

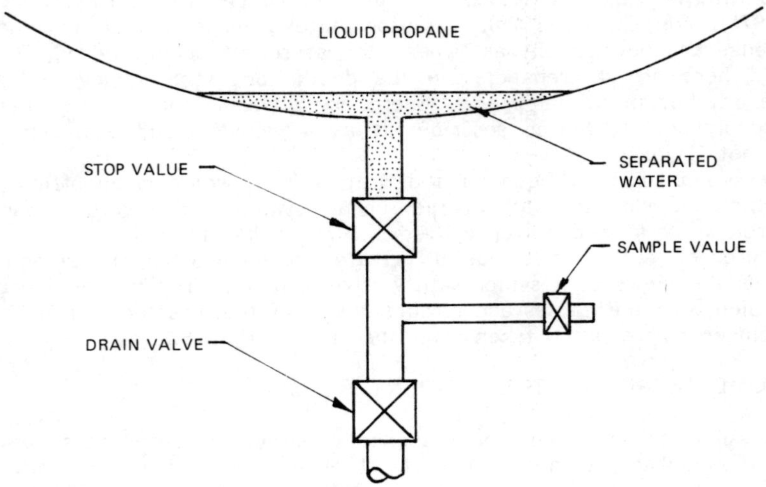

Fig. 9.4. RUN-OFF SYSTEM AT FEYZIN.

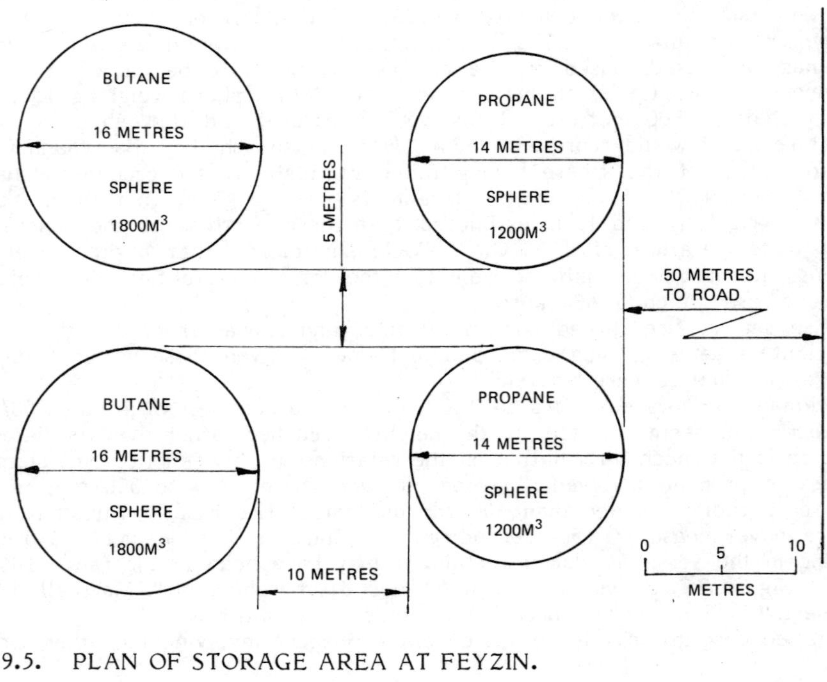

Fig. 9.5. PLAN OF STORAGE AREA AT FEYZIN.

Fig. 9.6. THE WRECKAGE AT FEYZIN.

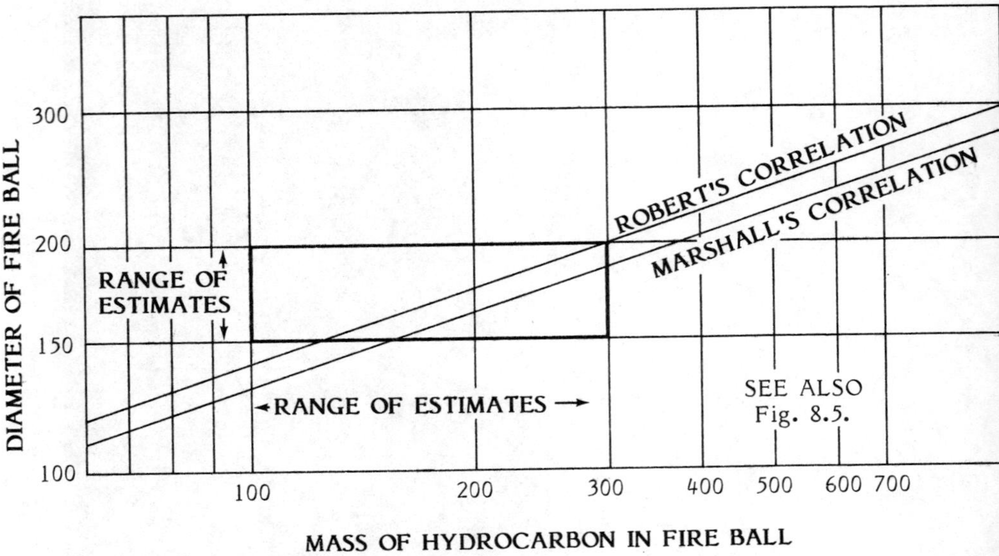

Fig. 9.7. RANGES OF FIRE BALL DIAMETER FOR VARIOUS SPILLAGES.

is lacking. There were three further propane spheres and four larger (2000 m^3) butane spheres. ANON (1966) says four further spheres exploded, Leonetti and Brats (1966) say two were affected. Neither indicate whether they were propane or butane spheres. Figure 9.6. shows "petal fractures" to two of the smaller spheres, i.e. those containing propane. No injuries followed on these later incidents. There is no evidence that there was a vapour cloud explosion at Feyzin. Blast would clearly attend the pressure rupture but the photograph shows no evidence of a Flixborough type event. The photograph referred to above shows clear evidence that three at least of the spheres suffered damage to their tubular legs. A feature which is clearly evident is that the joint between the legs and the shell was especially weak.

There is no information on the location of those killed and injured at the time of the rupture. Leonetti and Brats (1966) speaks of 150 people as working in the vicinity. There is no indication of what measures of evacuation, if any, were instituted. This is an area of information which, were it available, would shed valuable light on the hazard radii of fire balls.

9.3.4. ANALYSIS OF CAUSES AND CIRCUMSTANCES.

The factors which, taken in combination, caused this to become a major disaster included:-

(1) The unsatisfactory nature of the draw off system in which the dislodgement of a cock key rendered the operators powerless to intervene. That the purge discharged directly under the vessel, instead of to the side of it, was a further factor, as was the lack of ability to close the valves by remote control.

(2) The apparent failure to keep motor vehicles from the scene. The source of ignition appears to have been a vehicle on a service road which was closer to the scene than the motorway. Both service roads and the motorway

should have been closed to remove ignition sources.

(3) The apparent failure to discharge the contents of the leaking tank to the other three storages which, at two-thirds full, could have accommodated the entire contents of the leaking tank.

(4) The failure of the occupiers to foresee the consequences of fire on the vessel. It seems clear that they saw the situation as one where it was only a question of time before the combination of liquid leak from the bottom, and vapour leak from the safety valves, would empty the sphere. In such a case there was no point in trying to cool the sphere.

However, this was a totally mistaken view. As has been discussed above, there were two heat transfer regimes appropriate to the case. There was the regime below the liquid level which is characterised by flame on the outside and a boiling fluid on the other. In such a case the temperature would be controlled by the boiling heat transfer coefficient and the temperature would be close to that of the liquid which, in this case, would be boiling at about 30°C - 35°C. There is a similar situation with a domestic gas cooker where the temperature of the bottom of a kettle or saucepan is but little above 100°C. It is possible, for this reason, to boil water in a paper bag. Above the water line, however, matters would be very different. There would no longer be a boiling liquid to conduct away the heat rapidly and the wall temperature would approach the temperature of the flame. (There is a detailed discussion in standard works on heat transfer such as McAdams (1954).)

As the propane level fell more of the surface of the tank would be affected by the heat transfer regime operational above the liquid level and the flame temperature would, if anything, be higher at the lower levels. The steel would be greatly weakened above 600-700°C and even though it might have been capable of withstanding 80 bars when cold, it would not do so when hot (see Fig. 9.8).

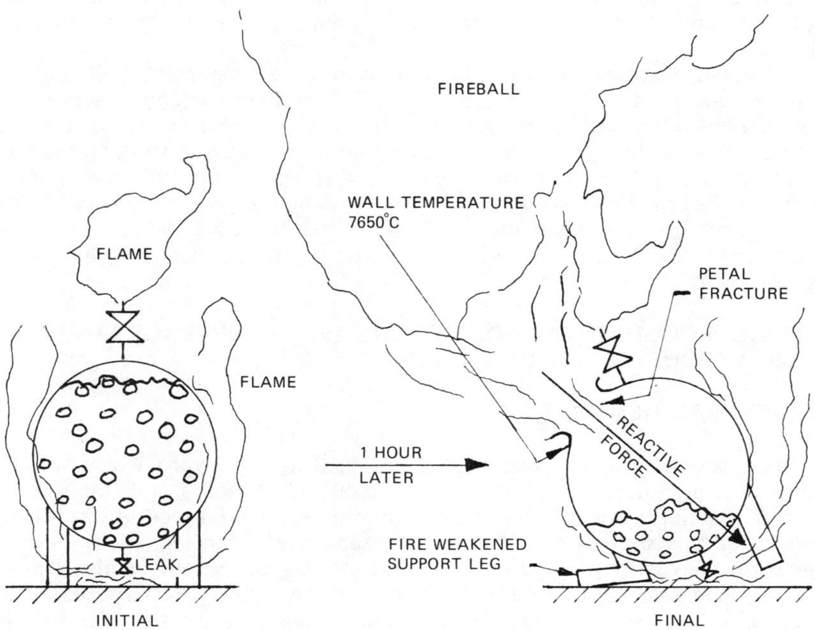

Fig. 9.8. THE DEVELOPMENT OF A BLEVE.

A further factor, which can be understood without recourse to considerations of heat transfer to boiling liquids and their vapours is that the support legs would be weakened by flame engulfment.

(5) There was a lack of any permanently installed water spray system on this and the other spheres which could have protected the walls from over heating. The legs could have been protected by encasement in concrete or by permanently installed water sprays.

(6) There was a total lack of liaison (according to Anon 1966) between the Lyons and other local fire brigades and the occupiers, such that there was no previously worked out plan. Had, for example, the fire brigades have known that the over-riding need to keep the tank cool by diverting their jets upon it, the disaster might have been averted.

9.3.5. CONCLUSIONS.

The Feyzin disaster, in common with many disasters in the chemical and process industries, was the result of a chain of mistakes. Had this chain been broken at any point the full effects of the disaster might not have been felt. Many lessons have been learned from it and some are incorporated in Health & Safety Executive (1976a) which gives guidance on LPG storage at factories. Photographs of large storages in Health & Safety Executive (1976a) show them to stand on low plinths. The legs are either encased in concrete or water drenched. The spheres are ringed with water spray pipes at eight or nine levels. The sprays may be actuated by excess temperature or excess internal pressure. Separation distances recommended for large adjacent tanks are given as one-quarter of the sum of the diameters of the adjacent tanks but not less than 1.5 metres. Applied to Feyzin this would have given a distance of 7 metres in place of 5 metres which was the actual distance.

Health & Safety Executive (1976a) does not recommend bunds which would hold all the tank contents, but does recommend low walls of maximum height, 0.60 metres.

The Feyzin disaster focussed attention on the importance of close liaison between occupiers of hazardous plant and the public emergency services. The primary responsibility for understanding the hazards associated with large-scale LPG stores lies with those who operate them. It is these latter who should point out to the public fire brigades the strategy which must be adopted to deal with a Feyzin-type situation. It is completely unreasonable to expect a fire brigade to improvise their response to situations which lie outside of their ordinary experience as they arrive at the fire ground. In the case of Feyzin, having to do this cost firemen's lives.

9.4. FLIXBOROUGH, ENGLAND, 1st JUNE 1974. CONFLAGRATION FOLLOWING VAPOUR CLOUD EXPLOSION.

9.4.1. GENERAL DESCRIPTION.

There was a severe vapour cloud explosion (VCE) at the Nypro Works, Flixborough, Humberside, England, on 1st June 1974 at about 16.50 hours and this was followed by a major conflagration. The VCE will be examined in detail in Chapter 13, the purpose of this section is to discuss the subsequent conflagration.

The area devastated by fire (in addition to damage by explosion) was the south east quadrant of the site; the fire covered about 6 hectares (15 acres). The area affected is shown on Figure 9.9., which is a sketch plan of the site.

There were large stocks of highly flammable liquids on site and the Official Report [Flixborough (1975) paragraph 194c] listed the following, (the quantities are approximate):-

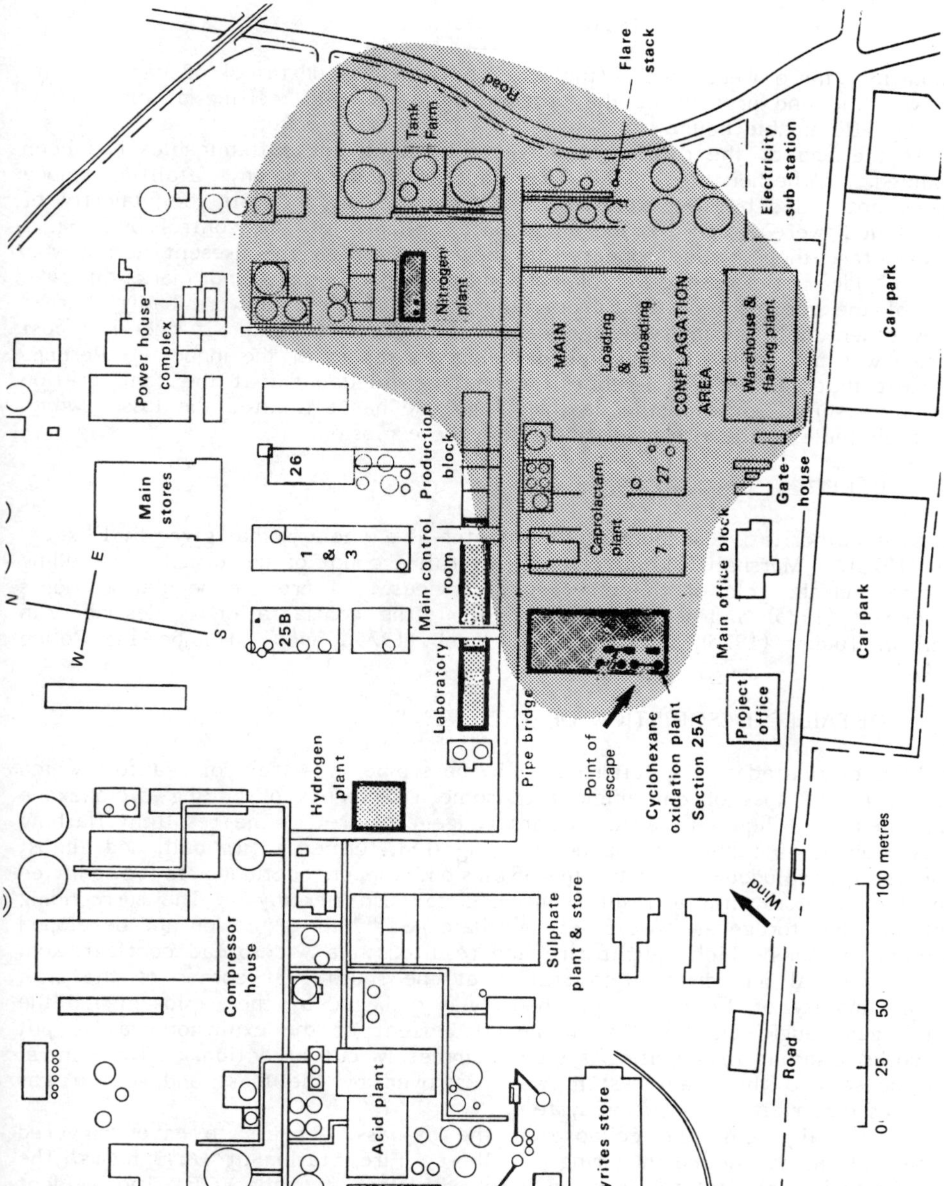

Fig. 9.9. SKETCH PLAN OF NYPRO SITE SHOWING DAMAGED AREAS.

Cyclohexane	1,250 tonnes
Naphtha	250 tonnes
Benzene	100 tonnes
Toluene	40 tonnes
Total ca	1,650 tonnes

In addition, there were stocks (the level of which is unstated) of caprolactam. This was the end product of the process and is a solid melting at circa 79°C to yield a flammable liquid.

At the end of the conflagration virtually all of these flammables had been consumed. Whether or not this fire was responsible for any fatalities among persons not killed by the explosion is difficult to say. That the majority of those killed were killed by the immediate collapse of the control building is not disputed; their bodies appeared to bear, so far as the present author was able to judge from inspection of post mortem photographs, no marks of fire. Some of the corpses recovered from other parts of the site were badly charred but whether the charring was the cause of death or whether it occurred post mortem was not made clear by the pathologist's report at the inquest. Perhaps the most that can be said is that it cannot be ruled out that the conflagration may have caused some of the deaths. No one has estimated the losses which are attributable to the fire as distinct from the explosion.

9.4.2. REFERENCE SOURCES.

The most authoritative account of the disaster as a whole is that given by Flixborough (1975). Marshall (1979a) gives an overall account of the disaster, including the fire and the role of the emergency services. There are popular accounts in Kennett (1975) and in Canning (1976). The conflagration is described in detail in Tucker (1975), in Tucker and Webb (1975), and in Humberside Police (1974).

9.4.3. DETAILED DESCRIPTION OF THE EVENT.

The VCE originated in the failure of a by-pass pipe in a train of reactors which resulted in the loss of containment of some 125 tonnes of boiling cyclohexane maintained as a liquid at about 8 bars gauge pressure. The resultant flashing released about 40 tonnes of vapour which ignited, formed a fire ball, and almost immediately exploded. After the explosion, vapour continued to be emitted from the reactor train and this continued to burn fiercely. The surrounding plant and the storage area were severely damaged by the explosion and developed innumerable leaks which caught fire and resulted in a widespread conflagration.

The natural gas distribution station at the south east corner of the site, with a capacity of 14,000 m^3 per hour, was damaged by the explosion and the leaking gas caught fire. One immediate effect of the explosion was to put the works' internal fire-fighting system completely out of action. The central pump house and the foam system was destroyed by the blast and so too was the works' electricity distribution system.

The initial event, the collapse of the by-pass, appears to have triggered the fire alarm at the neighbouring Scunthorpe Fire Station at 16.43 hours. The brigade was leaving the station when it felt the blast from the VCE. The incident was soon seen to be very serious and within 40 minutes, 30 pumps had been mobilised. Figure 9.10. shows the rate of build up of appliances.

Virtually no water supply was available at the site itself, but there was ample water available from the tidal River Trent near the works boundary where there were a number of lagoons to supply cooling water to a nearby steel works.

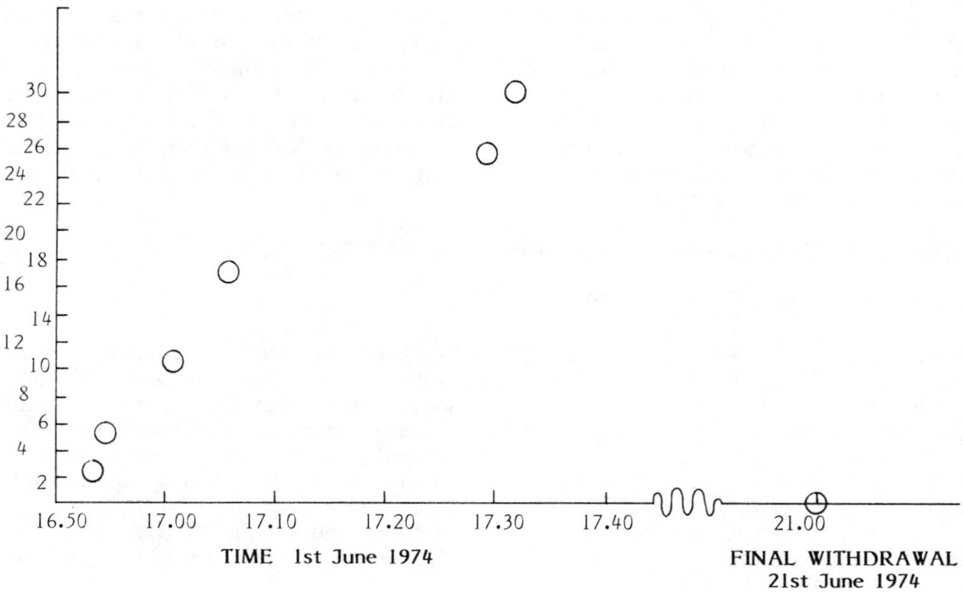

Fig. 9.10. RATE OF BUILD-UP OF FIRE APPLIANCES.

[The 0.75 m cooling water main to the steel works was ruptured by the explosion, presumably due to earth shock, and the fire service had to provide 10 pumps at the steel works to maintain its cooling system]. Fire fighting was largely completed within 48 hours by which time the pumps had been reduced to 5 and all but a few controlled fires had been extinguished. There were two explosions at about 72 hours after the fire started. The controlled fires were extinguished after 11 days and after 20 days in attendance the fire brigade eventually withdrew. The attendance included brigades from five fire authorities, a total of 50 appliances being involved with 139 full-time and 105 part-time personnel. 16 fire service personnel were treated for injuries.

9.4.4. GENERAL CONCLUSIONS.

One of the problems of siting is the relative location of the process units and of the storage units. In general, on a site handling flammable substances, the storage area will carry a much higher inventory of flammables than will the process units which it serves. On the other hand, the process units represent a much higher investment than the storage units. Also a single stream process possesses little redundancy. Storage facilities are fairly versatile, whereas process plant is custom-built and requires a much longer time to replace than does tankage.

The risk of fire spread from storage to process is minimised, though obviously not eliminated, by siting the storage area so that there is the least probability of the wind blowing from storage to process. This implies that, if possible, storage in the UK should be sited to the east of process. This was done at the Nypro works. There was, in addition, a clear separation there between the two areas, the loading and unloading took place in the area between the process and the storage areas, leaving a 60 metre gap between them. Thus the siting decisions at Flixborough were sound and accorded with generally established practice. It appears also that the provision of fire fighting equipment on the site was sound and adequate.

But the most general lesson to be learned from this conflagration is that the destructive effect of VCEs is massive and that it can totally disable even a well-planned conventional fire control system. It would be going beyond the scope of this book to examine in detail the steps that could or should be taken to "harden" fire fighting facilities on a site where a VCE is apprehended. Such measures could include impounding emergency fire fighting water in a lagoon rather than in a tank and a policy of non-reliance on mains electricity to drive fire pumps or to disperse foam.

9.5. SAN CARLOS, SPAIN, 11th JULY 1978. A FIRE BALL ?

9.5.1. GENERAL ACCOUNT OF DISASTER.

There was a disaster at the Los Alfaques (The Sand Dunes) camping site at San Carlos de la Rapita on the Mediterranean coast of Spain at approximately 14.30 hours on 11th July, 1978. A road tanker loaded with 23.5 tonnes of liquid propylene burst whilst travelling along the road past the camp and released a cloud of flammable vapour. The cloud ignited and the resulting conflagration killed 215 persons, almost entirely as the result of burns, about 100 of them instantly. A further 67 people required hospital treatment.

In addition it caused severe damage to fixed and mobile property over an area of some 50,000 m^2 (5 hectares or 12.5 acres). The scene is shown as a sketch map in Figure 9.11.

9.5.2. REFERENCE SOURCES.

The present author had access to the following reference sources:-

Spanish Ministry of Justice (1982), which is the judgement of the Tarragona Court of 27/1/82. This judgement combines criminal and civil jurisdiction in the matter.

Carrasco (1978), which is a report and technical study by a Spanish engineer which formed the basis of the technical advice to the Tarragona Court.

Scilly (1978), which is a report with photographs to the HSE by (at that time) an Explosives Inspector of the HSE who visited the scene.

Stinton (1978), which is a report, with photographs, by a UK fire officer who visited the scene.

Hymes (1982), a draft report by the Safety and Reliability Directorate (SRD) on the physiological and pathological effects of thermal radiation.

Moodie (1982). This is a critique, with detailed calculations, of a document compiled by the present author on the failure of pressure vessels by hydraulic pressure. This document, in its final form, has been given above (Table 6.5, P.82).

Davis (1979), which gives an informative non-technical account of the disaster.

Jones and Sands (1981), which is written by two authors with scientific qualifications who are not safety professionals but which, nevertheless, contains useful information and original thinking.

Stern (1978). A German newspaper account of the disaster.

The present author also had in his possession several dozen photographs of the wreckage and of the bodies of victims.

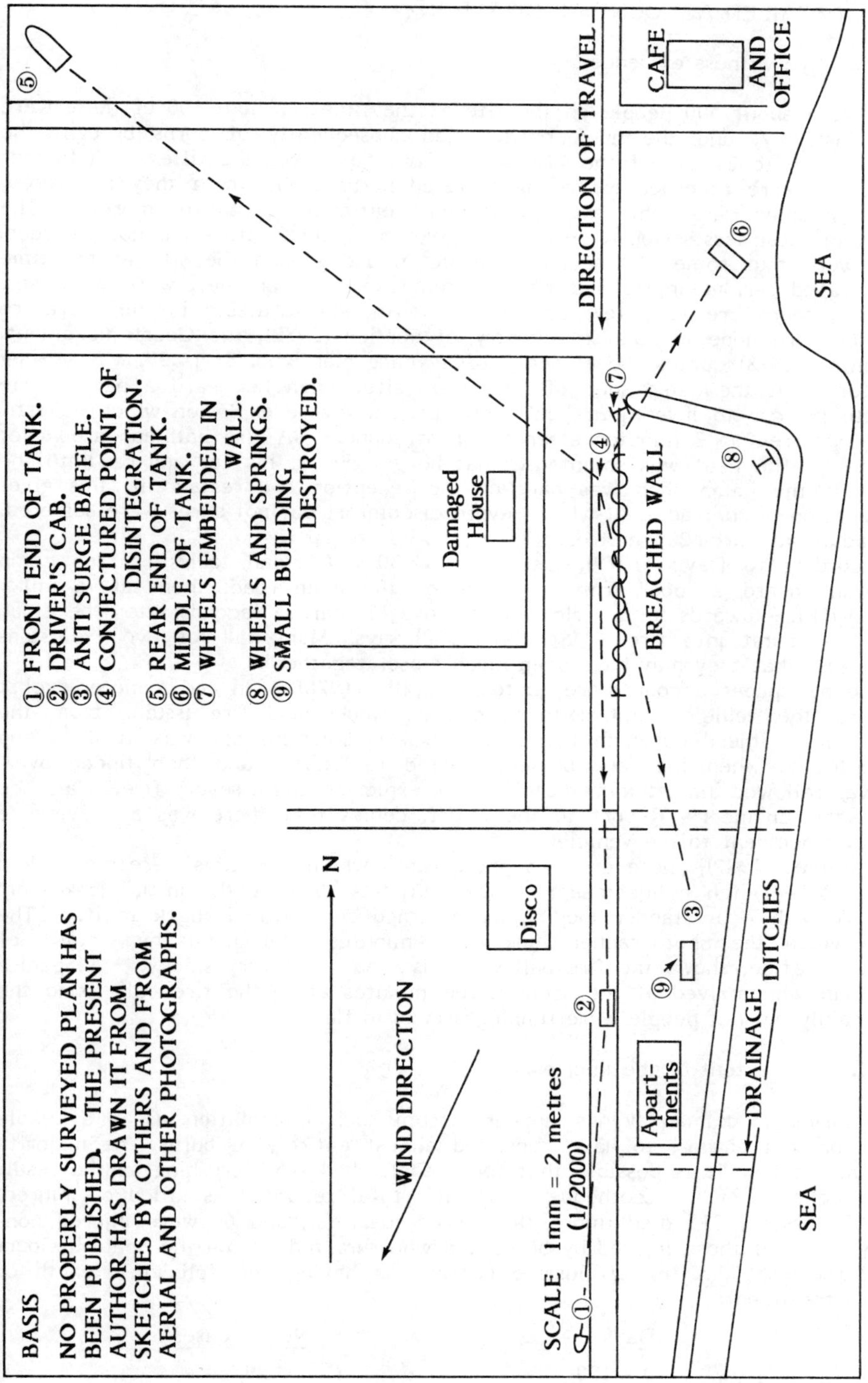

Fig. 9.11. SKETCH OF SAN CARLOS CAMP SITE.

9.5.3. DETAILED ACCOUNT OF DISASTER.

9.5.3.1. Eye witness evidence.

There were about 500 people on the site at the time. About 100 of these were killed instantly, and the number who died subsequently of burns brought the death toll up to 215. Of the 215, all but two have been identified. A further 67 persons were recorded as having suffered burns from which they recovered, though in some cases only after prolonged treatment (a year or more). The number of those less seriously burned and given first aid treatment is not recorded. There were thus some 200 or more people in and around the site at the time who escaped serious injury and who, in some sense or another, were witnesses. However, there are very few accounts of witnesses available in the literature and there are none in Spanish Ministry of Justice (1982), the Court judgement.

Stinton (1978) quotes the evidence of a young man who, at 14.29, was serving a customer at the camp shop off the main site, when he heard a bang. He got into his car to investigate and then heard a severe explosion which he estimated occurred at 2 minutes after the first sound. A fire ball appeared over the site. The heat was so intense that he got out of his car and ran into the sea. If the camp shop was part of the reception and restaurant, this gives an indication of the radius at which severe discomfort but not burns were suffered. This radius was circa 200 metres.

According to Davis (1979), just after 14.30 a German tourist in the camp restaurant heard a "pop" from a tanker on the main road, then saw a milky cloud drifting towards him. He ran to move his car. Seconds later the cloud flared and burst into fire. Stinton (1978) says "Many talk of two explosions but suggest that they may have been much closer together".

Two newspaper accounts are quoted in Scilly (1978). In one a motor-cyclist following the vehicle claimed to have seen smoke and fire issuing from the tank. In another, a French tourist who was walking his dog was at the camp site entrance when the tank passed, started to "snake" and then tipped over. This was followed almost immediately by an explosion and a severe fire. Carrasco (1978), the Engineer's Report to the Court, denies that there was any evidence of a prior accident to the vehicle.

In Hymes (1982), there are two photographs which it claims were both taken by the same Dutch holidaymaker. One purports to show the initial leak from the tank with a by-stander looking unconcerned and with his back to it. The other, which the photographer took some minutes later from the vicinity of the camp office, shows the fire ball which is seen to be very smoky. A Spanish policeman who arrived at the scene three minutes after the fire ball found the fires mainly out but people were running around in flames.

9.5.3.2. The extent of the injuries.

The injuries to human beings appears, from such information as is available to the present author, to have consisted almost entirely of burns, the majority of them fatal. It is possible that four people died from crushing as the result of the collapse of a discotheque. Of the total recorded as killed or injured, i.e. 282 persons, 215 died (more than three-quarters), and 67 were injured non-fatally. 61 of those injured by burns and who survived are on the Spanish Court record [Spanish Ministry of Justice (1982)], as having the following length of medical treatment:-

Days	No. of cases
0 - 100	29
100 - 199	8

200 - 299	12
300 - 399	4
400 - 499	5
500 - 599	2
600 - 690	1

These represent the numbers of persons making claims for injury.

9.5.3.3. Damage to property.

The following buildings were damaged:-

Outside the camp site:-

A discotheque completely collapsed.

An apartment block which appeared to have been damaged by an internal explosion.

A two-roomed house.

Inside the camp site:-

A toilet block.

A small building identified by the present author from photographs.

74 cars were damaged, of which 23 were specifically stated by the Court to have been totally destroyed.

5 motor-cycles or scooters were damaged.

The Court recorded numerous other claims for property damage.

Both Scilly (1978) and Stinton (1978) agree reasonably well that the area of damage and injury is that shown in Figure 9.11. This seems to indicate an area of some 50,000 m^3 (5 hectares); the radius of an equivalent circle would be 125 metres. However, doubt exists as to how far severe thermal effects spread over the sea. There are indications that some people were injured in the sea. Such a circle would be greater in radius than the fire ball which gave rise to the damage or injury, on account of the effects of thermal radiation outside of the fire ball envelope. This question will be further discussed below.

9.5.4. THE TANKER AND ITS LOAD.

9.5.4.1. Description of the tanker.

The road tanker which was involved in the accident consisted of a tractor (owned by the driver, who was killed) and an articulated semi-trailer which was leased to the driver [Spanish Ministry of Justice (1982)]. The semi-trailer, which was constructed in 1973, consisted of a rear truck supporting a pressure vessel (referred to below as the "tank") with a set of small wheels at the front end to allow the semi-trailer to be parked independently of a tractor unit. The tank was not a simple cylinder; the front portion, which fitted onto the trailer, tapered to a smaller diameter than the main rear portion which was 2.3 metres in diameter. The tank was thus made up of a cylindrical section with a tapering section at the driver's end. The roof formed a straight line about 10 metres long, but the bottom of the tank included an obtuse angle with the tapering portion. The tank ends were closed by approximately hemispherical dishes. There were at least two internal baffles welded to the walls of the tank, the principle purpose of which would be to prevent the load from surging. There seems some lack of agreement as to the wall thickness but it was not less than 8 mm. The design pressure of the tank was 18 bar and the test pressure was 30 bar. [At 21°C for example, the vapour pressure of propylene is ca 10 bars]. No relief valve was fitted nor was one required by Spanish law at the time [Spanish Ministry of Justice (1982)].

The main body of the tank was made from a low carbon steel T1 A (ASTM A 517 - 72a) and the ends were made from BH 51. The ultimate tensile stress of the TL steel in the tanker was given as 81.2 kg mm^{-2} and of the BH steel as 67.1 kg mm^{-2} in the documents provided by the suppliers of the tank [Carrasco (1978)].

The nominal volume of the tank was 45,000 litres. This is equivalent to 24,390 kg of liquid propylene at $4^{\circ}C$ with the tank completely filled with liquid. The quantity permitted to be loaded into the vehicle was 19,350 kg.

9.5.4.2. The loading of the tanker.

The vehicle was loaded with liquid propylene at the ENPETROL S.A. refinery on 11th July 1978 at about 12.00 noon when, despite the permitted load being only 19,350 kg, which would have occupied 80% of the total volume of the tank, the driver was provided with, and accepted, 23,470 kg of liquid propylene as measured by ENPETROL's weighbridge. This latter load, at the storage temperature of $4^{\circ}C$, which is the temperature at which it is believed the propylene was loaded, would have occupied 96% of the volume of the tank. However it seems likely that the bulk temperature at the time of setting off would have exceeded this. The facts regarding this loading are not in dispute by any party and are recorded in Spanish Ministry of Justice (1982). Some two and a half hours after leaving the refinery, and after covering 102 kilometres, the road tanker was passing the Los Alfaques camp site when the tank disintegrated, substantial portions of it travelling up to 300 metres in a number of directions from the point of disintegration (see Figure 9.11). As to the precise circumstances of the disintegration, the judgement of the Court, [Spanish Ministry of Justice (1982)], gives no information. What the Court did say, and without equivocation, was that the accident was "Due solely to the overfilling of the tanker, which means that the obligatory and necessary ullage to allow for the natural expansion of the propylene because of the heat, which in the time referred to was that normal on the coast of this province, was not available . . .".

9.5.5. THE NATURE OF THE DISASTER.

The important question to be decided here is whether this disaster was predominantly a fire or whether it was predominantly a vapour cloud explosion or whether it partook of elements of both. To decide this question, it is necessary to examine the nature of the damage to the surroundings and the nature of the injuries to human beings.

The evidence for blast effects rests on these facts:-

(a) A number of witnesses testified to hearing an "explosion" immediately before the disaster.
(b) Several buildings suffered severe damage of a mechanical nature, including a discotheque which was totally wrecked. Also a length of wall was pushed over near the presumed point of disintegration of the tanker.
(c) The various portions of the tanker were found, in some cases, hundreds of metres from the point of disintegration, as is disclosed in Fig. 9.11.

It may be said at the outset that non-technical witnesses are unlikely to make distinctions between pressure ruptures, VCEs, and the detonation of high explosives.

There can be no doubt whatsoever that of those who died immediately, many were burned beyond recognition. The photographs show some victims as partly cremated. This cannot happen solely as the effects of an explosion which are over in a fraction of a second. Domestic gas explosions, for example, hardly scorch furnishings. The ratio of killed to injured at San Carlos, 215 - 67, is not typical of explosions. The VCE at Ludwigshafen, 1948, for example, killed 208 but the non-fatally injured numbered 3,818. The VCE at Flixborough killed 28 and injured 89. The 61 injured people at San Carlos who survived after hospital treatment, were all treated for burns.

The damage to buildings and walls could perhaps, in some cases, be explained by collision with parts of the disintegrating tanker, some pieces of which, weighing several tonnes, travelled several hundred metres before coming to rest. Some of these pieces certainly collided with structures. An apartment house showed signs of internal overpressure which could have been caused by an inleakage of propylene vapour which was then ignited internally or from the main fire ball. The discotheque, which collapsed completely, could similarly have been destroyed by internal explosion. Further confirmation is provided in Stinton (1978), "The blast damage was an interesting feature. About 75 metres from the explosion (i.e. the point of disintegration of the tanker, VM), a single-storey building of substantial construction used as a discotheque, was completely demolished causing the death of four people. On the other hand, a motorcycle was still standing on its footrest only about 20 metres from the blast area, but (it was) completely burned out".

This leads into the question of the origin of the energy which was available to disintegrate the tanker. In the view of the present author, this arose solely from the internal energy of its contents. Prior to disintegration the contents occupied a volume of the order of 45 m^3. After expansion, the vapour flashed off would, theoretically, have a volume of around 4,000 m^3. The energy so released would be appreciable. Moodie (1982) estimates that 24 tonnes of liquefied propane at 20°C would have a theoretical energy release equivalent to 260 kg TNT. Not all of this energy would be available for the projection of projectiles; High Pressure Technology Association (1975) gives, for complete shattering due to brittle failure of a pressure vessel, the following partition of energy:-

Fragment kinetic energy	=	0.2 Es
Shock wave energy	=	0.8 Es
where Es	=	system energy

There have been a number of incidents in the past in which vehicle fragments have been hurled considerable distances following the rupture of pressure vessels containing liquefied gases. One of these, with close similarities to the vehicle at San Carlos and its load, was the LPG tanker disaster at Eagle Pass, Texas on 29/4/75 which is described in National Transportation Safety Board (1976). The salient features of this event were as follows:-

A tanker containing some 18 tonnes of LPG and of dimensions similar to that involved at San Carlos was involved in a traffic accident in which the rear end of the tank collided with a reinforced concrete abutment of a bridge. The accident catastrophically ruptured the rear end, and the front end, which at the instant of disintegration contained about 0.80 of the load, travelled some 500 metres from the point of collision with multiple impacts en route. Other, and much smaller portions, travelled up to 250 metres.

At the Laurel, Mississipi, rail accident of 25/1/69, which is described in National Transportation Safety Board (1969), a major portion of a tank car travelled 500 metres, clearing the roofs of houses, a major portion of a different car travelled some 350 metres with multiple impacts.

Similarly, at the New Jersey Turnpike on 21/9/72, described in National Transportation Safety Board (1973), one portion of a tanker travelled 400 metres, another portion travelled 180 metres and yet another portion travelled 70 metres from the point of collision.

This evidence shows that the disintegration of the San Carlos tanker was not an isolated incident; on the contrary, it was to be expected in such a case.

There is nothing chemical about such explosions; they arise purely from the reactive forces which attend the release of flashing liquids. This is especially true where the fragments include cylindrical portions with length to diameter ratios of, say, 3 or 4 to 1; such portions behave like crude jet or rocket engines. They would behave exactly the same way were the cargo to be a non-flammable liquefied gas of equivalent TAFF such as a Freon.

Study of the documents leads to a general consensus that the San Carlos disaster was primarily a fire but that it was attended by some blast effects arising from internal explosions in buildings, and explosions caused by the effects of heat on camping gas cylinders and the petrol tanks of motor vehicles. However, the blast effects due to the flashing of the liquid after disintegration seem largely to have been ignored by the other investigators.

9.5.6. THE DISINTEGRATION AND ITS CONSEQUENCES.

9.5.6.1. Reconstructing the disaster.

It may be said at the outset that any attempt to construct a detailed scenario of the accident must end in failure because the information needed for this does not exist. What is attempted below is to reconstruct the main features of the disaster using the principle of "Occam's razor", i.e. that an explanation be sought which requires the fewest hypotheses. A method of doing this would be to "run the film backwards", i.e. to try to retrace the path of the various parts of the vehicle from the points where they ended up back to the point of disintegration.

9.5.6.2. The trajectories of the parts of the wreckage.

There seem to have been three main trajectories. The first to be considered is that of the rear portion which comprised about half of the tank and thus carried about half of the load, i.e. 12 tonnes. It was found upside-down and embedded in the wall of a restaurant on the west side of the road. Some of the commentators have claimed that it collided with overhead cables on the way. Scilly, in a sketch, estimates that its trajectory lay at about 135° to the original line of travel and that it travelled about 250 metres. The second trajectory is that of the front end of the vehicle which comprised, at the moment of disintegration, the front, tapering, portion of the tank which probably carried about one-sixth of the load, about 4 tonnes, and the cab. The front of the tank was found by the road, on its west side, about 300 metres from its presumed point of disintegration. The cab, according to Stinton, was found on the road in its direction of travel but upside-down. The photographs in the present author's possession show it near the roadside on the west side of the road to which it had, presumably, been moved to clear the road. It had travelled about 100 metres from the presumed point of disintegration. The trajectory of the front portions seem thus to have been along the line of travel. The mid-section, which comprised about a third of the tank and thus carried about 8 tonnes, was found in the site upside-down about 50 metres from the presumed point of disintegration and its apparent trajectory lay at an angle of about 120° to its line of travel. It had opened up along its bottom seam and had a semicircular section.

Other significant portions found included an internal baffle which was found embedded in the engine compartment of a car in the site about 70 metres away and at an angle of 45° to the line of travel. A rear axle with two wheels was embedded in the camp wall to the rear of the presumed point of disintegration and the other rear axle was found in the camp about 40 metres away and at an angle of roughly 90° to the angle of travel.

The presumed trajectories are shown in Fig. 9.12a, and a reconstruction of the disposition of the various portions at the moment they parted company is shown in Fig. 9.12b.

9.5.6.3. The collapse of the wall.

Thus, from the moment of disintegration, the rear portion would be rocket propelled and would discharge a jet of flashing liquid across the road, which would strike the wall at about 45°. There seems nothing implausible about assuming that the impact of 12 tonnes of a high velocity two phase system could have brought about this breach in the wall. Photographs of the collapsed wall show it to be of light construction and to have been built of hollow terracotta blocks on a concrete base. The damage does not seem consistent with impact from a vehicle. Lengths of it seem to have come down in one piece with longitudinal cracks. Mechanical impact might have been expected to have punched a hole in it and have damaged the footing also. This does not seem to have happened.

9.5.6.4. Did the vehicle enter the camp ?

There was a ditch about 1 metre deep and two metres wide between the road and the wall. Had the vehicle ploughed across this ditch and hit the wall, it might have been expected that there would have been tyre marks and heavy scoring. In addition, along most of the length, and just inside the wall, there were trees still standing. There should also have been cars in the camp which showed signs of collision damage. The only report which gives the credence to the theory that the vehicle ran into the camp is that in Stern magazine which shows an aerial photo with annotations "Here the lorry broke through the wall", and "Here lay part of the driver's cab", showing a point near the internal camp road for the latter. The article did not show a photo of this part of the cab. Stern also claimed that the body of the driver was found in his cab, and that it was severely lacerated. "Shreds of skin" was the expression used.

9.5.6.5. The jets.

On the same basis as the above arguments, it might have been expected that the front portion would discharge a jet, but for a much shorter time compared with that of the rear portion, in a direction opposite to the line of travel.

The middle portion would have discharged both downwards onto the road, and from both ends. It would be the reactive force which partly flattened it. The present author has seen a photograph of the remains of an LPG tanker which split along its top seam. The shell was flattened to the ground and the under parts, wheels etc, were crushed as if they had been in a vehicle crushing press in a breaker's yard. The reactive forces in such disintegrations are enormous. The thrust, at the moment of disintegration of the rear portion, would have been equal to that of a space shuttle engine, i.e. about a gravitational force of 375 tonnes.

It is the present author's conclusion that the vehicle disintegrated in such a manner that the resultant of the jets which emerged was a widely diverging jet which blew down the campsite wall and swept across the site and possibly out to sea.

9.5.6.6. The quantity which flashed.

A question not previously discussed by any writer is what proportion of the

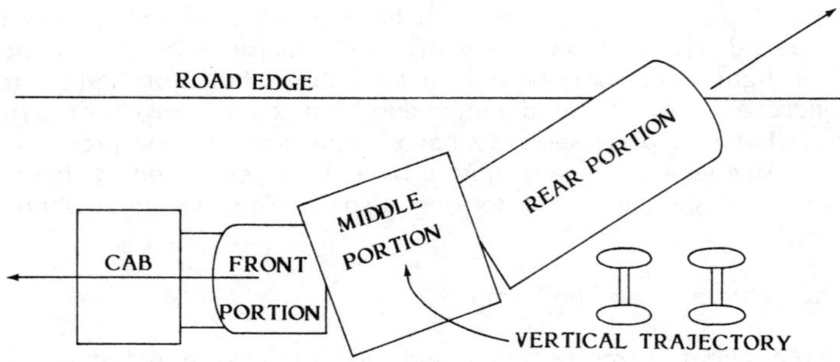

Fig. 9.12a. PRESUMED TRAJECTORIES OF DISINTEGRATION.

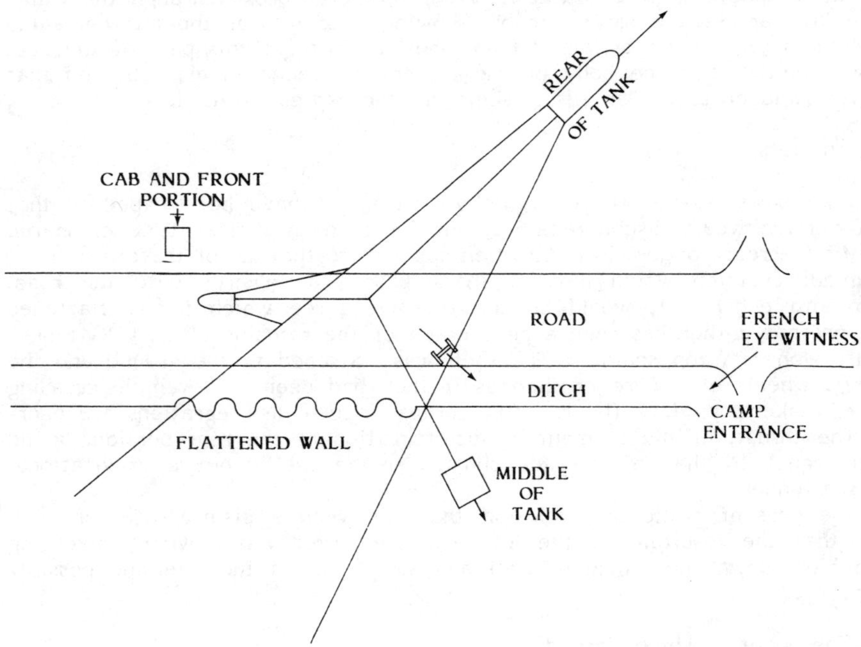

Fig. 9.12b. DISPOSITION OF PORTIONS OF WRECKAGE.

liquid flashed. The theoretical adiabatic flashing fraction (TAFF) may be calculated from the following relationship where the final state is the atmospheric pressure boiling point:-

$$E_{IL} = (1\text{-}x)E_{FL} + (x)E_{FV}$$

where
x = TAFF
E_{IL} = Enthalpy of original liquid
E_{FL} = Enthalpy of final liquid
E_{FV} = Enthalpy of final saturated vapour

The necessary data have been obtained from the Physical Properties Data Service of the Institution of Chemical Engineers as below. Enthalpies are expressed in MJ per kg mole, and density is expressed in kg per cubic metre. The data at 6°C are used for another purpose later.

TABLE 9.2. THERMODYNAMIC PROPERTIES OF PROPENE (PROPYLENE).

	-47°C	6°C	15°C	20°C
Enthalpy of liquid	-22.90	-17.718	-16.760	-16.20
Enthalpy of saturated vapour	-4.45	-2.216	-	-
Density of saturated vapour		14.66		

The exact temperature of the liquid at the time of discharge is not known and instead a range of values has been calculated for temperatures between 15°C and 20°C. This gives a TAFF in the range 0.33 to 0.36 which is fairly narrow and from which a figure of about one-third seems, in view of the uncertainties, to be a reasonable approximation.

9.5.6.7. The victims drenched in liquid propylene.

If the rear portion of the tank contained 12 tonnes of liquid initially, the discharge would have been a two-phase system consisting of about 8 tonnes of liquid and 4 tonnes of vapour. The unfortunate victims in the path of this jet would have been subjected to a Siberian blast at ca -47°C and they and their belongings would have been drenched in liquid propylene. This drenching may account for the extreme charring to which many of the bodies had been subjected.

It may be, mercifully, that many died of shock and asphyxiation before the flames reached them.

9.5.7. DETAILED ASSESSMENT OF THE CONSEQUENCES OF THE DISINTEGRATION.

9.5.7.1. Introduction.

It is possible, taking account of the circumstances of the disintegration discussed above, to provide more detailed speculation as to the sequence of events following the disintegration.

9.5.7.2. The geometry of the cloud.

The various parts would be propelled by reactive forces. The front and rear portions, after their contents were expelled, probably were then less stable in flight being without fins to guide them. They may have spun in flight, Scilly comments on spiral scoring of some of the pieces. The existence of two pieces with length to diameter ratios greater than 1.0 would disrupt the tendency for the vapour to spread evenly around the point of disintegration. It has been suggested above that the spillage pattern was dominated by the discharge from the rear portion which spread in a diverging jet across the camping ground with an axis at about 45° to the direction of travel of the vehicle. It seems likely that the discharge was almost entirely at low level.

The centre of the cloud would be intensely cold and would contain two parts by weight of liquid to one of vapour. On the fringes of the cloud, chilling would produce a mist, or even hoar frost, by condensation of water vapour in the air.

9.5.7.3. The nature of blast effects.

There seems no evidence, as Scilly (1978) has pointed out, to indicate that there were generalised blast effects of the kind expected in a vapour cloud explosion. This is in spite of conditions which today are currently considered to favour vapour cloud explosions and which are discussed in Chapter 12. These include the presence of obstructions and of features which favour semi-confined explosions. Propylene vapour is also a fairly reactive gas.

But there is evidence of internal explosions in at least three buildings. The disco building was totally demolished though Scilly comments that a motorcycle nearby was still on its footrest though burned out, which argues strongly against external blast. Photographs of damage to apartment buildings to the south of the site are strongly indicative of damage due to an internal explosion. Examination of photographs by the present author seem to show a demolished small building to the south end of the site which is not commented on by either Scilly or Stinton. The position of this is indicated on Fig. 9.11, the sketch map of the site. It may well have been a toilet building intended to serve the south end of the site. Many of the cars have their engine compartment lids blown up and it would be likely that fuel tanks and camping gas cylinders would have exploded.

9.5.7.4. The time scale.

If internal explosions had occurred as far as two hundred metres from the point of disintegration, this means that vapour must have reached these points before it caught fire. The wind was blowing from the sea and may have blown vapour off the sea to be sucked into the air conditioning system of the apartment houses, the damage was at the back of the houses away from the road. This suggests a minute or two would be needed from the moment of disintegration to the onset of fire. It is of interest that there is no report that suggests that any other vehicle was involved on the road by running into the gas cloud. Photos taken the following day suggest that the road was fairly busy and had the interval between discharge and ignition been more than a minute or so, it might have been expected that other vehicles would have run into the cloud. Experience of other incidents on roads suggests that engines stall in vapour clouds because of the over-rich mixture which enters the engine and that it would be extremely unlikely, therefore, that any vehicle could have run through the cloud.

It seems not unlikely, therefore, that the sequence reported by some eyewitnesses that there was a "pop" followed by the appearance of vapour followed by a loud explosion followed by fire, would accord with the sound of the disintegration of the tanker being followed by the spreading of a vapour cloud followed by an internal explosion of one of the buildings, probably the disco, and succeeded by the fire. One factor which may have delayed the onset of fire was that, in the area mainly affected, the vapour was too rich to burn and would actually extinguish sources of ignition. The upper flammable limit for propylene, according to Harris (1983), is 10.3% vol/vol. Propylene has a maximum laminar flame speed, according to the same source, of 5.1 metres per second. It is not likely that this speed would be achieved and the time scale may be consistent with ignition starting in several places at the fringes, where the propylene/air mixture was below 10.3% propylene, and coalescing into a major conflagration. Any idea that a single ignition source could ignite the whole mass in seconds can be dismissed on account of the limited burning velocity of the vapour.

9.5.7.5. Was San Carlos a fire ball ?

In accordance with the question mark in the title given to this case history, some doubt arises as to whether the San Carlos incident was truly a fire ball as defined earlier in this book. It was certainly an incredibly fierce flash fire, but whether there was a fully developed fire ball seems open to some doubt. It may have been the circumstances of the disintegration which spread the cloud for great distances along the ground which militated against lift-off.

9.5.8. WAS THE DISINTEGRATION CAUSED BY HYDRAULIC PRESSURE ?

9.5.8.1. The judgement of the Court.

The Spanish Court, in Spanish Ministry of Justice (1982), was quite uncompromising in its conclusions, stating that the tanker failed **solely** as the result of overfilling. Though it is beyond dispute that the tanker was overfilled, this does not prove that a hydraulic rupture occurred in this instance. To prove the point, it is necessary to show that there was a rise in temperature of the contents sufficient to cause an expansion of the liquid to such a degree that it would be equal to the volume of the vessel at the point of disintegration.

9.5.8.2. The Engineers Report.

Carrasco (1978) set out to provide such a proof, but, as is shown below, he failed to furnish proof; at least he failed to establish conclusive proof. Table 6.5. (Page 82), sets out a comprehensive view of the general case of a vessel without pressure relief, and containing a liquid, and which is subjected to increasing temperature. A limiting case which arises is that in which the temperature (T_R), at which rupture would be predicted to occur, exceeds any feasible ambient temperature. In such a case rupture cannot occur and this represents the consideration which determines the theoretical minimum ullage which is to be allowed on filling a vessel. (Such minimum ullage is increased in practice by a safety margin to allow for abnormal ambient temperature and other factors).

Thus the first step is to determine whether it is feasible, under assumed conditions of ambient temperature, for rupture to occur. This being so, the next step is to demonstrate that the necessary rise in temperature can be achieved in a time span dictated by the circumstances of any particular case. These general points being established, it remains to consider Carrasco (1978) in detail. Carrasco states in his report:-

> "The author of the present report finds that the cause of the explosion of the tanker was an excess in the load of propylene, the volume of which, in liquid phase, at the moment of the explosion, was greater than the vessel which was containing it, which burst the vessel because of the almost total incompressibility of liquids in the same way that thermometers burst when the mercury or other thermometric fluid increases its volume as a result of a rise in the ambient temperature and does not have enough

room in the vessel which contains it".

Even allowing for problems of translation, this is not well put, Carrasco evidently means that the vessel cannot accommodate any further increase in the volume of its contents without rupture. Carrasco states further:-

"The fact that liquid propylene was transported in a completely enclosed vessel led us to consider the possibility that the liquid which had been loaded, when increasing its volume by increasing its temperature, reached a point where it took up all the available volume in the tank. Once the liquid exceeded this volume because of its temperature, the tank would have to burst, whatever its mechanical strength and whatever the material it was made of or the thickness of its walls".

(Present author's underlining).

The assumption that rupture will occur at the instant that the volume of the expanding liquid becomes equal to that of the free space in the empty vessel may be challenged at once. This assumption is based upon the concept that steel, under stress, does not exhibit strain, that is it has a Young's Modulus of infinity. Moodie (1982) considers that the ability of steel to stretch would require that the liquid would have to be heated by 3 to 6°C (according to the steel from which it was made) above the temperature at which the ullage becomes zero. This latter assumes also that propylene is totally incompressible; Moodie considers that the compressibility of propylene is so small that it may be ignored. The version of Carrasco (1978) in the present author's possession does not have the table of densities used by Carrasco, but he has obtained data from the Physical Properties Data Service, Institution of Chemical Engineers, Rugby, (Fig. 9.13).

According to Spanish Ministry of Justice (1982), the capacity of the tank was 45 m^3 (a statement which does not appear to have been checked by measurement of the fragments) and that it was charged with 23.47 tonnes of propylene. This statement though unchallenged has not been checked for accuracy. At a point at which the tank would be entirely filled (assuming that there was no nitrogen present) the density would have been $\frac{23.47}{45.00} = 0.521$. The temperature corresponding to this density = 14.2°C which is much lower than Carrasco. However, the accuracy of this figure is subject to the accuracy of a number of parameters. These include the accuracy of weighing, the accuracy with which tables of density reflect the density of the actual material which, as a commercial substance, may have had a higher or lower density than that of pure propylene, and finally the previously mentioned uncertainty as to the actual volume of the tanker.

As has been pointed out earlier, Carrasco ignores the increase in volume consequent upon the strain in the metal walls which arises from the internal stress. These factors are taken account of in Moodie (1982) where, for a pressure vessel of about 48 $metres^3$ capacity made of structural steel (1501 - 151 - 28A) or of high tensile steel (EN 25), the respective increments of temperature between disappearance of ullage and the onset of vessel burst (which would occur at 2.2×10^6 N $metres^{-2}$ for the structural steel, and 4.5×10^6 N $metres^{-2}$ for the high tensile steel) would be:-

Structural steel	ca 3°C
High tensile steel	ca 6°C

The burst pressure calculated by Carrasco is 57.3 kg/cm^2 which corresponds to 5.73×10^6 N $metres^{-2}$. The ultimate strength of the steel as used in calculations by Carrasco is 81.2 kg mm^2 or 8.12×10^8 N $metres^{-2}$, which corresponds to Moodie's EN 25. [It may be noted that there are economic factors which favour the use of high tensile steel in tank cars. Tanks made of such steel are only half the wall thickness of structural steels and hence half the weight. The reduction in tare weight is reflected in more economical use of fuel].

Thus Carrasco should have allowed up to 6°C extra, making the temperature

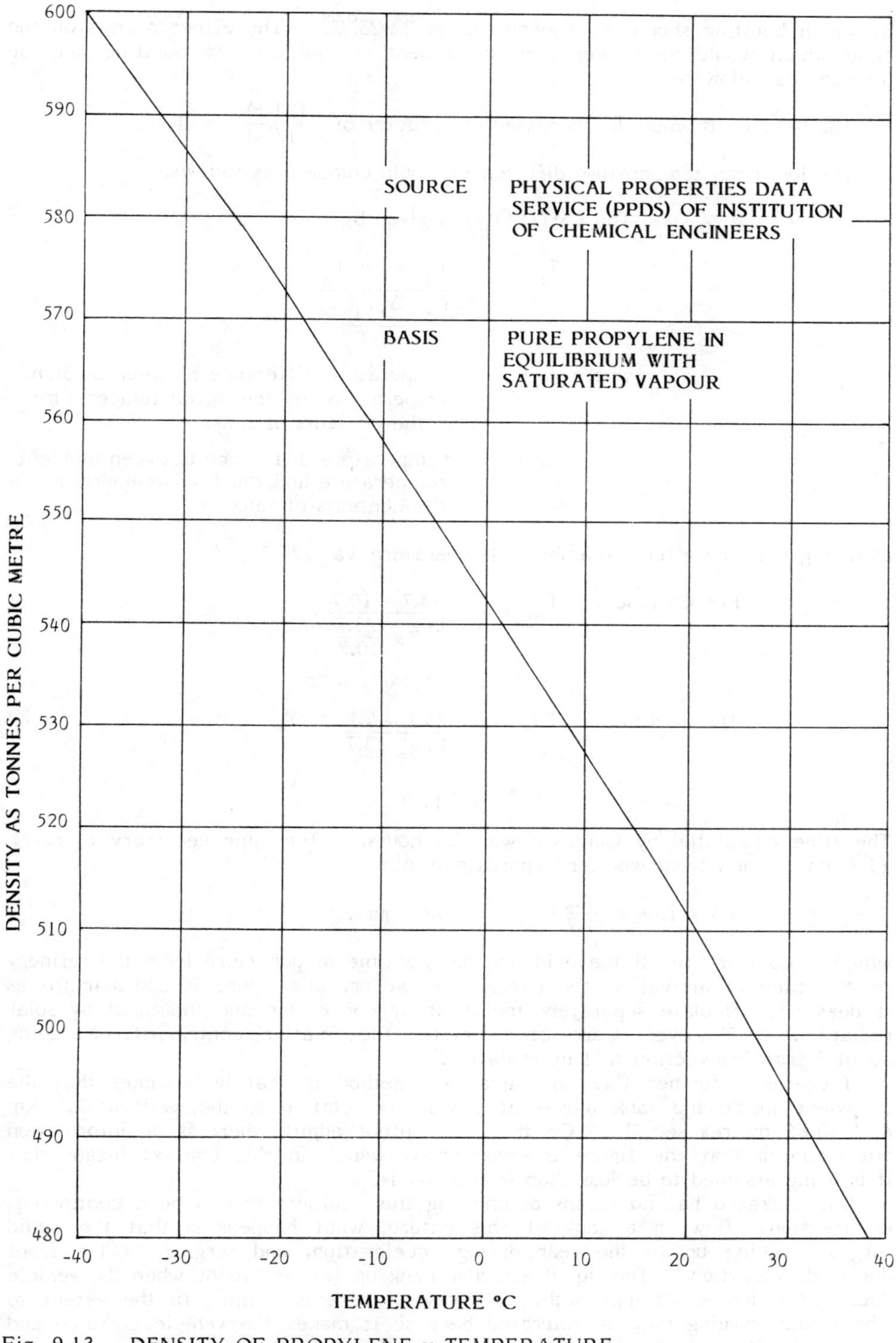

Fig. 9.13. DENSITY OF PROPYLENE v TEMPERATURE.

at which bursting should have occurred as 22-23°C. The effect of this on the time which would have been needed to heat up the tank to bursting may be assessed as follows:-

The heat input would be increased by a factor of $\frac{(23-4)}{(17-4)} = 1.46$.

The log mean temperature differences would compare as follows:-

The LMTD (T_m) is given by

$$T_m = \frac{\Delta T_1 - \Delta T_2}{\text{Log}_e \frac{\Delta T_1}{\Delta T_2}}$$

where ΔT_1 = temperature difference between ambient temperature and the initial temperature of the contents of tank

ΔT_2 = temperature difference between ambient temperature and the final temperature of the contents of tank

assuming that the effective ambient temperature was 27°C.

For Carrasco, $T_m = \frac{23.7 - 10.7}{\text{Log}_e \frac{23.7}{10.7}}$

$= 17.23$

The modified $T = \frac{23.7 - 4.7}{\text{Log}_e \frac{23.7}{4.7}}$

$= 11.7$

The time calculated by Carrasco was 2.5 hours. The time necessary to reach 23°C on the new basis would be approximately

$$2.5 \times 1.46 \times \frac{17.23}{11.7} = 5.4 \text{ hours.}$$

which is contrary to all the evidence on the time of departure from the refinery to the time of arrival at the camp. However, this figure is approximate as it does not calculate separately the input by convection and the input by solar radiation. However, Carrasco estimates the relative contribution as being about 7 from convection to 1 by radiation.

There is a further flaw in Carrasco's method in that he assumes that the propylene inside the tank moves at a velocity relative to the walls of 2.4 km hr^{-1} (0.66 metres sec^{-1}). On this that author admits there is no information but declares that the figure is conservative which in this context means that it is being assumed to be less than it actually is.

But Carrasco has no means of knowing this. There cannot be a continuous, unidirectional flow in a tank of this nature, what happens is that the liquid surges, backing up to the rear during acceleration, and surging to the front during deceleration. The liquid will also bank up to the outside when the vehicle does not follow a straight path. However, there is a limit to the extent to which such surging may be tolerated because it makes the vehicle unstable and internal baffles are fitted to minimise this effect. One such baffle was seen on the site after the explosion and Scilly (1978) has speculated that another may have been hurled into the sea. There is evidence that one landed part way along the path of the rear section.

The key point to note is that surging gets less as the tank becomes fuller and finally ceases at total occupation by liquid. A homely analogy is shaking a medicine bottle. It is difficult to mix when full and easy when half full. Thus the internal heat transfer mechanism between 17° and 23° when the tank was full of liquid would be solely by conduction and natural convection. Solar radiation would have little effect on the convective process as it would heat the top of the tank. Carrasco attempted to justify his hypothesis that the tank would have achieved 17° in 2.5 hours by loading a similar vehicle with ca 23.3 tonnes of water which it was claimed heated up from $19.5^{o}C$ to $23^{o}C$ at an ambient temperature of $34^{o}C$. The calculated time to heat up using the method applied to the propylene tanker was 2.29 hours and the actual time was 2.3 hours.

Those with experience of the errors in calculating heat transfer coefficients in the laboratory, let alone for those of road vehicles, may consider that this measure of agreement is purely coincidental. However, there must have been considerable differences in the internal heat transfer mechanism in a vehicle which was barely half full as compared with one in which the liquid was 0.95 to 1.00 fraction of the total volume. It would appear from the above that neither Carrasco's theoretical methods nor his experiment are immune from serious criticism.

9.5.8.3. An alternative justification for the theory of hydraulic rupture.

Though the discussion above is dismissive of the proof advanced by Carrasco, it does not follow that the hypothesis that the cause of the disaster lay in hydraulic rupture is thereby disproved. As will be shown below, a detailed analysis of the physics and thermodynamics of the system leads to the conclusion that hydraulic rupture was perfectly feasible and the most likely cause. Indeed no other contributory cause need be advanced.

The first question to be examined is the relationship between the temperature and the density of liquid propylene. Figure 9.14. shows that there are considerable differences between values quoted in the literature with the density values quoted by Physical Properties Data Service (1986) being lower than those for other sources including the source used by Carrasco. The PPDS figures are highly reputable and represent the most up-to-date knowledge. However, they do apply only to the pure substance and cannot be absolutely relied upon to apply to the commercial product involved at San Carlos.

There is also the question of the true volume of the tank. Stinton has claimed, without stating the source, that the tank's true capacity was only 99% of its nominal volume.

Carrasco assumed that the initial temperature of the tanker, presumably based upon the temperature at which the propylene was kept in bulk storage, was $4^{o}C$. Ens (1986) has cast doubt upon this, calling attention to the time it took to fill the tanker, namely about 2 hours on a hot morning. But there is a further factor. If the tanker was filled by pumping in liquid into a tank containing vapour, without venting back the vapour displaced to the storage, the condensation of this vapour during filling would increase the sensible heat of the liquid. Had the tank been initially full of vapour at 7 bars, which is the vapour pressure of propylene at $4^{o}C$, the temperature of the liquid would have been raised by $2^{o}C$ in filling through this cause alone. A further factor would have been work done by the pump on the liquid. So possibly the initial temperature of the load may have been around $7^{o}C$.

There is also the question of the possible presence of a non-condensible gas in the original vapour space. This could arise from the use of nitrogen

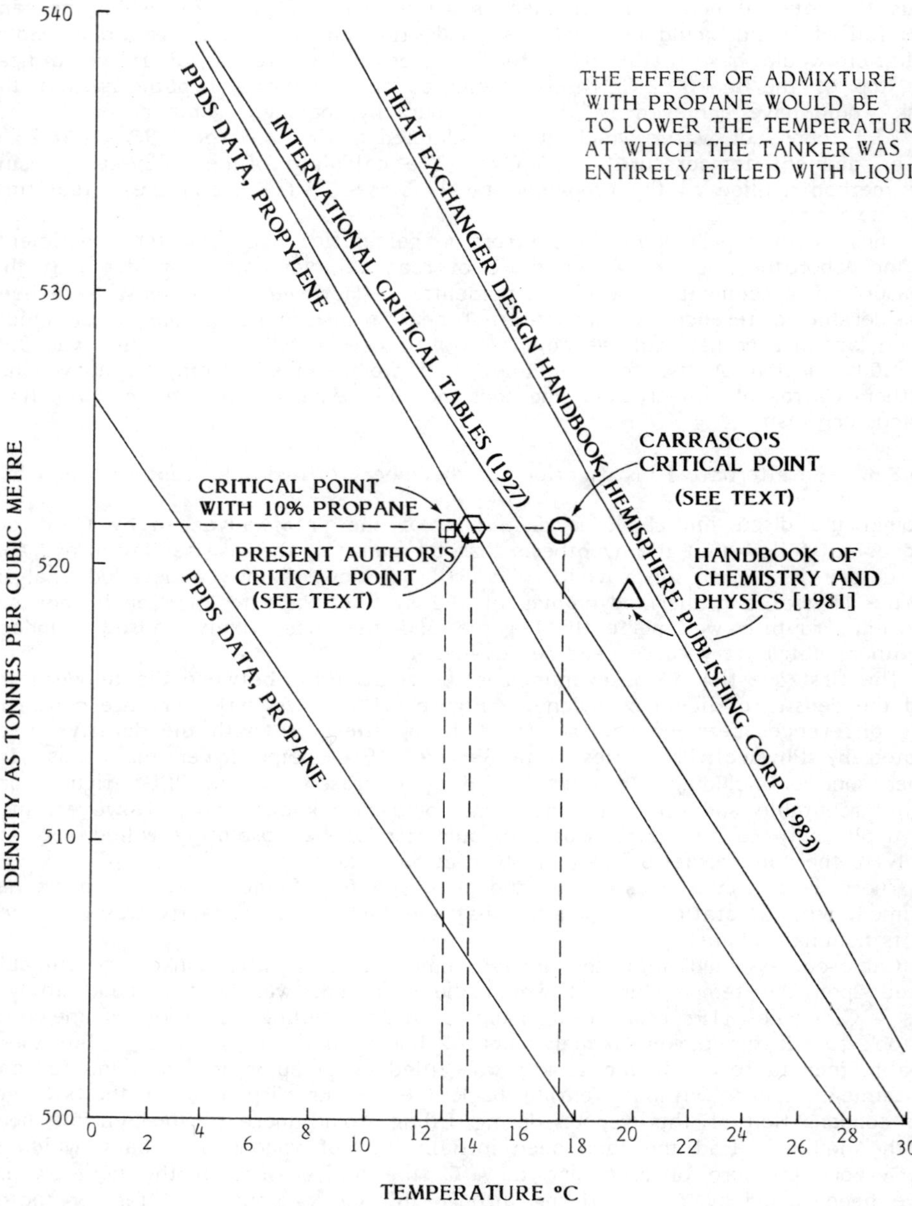

Fig. 9.14. DENSITY OF LIQUID PROPYLENE, VARIOUS SOURCES.

for purging. The influence is impossible to quantify as it is not known, and never will be, whether non-condensible gas was present. All that can be said is that the presence of such gas would increase the internal pressure above that calculated for the vapour pressure of the liquid. Even if the gas were soluble, and nitrogen is only sparingly soluble in organic liquids, it would still exert a pressure.

A different approach to that used by Carrasco is to estimate the volume of the tank at its bursting pressure thus making allowance for the stretching of the walls before burst. This can be done approximately by assuming that they were elastic up to the point of rupture and then using a standard formula for the specific increase in volume of a thin-walled cylindrical vessel when subject to internal pressure. This is given as follows, in consistent units:-

Specific increase in volume	=	$p\,r / E\,t\,(2.5 - 2/M)$
where p	=	pressure
r	=	radius
E	=	Young's Modulus
t	=	thickness of wall
1/M	=	Poisson's ratio, say 0.25

Taking Carrasco's burst pressure of 57.3×10^5 N/m^2 and assuming that r = 1.15 metres and t = .008 metres and that E for the steel was 206×10^9 N m^{-2}, the specific volume increase at the point of bursting would be .008 or roughly 1%. This seems to cancel out the 1% deficiency in volume. Thus the capacity at burst, but not allowing for the breakdown of the above equation beyond the elastic limit, would be 45 cu metres. The density at burst would be about 0.5216 which corresponds to about 14°C.

On the basis of this calculation, the heat input would be as follows:-

$$23{,}470 \times (14-7) \times \text{specific enthalpy}$$

The specific enthalpy is about 2,550 J/kg and the heat input is thus 4.18×10^8 J. The required overall coefficient to yield this input in ca 2.5 hours would be given as follows:-

$$U = \frac{4.18 \times 10^8}{2.5 \times 3600 \times \text{LMTD} \times \text{Area of vessel}}$$

$$\text{LMTD} = \frac{\text{ca}(27-7) + (27-14)}{2} = 16.5$$

Surface area of vessel = 80 m^2

Thus U = ca 35 W/m^2 $^{\circ}$C.

It was not possible to find evidence in the literature to support this value for a strictly parallel case, though it lies in the region of reported results for natural convection. However, tests conducted by the present author in a wind tunnel in Bradford University using a model of 1/20th scale, filled completely with water, gave values of ca 150 W/m^2 $^{\circ}$C. It would be expected that the internal heat transfer on the full scale would be less than the coefficient so achieved. Moreover neither these tests nor the calculations above make any allowance for the input of solar radiation.

Attempting exact calculations would be a futile exercise as the assumptions are too coarse. What can be done, at a minimum, is to indicate that the feasibility of hydraulic rupture cannot be ruled out. This has clearly been achieved. Further, the calculations suggest that such a rupture was feasible given a combination of unfavourable circumstances. After all, the vessel had a load which at the time of setting off had, at the most, 3% ullage compared with the 20% ullage customarily allowed.

Nothing has been said above regarding the theory that the metal had been weakened by ammonia embrittlement. The Court received a sworn affidavit from the owners of the refinery which had filled the tanker declaring that it had been used solely for transporting liquefied hydrocarbon gases. But, as they did not own the tanker, it is difficult to see how they could know. Obviously, had there been ammonia embrittlement, this would have added to the feasibility of rupture.

9.5.8.4. Are any other explanations necessary ?

The above calculations suggest that theories by other investigators that the disaster was caused by prior collision perhaps initiated by a burst tyre, or that it arose through a so-called BLEVE, are unnecessary. One theory advanced is that a tyre blew and caught fire. This caused the vehicle to cross the road and to career through the camp wall. The mechanical damage which ensued stressed the vehicle to rupture point. Discussion above has discounted the theory that the collapse of the wall arose from mechanical impact. Carrasco, who examined the wreckage, has declared that there was no evidence for a collision prior to the disintegration of the vehicle. As to the BLEVE theory, the hypothesis that the tank was completely filled rules out the possibility of a dry wall.

9.5.9. SOME GENERAL LESSONS.

9.5.9.1. The judgement of the Court.

It seems completely reasonable to agree with the judgement of the Spanish Court that the accident was caused by overfilling, but this would be on the balance of the evidence rather than absolute proof which was, and still is, lacking.

9.5.9.2. Supervision.

Supervision by the refinery seems to have been practically non-existent. No disclosure was made as to the method of payment but if this was based upon payment by the tonne carried, this would have given the owner driver every incentive to overload the vehicle. An upper limit of payment based on the safe load, or even penalties for overloading, might have prevented this accident.

9.5.9.3. The need for a safety valve.

A vehicle of this character should have been fitted with a safety valve. Stinton (1978) alleges that there was a point provided for fitting a relief valve but that it was blanked off. There are dangers from relief valves blowing off flammable liquids but these are far less than the consequences of disintegration.

9.6. MEXICO CITY (SAN JUANICO). A SUPER FEYZIN.

9.6.1. GENERAL ACCOUNT.

On November 19th, 1984 at 5.35 a.m. at this site there commenced a series of explosions and fires which lasted until 23.00 hours that day. These events killed at least 500 people and injured 7,231, of whom 144 died in hospital. Some 39,000 people were rendered homeless or were evacuated. It has been described in TNO (1985) which is summarised in Pietersen (1985) and in Skandia (1985).

The disaster occurred in the premises of the organisation PEMEX (Petroleos Mexicanos) a fairly modern plant built, or so it was claimed, to API standards, and largely manufactured in the USA. Its purpose was to take in LPG from refineries and to supply it to wholesale distributors. It was then widely sold for domestic heating in Mexico City, which is one of the world's largest cities with a population estimated at 16 million people. Figure 9.15. is a sketch of the installation and its environs.

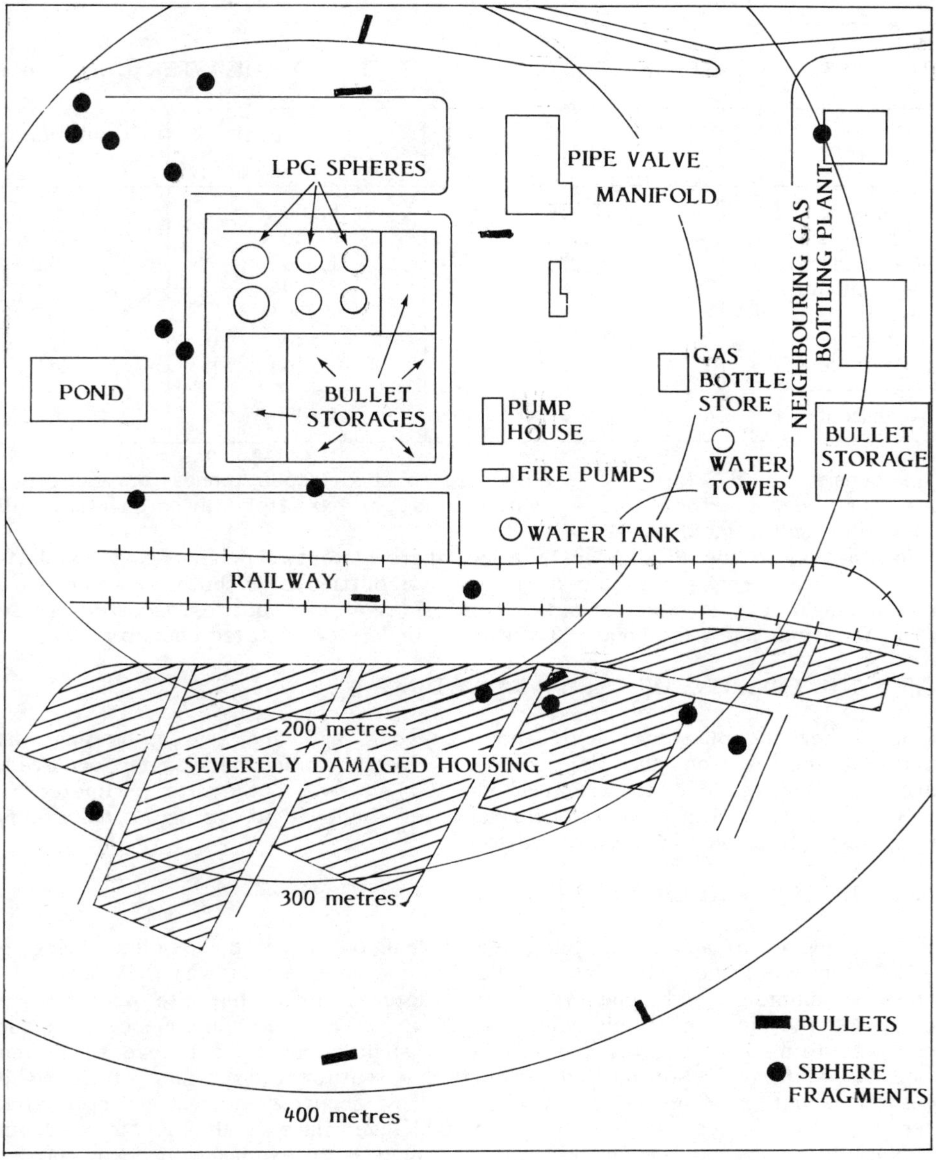

Fig. 9.15. SKETCH PLAN OF INSTALLATION AT MEXICO CITY.

9.6.2. TURNOVER AND QUANTITIES BEING STORED.

The throughput of the installation was 5,000 m^3 of LPG which, according to Skandia (1985), was 80% butane, 20% propane, but whether by weight or volume is not stated nor is it stated whether the butane was normal or iso butane. Such mixtures may, in any case, vary in composition from day to day. At, say, a density of 560 kg/m^3, this would have amounted to 2,500 - 3,000 tonnes per day. The total capacities of the storage tanks and their estimated contents at the time of the disaster are given in Skandia (1985). Table 9.3. is based on these data.

TABLE 9.3. CONTENTS OF STORAGES AT TIME OF DISASTER (All in m^3).

	Nominal Capacity (each)	Total Capacity	Contents m^3
2 spheres	2,400	4,800	4,320
4 spheres	1,500	6,000	3,000
4 cylinders	270	1,080	972
14 cylinders	180	2,520	2,268
6 cylinders	54	324	292
3 cylinders	45	135	121
21 cylinders	36	756	680
54 vessels		15,615	13,653

At a density of 560 kg/m^3 this amounted to about 6,500 tonnes, or about two days' turnover. The storage system was approximately three-quarters full at the time of the incident.

In addition to the PEMEX installation, there were two neighbouring installations which were engaged in the distribution of bottled LPG, both of which were badly damaged and in which countless explosions of LPG bottles occurred. These two installations had each about 20 storage cylinders of unstated capacity.

9.6.3. THE ENVIRONS OF THE INSTALLATION.

Though, when the plant was built about 25 years ago, the nearest housing was about 300 metres from the plant boundary, by the time the disaster occurred, housing had encroached to within about 100 metres of the site perimeter. The houses were of flimsy construction and were believed to average 5 inhabitants per house together with numerous livestock.

9.6.4. THE SEQUENCE OF EVENTS.

The initiating event appears to have been a leak on one of the pipelines bringing the LPG in from the refineries. The pipeline in question was 0.2 metres (8 inches) in diameter and appears to have ruptured inside the site near to the storages. The pipeline at this point was above ground. In consequence a vapour cloud was formed which was driven down a slight slope by a breeze estimated at ca 0.4 metres per second and blowing in a southwest direction. The cloud measured about 200 metres by 150 metres by about 2 metres in depth and, after 5 to 10 minutes, it ignited at a ground level flare about 100 metres from the point of escape. The ignition was described as an explosion by local people; according to TNO (1985) it was the first of nine to be recorded on the seismograph of Mexico City University some 30 kilometres away. The ignition created a fire ball which rose into the air. The original point of rupture now had a highly directional flame issuing from it which was directed towards one of the smaller spheres. The original ignition also left some 10 houses burning. Within 12 minutes the small sphere had burst creating a fire ball estimated at 300 metres in diameter.

During the next hour and a half, seven or eight large and violent ruptures occurred, the last of these being recorded at the University at 7.00 a.m. They were accompanied by numerous smaller explosions as most of the cylinders burst. Many of these were rocket propelled in a southerly direction as it was their northern ends which suffered most from flame impingement. There were also cases, probably numbered in hundreds, of small bottles rupturing in the nearby bottle-filling establishments. The last tank explosion occurred at 11.00 a.m.

9.6.5. THE CONSEQUENCES.

The considered judgement of the TNO investigating team, summarised in Pietersen (1985), is that there is little evidence to support the view that there was a vapour cloud explosion at Mexico City, though some overpressure effects were noted. Blast effects were localised and were due either to the flashing of the LPG on the rupture of containment or to confined explosions. A very large number of fragments of the vessels were propelled for distances of many hundreds of metres. Only 4 of the original 48 cylinders remained on their supports. One cylinder travelled for 1200 metres and 11 more were propelled for more than 100 metres. The 4 small spheres all exploded, some fragments travelling more than 400 metres. The 2 largest spheres did not explode but collapsed through buckling of their legs.

Damage to houses reached severe proportions, practically all houses within a 300 metre radius being severely damaged. There were many cases of internal explosions in houses. Skandia (1985) records that many people were sprayed with LPG droplets which led to severe burns.

Pietersen (1985) considers that current models seem to understate fire ball diameters as evidenced at Mexico City, though not to a very marked degree.

9.6.6. SOME LESSONS.

9.6.6.1. Strategic and tactical considerations.

It is of significance that the plant was built to reputable standards and largely manufactured in the US. But these were the tactics of the situation in which the strategy was bad. It was bad in that the site was very restricted in area and the entire fire fighting system was only capable of extinguishing small fires because it was far too close to the centre of the area in which a major fire would start. The pump house was only about 50 metres from the centre of the storage area.

The area of storage measured, according to the figure in Pietersen which shows the damage area, a mere 110 x 120 metres, i.e. some 14,000 sq. metres. The LNG loading was thus an astonishing ca 450 kg/sq. metre.

9.6.6.2. The lack of town planning.

As will be noted also when Bhopal is discussed in a later chapter, the close proximity of a densely populated area must have contributed greatly to the high death toll. Obviously PEMEX were in no way to blame for this as the site originally had no adjacent housing. A comparison with the Feyzin disaster of January 1966 may be apposite here and is given in the table below.

TABLE 9.4. COMPARISON BETWEEN FEYZIN AND MEXICO CITY.

	Feyzin	Mexico City
Quantity of LPG involved	6,400	12,000 cu metres
Number killed	17	500 +
Number injured	80 +	7,000 +
Distance of nearest housing from perimeter of storage area.	$\geqslant$ 1,000 ?	100 metres

It is clear from the above that though Feyzin was half the scale of Mexico City, it caused only 1/30th of the fatalities. The much greater distance of separation from housing must have played the major role in this.

9.6.6.3. The vessels and pipework.

Had there been a facility to shut off the pipelines at the site perimeter which could have been arranged to operate automatically in the event of leak, this tragedy could have been averted.

The lessons of Feyzin were not applied at Mexico City. Photographs in Skandia (1985) show that the spheres were not provided with water cooling sparge pipes and that the external legs of the spheres were not thermally insulated by encasing them in fire resistant concrete.

9.7. ANAEROBIC FIRES.

9.7.1. THE SINKING OF THE ROYAL OAK.

The British battleship, the Royal Oak, was torpedoed in the early hours of 14th October, 1939, in Scapa Flow, Orkney. The sinking is described in McKee (1972). The account is of relevance to this book in that it describes the development and effects of an anaerobic fire caused by the ignition of cordite propellant in one of the magazines.

The cordite was a mixture of collodion cotton, substantially cellulose di-nitrate with nitroglycerine, which would have been stored in canvas bags in the magazine. This material will not detonate but it "burns" fiercely producing a mass of incandescent gas at ca 1500°C. Such an incandescent gas is indistinguishable from a flame. McKee describes it as "seeking the open air" but this is merely a figure of speech. What was happening was that, as the cordite continued to burn, it produced an ever-expanding mass of hot, buoyant gas which penetrated into any space available to it and especially it travelled up through the ship's ventilation system. Witnesses saw it as an orange flame. In the words of one eyewitness "It was like looking into the muzzle of a blowlamp, the flame was bright orange outside and an intense blue inside".

Those caught in it were severely burned. Though many were seen to die, there were some survivors. One was surrounded by flame but, though requiring long hospital treatment, survived. He held his breath during exposure to the flame, so exposure could only have been for a minute or so. His survival was probably due to his wearing thick trousers and a jersey. There are no statistics on how many were burned in total nor can we know how many might have survived burning if they had not been drowned.

It might be anticipated that similar effects could occur if a large mass of nitrocellulose were to catch fire under industrial conditions.

9.7.2. FIRE AT ALFRED HARRIS LTD.

This fire broke out at a factory in Richmond, Surrey, England, on September 4th, 1947. The premises did not constitute a chemical works. As a result of an extremely fierce fire 7 persons living in nearby houses were burned to death, one of them in the street. The findings of a Public Inquiry are reported in Long (1948). The factory had at the time some 15 tonnes of celluloid (Xylonite) mainly in the form of scrap cinema film. The material is a nitrated cellulose containing about 10.5% nitrogen plasticised with camphor. Because of its high flammability, the material is no longer used for consumer goods, though some is still made for special purposes and stocks still exist in the UK in the form of archives of old films.

The Report notes the extreme severity of the fire and the rapidity of its spread without attributing it to the particular chemical nature of celluloid which renders it able to burn in an anaerobic way. The substance is, however, oxygen deficient and would be expected to burn also at the front of the expanding incandescent gas which is produced. The Inquiry noted, with an expression of surprise, that some wooden boxes containing the much less flammable substance cellulose acetate had survived the blaze. As the Advisory Committee on Major Hazards (1984), (Appendix 8), has commented, some cellulose nitrate materials stored in containers made of fibre board, presumably because of the low thermal conductivity of this material, are much less likely to give rise to cataclysmic fires than if the substances are contained in steel drums.

9.7.3. NITROCELLULOSE FIRE, STEVENSTON, AYRSHIRE.

On 23/5/73, there was one fatality, 3 serious injuries, and a further 30 persons had to be treated for burns, resulting from an intensive fire at this site. On the site were stored some 4,000, 227 litre drums of nitrocellulose, of which about 3,000 were involved in the fire.

In two-thirds of the drums, the nitrocellulose was wetted with isopropanol and the remainder were wetted with water. In all some 600-700 tonnes of material were involved.

The open air storage, in which the drums were stored on pallets, was 4,000 m^2 in area. There was an intense "ball of fire" with flames 30 m high. The fire eventually damaged 28 buildings and there was scorching and blistering of buildings up to 140 metres from the storage area. Damage was estimated at £600,000 (1974 prices).

A view of the scene after the fire, showing drums with their ends blown out, appears in Figure 9.16.

One of the first buildings to be involved, and one which was eventually totally destroyed, was the works fire station. This was engulfed in flames before all but one of the appliances could be driven out and three firemen and a clerk were burned in the building. One of the firemen died later in hospital.

The incident is reported in Fire Prevention (1973) and in Fire (1973). Neither of these references see fit to draw attention to the dangerous properties of nitrocellulose; one attributed the severity of the blaze to the presence of isopropanol as a wetting agent.

9.7.4. AMMONIUM NITRATE FIRE, HOECHST, NEAR FRANKFURT.

Clancey (1963), describes a fire at Farbwerk Hoechst on 2/3/61. A self-sustaining thermal decomposition, which had been started by friction in a conveyor belt, eventually propagated through a mass of some 8000 tonnes of a fertiliser mixture. The mixture contained about 1/3rd ammonium nitrate, the remainder being relatively inert fertiliser substances.

Fig. 9.16. SCENE AFTER FIRE AT STEVENSTON, AYR.

The fire took about 12 hours to extinguish and gave off copious nitrous fumes. The fire brigade had to wear breathing apparatus. There was no injury to the public as the fumes were relatively dilute before they reached populated areas. The fire was not accompanied by any explosion.

Chapter 10

EXPLOSIONS

10.1. DEFINITIONS.

Dictionary definitions of "explosion" are as follows:-

Shorter Oxford English Dictionary (1956) gives, inter alia,
(2) The action of driving out, or issuing forth with violence and noise,
(3) The action of going off with a loud noise, or of bursting under suddenly developed internal energy. (Latin "to drive off the stage by noise").

Chambers (1979) gives, inter alia,
"Act of exploding; a sudden violent burst with loud report, a great and rapid increase or expansion".

Uranov and Chapman (1964) gives, inter alia,
"A violent and rapid increase in pressure in a confined space. May be caused by an external source of energy (e.g. heat) or by an internal exothermic chemical reaction in which large volumes of gases are produced . . . .".

The common features of the above seem to lie in the concept that explosions are sudden events in which internal energy is released and in which pressure is developed. The process is accompanied by a loud bang or noise.

Robinson (1944) gives,
"In a general sort of way, an explosion consists of the very sudden production of large quantities of gases, usually hot, from a much smaller amount of a solid, a liquid or a gas. This is usually accompanied by a loud noise or report".

(Robinson seems to be in error in using the word "amount"; volume is what he should have said).

Biasutti (1981) gives,
"A rapid, self-propagating chemical decomposition accompanied by a large amount of heat and gas".

Baker (1973) gives,
". . . a process by which a pressure wave of finite amplitude is generated in air by a rapid release of energy".

The reference goes on to say:-
"Some widely different energy sources can produce such pressure waves and thus be classified as explosives by our definition. The stored energy

in a compressed gas or vapor, either hot or cold, can be such a source. The failure of a high pressure gas storage vessel or steam boiler or the muzzle blast from a gun are therefore examples of explosions. Release of electrical energy or rapid vaporisation of a fine wire or thin metal film can produce strong pressure waves in air and thus be classified as an explosion source. The more usual energy sources for explosions in air, however, are either chemical or nuclear materials capable of sustaining violent reactions when properly initiated".

Canvey (1981) gives,

"A rapid release of energy, which causes a shock wave, moving away from the source. An explosion can be produced by the detonation of a high explosive, the rapid burning of a flammable gas cloud or the sudden failure of a pressure vessel containing high pressure gas".

McGraw-Hill Encyclopedia of Science and Technology (1982), Vol.5. gives,

"An explosion is a violent and noisy outburst . . . This discussion is limited to events associated with the build up and release of pressure".

It seems as if one of the difficulties in arriving at agreed definitions lies in the remoteness of the meaning of the source from the modern technical meaning of the term. As a result the popular meanings of explosion are somewhat varied. For the purposes of this book, and taking account of the above, an explosion is defined as follows:-

"An explosion is a sudden release of energy accompanied by the generation of a pressure wave and by a loud noise. It has a potential for injury to human beings and for damage to property. Explosions may be physical or chemical. In the former category are nuclear explosions (which are excluded from detailed discussion in this book) and such events as the rupture of vessels containing pressurised or liquefied gases.

Chemical explosions may take the form of dense explosions in which there is an exothermic chemical reaction in a solid or liquid, or gas phase reactions in which there is a redox reaction as, for example, between the oxygen in air and flammable solids or liquids in finely divided form, or with the molecules of a gas or vapour mixed with the air. Gas phase explosions may be confined or unconfined".

Within the context of this book, the explosions principally studied are those with the potential to cause a major accident. Explosions of such a magnitude are sometimes the end product of a series of events which include explosions of a lesser character and the discussion, therefore, will devote some attention to such lesser events.

10.2. SELECT LIST OF REFERENCE SOURCES.

There have been numerous treatises on the subject of explosives and it is not possible to provide an exhaustive list of them here. The list that follows is that of the general works on the subject which have been consulted in the course of writing this book.

An early study of the phenomenon of explosives and their effects is that of Robinson (1944). By today's standards, Robinson's treatment is elementary but it provides a useful list of accidental explosions. A much more detailed analysis is that of Cook (1966), which is concerned more with the explosives themselves than with their effects. Baker (1973) has the opposite approach, being concerned not with explosives per se but with the production and properties of blast waves in air. Stull (1977) is a monograph which has the aim of teaching the fundamentals of explosions to chemical and other engineers in charge of chemical plant. Bodurtha (1980) is a monograph with a similar aim.

There are also a number of encyclopedia articles, such as Kirk-Othmer (1985).

The subject of accidental explosions involving conventional explosives has given rise to a number of monographs. An early example is Assheton (1930) which lists 140 explosions. Robinson (1944) lists 125, and Healy (1965) analyses 24 incidents in detail. A recent compilation, Biasutti (1985), lists around 1000 such incidents occurring throughout the world, but whether this represents the full total is much open to doubt. It should be noted that there is a considerable degree of overlap of the data in succeeding compilations. Strehlow and Baker (1975) is an analytic evaluation of the subject of accidental explosions. Dealing with specialised areas are the monographs on dust explosions, Palmer (1973) and Field (1982); on vapour cloud explosions, Gugan (1979); and on ammonium nitrate, Clancey (1963).

10.3. A TAXONOMY OF EXPLOSIONS.

10.3.1. LIMITATION OF THE AREA OF DISCUSSION.

Though certain purely mechanical events such as the disintegration of a flywheel are loosely described as "explosions", such events will not be discussed here. The explosions under discussion will be those in which there is a sudden release of energy in the form of pressurised gas.

10.3.2. ENERGY AND POWER.

The concept of a sudden release of energy implies two things:-

(a) The system contains locked up or potential energy.

(b) This energy is capable of being released at such a rate that it can cause injury or damage through blast.

It can be amply demonstrated that the energy released by a given mass of a great number of organic substances on combustion, exceeds the energy released by the detonation of an equal mass of TNT by a factor of about 10. However, it may be noted that this discrepancy is considerably reduced if account be taken also of the mass of oxygen equivalent to the hydrocarbon without which the energy release could not occur. Thus:-

$$\frac{\text{Energy released by combustion of X kg of stoichiometric mixture of typical paraffin and oxygen}}{\text{Energy released by detonation of X kg of TNT}} = \text{ca } 2.25$$

The question of what constitutes a "sudden" release requires quantification. The speed at which detonation spreads through a solid or liquid explosive (referred to below as a "dense" explosive) is roughly the velocity of sound in the material. Kirk-Othmer (1980) Vol. 9, gives a range of 2×10^3 to 9×10^3 m sec^{-1} for this detonation velocity. With gas explosions due to chemical reaction or the physical explosions in which pressurised gas is released, the pressure waves move outward at roughly the velocity of sound in the medium, which can be taken as air. Sonic velocity in air is about 330 m sec^{-1}. On the other hand, flame velocities in ordinary combustion are of the order of 0.5 m sec^{-1}. Taking combustion velocity in flames as equal to 1, suddenness as reciprocal time has the following values.

Gas explosions	=	10^2
Dense explosions	=	10^3

However, this comparison is by no means the whole story. Suddenness is perhaps better expressed by the concept of "power density" or energy released in unit time per unit volume.

The three cases, detonation of dense explosives, gas explosion and flame propagation, are compared in Table 10.1. which is based on a cube of 1 metre sides, where it is assumed that reaction commences across an entire face.

TABLE 10.1. COMPARISON OF ENERGY DENSITIES.

Considered Case	Mass Involved kg	Energy released J/kg	Energy released J/m^3	Velocity of propagation metres/secs.	Time to traverse 1 metre cube secs.	Power Density W/m^3	Likely maximum overpressure developed
TNT (1)	1,650	4.23×10^6	7×10^9	7×10^3	.00014	5×10^{13}	ca 100,000 bars
Methane/air explosion (constant volume)	1.2(2)	-	3.3×10^6	3.33×10^2	.003	1.1×10^9	ca 6 bars(3)
Methane flame (constant pressure)	1.2(2)	-	3.3×10^6	ca 0.45 (3)	2.22	1.48×10^6 (4)	

(1) Data from Kirk-Othmer (1980), Vol. 9.
(2) Mass of stoichiometric mixture.
(3) From Bradley (1978).
(4) Note that this is an order of magnitude greater than that deduced for large pool fires.

10.3.3. PHYSICAL EXPLOSIONS AND CHEMICAL EXPLOSIONS.

The discussion will be concerned with two types of explosion which are often confused in the minds of the public and the mass media.

The first is the purely physical explosion in which the energy released is the internal energy of a compressed gas or a liquefied gas (more strictly, a liquefied vapour). In such explosions, the violence of which depends upon the internal pressure, damage may be caused by a blast wave from the expanding gas or by portions of the disrupting container. As has been referred to in the previous chapter, there have been a number of incidents in which there were physical explosions arising from the catastrophic failures of tank cars. Depending on circumstances, parts of such a vessel may be jet propelled for hundreds of metres. This can happen too on a smaller scale with portable gas cylinders if they fall and knock off a reducing valve. There is ample evidence which demonstrates that the release of pressure on a liquefied gas (which seldom exceeds 40 bars as this is the critical pressure of most common liquefied gases) has been attended by substantial damage from the physical explosion alone.

Chemical explosions may also be of many different kinds and are discussed in detail later. The molecules of some explosives contain nitrogen which separates oxygen atoms from carbon and hydrogen atoms elsewhere in the molecule. In other cases flammable materials are dispersed in air as gas mixtures or in dust or aerosol clouds.

Other circumstances giving rise to explosions lie in "run-away" reactions in which there is an uncontrollable pressure rise in a reaction vessel which, if adequate relief arrangements are not provided, may cause the rupture of the vessel succeeded by a blast wave and flying missiles.

10.3.4. CLASSIFICATION ON THE BASIS OF DENSITY.

10.3.4.1. The two basic types.

There are two basic types of explosions under this heading, dense and rarefied.

10.3.4.2. Dense Explosives.

These include those arising from all solid explosives and the relatively few liquid explosives which include nitroglycerine. Such explosives commonly have densities between 1.5 and 1.80 g/cm^3 (i.e. between 1,500 kg and 1,800 $kg/metre^3$). However, primary explosives containing lead or mercury have much higher densities.

10.3.4.3. Rarefied explosions.

Rarefied explosions arise from mixtures of air and some oxidisable substances in the form of dust, aerosol or vapour. Such mixtures have a density barely distinguishable from air. Rarefied explosions may be divided into two classes, confined and unconfined.

All but one of the dust explosions of which the present author has details have been confined explosions. Bartnecht (1979) shows a photograph of an unconfined explosion of 4 Kg of pigment dispersed in air but the reference does not, in its text, discuss unconfined dust explosions. It may be assumed, from lack of evidence to the contrary, that unconfined dust explosions are of little technological significance.

There is little evidence, with one major exception, of serious explosions associated with aerosols, though comparatively small aerosol explosions are not uncommon in compressed air systems where the aerosol is a mist of oil droplets. The one exception to this is the Pernis explosions of 20/1/68 where, according to Pernis (1968), a severe explosion at an oil refinery was due to the ignition of an aerosol/vapour cloud containing about 50 tonnes of hydrocarbon.

Gas or vapour explosions may be either confined, which the vast majority of them are, or unconfined when the quantity of gas/vapour is likely to be of the order of at least three tonnes.

10.3.5. SUSCEPTIBILITY TO SHOCK.

Dense explosives may be classified according to their susceptibility to shock as:-

(a) Propellants. These are very insensitive to shock and are relatively slow burning. Nitrocellulose is a typical example of this class.

(b) Secondary Explosives. (Also known as high explosives). TNT is a typical member of this class and is often used as a reference substance.

(c) Primary Explosives. Though not so powerful as secondary explosives, they are easily detonated by mechanical shock. Lead azide and mercury fulminate are the best known examples of this class.

The firing and eventual explosion at a target of a high explosive shell gives examples of the use of all three classes. When in a gun there are four explosive components present. There is a cap consisting of a relatively small quantity of primary explosive which is detonated by mechanical shock, thus igniting the propellant which propels the shell. The propellant has to generate a high pressure, circa 10^3 bars, but without shattering the gun. The shell consists of a casing containing high explosive and fuse which on impact detonates a relatively small quantity of primary explosive which, in its turn, detonates the high explosive charge.

For some military applications, a high degree of brisance or shattering power is required. Propellants are low brisance explosives, high explosives are high brisance explosives. Brisance is a function of rate of release of energy, i.e. power.

10.3.6. CLASSIFICATION BY CHEMICAL NATURE.

10.3.6.1. Decomposition reactions.

The simplest case of an explosion is that involving simple decomposition which gives rise to gaseous products. One example is that of hydrogen peroxide which decomposes with considerable heat of reaction to give steam and oxygen.

$$2H_2O_2 \rightarrow H_2O + O_2 \quad - \quad 23.44 \text{ kcal/mol}$$

As a domestic product, hydrogen peroxide is sold as "10 volume" or 3% H_2O_2 and presents little danger though it yields 10 times its volume of oxygen. Matters are very different with "high test" hydrogen peroxide which is 90% or more H_2O_2. The decomposition of high test H_2O_2 is catalysed by a variety of materials and it is used as a rocket fuel or in a gas turbine to pump fuel to the main motors.

A second example is that of lead azide which is readily decomposed by friction or shock.

$$PbN_6 \rightarrow Pb + 3N_2 \quad - \quad 115.5 \text{ kcal/mol}$$

Lead azide is a primary explosive or detonator. The great majority of explosives, after detonation, yield a variety of products. For example high density TNT, (1.59 g/ml) on being detonated, yields, according to Cook (1966), Page 286, the following products of decomposition:-

TABLE 10.2. DECOMPOSITION PRODUCTS OF TNT.

Product	CO	CO_2	H_2O	N_2	NH_3	CH_3OH	HCN	C
mol / kg	0.6	10.0	0.8	6.0	0.4	4.7	1	14.9

It will be noted that TNT is "oxygen deficient" and therefore yields carbon as one of its principal products. TNT explosions are thus very smoky. Most high explosives (nitroglycerine is an exception) are oxygen deficient, i.e. they have insufficient oxygen atoms in the molecule to yield 1 mol of CO_2 for every atom of carbon present in the molecule and 1 mol of H_2O for every two atoms of hydrogen. A number of commercially important substances, not classified as explosives, behave in this way. They are listed in European Community Directive (1982) and, in many cases, they are organic peroxides.

It is possible, using the principles of thermochemistry, to demonstrate whether or not a given compound has the potential for exploding. These principles are conveniently set out in Stull (1977). If, in a given reaction, the sum of the heats of formation of the products has a lower value than the heat of formation of the decomposing compound, then there is a potential for explosion.

Thus for A $\rightarrow$ B + C + D

then $\Delta f A$ must be greater than the sum $\Delta f B + \Delta f C + \Delta f D$,

where Δf = heat of formation.

Δf, in the convention used here, has a positive value for compounds which absorb heat during formation (known as an "endothermal" reaction) and, conversely a negative value for compounds which give out heat during formation (known as an "exothermal" reaction).

e.g. $$2C + H_2 \rightarrow C_2 \equiv H_2 \quad + 54.2 \text{ kcal/mol}$$

Acetylene

$$C + O_2 \rightarrow CO_2 \quad - 94 \text{ kcal/mol}$$

Δf for the elements in their normal molecular state, for example, O_2, N_2, H_2, (but not O_3) is taken as zero.

It should be made clear that thermo chemistry discloses only the potential for explosion, it is the reaction rate which determines power, i.e. the violence of the effect. The reaction between candle wax and oxygen is highly exothermal but does not give rise normally to explosion.

10.3.6.2. Redox reactions.

Redox reactions, in which air or oxygen reacts with a reducing agent, are very common and form the basis of all combustion reactions. Where the reducing agent is a massive solid or a liquid, combustion reactions are insufficiently rapid to create an explosion. If, however, a solid is finely divided or a liquid is in the form of droplets, rapid pressure rise is possible. This, in general, can give pressure rises up to about 8 bars under conditions of total confinement.

Flames have been discussed in Chapter 8. Steady flames give rise to only minor pressure effects; some degree of pressure rise is inescapable to provide the driving force whereby the combustion products flow from the flame into the surrounding atmosphere.

10.3.6.3. Polymerisation, isomerisation and condensation reactions.

A number of substances can polymerise more or less spontaneously and it is usual for polymerisation reactions to be exothermal. If the monomer is volatile, which it often is, a stage is reached at which there can be a dangerous rise in pressure. Sometimes polymerisation will proceed only at elevated temperatures, but with some substances such as ethylene oxide, polymerisation may be initiated at room temperature particularly if they become contaminated with substances which catalyse the polymerisation. Ethylene oxide may also isomerise to acetaldehyde exothermally:-

- 27.10 kcal/mol

Stull (1977) discusses this reaction and the subsequent reaction in which acetaldehyde decomposes to yield two mol of permanent gases.

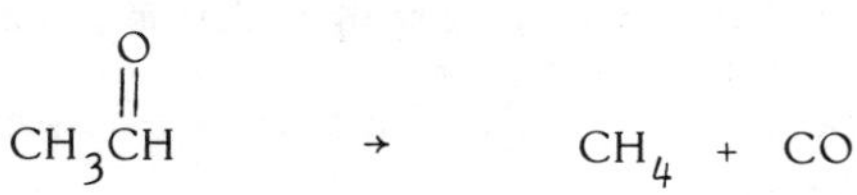

- 4.55 kcal/mol

Condensation reactions are very common. They feature prominently in the paint, varnish and resin industries where they are carried out batchwise in stirred reaction vessels with coils for heating or cooling. There have been many examples of "run-away" reactions which are rooted in the problem that the rate of heat transfer in such vessels is a linear function of the temperature difference between the reaction mass and the coolant, whereas the rate of reaction is an exponential function of the reactant temperature. However, the rate of production of heat is a function also of the concentration of the reactants which, as the reaction proceeds, become less and less, to some extent off-setting the effect of increasing reaction rate.

10.3.6.4. Reactions of mixtures.

The most obvious example of a mixture which reacts to give an explosion is the mixture known, originally, as "black powder" and, later, as "gunpowder". A typical composition for gun powder is:-

Potassium Nitrate (nitre) 75%, Charcoal 15%, Sulphur 10% [Read (1942)]

In this mixture the oxygen rich "nitre" oxides the carbon and sulphur.

Similar effects attend the detonation of mixtures of other oxidisers such as perchlorates or chlorates with organic compounds. Some organic materials, if soaked in liquid oxygen (LOX) and suitably detonated, give rise to explosion. Wood soaked in LOX has been used as an explosion in mining.

It is of interest to note that the reaction:-

$$2Al + Fe_2O_3 \rightarrow Al_2O_3 + 2Fe$$

though highly exothermal, does not, of itself, give rise to explosion as it yields no gaseous products.

10.4. THE TAXONOMY AND MAJOR CHEMICAL HAZARDS.

By no means all the explosions listed have significance as major chemical hazards. The manufacture of conventional explosives will not be discussed here in detail, as their manufacture entails highly specialised technologies and, in the UK, is governed by special legislation. The explosives industry is discussed in Shreve and Brink (1977). However, it is relevant to consider the **behaviour** of conventional explosives, as they constitute a model for the behaviour of unstable substances not classified as explosives and also ammonium nitrate which represents a special case as, though it has been, and still is, used as an explosive, it finds its principal use as a fertiliser. It is, in the pure state, very difficult to detonate.

Redox reactions are important as the source of major chemical hazards, in the form of dust explosions and gas explosions, whether confined or in the open air. They will be further discussed with especial reference to vapour cloud explosions.

10.5. THE EFFECTS OF DENSE EXPLOSIVES.

10.5.1. THE DETONATION OF A DENSE EXPLOSIVE.

The events referred to are explosions which occur in air, at or near ground level. Explosions which occur in water or at high altitudes may differ considerably in their effects from those described here.

The sub-section is concerned with primary and secondary explosions but not with propellants. In the case of primary explosives detonation may be initiated by impact or friction, in the case of secondary explosives detonation will typically be initiated by the detonation of a primary explosive in contact with the secondary explosive.

Though there is no practical method of encasing a dense explosive so as to contain it in its initial volume, a stout casing will temporarily contain the initial pressure and transmit pressure to the detonating material and thus enhance detonation. The shock wave from the detonation then spreads through the mass. There seems to be a critical charge diameter below which detonation falls away, this diameter is 2 to 7mm depending on the explosive and its density. Above this diameter the detonation velocity, depending on the explosive, may be 5 - 9 km sec^{-1} [Stull (1977); Baker et al (1983)]. The gases initially produced are so compressed that ideal gas laws do not apply and an equation of state has to be selected. Explosions are so rapid that the products are not in chemical equilibrium.

10.5.2. EFFECTS ON SURROUNDINGS.

The explosion initially produces a fire ball at a temperature exceeding 2000°C and an initial pressure which may theoretically approach 20 - 40 GPa or 200,000 to 400,000 bars [Baker et al (1983)].

From the site of the explosion the products expand, cooling as they go. The initial velocity of this outward expansion (described sometimes as a "wind" or "hurricane"; both descriptions are grossly inadequate) is of the order of several km sec^{-1}. But this effect is relatively localised; Robinson (1944) suggests about 8 metres for 100 kg of high explosive. The pressure of the expanding gases produces a shock wave which, though it starts out at the same velocity as the expanding gas front, soom decelerates to the sonic velocity in air, i.e. ca 1/3 km sec^{-1}. Though the "wind" is short range, the shock wave travels outward far beyond the range of the "wind", degrading eventually into a sound wave. Man-made explosions have been heard tens of kilometres away; the explosion of the volcano Krakatoa in 1883 was heard as a "boom" 5000 kilometres away and about four hours after the event [Houwink (1976)]. One effect of this wind is to scour out a crater in the case of explosives resting on, or near, the surface of the ground.

Robinson (1944) gives a figure which relates crater volume and diameter to the mass of an explosive charge. It may be expected that there would be a considerable degree of scatter on account of the variation in the nature of the ground. Robinson's figures suggest the following mean approximate relationships.

1 tonne explosives ≡ 60 $metres^3$ of crater volume

and that $D = 3\left(\frac{V}{h}\right)^{0.5}$

where D = diameter (metres)

V = volume ($metres^3$)

h = depth (metres)

Thus for 1 tonne:-

$$D = 3 \times \frac{7.74}{h}$$

Robinson attributes to Olsen (no reference given) the observation that $h = \frac{D}{2}$.

Thus
$$D = \frac{3V^{0.5}}{0.5D}$$
$$0.5\ D^2 = 3\ V^{0.5}$$
$$D^2 = 6\ V^{0.5}$$
$$D = 6\ V^{0.25}$$

Thus for 1 tonne
$$D = 6\ 60^{0.25}$$
$$= 16.7 \text{ metres.}$$

M, the mass of explosive, in the above equations, strictly only relates to a standard explosive of a standard "brisance". The concept of brisance has in recent times been quantified by the concept of detonation pressure which is usually expressed in gigapascals or GPa (1 Gpa = 10^4 bars). Detonation pressure is theoretically calculated, as it is not practicable to measure it. [See Cook (1966) P.32; Kirk-Othmer (1980) Vol.9; and Baker et al (1983) P.110].

The data in Cook (1966), (Fig. 2.8 in that reference), suggests that the hole volume in a steel plate (a form of crater) may be expressed by the equation:-

$$V = K\ P^{0.46}$$

10.5.3. THE FORM OF A BLAST WAVE.

The subject of blast waves is treated in Baker (1973) in great depth and what follows is, by comparison, only the merest outline of the subject.

Figure 10.1. shows the form of an ideal blast wave. From this it will be seen at once that there is a positive phase (where there is an overpressure) succeeded by a negative phase (where there is an under-pressure). There is a general view that the algebraic sum of the integrals of the two phases is zero. Since underpressure is only, and can only be, a fraction of an atmosphere it follows that if, for example, overpressure exceeds 1 atmosphere, for the integrals to be algebraically equal the duration of the negative phase must exceed the duration of the positive phase.

i.e. $I_s^+ + I_s^- = 0$

where I_s^+ and I_s^- are known as the positive and negative impulses, and

$$I_s^+ = \int_{t_a}^{t_a + T^+} [P_o - P(t)]\ dt$$

$$I_s^- = \int_{t_a}^{t_a + T^+ + T^-} [P(t) - P_o]\ dt$$

Symbols are as Figure 10.1.

Various authorities have attempted to derive functions which relate amplitude to time. These are discussed in Baker (1973) but do not appear to be relevant here. Baker points out that most of the work has been concerned with the positive phase and that most authorities consider the negative phase to be unimportant. In practice, it appears that after the passage of the negative phase, there may be repeated positive and negative phases of lower and rapidly declining amplitude.

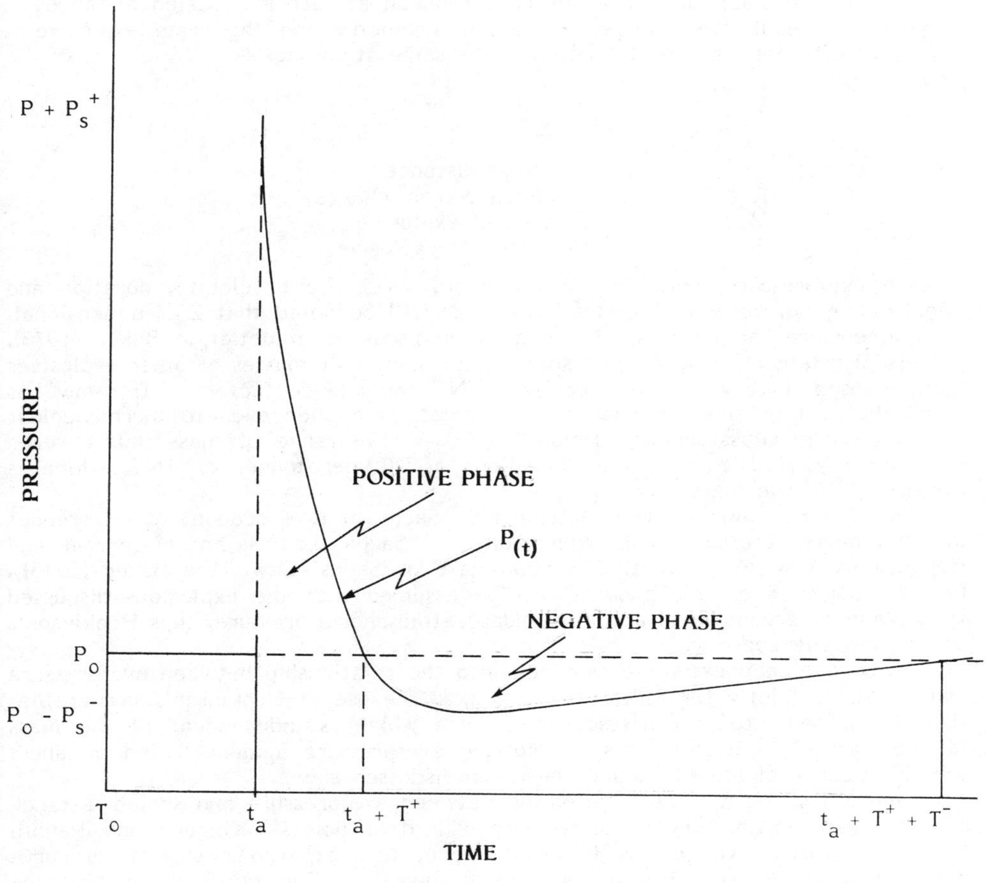

Fig. 10.1. THE FORM OF A BLAST WAVE.

10.5.4. OVERPRESSURE AND ITS MEASUREMENT.

Overpressure is one of the key parameters in the analysis of explosions and is conveniently handled as expressed "side-on, peak overpressure". "Side-on" signifies that the pressure is measured with the face of the transducer parallel with the direction of propagation of the blast wave. (The use of the term "static" to describe this variable has now been discontinued).

If the overpressure be measured by a transducer at 90° to the direction of propagation of the blast wave, then the physical property measured is known as the "reflected" overpressure. Reflected overpressures may have values ranging from 2 to 8 times the side-on overpressure. They will be treated in more detail in a later chapter when the subject of the resistance of buildings to blast is discussed. Where the word "overpressure" is used below on its own, the meaning of side-on, peak overpressure is implied.

The relationship between mass of explosive, overpressure and distance is greatly simplified by the use of Hopkinson's scaling law [see Hopkinson (1915)]. The law may be stated as:-

> "Self-similar blast (shock) waves are produced at identical scaled distances when two explosive charges of similar geometry and the same explosive but of different size are detonated in the same atmosphere".

Thus

$$Z = \frac{R}{M^{1/3}} \quad \text{or} \quad \frac{R}{E^{1/3}}$$

where Z = scaled distance
R = distance from charge
M = mass of explosive
E = energy of explosive

Side-on overpressure, reflected overpressure, shock front velocity, duration and impulse may all be scaled in this way. It will be noted that Z is dimensional. The dimensional analysis of blast waves is discussed in detail in Baker (1973).

As M relates to a mass of standard explosive, TNT masses of other explosives may be used if they are adjusted by a TNT equivalence factor. This law has stood the test of time and has been successfully applied even to thermonuclear explosions [see Glasstone and Dolan (1980)]. The range of mass thus covered is from, say, 100 g to the equivalent of 20 megatonnes of TNT, which is a range of 10^{11} in mass.

Hopkinson's Law has been extended by Sachs to take account of differences in atmospheric pressure and temperature. Sachs' Law is more general and Hopkinson's Law is, in fact, a special case of Sachs' Law [see Baker (1973)]. For the purposes of this book it will be assumed that the explosions discussed take place at ground level and at standard atmospheric pressure, thus Hopkinson's scaling law will apply.

There has been extensive research into the relationship between overpressure and distance, chiefly for military purposes. The use of Hopkinson's law enables this relationship to be expressed in a form which is independent of the mass of the charge. It does this by plotting overpressure against scaled distance; the significance of scaled distance has been discussed above.

Figure 10.2. shows the relationship between overpressure and scaled distance and is based on a curve incorporating 273 data points [Kingery and Pannill (1962)]. This curve relates to ground level test explosions, whereas a curve given in Baker et al (1983) is for free air bursts. This curve shows differing values of overpressure at a given scaled distance than does Kingery and Pannill.

It must be strongly stressed that there is a great deal of variance in experimental results and that the graph represents "best fit" relationships.

10.5.5. DURATION OF IMPULSE.

From the equation given above it will be clear that impulse is a function of both overpressure and time. However when damage potential is considered, it may be argued that not only overpressure but also duration is significant. Thus the overpressure required to produce a given level of damage may be different in cases where the duration is different. According to Hopkinson's scaling law, for any given explosion, the duration, and hence the impulse, is fixed for a given level of overpressure. In principle this may not be true between one explosive and another. However, for dense explosives, there seems no great variation.

This may, however, not be true when comparisons are made between dense explosives on the one hand and either nuclear or vapour cloud explosions on the other. It seems from Glasstone and Dolan (1980), Page 119, that the duration

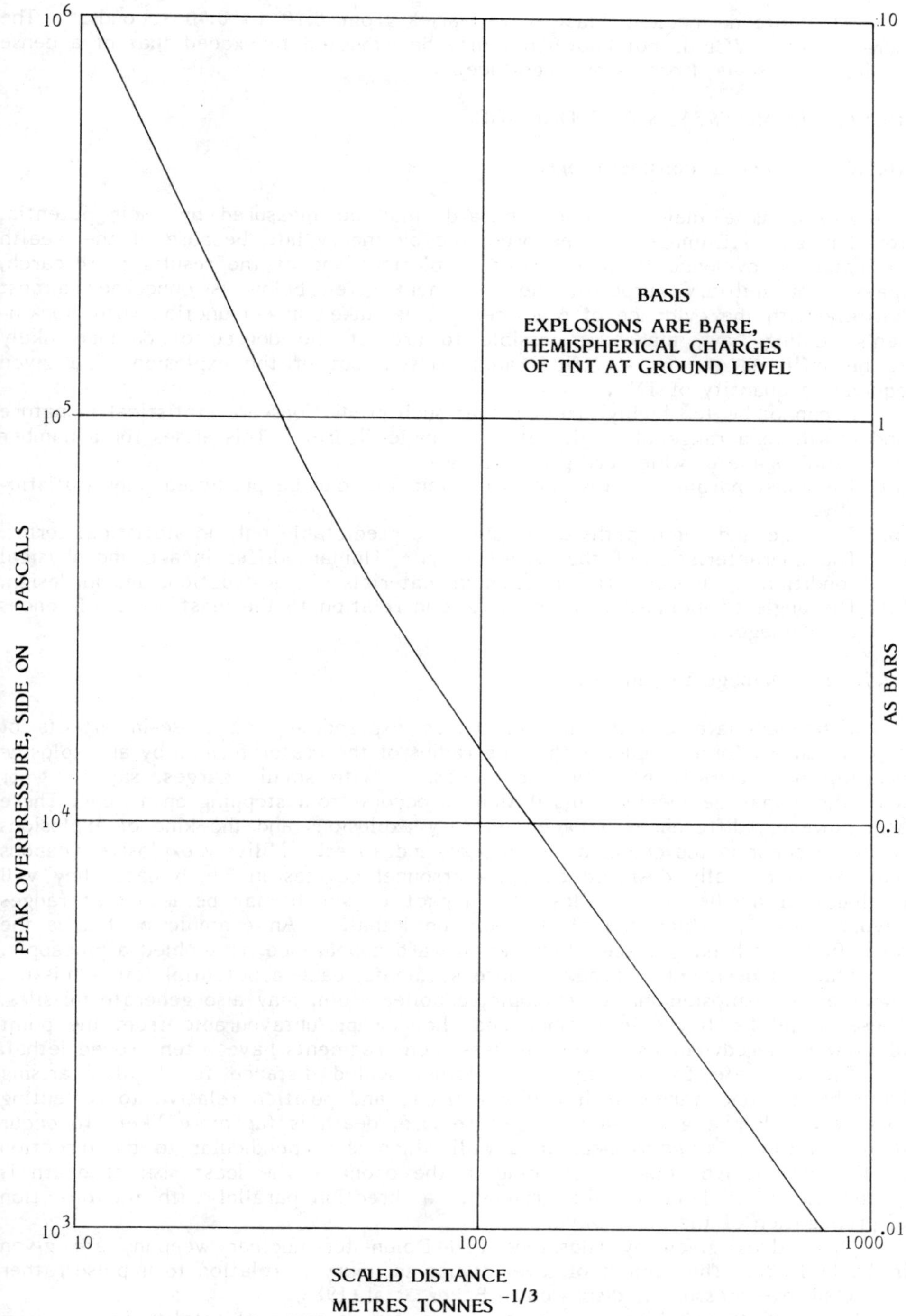

Fig. 10.2. PEAK OVERPRESSURE VERSUS SCALED DISTANCE.

of a 1 kilotonne nuclear blast wave varies from 0.10 to 0.40 seconds. The duration of a VCE is not known but may be expected to exceed that of a dense explosion by several orders of magnitude.

10.5.6. BLAST WAVES AND DAMAGE.

10.5.6.1. General considerations.

Explosions, as a major chemical hazard, may be measured by their potential for damage. Damage may be predicted by theory but, because of the wealth of historical evidence from accidental explosions and of the results of research, mainly for military purposes, the treatment given below is concerned almost entirely with the evidence of practice. This, taken in conjunction with Hopkinson's Scaling Law, makes it possible to predict the degree of damage likely to be inflicted at any specified radius arising out of the explosion of a given equivalent quantity of TNT.

It cannot be too highly stressed that such predictions are statistical in nature and relate to a range of radii, not to a single figure. This arises for a number of reasons, some of which are given below:-

(a) The blast parameters (overpressure, impulse) can be predicted only statistically.
(b) The size and flight paths of missiles are predictable only in statistical terms.
(c) The characteristics of the target vary. Humans differ in age and physical condition. Houses differ in size, in materials of construction, and in design.
(d) The angle of inclination of the target in relation to the blast wave influences the damage.

10.5.6.2. Damage to humans.

As with inanimate objects, human beings exposed to the close-in effects of explosions, as, for example, within the radius of the crater formed by an explosive charge, may literally be blown to pieces. With small charges, say 1 kg or less, there may be localised mutilation as occurs from stepping on a mine. There are, however, differences between military explosions and the kind of explosions liable to occur in the chemical and process industries. Military explosive weapons may be specifically designed as anti-personnel devices in which case they will discharge a number of missiles, the impact of which may be lethal at ranges beyond those at which the blast would be lethal. An example of this is the Mills Bomb or hand grenade which, in outward appearance, resembled a pineapple, and which is designed to break up into segments, each a potential lethal missile. However an explosion in, for example, a boiler drum, may also generate missiles. These would be fewer in number and their shape unfavourable from the point of view of aerodynamics. Nevertheless such fragments have often proved lethal.

The estimate for the maximum lethal scaled distance for fatality arising from blast alone varies with bodily attitude and position relative to reflecting surfaces. For a given side-on overpressure, death is far more likely to occur if the victim is standing near to a wall which is perpendicular to the direction of the blast wave, than if standing in the open. The least risk of death is if the victim is lying on the ground in a direction parallel with the direction of propagation of the blast wave.

The values given by Glasstone and Dolan for nuclear weapons are given in Table 10.3. The subject of level of human injury in relation to impulse rather than peak overpressure is discussed in Baker et al (1983).

However, though blast may be the predominant cause of fatality for a victim caught in the open, other factors operate in the built environment. The first of these have been noted above, namely that overpressure may be considerably

TABLE 10.3. OVERPRESSURE AND HUMAN INJURY.

100% Lethality	5	-	8	bars	
50% Lethality	3.5	-	5	bars	
Threshold Lethality	2	-	3	bars	
Severe Lung Damage	1.33	-	2	bars	
50% Eardrum rupture	2	-	2 1/3	bars	(under 20 years of age)
50% Eardrum rupture	1	-	1 1/3	bars	(over 20 years of age)

[Note that Glasstone & Dolan's figures have been rounded off].

enhanced by reflections from walls or vessels. Fatality may result from the collapse of buildings. Such collapses can lead to asphyxiation, crushing, or burns, if the building catches fire. In the built environment, secondary missiles may arise such as bricks, roofing tiles or, perhaps most serious of all, flying glass. Or the victim may become a missile and be hurled against a structure.

10.5.6.3. Damage to buildings.

The subject is sufficiently complex as to merit a Section to itself. It will be considered in Chapter 21.

Chapter 11

CASE STUDIES OF INCIDENTS INVOLVING DENSE EXPLOSIONS

11.1. OPPAU AMMONIUM NITRATE EXPLOSION.

At 7.30 on 21/9/21, there was an explosion at the Oppau, near Ludwigshafen, works of the Badische Anilin und Soda Fabrik (BASF). Accounts of this explosion vary greatly and it has been described by many.

The explosion killed 561 people including 4 who were killed in Mannheim 7 kilometres away. There were 1,500 injured and in the town of Oppau 75% of the houses were destroyed. A total of around 1,000 houses in all were destroyed.

There seems general agreement that 4,500 tonnes of a mixture of ammonium nitrate with ammonium sulphate blew up and that the material was detonated by explosive charges which were customarily used to break up the caked material. Clancey (1963) claims that the material was a co-crystallised mixture of ammonium nitrate with ammonium sulphate in the molecular proportions of 2 : 1, which would be about 58% ammonium nitrate by weight. Biasutti (1985) claims that the mixture which blew up was 50% by weight of nitrate and sulphate. The same reference claimed that it was proved impossible to detonate any mixture containing more than 40% sulphate, but that the mixture was not homogeneous and may have contained up to 70% nitrate in some places. If, however, the mean content was 50% nitrate, some places must have been leaner than this, and it is difficult therefore to see why the whole heap went up. Both Clancey and Biasutti are agreed that the decision to use explosives was only taken after exhaustive tests and that many thousands of actual blastings had been conducted.

There are varying accounts of the size of the crater. BASF (1958) gives a depth of 10 metres, breadth 75 metres and length 115 metres. Within a radius of 250-300 metres, all buildings were totally destroyed or so badly damaged as to require demolition. The scene after the explosion is shown in Figures 11.1 and 11.2.

11.2. TEXAS CITY AMMONIUM NITRATE EXPLOSION.

In this disaster, which occurred on April 16th and 17th, 1947, a ship, the Grandcamp, was carrying some 2,300 tonnes of ammonium nitrate in its hold No. 4. According to Nash (1976), a fire started among the ammonium nitrate which was being carried in paper sacks. The ship's master decided not to extinguish it with water to avoid ruining the cargo and tried to put it out by battening down the hatches and admitting steam. [This would have made the situation worse as it intensified the anaerobic fire by heating up the ammonium nitrate]. The fire started at 8.00 a.m. and at 9.12 the ship exploded killing, among others,

Fig. 11.1. SCENE AFTER EXPLOSION AT OPPAU.

Fig. 11.2. CLOSE-UP VIEW OF OPPAU CRATER.

over 200 people who had crowded to the dockside to watch the fire. According to Nash, those killed included the captain of the ship and 32 crew, and 4 people in two aircraft who were circling overhead.

Severe damage and widespread fires broke out in the nearby Monsanto complex and in the tankage of a number of oil companies. Pieces of wreckage travelled up to thousands of metres, a 1 tonne piece of propellor shaft travelled for 4 kilometres.

At 1.10 p.m. on April 17th a second ship, the Highflyer, which had caught fire the previous day as a consequence of the explosion on the Grandcamp, also exploded. This ship was also carrying ammonium nitrate and, according to some reports, sulphur.

The total death toll has been estimated at 552 with 3,000 injured and 200 missing. Damage was in excess of $100 million (1947 prices).

There is a photograph of the fire in Nash (1976) and the scene after the explosion is shown in Brasie and Simpson (1968).

The Texas City disaster shows that, in some circumstances, ammonium nitrate anaerobic fires may escalate to a detonation even when barely confined, as the hatches of the hold must already have been destroyed or displaced by the fire. The quantity of ammonium nitrate which actually exploded cannot be stated with certainty. Some of the nitrate in both ships must have "burned" before the explosion occurred. The order of magnitude of both explosions might have been about 2,000 tonnes of ammonium nitrate each. This would have been equivalent to about 1,000 tonnes of TNT in each case. The explosions led to pronounced "domino" effects by starting fires in nearby chemical plant oil storages and dry goods warehousing.

It seems clear that the captain of the Grandcamp had had no training in how to deal with an anaerobic fire and also that the harbour was inadequately equipped to deal with a major fire. The death toll was undoubtedly considerably aggravated by the phenomenon of "negative evacuation", that is people flocked to the scene to witness the fire. Had the authorities acted (or been able to act) to evacuate the area, the death toll could have been far less. Texas City is not the only example of people being killed as spectators at the scene of a disaster.

Chapter 12

SPECIAL PROBLEMS OF RAREFIED EXPLOSIONS

12.1. INTRODUCTION.

Rarefied explosions will be discussed under two main headings, dust explosions and vapour cloud explosions. Whereas virtually all dust/air explosions have taken place in confined spaces, vapour/air explosions may occur both in confined spaces and in the open air. In the latter case, they will be described as "vapour cloud explosions" or VCEs.

The Chapter will, accordingly, have three parts, dust explosions, confined vapour (which includes gas) explosions, and vapour cloud explosions.

12.2. DUST EXPLOSIONS.

12.2.1. INTRODUCTION.

The subject of dust explosions has led to extensive study and has been the subject of many papers and of several monographs. The topic may be considered as having two quite distinct aspects, dust explosions in galleries and dust explosions in equipment and in buildings.

The former are distinguished by shock waves which travel down long galleries which may be several kilometres in length. The author regards the study of dust explosions in coal mines as lying outside the main theme of the book, but will comment on them briefly below.

12.2.2. DUST EXPLOSIONS IN COAL MINES.

The history of the growing realisation of the role played by coal dust in mining disasters is charted in Bryan (1975), who points out that the first time that it was appreciated in the United Kingdom that coal dust had played a role in a mine explosion, was in 1803 in a report on the Wallsend, Northumberland, Colliery explosion of that year. However, realisation of the role played by coal dust was incomplete. Later in the century, following the Haswell, Co. Durham, disaster (1844), in which 95 were killed, the distinguished scientist Michael Faraday was invited to conduct an investigation. Though the Report on this disaster recognised that it could not entirely be due to firedamp (methane), it was not till more than fifty years later that Professor W. Galloway (Professor of Mining at Cardiff) recognised that it was coal dust which was chiefly responsible for the more serious explosions.

The mechanism which is now recognised is that an initial, and relatively minor, explosion of methane can create such turbulence as to create a dust cloud from the coal dust in the galleries and that it is this dust cloud which, on ignition, sends shock waves which disturb more dust to contribute to what is, eventually, a devastating explosion. Galloway (1898) estimated that no fewer than 645 explosions occurred between 1835 and 1850 in the UK alone, many of them involving dust explosions. Cybulski (1975) has estimated that, world-wide, 135 disasters in which 50 or more were killed, occurred between 1900 and 1951; there was a mean death toll of 151 per incident. Between 1931 and 1955, according to the same source, an average of 117 persons per year were killed in US pit explosions.

The strategy for preventing dust explosions in coal mines, or at least for limiting their severity, has the following elements:-

(a) The prevention of initiating explosions by practising methane drainage and by the rigorous elimination of ignition sources.
(b) As far as practicable, limiting the amount of dust lying in galleries.
(c) Keeping the coal dust wet.
(d) Providing inert powder as a supressant.

This inert powder is a non-siliceous dust and is usually limestone. Today, rather than simply mixing it with the coal dust, as was the former practice, it is more usual to charge the dust into trays set in the gallery roof. These tilt when disturbed by the precursor of an explosion and discharge a curtain of the inert dust which mixes with the air borne coal dust. The stone absorbs heat of combustion, and thus cools the flame. This, in its turn, reduces the flame velocity. It is possible also that the limestone undergoes an endothermal decomposition which further cools the gas.

Figure 12.1. is a chart of the numbers killed each year in UK mining disasters, in which 20 or more persons were killed. It will be seen from this that such serious accidents in UK mines are largely things of the past. However, it should be borne in mind that coal production has considerably declined and, due to higher productivity, manpower has declined even more rapidly over the period in question. However, the proportion of serious accidents involving dust explosions has declined even more markedly than either of these.

12.2.3. DUST EXPLOSIONS IN EQUIPMENT AND IN BUILDINGS.

12.2.3.1. Introduction.

The subject matter of this section lies closer to the area covered by the title of the book than does the previous section, but nevertheless only a portion of dust explosions in equipment and in buildings can be regarded as occurring in the chemical and process industries. For example, a significant number of the most severe explosions have occurred in grain elevators or in flour mills. The basic causes and the mechanism of dust explosions in industry have a generic similarity and the lessons to be learned are the same whether the explosions occur in handling grain or in, say, handling dyestuffs in powder form.

12.2.3.2. Literature survey and historical review.

The first recorded dust explosion appears to be that which occurred in a flour warehouse in Turin, Italy on December 14th, 1785 [Morozzo (1785)]. It appears from the reference to it in Field (1982), that Morozzo recognised what are today seen as characteristic features of dust explosions, though the general recognition of these characteristics did not occur for over a century. There seems to be only one major historical study of dust explosions, National Fire Protection Association (1957). This surveys the field of dust explosions in the USA and Canada for the years 1860 to 1956 and gives individual accounts of 75 explosions which occurred between 1932 and 1956.

The Report gives a total of 1,120 explosions of which 391 involved fatality.

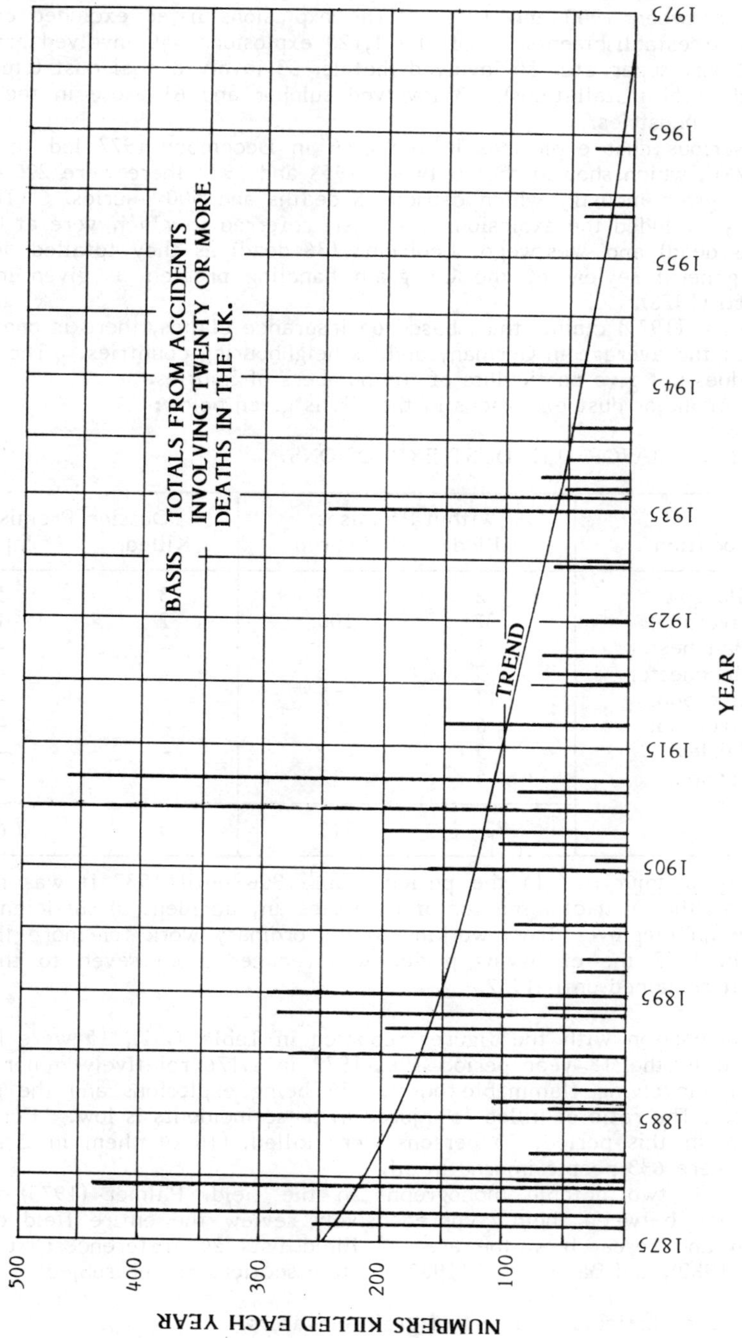

Fig. 12.1. NUMBERS KILLED EACH YEAR IN UK MINING DISASTERS.

There was a total of 676 deaths and of 1,770 injuries, the overwhelming majority occurring between 1900 and 1956. The explosions listed excluded coal mines and defence establishments. Of the 1,120 explosions 540 involved grain, other cereals, flour, sugar etc, 80 involved metals, 63 involved coal dust (chiefly from pulverised fuel installations), 33 involved sulphur and 61 arose in the chemical and process industries.

Two serious dust explosions in the USA in December 1977 led to a report, [Hall (1978)], which showed that between 1958 and 1977 there were 220 explosions in the US grain industry which caused 48 deaths and 500 injuries. (This report presumably excluded the explosions previously referred to which were at Galveston, Texas (18 dead) and Westwego, Louisiana (36 dead) as they totalled 54 deaths). A more general review of the US grain handling problem is given in Verkade and Chiotti (1978).

Bartnecht (1979) claims that, based on insurance claims, there is one explosion per day on the average in Germany and its neighbouring countries. The reference however does not give particulars of the numbers of injuries.

A list of major dust explosions in the UK is given below:-

TABLE 12.1. MAJOR U K DUST EXPLOSIONS.

Year	Location	Within Premises Killed	Within Premises Injured*	Outside Premises Killed	Outside Premises Injured*
1911	Glasgow	2	3	3	5
1911	Liverpool	37	100	2	1
1911	Manchester	3	5	-	-
1913	Manchester	3	5	-	-
1930	Liverpool	11	32	-	-
1941	Liverpool	6	40	-	-
1964	Paisley	5	2	-	-
1965	London	5	32	-	-
Totals		72	219	5	6

*Definition of injury: In the period from 1906 until 1937 it was mandatory to report to the District Inspector of Factories any accident of sufficient severity to disable an employee from working at his ordinary work for more than seven days. In 1937 the qualifying period was reduced from seven to three days, at which it remained until 1982.

For comparison with the figures reported in Table 12.1., 15 were killed and 599 injured in the 18-year period 1962-1979 in 1,176 relatively minor incidents reported as involving flammable dusts, 474 being explosions and the remainder were fires. The ratio of killed to injured in these incidents is low. For example, in the UK in this period, 25 persons were killed, (10 of them in 2 incidents), and there were 633 people injured in all.

There are two notable monographs in the field, Palmer (1973) and Field (1982), which between them comprehensively review the entire field of causes, prevention and research in the area. Field lists 247 references; Stull (1977), Bodurtha (1980), and Baker et al (1983) contain sections on the subject.

12.2.3.3. The particular problem of grain elevators.

Grain elevators seem to present the most serious single dust explosion hazard in the United States and Canada. However, in discussing the problem, however briefly, it is necessary to clarify the meaning of the term "elevator". This has two meanings in American English. It can mean a piece of machinery for lifting powdered solids as in the United Kingdom, but its more important meaning in the USA is:-

"A building used for storing or discharging grain"

[Merriam-Webster (1958)]

Such buildings usually take the form of a number of silos, fed by equipment which lifts up the grain from rail or road cars or from ships. The grain is then fed to the silos by conveyor belt and there is also a conveyor system for extracting the grain from a silo either to be transferred by rail, road or sea or, sometimes to transfer it to another silo. Particularly when freshly harvested, the grain readily generates dust by the attrition which occurs when it is handled.

Such elevators are far larger than anything of a similar nature to be found in the UK. Guinness Book of Records (1971), states that the largest single unit elevator is that of C.G.F. Grain Company at Wichita, Kansas. It consists of 246 silos arranged in three parallel rows occupying a site about 825 metres long by 30 metres wide. Each silo is about 35 metres high by 9 metres internal diameter. The total storage volume is about 700,000 m^3. The machinery for lifting the grain may have, for a single unit, a capability of handling 2000 tonnes per hour [Science and Invention Encyclopedia (1977), Page 892].

In contrast, the corresponding installations in the UK are far smaller. No.2 Granary, Manchester Ship Canal, for example, has a capacity of about 40,000 m^3 [Encyclopedia Britannica (1951), Vol.10, Page 624]. It is noted in this reference that the grain has largely been dedusted in the handling before it reaches the UK. The length of the passages and the large volume of North American elevators all tend to produce very severe effects should an explosion be induced.

12.2.3.4. The causes and circumstances of dust explosions.

In broad terms, a given volume of combustible dust mixed with air is capable of releasing a quantity of energy comparable with that released by the same volume of flammable vapour/air mixture. However, the power (energy released per unit time) may be less.

In a given volume of dust laden air the limiting factor is not the mass of the dust, it is the mass of the oxygen present. Provided that the mass of the dust is stoichiometrically equivalent to this mass of oxygen, or exceeds it, the energy released by organic dust clouds will be roughly equal to that released by organic vapour clouds. However, more or less any solid which will burn, if sufficiently finely divided, can give rise to a dust explosion. Thus metal dusts such as those of aluminium, or finely divided elemental sulphur, can give rise to dust explosions.

The power of the explosion (the rate of energy release) is related to an important parameter, the rate of pressure rise. Unlike the case of vapour cloud explosions the combustion of dusts takes place at a solid/gas boundary and, other things being equal, the more finely divided the solid, the more rapidly it will be consumed.

Generally speaking, a dust explosion will occur when the material contains a substantial fraction of material passing a 200 British Standard mesh sieve, i.e. $\leqslant 76$ microns [British Standards (1958)]. Baker et al (1983) points out that the concentration of dust necessary to sustain an explosion is so high as to be intolerable for long to humans; such clouds are essentially opaque. Though clouds of this character may exist in the interior of equipment such as elevators, or in machinery for crushing and grinding, they cannot exist, except for a very short time, in buildings.

In practice, the most devastating explosions are those which arise by a mechanism similar to those in coal mines; there is an initial explosion which disturbs the dust and makes a second and much more powerful explosion possible.

Palmer (1973) discusses the rates of propagation and concludes that most industrial explosions are deflagrations rather than detonations and that rarely does the rate of propagation approach sonic velocity. Such detonations may occur in the galleries of coal mines but these may be hundreds of metres long. Detonations may occur also in grain storages in the USA where conveyors and elevators may have similar lengths of travel.

There seems general agreement that pressures of up to 8 bars may be generated in laboratory equipment in some instances. Probably the most important examples of these are those associated with powdered foodstuffs such as cornflour, wheat flour and yeast. Many substances listed as explosible in Palmer (1973) and Field (1982) have maximum explosion pressures around 5 bars. Since Palmer (1973) claims that in practice the peak pressure reached corresponds closely to that predicted for adiabatic conditions, it would seem likely from this that the adiabatic peak temperature may also be attained. This has been confirmed by Cashdollar and Hertzberg (1983) who report maximum temperatures for dust explosions comparable with those for hydrocarbon vapours.

Both maximum explosion pressure and maximum rate of pressure rise may be determined by use of the Hartmann bomb as is described in Field (1982) [sub-section 3.3.6], in which a transducer produces a plot of pressure versus time. Initial pressure in the bomb is slightly higher than atmospheric because of the introduction of an initial pulse of compressed air to disperse the dust. If this pressure be denoted P1 at time T1 and the maximum pressure reached denoted by P2 at time T2, then:-

$$P\ max = P_2 - P_1$$

and

$$dp/dt\ (average) = \frac{P_2 - P_1}{T_2 - T_1}$$

dp/dt (max) may be obtained by inspecting the graph and determining the maximum slope.

Some typical figures are given in Table 12.2., taken from Field (1982), Appendix 1, and including minimum ignition temperatures.

Field (1982) summarises the work conducted in Switzerland under the leadership of Bartnecht at the Ciba-Geigy Organisation, Basle. Much of the previous work in the field had been conducted in the Hartmann bomb of 1.2 litres capacity. Bartnecht and his co-workers have compared the results from the Hartmann with results in larger vessels of up to 60 m^3. Bartnecht has concluded that a 20 litre sphere (.020 m^3) is the smallest vessel which yields reliable results and a version of this is currently available. In this apparatus, the vessel is partially evacuated to 0.4 bar absolute before a measured quantity of compressed air is admitted to disperse the dust. The quantity is that which just restores the internal pressure to 1 bar absolute. The dust is ignited by a 10^4 joule chemical detonator. Bartnecht has developed the following cube root law for vessels of similar shape:-

$$dp/dt\ (max) \times V^{1/3} = K_{ST}$$

dp/dt (max) = maximum rate of pressure rise (bars sec^{-1})

V = volume of vessel (m^3)

K_{ST} = an index with dimensions (bars metres sec^{-1})

(ST = Staub i.e. dust)

TABLE 12.2. EXPLOSION PARAMETERS OF EXPLOSIBLE DUSTS.

	Minimum Explosible Concentration $g\ m^{-3}$	Minimum Ignition Temperature ^{o}C	Max. Pressure Bars	Max. Rate of Pressure Rise Bars sec^{-1}
Aluminium (Flake)	45	610	8.8	1380
Calcium Stearate	25	400	6.7	690
Cellulose	45	410	8.1	552
Coal (Std Pitts-burg)	55	610	6.2	159
Coffee (Instant)	150	490	4.4	38
Cork	35	400	6.7	518
Epoxy Resin	12	490	5.4	902
Flour	40	390	7.1	141
Iron	200	510	3.3	145
Magnesium	30	560	8.0	1035
Nylon	30	500	6.6	276
Soap	20	430	5.4	194
Sulphur	20	190	5.4	324
Titanium	45	330	5.9	759
Wheat Flour	50	380	7.6	256
Wheat Starch	45	430	6.9	449
Wood	no data	360	6.2	393

The above are all derived from small-scale tests with dusts with a mean particle size of less than 75 microns.

Bartnecht (1979) classifies dusts as below:-

TABLE 12.3. EXPLOSIBILITY OF DUSTS.

Parameter K_{ST}	Class	Corresponding Hartmann Data (After Field (1982) Bars sec^{-1})
0	St 0	0
$\geqslant 0 \quad \leqslant 200$	St 1	$\leqslant 500$
$\geqslant 200 \quad \leqslant 300$	St 2	$\geqslant 500 \quad \leqslant 1500$
$\geqslant 300$	St 3	$\leqslant 1500$

12.2.3.5. Prevention of dust explosions.

As will appear from the above, though dust explosions in equipment may be very damaging, secondary explosions which involve entire buildings may be much worse.

The primary strategy is to avoid accumulations of dust which can be stirred up to create the circumstances for a secondary explosion. This strategy may be more difficult to put into practice in the case of grain elevators, grain silos or other silos which, in general, are likely to present more danger when practically empty than when they are full, as the severity of an explosion is likely to be a function of the free volume. Prevention is discussed in Palmer (1973), Field (1982), Bartnecht (1979), and in Baker et al (1983). The principal elements are:-

(1) Basic building design.
(2) Good housekeeping.
(3) Minimising the volume in which explosion can occur.
(4) Avoidance of ignition sources.
(5) Venting.
(6) Inerting.
(7) Suppression.
(8) Effective fire fighting methods.

Basic building design may have among its object the minimising of dust accumulation. For example, roof trusses which can permit significant build up of dust on flat surfaces must be avoided. Regular cleaning of plant is essential though this is related, to a degree, to basic plant design. Minimising the volume includes such measures as breaking up the volume into discrete parts as, for example, in a multisilo installation, isolating each silo from the others.

The subject of the avoidance of ignition sources is too lengthy to treat in detail, but attention is called to a few elements. Electrical equipment needs to be dust proof and especial care needs to be taken to avoid the dangers associated with the build-up of heat. Motors whose air cooling vents become blocked by dust will overheat. Lamp bulbs may become coated with dust and, in consequence, they may overheat. In at least one case, a major explosion was brought about by a handlamp becoming buried in dust. Many of the explosions recorded in Verkade and Chiotti (1978) arose through friction in pulley belts and elevators. In crushing and grinding operations, the introduction of "tramp" metal may cause sparks and overheating.

The standard method of mitigating the effects of explosions is by explosion relief (venting) which takes the form of providing weak panels in equipment or in buildings. A simple approach is to seek a ratio, m^2 of vent per 1 m^3 of volume of building or equipment (which has the dimensions L^{-1}), but Palmer (1973) shows that this ratio will depend on the explosion characteristics of the dust. Palmer puts forward ratios of 1:20 for low explosivity and 1:10 for high explosivity. Field (1982) recommends the use of the cube root law which is now embodied in VDI (1979). The fundamental equation is:-

$$f_1 (V_1)^{1/3} = f_2 (V_2)^{1/3}$$

where f_1 and f_2 are the vent ratios required to protect volumes

V_1 and V_2 respectively.

or

$$\frac{A_2}{A_1} = \frac{V_2^{\;2/3}}{V_1^{\;2/3}}$$

where A_2 and A_1 are the appropriate vent areas of volumes V_2 and V_1.

Thus if the vent area to protect a given volume for a given dust is known, the vent area for any other volume may be deduced. Field (1982) gives a series of nomographs for various dusts and ignition sources.

Reference has been made above in section 12.2.2. to the use of inert dusts in the suppression of coal dust explosions. This technique is seldom feasible in manufacturing industry. However, inert gases may sometimes prove useful in equipment though obviously not in buildings. Pure nitrogen does not contain undesirable impurities but is expensive. Inert gas, manufactured in commercially available units, is appreciably cheaper and recycled chimney gases are cheaper still but they may be heavily contaminated and also they are humid. All such inerting carries dangers of asphyxiation to operators particularly during maintenance or when attending to blockages.

Suppression is a technique in which a suitable transducer, e.g. a pressure rise, or infra-red, detector actuates a system which rapidly releases an inerting gas such as CO_2 or a halogen-type suppressor into a piece of equipment in which explosion is in the incipient stage.

Fire fighting has to be conducted with special care in dusty buildings, as impingement by hose jets may stir up dust and give rise to explosion. Hence a fine spray is to be preferred. However, with metal dusts such as magnesium or aluminium, fine sprays of water may be very dangerous.

12.2.4. CASE HISTORIES OF DUST EXPLOSIONS.

12.2.4.1. Sources of information.

Case histories of dust explosions are given in a number of reference works and there is usually some overlap in that later works take some of their case histories from earlier works. The chief source of case histories are given in Table 12.4.

TABLE 12.4. SOME SOURCES OF CASE HISTORIES.

Source	National Fire Protection Association (1957)	Palmer (1973)	Verkade & Chiotti (1978)	Field (1982)
Description of information offered	75 case histories of explosions of various dusts.	No detailed case histories. 14 Photos	Compilation of 189 explosions in cereal handling facilities with causes where known. Full statistical analysis.	20 case histories. 17 photos

Three further reference sources, not consulted by the present author, are:- Price and Brown (1922) [for early records], Aldis and Lai (1979) [grain dust], and Cardillo and Anthony (1979) [dusts in general]. A case history of the Westwego, Louisiana explosion and other dust explosions is given in Chapter 13.

12.2.5. DUST EXPLOSIONS AS MAJOR CHEMICAL HAZARDS.

It will be clear from the discussion above that dust explosions are events of frequent occurrence in industrial installations and that some of these have had serious consequence. However, the most serious explosions have occurred outside the chemical and process industries in "dockside" industries such as grain storage, flour milling and in the processing of human and animal foodstuffs.

One characteristic of the grain handling industry is that it produces, by attrition, a non-toxic dust of low, or indeed, negative value and that there is little or no economic incentive to collect it for sale or for recycling. By contrast the dusts produced by the chemical industry are usually obnoxious to humans or commercially valuable, or both, and therefore there are strong incentives to maintain a high level of containment on grounds either of occupational health or of economics or both. A further contrast is that, unlike the grain handling industry, which is a traditional industry (and, indeed, is an off-shoot of agriculture), the chemical industry is accustomed to such techniques as the use of inert atmospheres and to the technology of instrumentation and control. Moreover, the understanding of powder technology, including the two-phase flow of gas/solid mixtures and of gas filtration, forms a component part of chemical engineering. For these reasons, dust explosions in the chemical and process industries are much less likely to give rise to devastating secondary explosions.

It is the author's conclusion that dust explosions in the chemical and process industries are a borderline case of a major hazard. The potential for serious consequences exists and chemical engineers must be continually aware of this. Therefore there does not appear to be a case for the institution of special new measures of legislative control in what is today a long-recognised area of hazard, in which the necessary techniques for reducing risk are well understood. This was the view expressed in Advisory Committee on Major Hazards (1984).

12.3. CONFINED GAS EXPLOSIONS.

12.3.1. INTRODUCTION.

The term "gas explosion", as used in this section, includes the meaning "vapour explosion" as, for all practical purposes, chemical explosions which involve gas/air mixtures are identical with those which involve vapour/air explosions.

Confined gas explosions may occur in equipment (including store tanks) or they may occur in buildings. Explosions in buildings may arise from internal leakages or they may result from the penetration of buildings by gas clouds which have originated outside of the building.

Theoretical calculation shows that gas explosions are capable, under adiabatic conditions and starting at atmospheric temperature and pressure, of yielding overpressures of up to about 8 bars. For a given mixture the maximum overpressure is that yielded by a mixture which is slightly richer than the stoichiometric composition. The maximum rate of pressure rise for gas explosions appears to be very similar to that recorded for the worst cases of dust explosions [Baker et al (1983)].

12.3.2. LITERATURE SURVEY AND HISTORICAL REVIEW.

That mixture of air with a flammable gas can ignite and produce dangerous overpressures has long been recognised. Nash (1976) records 10 pit explosions, all in England, 9 of them in County Durham, as having occurred in the 18th century and having killed a total of 361 persons.

During the next century such explosions were recorded worldwide and their numbers rapidly increased, [Nash (1976) needs four pages to list them].

At the same time flammable gases became handled commercially. Following the pioneering work of inventors such as Murdock, in the latter years of the 18th century, gas for purposes of illumination became commercially available in London from 1812 onwards, when the Gas Light and Coke Company was incorporated.

It would have been a matter for surprise, considering the general level of technology, and the general state of ignorance, had such a development not been attended by explosions at consumer's premises and at gas works. Everard (1949) states "Regrettably explosions were not as unusual as might have been wished, and the Company were on another occasion forced to meet a large demand for damages when the Recorder of Grantham was blown up in the wine cellar of his London house". The same work cites other serious explosions in London during the 19th century. Williams (1981), in a history of the British gas industry, has little to say on the subject, but there is a reference to the Silvertown, London explosion of January 1917, which occurred in a munitions works and caused a nearby gas holder with a capacity of ca 250,000 m^3 to collapse and to give rise to a fire ball.

This incident, however, sheds little light upon the intrinsic safety of the manufacture of town gas. Furthermore it is reasonable to argue that the character of the industry has so greatly changed in the last few decades as to render the examples given above invalid in relation to present-day circumstances.

In broad terms, there have been three main eras of gas production technology in the United Kingdom; it would, however, be going beyond the scope of this book to describe them in detail. In the first of these, virtually all the gas distributed was made from coal and it was known as "coal gas". It had a typical composition as below:-

CH_4	0.50	mol	fraction
H_2	0.33	"	"
CO	0.07	"	"
Inerts	0.10	"	"

The presence of hydrogen made the gas low in density and produced a high flame velocity. The presence of the CO made the gas highly toxic, a circumstance which will be commented on in Chapter 15.

In the second era, which commenced about 1960, coal as a raw material began to be replaced by light distillates from the oil industry. Oil was then exceptionally cheap and the use of this new feedstock led to a considerable reduction in the price of the raw material. The new processes, which owed their technology to the chemical and petrochemical industries, used equipment which was very compact, Williams (1981) suggests that an equivalent amount of gas could be made on a site which was only one-tenth of that which would be required for making it from coal.

The basic strategy employed was to produce a gas which could still be used on existing burners, as a consequence the gas produced in this way did not differ significantly in its properties from the coal gas which it had superseded.

It was during this period that British gas supplies began to be supplemented by the importation of liquefied natural gas, (LNG). This came in the first instance from Louisiana, but Algeria soon became the predominant source of supply. From this beginning, LNG was destined to become a major international trading commodity and the confidence engendered by the success of British Gas in handling LNG on a large scale led to the redevelopment of confidence in the United States, which had been shattered by the Cleveland 1944 incident which has been described in Chapter 14.

The third era was that of the exploitation of natural gas from the North Sea which, in the decade from about 1966 to the completion of the conversion in 1976, resulted in a complete changeover to a gas which is substantially methane with minor quantities of other hydrocarbons.

It was realised that there would be no point in trying to make North Sea gas compatible with coal gas or gas derived from light distillates and, accordingly, a total change-over of gas appliances to burners suitable for burning a gas of higher density and lower flame velocity was instituted. Any doubts that the new gas would carry greater dangers than the old appeared to be dispelled by the Morton Report, [Morton (1970)], which took account of a disaster at Ronan Point, the severity of which seems to have been due more to the method of construction of the flats than to that of the explosion itself.

However, a series of explosions during the winter of 1976/77 led to a loss of public confidence and the King Inquiry was set up. The findings of this Inquiry [King (1977)], which will be discussed below, provide useful statistics on the explosion hazards associated with present-day public gas supplies.

It is far from true that the gas industry is the sole source of confined gas explosions. It may be the principal source in the United Kingdom of gas explosions in buildings, but there are many examples of serious explosions in storage tanks. A common cause of such explosions is the ignition of vapours in store tanks which have been inadequately purged prior to welding operations, whether such welding is taking place inside or outside of the tank. In some cases the vapour is generated by the heat which accompanies welding operations.

Probably the most serious of "store tank" explosions are those which have arisen, or are presumed to have arisen, in the cargo spaces of supertankers. These ships, which may be regarded as floating store tanks with capacities in the range 100,000 to 300,000 cubic metres, and with a single cargo holds with volumes equal to perhaps one-tenth of these volumes, present an unusually large hazard. Hazardous Cargo Bulletin (1980) gives a table of ships showing six which exploded when in ballast. It is believed that some, if not all, exploded through the ignition of vapour in otherwise empty cargo space. One theory is that ignition was brought about by static charges generated by water jets during cleaning. Lessons have been learned from these losses and it is now usual for such large tankers to be "inerted" by having the free space filled with treated flue or exhaust gas from the ship's propulsion system in place of air.

12.3.3. CONFINED GAS EXPLOSIONS IN THEORY AND IN PRACTICE.

12.3.3.1. Introduction.

The subject of confined gas explosions has recently been greatly advanced by the publication of "Gas Explosions in Buildings and Heating Plant" by R.J. Harris of the British Gas Midlands Research Station. [Harris (1983)]. The discussion which follows is based, to a considerable extent, upon this book (unless otherwise referenced).

12.3.3.2. Statistics of gas explosions.

The statistics quoted below relate to explosions occurring as a consequence of the ignition of natural gas supplied to premises by the British Gas Corporation. They derive in part from King (1977) and from Yusuf (1982). Means and standard deviations have been calculated by the present author.

In Table 12.5. below, the fact that the number of fatal accidents closely coincides with the number of fatalities shows that during the period, gas explosions seldom resulted in multiple fatality accidents in Great Britain. However, this may not be a valid general conclusion. The October 1968 gas explosion at Clarkston, near Glasgow, in a shopping centre killed 21 people. The explosion of May 1968 in a block of flats at Ronan Point, East London, which has been previously referred to, killed 4. In this latter case, the death toll would have been much higher had not most of the people living there been in bed at the time. It has been argued, however, that the collapse was mainly due to a faulty design of the building which caused the entire corner of the building to collapse like a house of cards as the result of an explosion in the kitchen of just one of the flats.

An alternative way of viewing the risk associated with gas explosions is given in Harris (1983) and is set out in Table 12.6.

TABLE 12.5. GAS EXPLOSIONS IN GREAT BRITAIN.

Year	Serious explosions	Severe explosions	No. of explosions involving fatalities	No. of fatalities
72/73	104	26	8	8
73/74	98	26	10	11
74/75	100	28	14	17
75/76	128	27	10	11
76/77	111	37	9	10
77/78	101	44	3	3
78/79	98	36	13	13
79/80	99	31	7	7
80/81	83	31	11	11
81/82	97	41	12	14
MEAN	102	33	10	10.5
STANDARD DEVIATION	11	6	3	3.7

Note:- A "severe damage" explosion is defined in Yusuf (1982) as one which results in the demolition of, or serious damage to, outer or party walls, roof timbers, floors, structural woodwork or metalwork or other load-bearing element. A "serious explosion" is defined as one which results in the demolition of, or serious damage to, internal partition non-load bearing walls, windows and/or doors, or industrial plant, equipment or vehicles. The involvement of a fatality or serious fire damage also falls within this definition.

TABLE 12.6. RELATIVE RISK FROM DIFFERENT HAZARDS IN THE UK 1975-78.

Hazard	Fatalities per annum per 10 million at risk
Road travel	1000
Fire in the home	100
Electrocution	10
Gas explosions	2
Lightning	1

12.3.4. THEORETICAL ASPECTS.

12.3.4.1. Pressure Rise.

A gaseous explosion arises from the sudden release of energy resulting from a "redox" reaction. This energy release heats up the gas and, under conditions of confinement, a rise in pressure by a factor of about eight may occur. Thus for explosions of gas/air mixtures in which the starting pressure is one bar absolute, the final pressure may approximate to eight bars absolute or seven bars gauge. Where the initial pressure is different to that of atmospheric pressure the final pressure, measured in absolute terms, will still approximate to eight times the initial pressure.

The well known Charles' Law states that, at constant pressure, all gases expand by 1/273 of their volume at 273 K(0°C) for each 1K rise in temperature. Taken in conjunction with Boyle's Law which holds that the pressure and volume of a given mass of gas are inversely related it follows that, under constant volume conditions, the pressure of a given mass of gas will increase by 1/273 of its initial pressure for each 1 K rise in temperature. These two laws may be combined to give the well known Gas Law:-

$$PV = RT$$

Where R is the universal gas constant with the value of:-

$$8.314 \times J/mol \times deg\ K.$$

When applied to explosions in gas/air mixtures, the Gas Laws given above yield only approximate answers. Firstly the gases are not "ideal" and secondly, account has to be taken of possible changes in the number of molecules during combustion. Typical changes are shown below:-

$$2\,H_2 + O_2 = 2\,H_2O$$
No. of moles 2 + 1 → 2

$$CH_4 + 2\,O_2 = CO_2 + 2\,H_2O$$
No. of moles 1 + 2 → 1 + 2

$$2\,C_2H_6 + 7\,O_2 = 4CO_2 + 6\,H_2O$$
No. of moles 2 + 7 → 4 + 6

$$2\,C_6H_6 + 15\,O_2 = 12\,CO_2 + 6H_2O$$
No. of moles 2 + 15 → 12 + 6

From this it will be seen that the combustion of hydrogen results in a reduction in the number of molecules in the mixture, the combustion of methane results in no change, the combustion of ethane or benzene increases the number of molecules in the mixture. However, the ratio between the number of moles in the products of combustion and the number of moles in the unburned gas/air mixture is much closer to unity than the examples given above would suggest as account has to be taken of the nitrogen present. The mole number ratio for hydrocarbon gases in air lies, in most cases, between 1.00 and 1.05.

For many purposes it is sufficiently accurate to ignore both deviations from ideality and also the change in the number of moles and to assess the adiabatic pressure ratio approximately as the ratio between the theoretical flame temperature and the initial temperature. For propane, assuming an initial temperature of 15C (288 K), the ratio is 2198/288 or 7.6. The maximum theoretical overpressure for propane is thus 7.6 - 1.0 = 6.6 bars. For acetylene the ratio is 2598/288 or 9.0 giving a maximum theoretical overpressure of 8.0 bars. For the 14 flammable gases tabulated in Harris (1983), the mean value for the ratio is 7.7 and hence the mean overpressure is 6.6 bars. It is reasonable to expect, therefore, that hydrocarbon gases under conditions of total confinement and assuming adiabatic conditions will achieve an overpressure of circa 6.5 bars.

12.3.4.2. Preconditions for the explosion of gas/air mixtures.

For an explosion to occur, there are three necessary preconditions. These are:-

(1) The presence of a flammable gas (the reducing agent).
(2) The presence of oxygen (the oxidising agent).
(3) The existence of a sufficiently high temperature.

12.3.4.3. The influence of the concentration of reactants.

The flammability range for a number of gases or vapours mixed with air is provided in Fig. 12.2. Harris (1983), table 1.2, gives further data and also further sources of data. However, the range of flammability is a matter not only of the concentration of the reducing agent, but also of the concentration of the oxidising agent. Harris (1983) deals only with mixtures with air as this is the problem with which British Gas is chiefly concerned. It may be said that with oxygen concentrations greater than the 21% found in air, the flammability range will be extended as will also be the overpressure as a consequence of the higher flame temperature which will be achieved. Conversely with oxygen-deficient air, the flammability range will be reduced with the lower flammability limit being increased. For any given gas there will be a level of oxygen concentration below which ignition becomes impossible; this is the basis of "inerting" as a means of protection against fire or explosion. In ships for example, inerting may be brought about by filling the space above the cargo by the introduction of flue gases either from the ship's propulsion machinery or from a custom-built inert gas producer. Clearly, in oxygen-depleted air, the maximum flame temperature will also be reduced with a corresponding reduction in the maximum overpressure.

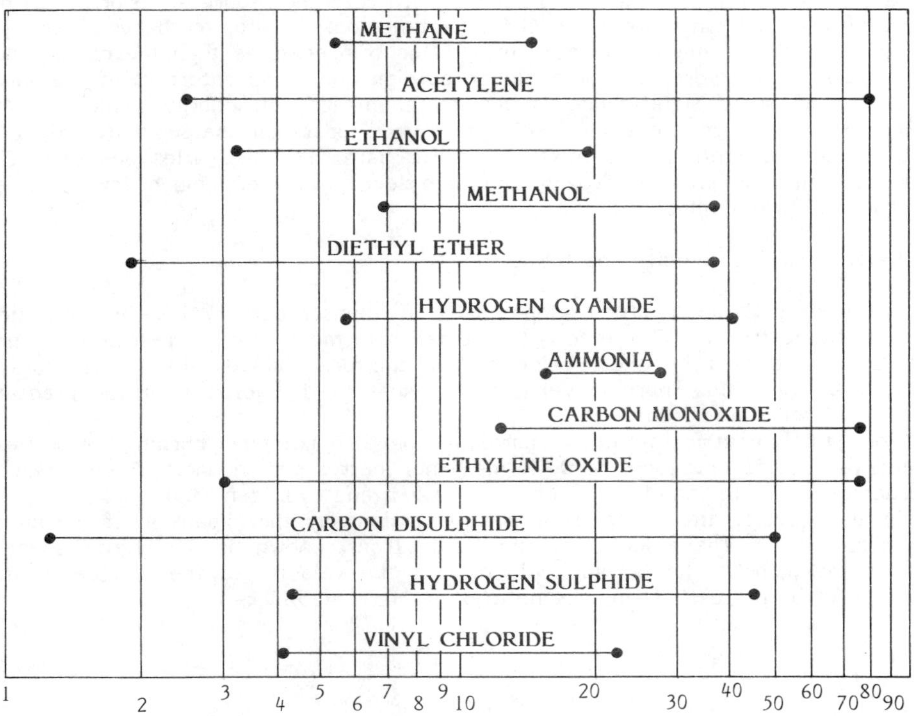

Fig. 12.2. FLAMMABILITY LIMIT OF METHANE COMPARED WITH SOME VAPOURS.

Characteristics of inert gas and of inert gas producers are given in Peabody Holmes (1978). This describes a standard range of inert gas producers of capacity from 100 to 4,000 cubic metres per hour with the possibility of producing equipment capable of providing up to 20,000 cubic metres per hour to special order. The gas has the following composition:-

CO_2	15% Max.
CO	0.01%
O_2	0.5%
H_2O	Equivalent to saturation at discharge temperature.

(It may be noted that inert gas is a very dangerous asphyxiant. There have been a number of fatal accidents consequent upon its use; the worst recorded to date was at Singapore where 7 lost their lives. This was a case, of which there are many other examples, where one man was overcome, a second was overcome attempting to rescue the first, a third was overcome trying to rescue the first two, and so on).

12.3.4.4. The influence of temperature.

There is for any gas/air mixture within the flammable range, a minimum temperature, known as the autoignition temperature, below which a self-sustaining combustion reaction cannot occur. Values of this temperature are given in table 1.2. of Harris (1983) and in other standard reference sources. For paraffins, the values quoted range from 214°C for heptane to 540°C for methane. Generally the olefines show somewhat lower autoignition temperatures than the corresponding paraffins. Hydrogen has a higher autoignition temperature than methane. A factor of some significance is the minimum ignition energy, which is the energy stored in an electric circuit prior to a spark discharge test, which is just capable of igniting the mixture. For paraffins this varies between 0.25 and 0.29 millijoules. Hydrogen and acetylene show very much lower values of around .02 millijoules.

12.3.4.5. Burning velocity and flame speed.

In a stationary flame, as, for example, that of a bunsen burner, there is a reaction zone or flame front. This reaction zone tends to move into the reaction mixture but this is opposed by the motion of the mixture towards the flame front. In such a flame this "burning velocity" is equal to the velocity of the reaction mixture at the flame front.

For stoichiometric mixtures, paraffins have maximum "burning velocities" in the range 0.52 (heptane) to 0.45 (methane) metres per second. Higher values are achieved by ethylene (0.83), acetylene (1.58), and hydrogen (3.5).

In an explosion the flame front is pushed along by the expansion of the burnt gas behind it. The velocity of the flame front, known as the "flame speed", is thus the sum of the burning velocity and the velocity of the surface of the envelope of burnt gas. Using the notation in Harris (1983) :-

$$S_f = S_o \times E$$

where S_f = flame speed

and S_o = burning velocity

and E = expansion factor

$$E = (T_f / T_i) \times N_m$$

where T_f = flame temperature
and T_i = initial temperature
and N_m = Ratio of moles present after and before combustion.

The value of S_f is dependent upon the geometry of the system in which the explosion occurs. For explosions in tubes of constant cross-section, open at one end only, it is approximately equal to 8 x S_o. For a completely unconfined situation it will be equal to approximately 2 x S_o. Values of maximum S_f for paraffins are of the order of 3.5 (methane) to 4.0 (heptane). There are higher values for acetylene (14.2) and hydrogen (28.0). It will be demonstrated below that these values are far below the velocities associated with the development of appreciable overpressures.

12.3.4.6. Scaling Laws.

Paragraph 12.2.3.4. of this book discussed, inter alia, the work of Bartnecht and his co-workers in developing a cube root law for predicting the rate of pressure rise in dust explosions. Harris (1983) advances a similar law for the early stages of a gas explosion which are those stages which occur before venting operates. The reference demonstrates that the rate of pressure rise dp/dt, is a function of the cube of the burning velocity. It also demonstrates that:-

$$t_2 / t_1 = (V_2 / V_1)^{0.333}$$

where t_2 = time to reach a given overpressure in a vessel of volume V_2.
and t_1 = time to reach a given overpressure in a vessel of volume V_1.

12.3.4.7. Considerations of mixing.

Most flammable gases have densities which differ markedly from that of air. Only ethylene (molecular weight = 28) and ethane (molecular weight = 30), have densities close to air which may be regarded as having a mean molecular weight of 29. Methane is much lighter, and hydrocarbons containing three or more carbon atoms are clearly much heavier than air. Even fairly small differences in density may produce considerable stratification as witness the way that atmospheric air may be stratified as the result of temperature differences.

As a consequence, even though a given leakage of gas may be calculated to produce a concentration of X, unless there is a considerable lapse of time or the contents of the confined space are agitated, some parts of the space will have a concentration greater than X and others a concentration less than X. This can mean that though a given quantity, if averaged out over the whole volume, may appear to produce a concentration below the flammable limit, there may be part of the volume where a mixture exists which lies inside the flammable limit. A consequence of considerable importance which arises from this is that even in the early stages of a leak, there may be a volume of a confined space which has the potential for an explosion even if only a minor one. For light gases, this dangerous volume will lie near the roof and, for heavy gases, near the floor.

However, much depends on such factors as the geometry of the system, i.e. whether the leak is at high or low level, and upon the source term, as, for example, whether the leak is in the form of a high pressure jet. Harris (1983) suggests that for leaks of natural gas, a fairly uniform concentration is fairly quickly achieved above the point of inleakage, thus a gas leak near the floor will fairly soon achieve a uniform concentration. The implications of this are probably more significant for heavy gases as the circumstances which can give rise to leakage at low level are much more numerous than the circumstances which can give rise to high level leakage of light gases. This point is not, however, made in Harris (1983), which is not concerned with gases other than natural gas.

12.4. VAPOUR CLOUD EXPLOSIONS.

12.4.1. INTRODUCTION.

It has to be said at the outset that the term "Unconfined Vapour Cloud Explosion" which has been much used in the past is now regarded by many investigators as a misnomer. This is because these investigators now take the view that such an open air explosion only occurs in circumstances in which there is some degree of confinement, even if this is no more than that afforded by pipe bridges or by the spaces between buildings. In this book the more general term "Vapour cloud explosion" will be used and it will be abbreviated to "VCE". A vapour cloud explosion is defined by the present author as follows:-

> "A vapour cloud explosion is a deflagration accompanied by blast effects which occurs in the open air as the consequence of the ignition of a cloud containing flammable vapour".

As has been noted in an earlier chapter, by no means all clouds containing flammable vapour, if ignited, give rise to a VCE; in some circumstances the deflagration may give rise to a flash fire or to a fire ball.

Because of the large quantity of vapour which may be involved, the most typical circumstance in which a VCE may arise is that following the loss of containment of tonnage quantities of a liquefied gas which then flashes to produce a cloud of flammable vapour.

VCEs have been responsible for many very serious incidents involving severe property damage and considerable death tolls, in one case exceeding two hundred persons.

12.4.2. HISTORICAL REVIEW.

Until the early seventies the phenomenon of the vapour cloud explosion was not recognised. For example, Brasie and Simpson (1968), in discussing guidelines for the estimation damage from chemical explosions, make no reference to VCEs. The first paper which gave a clear overview of the problem was Strehlow (1972) and this latter reference is thus of considerable historical importance. But Strehlow's list of VCEs is short and it omits the most serious of all, the Ludwigshafen explosion of 1948 which is described in the case histories below. Unfortunately it is very difficult sometimes to distinguish those incidents which gave rise to blast effects, the true VCEs, from those which gave rise to fire balls or to so-called BLEVEs. In consequence, many of the compilations of VCEs which have been published, including that by Strehlow, list such events as the Cleveland 1944 disaster (described above in Chapter 9), which was not a VCE.

Table 12.7. is a list of compilations of VCEs from various sources, which the present author does not claim to be comprehensive. It is his view, based on a detailed analysis, that the compilations display a considerable degree of overlap, but that the more serious incidents appear in practically all of them. But no one, up to now, has published a really comprehensive list in which all the known incidents are critically examined. Unfortunately there is a strong tendency for a compiler simply to accept, uncritically, the classification of any particular incident which has been adopted by earlier compilers and this constant reiteration may serve to consolidate errors where these exist.

TABLE 12.7. SOME COMPILATIONS OF VCEs.

Reference	No. of incidents	Period covered	Notes by present author
Strehlow (1972)	108	1930-72	Many are clearly not VCEs. 9 case histories.
Strehlow & Baker (1975)	12	1962-72	
Commercial Union (1976)	25	1948-75	Many not VCEs. Financial losses are given.
Davenport (1977)	43	1921-76	17 cases without overpressure and 7 without ignition also listed.
Marshall (1977b)	9	1959-75	Mortality indices calculated.
Slater (1978)	32	1948-77	Some were not VCEs. 70 other vapour cloud incidents listed.
Gugan (1979)	100	1919-77	Though listed as VCEs in only 25 cases does author claim that there was blast.
ACMH (1979)	19	1948-78	Vapour cloud explosions, some confined.
Kletz (1979)	220	1970-79	Serious accidents involving flammables. Not listed individually.
Davenport (1984)	71	1921-83	All are claimed to be VCEs. Financial losses, corrected for inflation, are given.
Wiekema (1984)	165	1921-80	86 VCEs, 64 fires, 15 uncertain

It is difficult to be sure which was the first VCE to be recorded. There is a reference in a nineteenth-century work, Berthelot (1892), to:-

"The clouds formed by petroleum vapours and other volatile hydrocarbons have also given rise to similar (i.e. similar to those produced by dusts - VM) explosions in cellars or magazines or even in the open air".

Whether this refers to explosions which gave rise to blast, or which merely gave rise to a loud bang, is impossible to determine.

Davenport (1984) listed an incident in 1921 in which a hydrogen-filled dirigible caught fire over the River Humber and gave rise to blast effects. Airship conflagrations will be discussed below, where it will be argued that few gave rise to VCEs. A later incident involving butane vapour at Newark, NJ, USA, on 2/1/39, which took place in a courtyard, is recorded by Davenport as having caused doors and windows to be blown in, though not any fatalities. Gugan (1979) recorded these circumstances but put a query in his column headed "Blast".

There seems no doubt that there was a serious VCE in 1943. This occurred at the BASF works at Ludwigshafen, Germany, on 29/7/43. This incident, which is discussed below under "Case Histories", may be regarded, in the absence of any evidence to the contrary, as the first industrial VCE which had serious consequences.

Davenport (1984) lists altogether 69 VCEs as occurring between 1943 and 1983, or roughly one every seven months worldwide. This figure is probably too low if only because of under-reporting of incidents in the Eastern Bloc; Davenport only lists two of these. (For a general discussion on under-reporting see Chapter 3). Davenport (1984) appears to be the most reliable and comprehensive compilation to date. Nevertheless, there remains a need to produce an authoritative and comprehensive incident file in which the individual incidents are written up in such a way as to disclose disagreements as to causes and circumstances and to put the salient features of the incidents onto a computer file to facilitate statistical analysis.

The latest compilation is that of Wiekema (1984). This comprises 165 incidents occurring between 1921 and March 1980. The incidents are tabulated in chronological order and are analysed for 10 factors including mass, reactivity of material, whether blast occurred, degree of confinement and the numbers killed or wounded. The incidents are only treated statistically and details of individual incidents are not given.

12.4.3. THEORETICAL CONSIDERATIONS.

12.4.3.1. The general problem.

The general problem may be stated as follows. A steady, i.e. constant volume, flame will not produce significant pressure effects. This is not to say that pressure effects do not exist. They are, however, small and are incapable of producing damage or human injury. Because flames are buoyant, the pressure field around them cannot be totally symmetrical. Figure 12.3. indicates that for a vertical flame, the pressure at the lower part of the flame is directed inwards towards the flame, whereas above it, the pressure is directed away from the flame. Were such pressures to be measured by a Pitot tube, the relationship between pressure and gas velocity is given by the energy conservation equation:-

$$P = 0.5\,\rho\,V^2$$

where P = force per unit area
ρ = density of gas
V = velocity of gas

For a maximum flame speed of 6 m/s at the flame tip:-

$$P = \text{ca } 0.5 \times 0.2 \times 6^2 = 3.6 \text{ Pa}$$

This equation only holds true for velocities which are low in comparison with the speed of sound. At higher velocities, the compressibility of air (or any other gas) may be significant and this will increase the value of the density term. Figure 12.4. gives the relationship between velocity and pressure for air up to approximately sonic velocity.

Recent work, Brossard et al (1984), in which stoichiometric mixtures of acetylene, ethylene and propane with air were ignited in balloons showed that

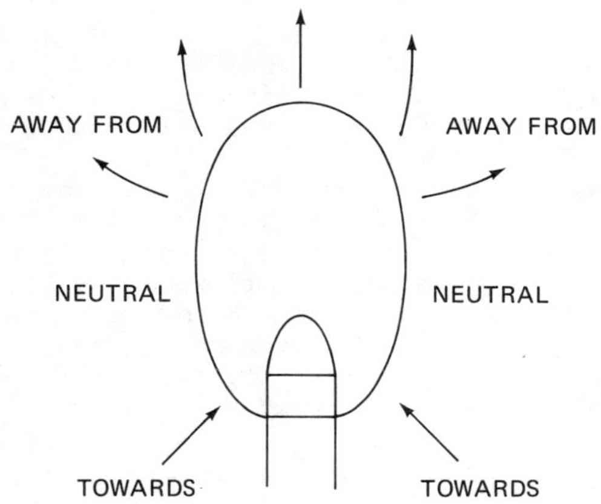

Fig. 12.3. FORCE FIELD NEAR A FLAME.

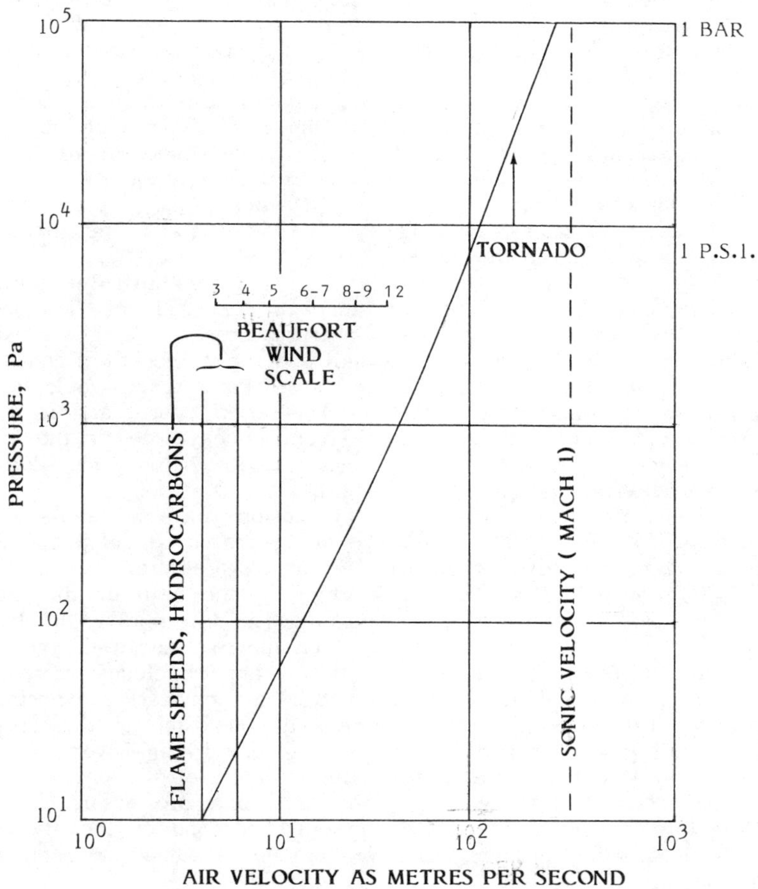

Fig. 12.4. PRESSURE v VELOCITY FOR AIR AT NTP.

the velocities of the flames produced were of the order of 11 to 16 m/s and the pressures generated were of the order of from those which might be deduced from Fig. 12.4. This latter work shows that when such mixtures are detonated by a solid explosive, pressures higher by orders of magnitude were obtained. This will be further discussed below.

The recognition that, in totally unconfined conditions, the production of significant overpressure would appear to presume values for the acceleration of flames which seem to be totally outside of experience to date, has led a number of investigators to seek for alternative explanations. Such explanations, and the theoretical difficulties which they create, are well reviewed in the appendix to Gugan (1979). The search for such explanations has now largely been abandoned in face of the growing conviction that there is no such thing as a totally unconfined vapour cloud explosion.

12.4.3.2. The influence of confinement.

As is well known, in conditions of containment, and where the flame has a volume which is a significant fraction of the volume of the container, very significant pressure effects may occur. This is the previously discussed case of the confined gas explosion. In neither case is it necessary to postulate flame speeds in excess of those quoted in the literature and which, for hydrocarbon gases, are of the order of 4 m/s.

The problem of the VCE is that it occurs in the open air so that the volume of the flame is negligible in comparison with its surroundings. Hence it would be expected to be able to expand freely to accommodate its increasing volume and all authorities are agreed that, at first sight, this should be able to take place without significant overpressure effects even for flame speeds appreciably in excess of 4 m/s. Conversely, to achieve values for the rate of increase in volume (dV/dT) of the vapour cloud in a VCE which would give rise to significant overpressure, it is necessary to postulate values for flame speeds which approach the speed of sound in the medium.

The discussion on fire balls in Chapter 8 clearly demonstrates that, under certain circumstances, it is possible for tonnage quantities of flammables to burn rapidly without producing significant overpressures. Using the equations given in Chapter 8 for the radius of a fire ball and for its duration, and assuming that ignition is initiated in the centre, the mean flame speed would be of the order of 29/3.8 or approximately 8 m/s. This latter figure, and it is only a rough approximation, is not greatly different from values quoted in the literature for the flame speeds of hydrocarbon gases and is far below the values which would be associated with significant overpressures.

These considerations have led to the conclusion that the explanation for the phenomenon of VCEs must lie in there being present a sufficient level of confinement to make possible the requisite level of overpressure.

Pikaar (1984) concludes "Structures, whether in the path of the dispersing cloud or upwind, can be of overriding importance in determining the behaviour of a vapour cloud . . . For dense clouds, which move close to the ground, the influence of structures is still more important than for clouds carried in an airstream . . ." After giving data obtained from combustion experiments in open terrain, the reference goes on "The overall conclusion of this large body of evidence is that clouds of LNG or propane vapours drifting over open terrain will, if ignited, burn with insignificant pressure effect, of the order of 1 mbar . . . recent findings support the view that the cause of explosions is to be sought in the flow related effects created in the neighbourhood of the flame front, as combustion in partially confined spaces pushes unburnt vapour through restrictions and past obstacles".

Zeeuwen et al (1984), reporting on trials in which various obstacles were placed in the path of a drifting propane vapour cloud containing 1 tonne of propane, conclude that vertical obstacles, in this case sewer pipes, had no detectable effect on the pressures generated on ignition. Nor did horizontally disposed pipes. However, when the vertical array of sewer pipes was covered with steel plates, flame speeds of up to 66 m/s were observed, accompanied by pressures of about 20 mbar. The positive phase duration of the blast wave was 200 to 450 milliseconds which corresponded to the duration of the period of relatively high flame speeds. It was noted that flames on leaving the zone of confinement tended to decelerate. Both the flame speeds and the pressures generated appear modest compared with those achieved in some actual incidents.

Wiekema (1984) concludes " . . . the presence of houses, structures, walls and so on, is a necessary condition for the generation of an explosion. In other words, the presence of obstacles forms a necessary condition for the creation of flame acceleration".

Some confirmatory evidence that the presence of buildings and plant may enhance the effects of a VCE comes from a study conducted for other purposes by the present author, which was reported in Marshall (1980). In this study, the damage areas of the Decatur disaster of 19/7/74 and of the Flixborough disaster of 1/6/74 were compared to test a hypothesis, advanced by others, that the yield of the Decatur explosion was some ten times greater than other well-documented explosions, such as Flixborough. An analysis of the two explosions is given in the next chapter but, for present purposes, attention is called to Figure 13.22. which shows the deduced damage pattern for Flixborough. This is shown to be roughly elliptical and to have a maximum range to the north east which is about 2.4 times the minimum range which is to the south west. Figure 9.9. previously given shows that at Flixborough the greatest density of plant and equipment lay to the north east of the point of release, whereas there was a "window" of relatively bare ground to the south west.

In addition, and in general support for the hypothesis that semi-confinement enhances VCEs, it may be noted from Figure 12.5. that the Flixborough reactor train formed a cascade. It was mounted on a reinforced concrete structure which provided a considerable volume of semi-confinement beneath it. However, the hypothesis that some measure of confinement is a necessary condition for the development of a VCE is not yet proven. It is further discussed in the final paragraphs of this chapter.

12.4.3.3. Quantifying VCEs.

There is little or no record by scientific instruments of either the overpressure generated or the duration of the positive phase of any accidentally occurring VCE and no experimental VCE which did not involve a detonation (a subject discussed below) has ever generated really significant overpressures. There have been examples of pressure pulses being picked up by recording barographs used for meteorological purposes but such instruments are not suited, either in their speed of response or in their range, for accurate recording of such data. In the case of the Flixborough explosion (described below), a nearby glider picked up a trace on a barograph [Flixborough (1975)]. Furthermore, signals believed to emanate from the Flixborough explosion were picked up by the University of Leicester where the Physics Department was studying the scattering of radio waves from the ionosphere. These traces certainly helped to place the timing of the incident and showed a disturbance with a duration of about one minute [Jones and Spracklen (1974)]. However, the evidence from such sources is not adequate to enable conclusions to be drawn as to the shape of the blast wave.

Instead, virtually all the evidence on the nature of VCEs is derived from study of the damage which they cause. By comparing the damage caused by

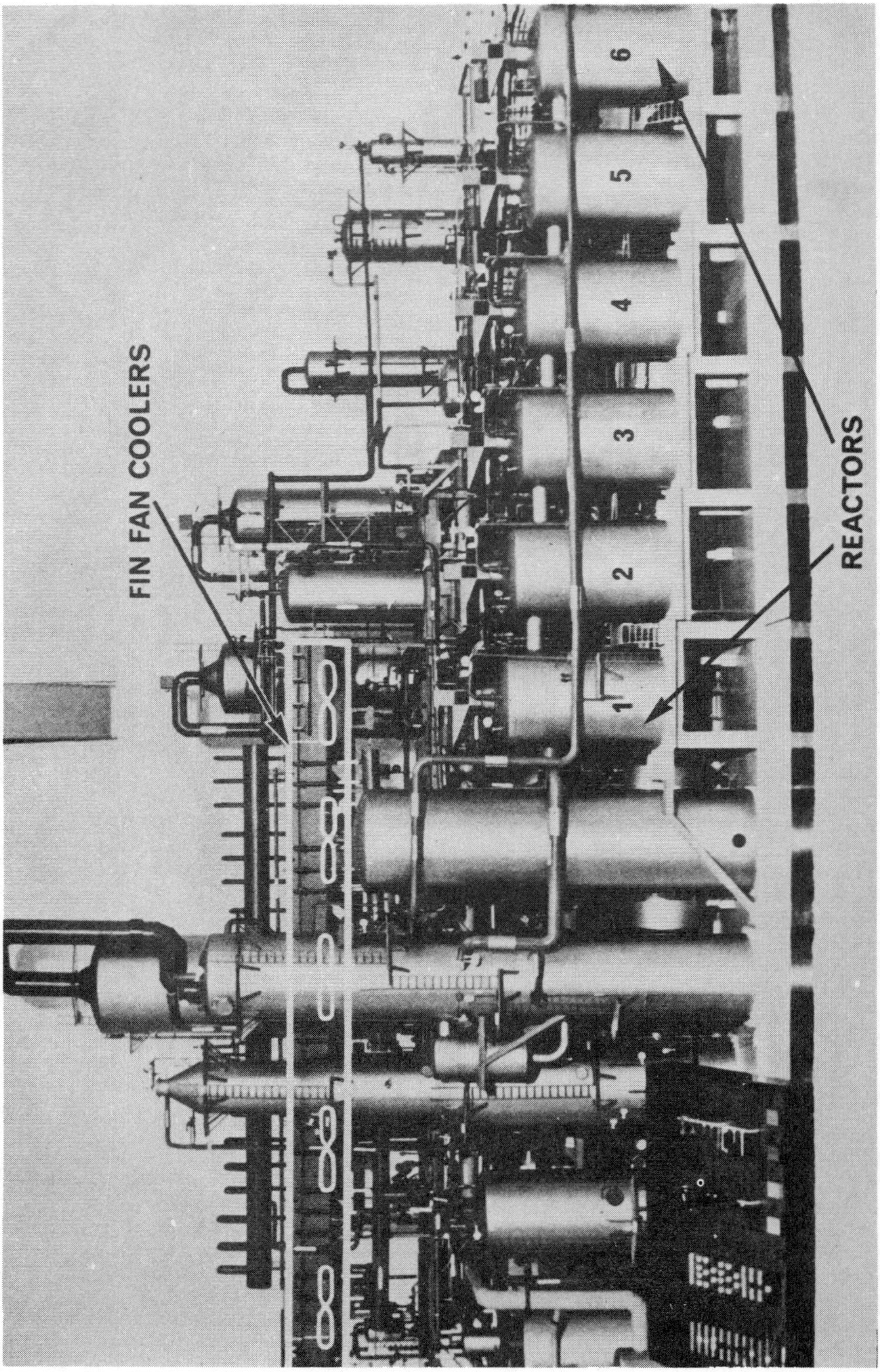

FIG 12.5. MODEL OF FLIXBOROUGH REACTOR TRAIN. THIS WAS KEPT IN THE CONTROL BUILDING AND WAS DESTROYED IN THE EXPLOSION.

a VCE with the damage caused by conventional explosions, conclusions may be drawn as to the energetics of a VCE.

It is necessary, therefore, to examine the chain of reasoning which leads to an estimation of these energetics. Firstly, there exists a vast amount of data on the damage effects of high explosive. This partly arises from actual experience of warfare such as from air raids. For example, some 2,300 V.1s (Flying Bombs) landed in London in 1944. These weapons did not penetrate the ground deeply and were thus ground-burst weapons. They all contained the same quantity of high explosive (ca 0.8 tonne). Extensive study of their effects was made. Alternatively, there is a considerable amount of unclassified material on research by defence establishments into the effects of high explosives in which damage effects have been correlated with scientifically measured overpressures and durations.

But there are limitations to such studies. There is a fair degree of variance even when overpressure measurements are correlated with scaled distance on flat, open ground, so that even this, the most scientific area of the study, has to be treated statistically. This is demonstrated by figures 3.2, 3.3, and 3.5 of Baker (1973). When the problem is extended to correlating damage with overpressure, the difficulties multiply. A house, for example, can hardly be considered to be a calibrated scientific instrument though it is true that it contains a variety of structural elements with differing resistance to overpressure. Unfortunately, actual houses can differ considerably in structural standards. Furthermore, there may be considerable differences between a house standing in isolation on a military proving ground and one standing in a street in a town. Then how is damage to be expressed ? During the Second World War, damage to dwelling house was classified on a scale from A (Almost complete demolition) to D (Damage called for urgent repairs but not rendering the house uninhabitable), [Healy (1965)]. These categories did not include houses with only window damage. Window damage represents a class of damage which enables lower values of overpressure to be evaluated. Such damage categories as those discussed above are far from exact measures and include a degree of subjectivity. Moreover, they do not correlate very precisely with overpressure levels. At best they are nodal points on a continuum.

All of this shows that there are considerable difficulties even when it is the damage effects of high explosives that is being studied. When the attempt is made to correlate damage for VCEs, there is a further complication. If a building has suffered a degree of damage attributable to a given level of overpressure for high explosive, then this has associated with it a characteristic duration for the positive phase of a dense explosive, i.e. a certain impulse. It by no means follows that a VCE is characterised by the same positive phase duration as a conventional explosive. Indeed such evidence as is available suggests that blast waves from VCEs have a longer duration than those from conventional explosives.

It might be fruitful to compare the time taken for the two kinds of explosion to take place. Taking TNT first, if it be assumed that the detonation wave travels at 7,400 metres per second [Table 2, Robinson (1944)], a hemisphere of TNT weighing 32 tonnes, and thus having a diameter of about 4.4 metres, if detonated at its surface would take a maximum ca 0.625 milliseconds for the detonation to travel to the farthest point. The Flixborough explosion has been rated at about 32 tonnes TNT equivalent and the cloud at about 200 metres in diameter. (For the justification for these figures, see the case history in Chapter 13). Even if the blast wave travelled at the speed of sound in air, the explosion would take about 650 milliseconds to travel. Put another way, an equivalent mass of vapour cloud is far larger and sound travels much more slowly in it.

According to Robinson (1944) "The push of an explosion against a wall is the product of the pressure (and) the time or duration . . . This product is called the impulse and is the most important factor in determining the resulting behaviour of the wall". Though this is a somewhat simplified view of the phenomenon, it is nevertheless basically true.

This suggests that a blast wave from a VCE, on account of its longer duration (or greater impulse), will do more damage for a given level of overpressure than would be inflicted by that level of overpressure from conventional explosives. Conversely this means that a given level of damage can be inflicted by a VCE at a lower level of overpressure than is required for high explosives. VCEs may, in this respect, show more resemblances to nuclear explosions, where the duration of the positive phase is of the order of about 250 milliseconds even for low yield bombs, [Glasstone and Dolan (1980), Table 3.76], than they do to conventional explosions.

If this be so then the difficulties presently associated with postulating mechanisms which could produce previously assumed levels of overpressure by VCEs, may be less than they appear at present because VCEs can do their damage at much lower levels of overpressure. However, because of lack of positive evidence, in the discussion below the levels of overpressure deduced from the damage inflicted by VCEs will be assigned the overpressures at which TNT would produce equivalent damage.

12.4.4. MODELLING VCEs.

12.4.4.1. The TNT Model.

Ever since the phenomenon was first clearly recognised, there have been attempts to model VCEs. The most popular model has been that which compares a VCE with the explosion of an equivalent amount of TNT; this approach was advocated in Strehlow (1972).

Though this approach has a certain measure of validity it may be said, at once, that such a model cannot possibly represent the state of affairs near the centre of a TNT explosion. This is because a VCE does not possess the brisance or shattering power of a dense explosive. In the immediate vicinity of a dense explosion, pressures in excess of 10,000 bars may be generated, [Cook (1966)], whereas, even under extreme conditions, the maximum overpressure generated by a VCE cannot exceed a few tens of bars. This is illustrated by comparing the scene after the Oppau explosion of 1921 which was caused by the detonation of some 4,000 tonnes of ammonium nitrate mixture and where there was a 10 metre deep crater, with the scene after the Ludwigshafen explosion of 1948 which was a VCE in which some 18 tonnes of dimethyl ether were involved. There was clearly no crater in this latter case, nor has one been noted as having been caused by a VCE in any other instance. Where craters have been noted at the scene of a VCE they have been attributable to an initiating event in which, say, a leak of liquefied gas has scoured out the ground in the immediate vicinity of the leak. It cannot be ruled out that a VCE may cause a shallow depression in soft ground detectable by surveying instruments but such a depression would lack the characteristic lip associated with the craters caused by conventional explosives.

Thus an overpressure versus scaled distance curve for a VCE would have a discontinuity for overpressures in excess of, at the most, a few bars. Such curves are illustrated as Figures 5 and 6 in Advisory Committee on Major Hazards (1979).

The question then remains as to how well the curve for a VCE for overpressures less than a few bars would then match the TNT curve given in Figure 10.2. It seems apparent that if equivalence has any meaning then if, in the near field, overpressures are lower than those for TNT, there must correspondingly be regions for which overpressures for VCEs are higher than those predicted for TNT. However, Phillips (1981), citing Luckritz (1977), is of the view that, in the far field, overpressures for VCEs are lower than would be deduced from the equivalent TNT curve. The Luckritz hypothesis would imply the overpressure/ scaled distance relationship of the form shown in Figure 12.6. The curve shown in the figure is notional. It is based on the concept that the sum total of damage in the entire area affected by a VCE is equal to the sum total of the damage brought about by that quantity of TNT which is designated as the equivalent quantity of TNT.

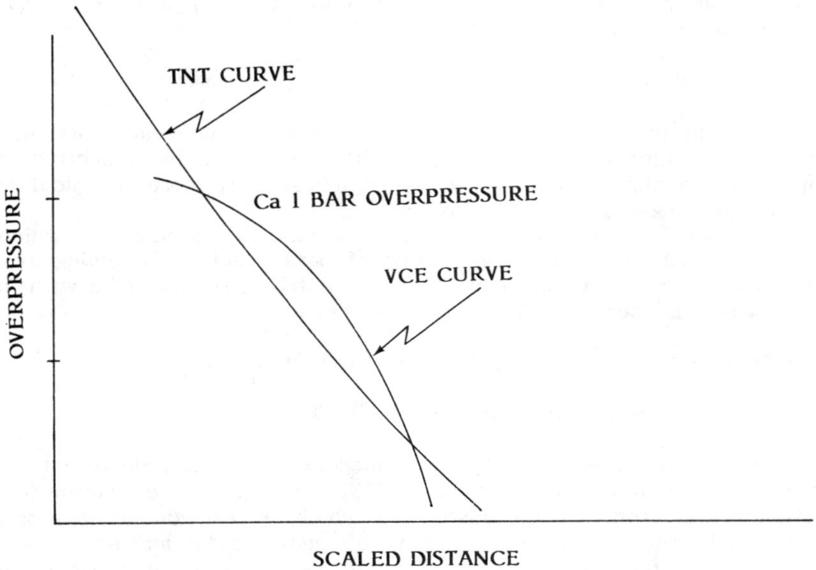

Fig. 12.6. OVERPRESSURE v SCALED DISTANCE, THE LUCKRITZ HYPOTHESIS.

A solution to the problem of producing a TNT model of what is essentially a "soft centred" explosion has been advanced by Sadee et al (1977). This proposes to model a VCE on an air burst TNT explosion. By distancing the centre of the TNT from the centre of the VCE the problem of brisance is avoided. This solution has not found favour. One reason is that it does not give a unique answer, closely equivalent damage patterns were shown in the paper to arise from 14 tonnes of TNT at 69 metres, 16 tonnes at 45 metres, and 18 tonnes at 21 metres above ground. A second reason is that an actual air burst produces reflections which have no counterpart in a VCE.

12.4.4.2. The Wiekema Model.

An alternative model has been put forward in Wiekema (1980) and in Wiekema (1984). This is a model based upon a theoretical approach which enables the author to produce a relationship between scaled overpressures, i.e. P/P_o, and a scaled distance for gas mixtures of varying reactivity assuming that the combustion is deflagrative. Methods are also given for detonative vapour explosions which, as will be demonstrated below, are most unlikely. The scaled distance used is a Sachs scaled distance which is $(E/P_o)^{0.333}$, [Sachs (1944)]. This model also provides a plot of dimensionless positive phase duration versus scaled distance. It has the advantage of not having to depend upon the damage criteria of the TNT model. However, its claimed justification lies in its ability to predict damage and this is checked in Wiekema (1984). Here it is claimed that the model agrees, qualitatively at least, with the following observations:-

(1) No one has been reported as having been killed by blast outside of the cloud in a VCE.
(2) Some people have been knocked down but not seriously injured by VCEs.
(3) Cars have never been recorded as having been overturned by a VCE.
(4) Houses have suffered damage in the vicinity of VCEs.
(5) Yield (further discussed below) increases with increasing size of the quantity of flammable material in the cloud.

12.4.4.3. The models compared.

In the view of the present author, the TNT model has some validity as one index of a VCE. Its chief merit lies in its ability to convey the magnitude of an explosion to the public. This it can do in a way that more technical models such as those of Wiekema are unable to do.

In the present author's view, the way forward for specialists, which will combine theory with the increasing amount of data which is becoming available from laboratory experiments and from large-scale trials, seems to lie with models of the kind which has been developed by Wiekema.

12.4.5. OTHER FACTORS IN THE EVALUATION OF VCEs.

12.4.5.1. Deflagration and detonation.

Reference has been made above to two models, one based upon deflagration and the other on detonation. Wiekema (1980) defines these two models as:-

Deflagration. A mode of combustion in which a reaction front propagates through a combustible mixture by means of heat conduction and convection from the burned to the unburned gas. (Not all deflagrations give rise to significant pressure effects, VM).

Detonation. A mode of combustion in which a shock wave is propagated of such strength that a chemical reaction starts directly behind the discontinuity.

Deflagration is a very well recognised phenomenon which has been extensively discussed above, and it is not proposed to add anything further to the discussion of it here.

The phenomenon of the detonation of mixtures of flammable vapours with air has been demonstrated in numerous trials in which such mixtures have been exploded by centrally placed charges of dense explosives. A description of such trials is given in Association of Petrochemicals Industries Japan (1974). These trials demonstrated that overpressures of up to about 20 bars could be generated when some 640 cu metres of 7.4% ethylene/air mixture were detonated by 20 gram charges of pentolite.

The question which has to be decided is whether or not detonations can occur in industrial circumstances in which the initiating mechanism is a relatively low energy source such as a naked flame, a cigarette lighter, or an electric spark of 200 to 400 volts. It is the general view today that detonation is extremely unlikely in such circumstances. To quote Pikaar (1984):- "Analyses of accidental explosions showed that the damage observed did not correspond to that produced by detonation. Furthermore, the development of cloud combustion theories which linked the development of pressure to the flame speed . . . showed that pressures of 0.3 bar, sufficiently large to account for damage observed in accidents, could be generated by flame speeds in the 150 to 200 m/s range.

So, gradually, scientific activity has moved away from a preoccupation with the detonation issue and is now concentrated on investigating causes of flame acceleration and assessing the magnitude and duration of the pressure waves which result . . ." In short, VCEs are deflagrations, not detonations.

The case of the VCE at Port Hudson in 1970 is one in which a number of investigators were convinced at the time that what occurred was a detonation. The Port Hudson incident is analysed below as a case history where it will be argued that the hypothesis of a detonation is not necessary.

12.4.5.2. The yield of VCEs.

By yield is meant the proportion of the total chemical energy present in the system which is released as blast energy. It has long been known that the energy of combustion of typical hydrocarbons exceeds the blast energy of typical high explosives by a factor of about ten. "There is more energy in a candle than in a stick of dynamite". To quote some figures, the heat of combustion of propane is about 5×10^7 kJ/kg [Weast (1981)], whereas the heat of detonation of TNT is about 4.2×10^6 kJ/kg [Kirk-Othmer (1980), Vol. 9, Page 586]. However, this is not strictly comparing like with like as the heat of combustion of TNT is much higher than its detonation energy at 1.5×10^7 kJ/kg (ibid), i.e. the detonation energy of TNT is only 28% of its heat of combustion. This, in itself, should be an indication that the energy content of a hydrocarbon gas cannot be fully realised as blast energy. In fact to postulate this is to postulate the existence of a perfect machine for converting chemical into mechanical energy, a machine where the final temperature is the same as the initial and where no heat is lost. Such a machine, though eagerly sought, has never been found. Wiekema (1980) suggests that results reported in Kogarko et al (1966) point to a 25% conversion of chemical energy into blast energy in a detonation. This might be taken as theoretical maximum yield from a VCE. Expressed in terms of TNT, this is roughly 2.5 to 3.0 tonnes of TNT per tonne of hydrocarbon. It seems reasonable to suppose that such a yield is unattainable in a deflagration.

Virtually all deflagrative experiments to date which have generally involved less than 1 tonne of hydrocarbon, have demonstrated either negligible overpressures or pressures of the order of a few millibars. In terms of yield this means that experimental trials have failed to produce any of significance. On the other hand, there have been numerous examples of VCEs which have produced a significant yield. These, in some cases, have been subjected to estimates based on damage analysis and TNT equivalences have been calculated. Fig.1 of Gugan (1979) plotted estimated yield against quantity of flammable material and Fig.2 of the same reference plotted estimated yield against a factor which combined the quantity of flammable material with the thermochemical properties of the material (heat of combustion, range of flammability and burning velocity). In neither case was there are apparent correlation. This may well have been due to the imprecise nature of the data, some of which seem very dubious. However, Wiekema (1984), using a plot of yield versus equivalent TNT, claims there is evidence that yield increases with size of explosion. In the crude sense, this must be so as it cannot be otherwise if, as has been amply proved, small releases have a yield of nil. However, Wiekema's chart shows yields of up to 60% which is twice the figure for detonations !!! The problem is only partly that estimating equivalent TNT is far from being an exact science. For, in addition to the problem of characterising damage, there is the problem of estimating how much material has been spilled, how much of the spillage is in the form of vapour, how much of the vapour is present in concentrations which lie within flammable limits, and how much of this vapour contributed to the explosion. The case history of Decatur 1974, given in Chapter 13, discloses some of the difficulties and demonstrates that misleading conclusions may be drawn unless the data are thoroughly and critically analysed. A rough estimate, based upon the present author's experience, would suggest that, for 10 tonnes to 40 tonnes of flammable material in a vapour cloud, a yield of 4 to 6% of the heat of combustion as blast energy is a reasonable assumption. This is equivalent to about 0.5 tonnes of TNT per tonne of hydrocarbon.

12.4.5.3. Maximum overpressure.

It is of considerable importance to have an estimate of the likely maximum value of overpressure which may be generated in a VCE. Marshall (1976) characterised a VCE as "soft-centred" and suggested a maximum overpressure of 1 to 2 bars. Later that year, in Marshall (1976a), this estimate of overpressure was reduced to 1 bar maximum. Since that time, other investigators have addressed themselves to the problem and a variety of answers have been put forward. Gugan (1979) put forward (Table 4 of his Appendix) 15 to 17 bars as being necessary to produce the crushing effects on two of the Flixborough vessels. Gugan adduces also (Page 67) a figure as high as 44 bars at Flixborough and 47 bars at Beek. These observations of Gugan have led to sharp controversy notably by Ale and Bruning (1980), and Fu and Eyre (1980). Ale and Bruning claimed that any deduced values of overpressure which exceeded the ca 8 bars which is attained as a final state by vapour/air mixtures under constant volume and adiabatic conditions, would violate the laws of thermodynamics. This statement cannot be true. There is ample proof, both theoretical and empirical, that in detonations of vapour/air, overpressures of the order of 18 bars can be attained [see Table 3 of Bradley (1978)]. Nor does it accord with the long known and well attested phenomenon of "pressure piling" in which compartments of a relatively small volume can build up pressures up to four times greater than are generated by a deflagration in a vessel to which they are connected, [see Fitt (1981) for a review of the subject]. As was pointed out in Phillips (1981), the laws of thermodynamics govern the maximum mean value of pressure generated in the system, not the maximum pressure in any of its component parts. In this paper, Phillips criticised Gugan's views but did not accept the more extreme criticisms of Ale and Bruning. Phillips considered that the methods adopted by Gugan to calculate the pressures necessary to crush the vessels were unsound; however this reference argued that, in the centre of the cloud, the negative phase of the blast wave returning the explosion centre might impinge upon itself to produce highly localised high pressure zones which might have produced damage to a drain cover which Gugan cited as being caused by overpressures well in excess of 10 bars.

Advisory Committee on Major Hazards (1979), in discussing the design of buildings, accepted the view of the UK Chemical Industries Association that buildings in the vicinity of possible VCEs should be designed to withstand 0.7 bar with a duration of 20 ms. Giesbrecht et al (1981) expresses the view that maximum overpressure at the edge of the cloud is of the order of 0.3 bar. Wiekema (1984) also expressed the view that, for medium reactivity, vapour overpressures of the order of about 0.3 bar are likely to be the maximum encountered. This may, however, reflect, without direct attribution, the views of Giesbrecht et al.

This subject is discussed below in Chapter 20 which deals, inter alia, with the protection of buildings.

12.4.5.4. The substances which give rise to VCEs.

Marshall (1979) analysed the causative agents responsible for 177 vapour cloud incidents, not all of them explosions. This analysis showed that hydrocarbons with 3 and 4 carbon atoms accounted for 40% of known incidents. Propane and butane and their mixture, LPG, accounted for three times as many incidents as petrol (gasoline) vapour. This is in spite of their much lower rate of production, ca 1,600,000 tonnes in 1979 as compared with 16,600,000 tonnes of petrol [see Central Statistical Office (1980)]. Though these figures relate to the UK only, they seem to be indicative of world ratio of LPG to petrol. On this basis these hydrocarbons are about 30 times more likely, tonne for tonne handled, to produce a vapour cloud incident than petrol.

Davenport (1984) lists 71 incidents which are considered by that author to be VCEs. Davenport comments "When Industrial Risk Insurers first began to study VCE incidents in the early sixties, we believed that . . . LPG type materials were the only materials of concern. At that time we were aware only of losses employing LPG flammables . . . After the Flixborough disaster it became apparent that explosions of clouds of (other) flashing flammable liquids were capable of producing damaging overpressures".

Wiekema (1984) divides materials into low, medium and high reactivity, without defining the categories or without giving examples of substances falling into the three categories. Out of a total of 136 cases, high reactivity substances (probably hydrogen, acetylene or ethylene) and low reactivity substances (probably methane) between them only account for six incidents, not all of which were VCEs. This analysis does not contradict the analysis given in Marshall (1979).

Two substances seem to merit further comment as possible causative agents. These are methane (natural gas), and hydrogen, and they are discussed below.

12.4.6. METHANE AND VCEs.

Methane deserves special comment because of the very large quantities handled industrially. The annual production by the UK of North Sea gas (ca 90% methane) in 1983 was of the order of 50 cu kilometres or about 35 million tonnes. Today the USA and the USSR each produce about ten times that quantity per annum. Most of the gas goes into heating and cooking but significant quantities are used for chemical synthesis. Considering the quantities handled there is little evidence of the involvement of natural gas in VCEs. Warner (1976) claimed " . . . that no case of an unconfined vapour cloud explosion with natural gas has been recorded. Methane is a very stable hydrocarbon and has been made to detonate only in mixtures with oxygen and (with) very large initiating charges of TNT . . ."

However, a case has since been made in Gugan (1979), Page 46, that there was a VCE at Raunheim, West Germany, with a yield of 0.5 to 1 tonne TNT equivalence from a release of 0.5 tonne of liquid methane at high level. Davenport (1984) takes this up "As is indicated by (the Raunheim incident) ignition of a spillage of methane in an industrial plant with partial confinement and bluff body interference from process equipment and structures can produce damaging overpressures and a $15.6 (1983) loss".

To summarise why methane has seldom, if ever, resulted in a VCE, the following factors, some of which have already been referred to, seem relevant.

(a) When spilled as a cryogenic fluid vaporisation is restricted by the rate of heat input from the surroundings unlike flashing liquefied gases where the energy for evaporation is drawn from internal energy.

(b) When released as a gas from pressurised containers at ambient temperature, methane is lighter than air and does not form ground-hugging clouds.

(c) Methane requires rather more energy to ignite it, and much more energy to detonate it, than do other common flammable gases.

(d) Methane has a low level of reactivity compared with other flammable gases.

12.4.7. HYDROGEN AND VCEs.

12.4.7.1. The general problem.

A number of the compilations listed earlier in the book give, as their first example, the explosion of a dirigible (i.e. steerable) airship, the ZR/2 or R38, over the River Humber, England, on 23/8/21. [See Gugan (1979) and Davenport (1984) - actually the date they give is wrong, it was 24/8/21]. From this it might appear that hydrogen is a possible cause of VCEs. There has, moreover, been a paper, Bulkley and Jacobs (1966), which listed a number of incidents involving hydrogen and which drew attention to the rapid growth in the scale of processes handling it, from which hydrogen is frequently vented off in considerable quantities. On the other hand, as will be shown below, there have been numerous examples of hydrogen-filled airships which have caught fire without significant blast effects being recorded.

The scale of hydrogen consumption in the USA is given in Shreve and Brink (1977), Page 98, as 2.5×10^{12} cu ft, which is equivalent to ca 70 cu kilometres or roughly the volume of natural gas consumed in the UK. One-third of the hydrogen is used in ammonia synthesis and a further important use is in hydrogenating vegetable oils. Davenport lists 3 incidents out of 71 VCEs as involving hydrogen which is small considering the scale of usage. Bulkley and Jacobs (1966) lists rather more minor incidents in addition to a substantial VCE at Jackass Flats, Nev., USA, which had been fully described in Reider et al (1965). (This latter reference includes "Unconfined . . . Explosion in its title). Gugan (1979) quotes from this reference extensively and concludes that about 90 kg of hydrogen gave rise to 27 kg of TNT equivalent or a yield of 1%.

12.4.7.2. Properties of hydrogen.

British Cryogenics Council (1970) summarises the relevant data on hydrogen from the point of view of safety. Hydrogen/air mixtures have a wide range of flammability (4% to 74%) and, when diluted with inert gas, hydrogen can still burn with only 5% oxygen compared with ca 11% for hydrocarbon gases. Hydrogen has a high burning velocity compared with hydrocarbon gases. It can be ignited by low energy sparks as it requires only about 1/10 of the energy that is required by hydrocarbon gases. Hence it is easily ignited by static electricity. (This may account for reports that hydrogen can ignite spontaneously, VM).

On the other hand, opinions differ as to its ease of detonation by high explosives. Atkinson et al (1980) claim detonation of hydrogen/air mixtures by milligram quantities of pentolite, but quote Matsui and Lee (1978) as claiming that hydrogen is harder to detonate by this means than any common gas except methane.

12.4.7.3. The literature of airship accidents.

Many general compilations of incidents involving flammable gas clouds refer to three examples of airships catching fire. The first of these is Dirigible ZR/2 (otherwise known as R38) of 24/8/21, the second is the R101 disaster of 5/10/30, and the third is the Hindenburg disaster of 6/5/37. Such an approach is anecdotal rather than analytical because the data on incidents involving the destruction by fire of airships filled with hydrogen are quite extensive. There is a considerable literature on airships, their history, their technical development and their eventual fates. But none of it is written from the standpoint of the study of the behaviour of large masses of hydrogen on ignition. However, the literature provides many eyewitness accounts and many photographs, some actually taken while the airship was still burning, and one film, that of the

Hindenburg disaster, which shows the whole incident from beginning to end.

The Imperial War Museum issues book lists of the material in its possession on a wide range of subjects and lists Nos. 1030, 1204, and 1332 are the most relevant to the matter under discussion. Because the literature is so vast, the present author will refer only to material on this subject which he has in his possession.

Encyclopedia Britannica (1872), [Vol 1, Pages 195 to 207], is a nineteenth century article on aeronautics which is concerned almost entirely with hot air and hydrogen balloons. It is a history of the first century of lighter than air flight. It points out that Charles (of Charles' Law fame) launched a hydrogen-filled balloon within weeks of the historical ascent of the Montgolfier brothers' balloon. During the period covered the hot air balloon seems to have been more popular than the hydrogen-filled balloon. Moreover, the reference claims that the use of hydrogen rapidly fell out of favour after about 1820, becoming superseded by coal gas which has a density about 0.4 that of air. (This compares with a value for hydrogen of about 0.07. Later, commercial, developments enforced the reinstatement of hydrogen on account of its superior lifting properties). The article is interesting in that it contains an estimate that, up till 1838, some 471 individuals had made balloon flights, many more than once, and only 9 were killed. Six of these were in fire balloons. In the period 1838 to 1870, according to the author, a further 3 died out of thousands to be added to the list of those ascending and one of these was attempting a parachute jump. This all adds up to a lack of any evidence of peculiar dangers from hydrogen or coal-gas-filled balloons. It falls under the heading of what historians call "The argument of silence", i.e. if we haven't heard about it, it is because it didn't happen. Alternatively, if it had happened we would have bound to have heard of it. In the matter under discussion in the article, there is no mention of explosions. The "argument of silence" is the more persuasive the greater the number of occurrences from which no evidence emerges. It is also more persuasive if the historian appears to have no axe to grind. It is difficult to see any evidence that the author of the Encyclopedia Britannica article would have had cause to suppress information on explosions.

12.4.7.4. An analysis of airship accidents.

The beginning of the twentieth century saw a sudden step forward in the development of the rigid dirigible airship under the guidance of Graf Ferdinand von Zeppelin. Whereas Charles made his ascent in a balloon with a volume of 325 cu metres, the first Zeppelin, the LZ1, had a volume of 11,400 cu metres. The final products of the Zeppelin works had volumes of ca 200,000 cu metres. The Zeppelins were always filled with hydrogen and the two largest ones, referred to above, each contained about 20 tonnes of the gas.

The author has conducted a statistical analysis of the history of rigid airships constructed in Germany and Britain during the twentieth century. This analysis took, as its starting point, Ventry and Kolesnik (1976), which is an outline history of all the world's airships but the analysis is confined to the rigid airships of Germany and Britain. The information in this reference has been checked against Meyer (1980) which gives fuller particulars for the German airships built by Zeppelin and by Schütte-Lanz. It has also been checked against the information in Morris (1969) for airships shot down or otherwise wrecked in air raids against Britain 1914-18 and against Deighton and Schwartzman (1978) which gives particulars of a number of airship disasters. The results are summed up in Table 12.8. The total number of airships studied was 130.

TABLE 12.8. STATISTICAL ANALYSIS OF AIRSHIPS AND THEIR FATES.

Eventual fate.	No. in category.	% of total.
Demolished or otherwise decommissioned.	44	34
Wrecked in flight or on forced landing, no fire.	35	27
Wrecked in shed or on airfield, no fire.	6	4.6
Caught fire, or set on fire by military action, in flight.	20	15
Caught fire, or set on fire by military action, in shed.	21	16
Exploded in flight.	3	2.5

Thus only one-third were destroyed by fire though the German airship fleet was heavily engaged in air raids on Britain and a number of other countries. Only one in forty are reported as having exploded.

There was a serious disaster at Ahlhorn, North Germany, on 5/1/18, when 5 airships were lost in their sheds. Meyer and Ventry both describe the event as an explosion, but Deighton and Morris both describe it as a fire. According to Deighton, it occurred when the gas bags of one of them were being filled. "Roaring along the gas lines from shed to shed, the fire reached across the long spaces between them. It burned out two of the giant double hangars and severely damaged two others". Deighton has a photograph of one of the gutted hangars which retains its structural framework and a substantial part of its cladding. Morris refers to the sheds as being "Razed to the ground", a claim which the photo cited above does not bear out. 15 men lost their lives in this disaster.

On 19/7/18, two airships were set on fire in their shed at Tondern, North Germany, by bombing. Meyer shows a photo of the burned out wreckage of one of them lying in what appears to be a largely undamaged shed. There is certainly no evidence of explosion to judge by the photo. There was a similar successful attack on 7/6/15 on an airship in its shed at Evere, near Brussels. Deighton gives a photograph of the shed with flames issuing from its open end.

None of these incidents give any clear indication of explosions, only of fierce fires.

There are clear accounts of airships being set alight by aircraft in Morris. It is clear from this reference that it took a considerable time for the air defences to realise that bullets were ineffective if fired into the gas bags of the airships and it was not until the development of exploding incendiary bullets that such tactics became successful. (This first occurred on 2/9/16. The airship was one with a wooden framework which burned for two hours). The descriptions of airships being set on fire as given by Morris bear a marked similarity to each other. Morris gives little information on the time taken for an airship to come down except (Page 167) "L48 did not come down with the rapidity usual in such catastrophes, the fall taking from three to five minutes". Morris refers (Page 152) to a pilot "his face was scorched by the heat of the flaming mass which he estimated to be some 300 yards away from him".

Some case histories of airships are given in Chapter 13.

12.4.8. SOME GENERAL CONCLUSIONS ON VCEs.

12.4.8.1. Confined or Unconfined ?

The wheel appears to have turned full circle. Prior to Strehlow (1972), there was extreme scepticism as to whether the ignition of a vapour/air mixture could give rise to significant overpressure. Then, from 1972, the term "unconfined vapour cloud explosion" came into common use. Now, in the eighties, many authorities deny that blast effects can arise without the presence of obstructions and of some measure of confinement.

Experimental evidence seems to confirm this as field trials suggest that, on open ground, pressure effects are negligible whereas, when obstacles are provided, there is an unmistakable development of overpressure. However, to date, none of these trials has produced overpressures comparable with those experienced at Ludwigshafen, Port Hudson, or Flixborough, all of which are described in Chapter 13. However, the quantities of vapour used in field trials have been much smaller than those involved in these disasters.

If the arguments of Pikaar, Zeeuwen, and Wiekema cited above are correct, it should follow that there should be no case in which a VCE occurred except in the presence of obstructions and of at least partial confinement.

There is, however, one case history which seems to cast doubt on the proposition and that is the Port Hudson VCE of December 9th, 1970. The official report on this incident, National Transportation Safety Board (1972), took the view that what had occurred was a "detonation". The circumstances were that a pipeline carrying propane burst. 24 minutes later there was a severe explosion which may constitute the largest such incident to date. Some 75 tonnes of propane vapour were involved. The ground over which it occurred was largely open but with some trees which were bare of leaves because of the time of year. There seems little doubt that the conclusion arrived at in Burgess and Zabetakis (1972), that the initiating event of the explosion was an internal explosion in a warehouse built of concrete blocks. The reference gives the dimensions of the building as 17 x 10 metres but the height is not stated. If the explosion occurred in the ground floor, this could be assumed to be 2.5 metres in height giving a volume of 420 cu metres. Burgess and Zabetakis estimated the flammable portion of the cloud to be 1,100,000 cu feet (31,000 cu metres). Thus the explosion in the warehouse was assumed to have been the sole causative agent for a VCE of about 60 times its volume. This cannot altogether be ruled out but the implications of the Port Hudson explosion do not give clear support to the hypothesis under discussion.

12.4.8.2. Case histories where VCEs did not occur.

It seems clear that there have been case histories in which the factors of obstructions and partial confinement were present but in which a VCE did not occur.

One of these is Cleveland 1944, which has been described above. Here there was a typical process plant landscape with many buildings, some of them houses, in the vicinity. There seems evidence of confined explosions but not of escalation to a VCE.

A second is San Carlos 1978 also described above. The terrain included many obstructions such as trees, parked cars, caravans and tents. There seems evidence of internal explosions in three buildings.

A third is Mexico City 1974 also described above. Here there was an archetypal industrial landscape together with close-packed housing and bearing similarities to the Cleveland 1944 locale. Here too there was evidence of confined explosions in houses.

12.4.8.3. The factors which may determine escalation to a VCE.

Though it seems apparent that the presence of obstructions and partial confinement play a major role in the escalation of a vapour cloud fire into a vapour cloud explosion, there are obviously other factors, some of which are set out below:-

(1) The clearly inescapable necessity is the presence of a vapour or gas which is flammable. But flammability may not, of itself, be sufficient as the controversy over natural gas shows. So reactivity of the vapour may play a part.
(2) Also an inescapable necessity is the presence of a source of ignition.
(3) At least some part of the cloud must have a concentration of vapour which lies above the lower flammable limit at the time of potential ignition.
(4) There must be a sufficient mass of vapour in the flammable range. This mass may be limited by the quantity spilled, or by the time which elapses to ignition in the case of otherwise sufficiently large spillages.
(5) The presence or absence of mechanisms for mixing are major factors which determine (2) and (3) above.
(6) Heavy gases or vapours are much more likely to give rise to persistent flammable clouds than are lighter gases such as hydrogen, methane and ammonia. (These latter two are both lighter than air when released at or near atmospheric temperature).
(7) The presence of obstructions and of possibilities for semi-confinement.

It may be argued, therefore, that (7) will only be operative if the other factors are present and that, even if they are, it does not necessarily follow that a VCE will occur.

Chapter 13

CASE HISTORIES OF RAREFIED EXPLOSIONS

13.1. INTRODUCTION.

The chapter provides case histories of dust explosions, confined gas explosions, and vapour cloud explosions.

Because of the importance of vapour cloud explosions as major chemical hazards, extended treatment is given to a number of such cases. Some case histories of airship accidents are given in which a VCE did not occur. These are included at this point to see if any conclusions can be drawn as to the circumstances in which hydrogen can give rise to a VCE.

According to Kauffman (1980), during an eight-day period in December 1977, five grain dust explosions occurred in the US causing a total of 59 deaths, 48 injuries and extensive property damage. Two case histories presented below were the worst of the five. It should not be thought, however, that five explosions in eight days represents a true picture of the rate at which such accidents occur in the US; there seem to have been peculiar weather conditions which had affected the nature of the grain being handled.

13.2. WESTWEGO, LOUISIANA, US. GRAIN DUST EXPLOSION.

This disaster involved a series of explosions during the morning of 22nd December, 1977. The death toll was 36. The installation was a grain elevator which consisted of 73 reinforced concrete silos each 35 metres high and with diameters ranging from 8 metres to 10 metres. In common with other such installations, the silos were closely packed together especially the group most recently erected which consisted of 28 silos of 10 metres diameter packed together on an equilateral triangular pitch. The total volume of the installation was thus ca 150,000 cu metres and was therefore, by US standards, a medium-sized installation.

At one end of the installation was a 75 metre high "head house" which accommodated the bucket elevators. The top 20 metres of this was fitted with light panels for explosion relief but the initial explosion brought down the top 35 metres. This crashed down onto a complex of buildings which housed offices, control room, laboratories, and the canteen. Next door to this complex of buildings was the maintenance block.

More than half of the silos were totally destroyed and others were severely damaged. The source of ignition has not been determined but attention has been drawn to the very dry weather with low relative humidity which had been experienced prior to the disaster and which may have made ignition easier.

13.3. GALVESTON, TEXAS, US. GRAIN DUST EXPLOSION.

This grain elevator disaster occurred during the night of 27th December, 1977, and involved a series of explosions. The death toll was 18 with 22 injured but this could have been much higher had the explosions occurred in the daytime.

The installation was very similar to that of Westwego with 60 silos of 8 metres diameter and 38 metres high. The total capacity was thus about 115,000 cu metres and the elevator was, therefore, somewhat smaller than Westwego.

The head house was 70 metres high and was fed by conveyor belts housed in a below ground tunnel. It is believed that the explosion was propagated through this tunnel, but the initiating source is unknown. There was virtually no explosion relief provided.

The destruction included the rail shed in which off-loading was taking place at the time of the explosion, the conveyor tunnel, and the central control room. The main office block and the head house were severely damaged.

13.4. BREMEN, WEST GERMANY. FLOUR MILL EXPLOSION.

This explosion took place at 21.24 on 6th February, 1979, and either destroyed or seriously damaged the greater part of the 40,000 m^2 Roland Mill complex in the docks area of Bremen. The death toll was 14 and 17 were injured out of about 50 people in the complex.

The explosion is described in detail with many photographs in Bremen (1979), which contains commentaries by the Fire Service and Police, by the Bonn Dust Research Institute, and by an insurance company.

A plan of the closely packed site is given in Figure 13.1., which shows the areas chiefly affected. There was almost total destruction of a seven-storey flour store, a six-storey mill, with a small mill, and of the southern part of the administration building. The centre part of the administration building and two silos were severely damaged. Lesser damage occurred to practically all remaining buildings on the site. A severe conflagration occurred which took about 24 hours to extinguish and then not completely as residual fires continued in the flour in silos up till March 10th, i.e. for more than four weeks.

As with many such occurrences, there is no completely certain scenario of the course of events. The conclusion as to the most probable series of events was that a fire of unknown cause started in the testing room in the basement of the waterfront store. This led to a primary explosion (possibly due to the ignition of gases given off by the pyrolysis (charring) of flour). This then stirred up dust on the upper floor which gave rise to secondary and powerful explosions on the upper floor. These were then propagated by the north west conveyor bridge, (about 40 metres long by 2 metres in width), into the flour store from which a series of explosions were generated in other parts of the complex leading to the massive fire.

The explosions shattered windows up to 2.0 km away and large lumps of masonry were projected for distances up to 100 metres, and a feed stuff store belonging to a different company and separated by a street was destroyed by blast.

A feature worth noting is that the pressure wave struck a parked lorry and threw it against a building. The contents of its load were spread into the air and are believed to have produced open air dust explosions.

Tests conducted by the Federal Institute for the testing of materials, established that the flour produced by the mill was mainly $\lesssim 63$ microns in particle size and that the dust was explosible producing experimental overpressures of up to 9 bars.

Of the 14 dead, there was no trace whatsoever of 7 of them, even though the debris was searched by tracker dogs. Some parts of the rubble showed

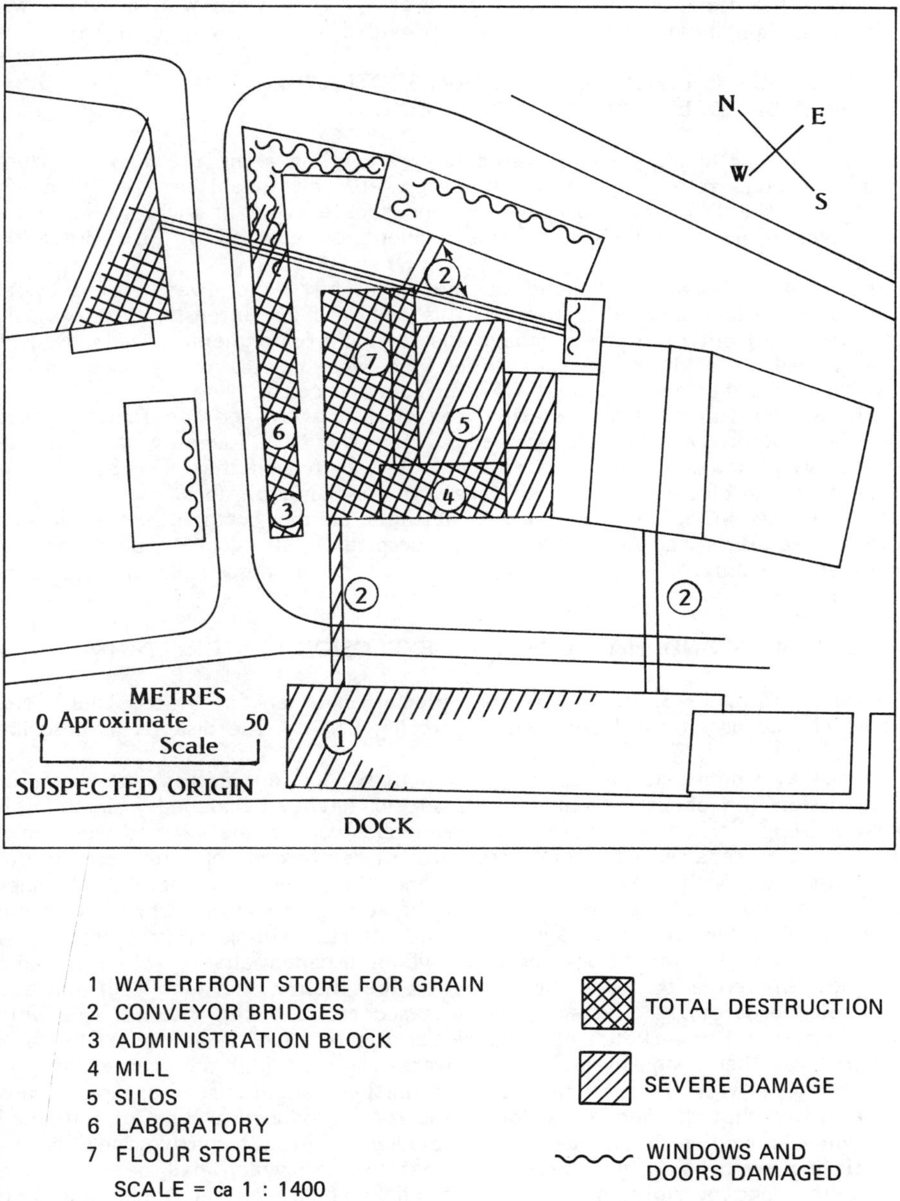

Fig. 13.1. PLAN OF DAMAGE AT BREMEN.

no trace of any combustible material even though the buildings had contained heavy wooden floors and joists. Temperatures of at least 1,000°C were maintained for several hours, which would be capable of totally cremating bodies.

The Report makes it clear that the buildings were fairly old, most being built during the time of the First World War. They were well provided with fire fighting equipment, but nothing was provided to deal with dust explosions.

13.5. COCKBURNS PATH, Near DUNBAR, SCOTLAND. POSSIBLE OPEN AIR DUST EXPLOSION.

On 2/3/82 at 8.29 hours, an articulated lorry travelling along the busy A1 trunk-road in Scotland, at a point at which there were few inhabitants, collided with a cottage. The lorry was carrying 19 tonnes of a resin in the form of a fine powder packed in plastic bags. The incident is reported in Fire Prevention (1982a).

The impact threw up a cloud of the resin dust which was ignited by the severed mains intake to the cottage. This caused a low intensity dust explosion which had the effect of scattering bags and setting fire to them. The subsequent fire gave rise to noxious fumes. The wind was blowing out to sea and the fumes affected fishermen in a boat 1.5 km from the coast.

The nearest full-time fire station was 46 km away and the fire had taken a good hold before the brigade arrived. The fire was fought with foam and, when this was exhausted, with water fog. It was feared that the use of water jets might stir up clouds of dust with danger of further explosions.

No one was killed, but the two inhabitants of the cottage, the driver and 24 other road users had to be treated in hospital. It would appear that the injuries were mainly, if not exclusively, due to the fumes. No fire personnel were injured.

13.6. STATEN ISLAND, N.Y. CONFINED EXPLOSION IN STORE TANK.

On February 10th, 1973, there was an explosion in a tank at Staten Island, New York, which had previously been used for storing LNG. The disaster is described in Davis (1979).

The tank, which had a capacity of about 95,000 cu metres, had been out of commission for about a year on account of having developed a leak on its internal lining. It had been purged of flammable vapour and 42 men were employed inside it on repairs. The tank itself was of reinforced concrete construction. It was insulated with polyurethane foam and the internal lining was of sheet aluminium laminate. An internal explosion lifted the steel roof which, according to Davis (1979), then crashed down into the interior killing or trapping the 40 men working on the tank floor, none of whom escaped alive. Two workmen inside the tank on a scaffold about twenty feet below the tank roof managed to scramble to safety. The explosion was succeeded by an internal fire which raged for hours. For a sketch of the tank see Figure 13.2.

There has been considerable controversy surrounding the causes of this disaster. One view is that the foam insulation caught fire, perhaps because of a spark, and that the decomposition of the foam produced a cloud of flammable smoke which mixed with air and then exploded. An alternative view is that the initiating event was the ignition of natural gas which had been occluded in the lining, and possibly in the earth embankment round the tank, and which had seeped out into the tank. A contributory factor here may have been the fall in atmospheric pressure which occurred prior to the explosion.

It seems to be generally agreed that the fire which succeeded the explosion was fuelled by the foam plastic lining and that fire precautions inside the tank during the repairs were non-existent.

In deciding between the two conflicting views, the time scale has to be taken into account. The evidence suggests that the initiating event was sudden and this strongly argues against the concept of the flammable vapour originating in the decomposition of the foam. That this can give rise to explosions is well attested by the experience of fire brigades in domestic fires, but it is well known that such explosions are preceded by the emission of clouds of choking and toxic smoke. This view that the explosion was too sudden to have originated in the thermal decomposition of the foam was put forward by Dr. M. Zabetakis of the U.S. Bureau of Mines in his testimony to the 1973 hearings.

It is impossible to ignore the circumstance that, following upon the public outcry consequent upon the Cleveland 1944 disaster, the use of LNG for peak shaving operations had been discontinued for nearly thirty years in the US. The Staten Island installation represented a return to this technology and now there had been a second disaster. To admit that it had been caused by NG would have intensified public criticism and might have led to the technology becoming outlawed. Thus a candid admission that NG had caused the initiating event was probably out of the question.

13.7. SHEFFIELD, ENGLAND. CONFINED EXPLOSION IN STORE TANK.

On 24th October, 1973, there was an explosion in a 45,000 cu metre underground tank at the Effingham Street, Sheffield, gas works. The tank, which had been out of use for eighteen months, was constructed of reinforced concrete. It was being converted by contractors and flame cutting operations were in progress. Three men were killed and twenty-nine people were injured, one of them seriously; the injured included work-people in nearby factories and members of the public. The body of one of those killed was recovered from a catwalk 30 metres above the ground. Damage was sustained over a radius of about half a kilometre; blocks weighing half a tonne were projected by the explosion and 60 cars were damaged. The concrete lid was blown off, it turned over and fell back into the tank upside-down.

The cause of the explosion is now believed to have been due to a failure to purge the tank thoroughly of its previous contents, primary flash distillate, which had been used to make towns gas. This material was found to be present in the debris and to have a low flash point. According to Her Majesty's Chief Inspector of Factories (1975), some 200 kg of distillate were subsequently recovered from the tank; the Report calculates that perhaps a further 350 kg may have evaporated to produce an explosive mixture in the tank.

The ignition source appears to have been a flame cutting operation which had been conducted by the man whose body was found on the catwalk.

13.8. BANTRY BAY, EIRE. CONFINED EXPLOSION IN A SHIP.

This disaster was primarily a major conflagration but it was accompanied by severe explosions. The event is fully described in the Report of the Official Inquiry [Whiddy Island (1980)] which, in the present author's view, is a model of its kind.

In the early hours of the 8th January, 1979, the Betelgeuse, a crude oil tanker of 62,000 gross registered tons (121,000 tons deadweight), which was moored at Whiddy Island, Bantry Bay, Eire, caught fire and later was subject to internal explosions. The entire crew, two passengers, and the jetty staff on duty, all lost their lives, making a total death toll of 50. The ship was a total loss and the jetty was severely damaged.

The cargo had been partly discharged and the ship had taken in ballast. The Inquiry concluded that the ship had been improperly ballasted with the result that it was more buoyant fore and aft than it was amidships. This circumstance, taken together with the corroded condition of the ship, (some of the scantlings were reduced by 50% or more), caused the deck to buckle and the ship thus failed by "sagging". The ship then caught fire from an undetermined ignition source and an uncontrollable conflagration developed. Shortly after the failure of the deck, there were explosions in both permanent ballast tanks which contributed to the ship breaking its back. Some half an hour after the initiating event, there was a massive explosion in one of the aft cargo tanks which was empty of liquid crude. A substantial piece of this tank, weighing about 500 kg, was subsequently found on Whiddy Island at a point 600 metres away. Figure 13.3. is a sketch showing the successive stages of collapse.

The blast from this explosion seems to have been highly directional. It shook houses up to four miles away yet some witnesses, close at hand, hardly noticed it. The present author visited and inspected the scene shortly afterwards and noted that there was no evidence of blast damage to the buildings on the jetty.

It is of interest that in May 1980, in Rotterdam, a tanker, the Energy Concentration, broke its back through "hogging" which is the opposite of "sagging". This was caused by incorrect ballasting as with the Betelgeuse. In this case there was no fire. It should be noted that, unlike the Betelgeuse, the Energy Concentration was fitted with inert gas protection.

13.9. CONCLUSIONS FROM CASE STUDIES OF STORE TANK EXPLOSIONS.

One conclusion which may be drawn is that the present-day scale of storage facilities is such that, when filled with flammable mixtures, they are capable of giving rise to explosions which can lead to multiple fatalities and to serious property damage.

Their potential for giving rise to these effects is highest when the tanks are empty, or nearly empty, of liquid. However, this only applies to fixed roof tanks. Floating roof tanks, provided that they are functioning properly, do not have this drawback.

A stoichiometric mixture of a typical hydrocarbon fuel gas could be expected to produce blast effects equivalent to about 20 kg of TNT per 1,000 cubic metres of vapour/air mixture assuming that 2.5% of the heat of combustion appears as blast energy. The subject of the TNT equivalence of gas explosions has been discussed in Section 12.4.4.

A second conclusion is that storage tanks, after being used for the storage of flammable liquids, may retain their dangerous properties for months or years, indeed dangerous properties may reappear after a period of dormancy. It is necessary, therefore, to apply the most stringent tests before such storages are entered or subjected to any form of hot work.

13.10. ETHYLENE EXPLOSION, EDMONTON, ALBERTA, CANADA. CONFINED EXPLOSION IN A BUILDING.

This explosion occurred on 18/4/82 and is reported in Marsh and McLennan (1985) and by High (1982). An ethylene leak of between 220 and 5,000 kg of ethylene took place as the result of the fracture, due to fatigue, of a 3mm pressure gauge connection under a pressure of about 1,000 bar in an ethylene compressor house. The period of leakage had a duration of several minutes. The building was approximately 30 by 15 metres in plan by 10 metres high, i.e. about 4,500 cu metres in volume. The ventilation rate was 8 changes per hour. There was an alarm system set at 50% of the lower explosive limit.

The building, which was only a light building, was destroyed. The neighbouring control room was wrecked and there was glass damage over a radius of up to 7 kilometres. Estimates of TNT equivalent range from 400 to 700 kg. Property loss was of the order of US $22 million.

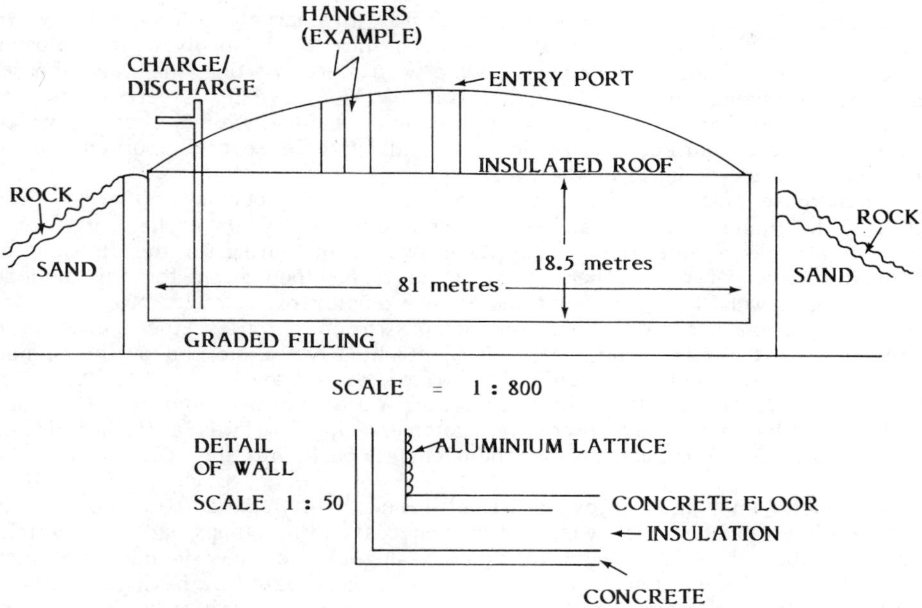

Fig. 13.2. THE STATEN ISLAND STORE TANK.

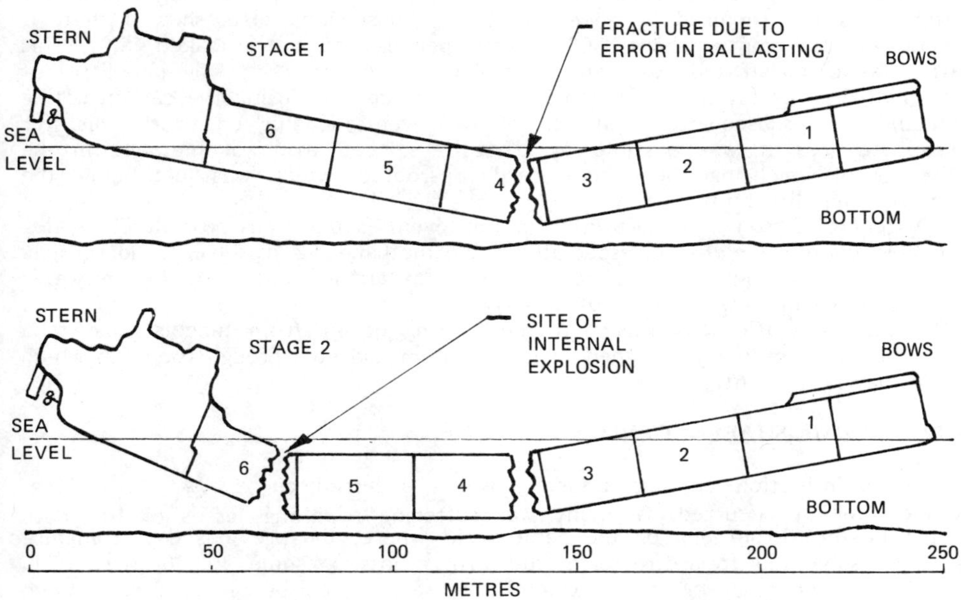

Fig. 13.3. THE DISINTEGRATION OF THE SHIP AT BANTRY BAY, EIRE.

13.11. ABBEYSTEAD, LANCASHIRE. METHANE/AIR CONFINED EXPLOSION.

This disaster occurred at 19.30 hours on 23/5/84 in a largely below-ground valve house by the side of the River Wyre in Lancashire. It involved an explosion of methane in air and it killed 16 people who were visiting the valve house. 8 died at the scene and 8 died in hospital. Of the dead, 3 were employees of the Water Authority. All of the remaining members of the party, which totalled 44, were injured, i.e. 28 in all. Many were severely burned, others were injured by the collapse of the roof.

The purpose of the visit had been to reassure local people who had alleged that a 12 km tunnel, which had been opened some five years earlier, and which drew water from another river, the Lune, was contributing to the flooding of the nearby River Wyre. (There was, in fact, no foundation for this belief). 36 of the party were local residents and 8 were employees.

The explosion occurred when, for demonstration purposes, the pumps were started up, for the first time for 17 days, at the River Lune end of the tunnel. If the connecting tunnel had been full of water, as it was supposed to be, water should have started to flow immediately. Water however did not flow and 18 minutes later, flow having not yet commenced, there was a flash followed by an explosion. The roof of the chamber fell back into the rooms where the party were.

Subsequent investigation by the Health and Safety Executive, reported in HSE (1985b), established that what had happened after the pumps had been switched on was that they had begun to expel a slug of methane/air mixture which had accumulated in the tunnel. This void had been created by having a washout (drain) valve open to drain away silt contrary to the established operating procedures. The design of the valve house was such that any air which accompanied the water was vented through a grilled floor into the valve house itself.

The cause of ignition has not been established, but there was no prohibition of smoking.

Inspection of the concrete tunnel showed it not to be water tight, but this feature was not required by specification. Subsequent tests showed that, if the tunnel was empty, about 1,000 tonnes per day of water leaked in. This water was demonstrated to contain dissolved methane (5 - 9 mg per litre) of geological origin. (That it was geological was established using the carbon dating method). A subsequent simulation of the events leading up to the disaster established that it was possible to create a concentration of methane in the valve house, which had been temporarily re-roofed, which was just below the lower flammability limit.

A general lesson to be drawn from this event is that it is possible for water to evolve dissolved methane, especially if the methane has dissolved underground under pressure, which gives a concentration in air in contact with the water which is above the lower flammable limit.

A more specific lesson is that the venting of air from tunnels must take place through a properly designed venting system and not through rooms in which people may be present.

13.12. LUDWIGSHAFEN, GERMANY. A VCE.

Detailed information on this disaster, which occurred on 29/7/43, is lacking. At the time it occurred, Germany was suffering appalling losses of life both on the Eastern Front and on the home front where, the previous day, Hamburg had been subjected to a firestorm raid with a loss of some 40,000 lives. In the circumstances an explosion which killed around 60 people, however much attention it would have attracted in peacetime, would be regarded as of secondary importance. Besides, there was a strict censorship which prevented such events from becoming public knowledge.

The earliest reference to it appears to be Stahl et al (1949), which is also an important reference for the next case history. It states (as translated by the present author):-

> "More recently, during the last war, on 29th July, 1943, the explosion of a tankcar containing 16.5 tonnes of a mixture of 80% butadiene and 20% butylene caused 57 deaths and injured 439. There was total destruction of buildings in the neighbourhood bounded by Buna Strasse in the north, Anilinfabrik Strasse in the east, Styrol Strasse in the south, and Diamin Strasse in the west".

The roads referred to can be clearly identified on the current factory layout plan published by BASF, Ludwigshafen (Figure 13.6). The dimensions of this block (C.500 on the current plan) are ca 350m (1100 ft) by 100m (330 ft) or 35,000 m^2 (370,000 ft^2). The block is about 700m (2300 ft) from the nearest point of the public domain in Ludwigshafen and about 1000m (3300 ft) from Mannheim across the river Rhine. Figure 13.4. shows the position of the centre of this explosion in relation to that of the 1921, Oppau, ammonium nitrate explosion, and the 1948 VCE which is described below. The area of severe damage is shown, along with that of the 1948 VCE, in Figure 13.6.

Giesbrecht et al (1981) discusses Ludwigshafen (1943), and plots the overpressure necessary to produce recorded damage to buildings and structures, using the criteria of Schardin (1954) and Glasstone (1964), against $r/E^{0.333}$ where E = heat of combustion times mass. For this explosion, the reference gives a value for E of 750 to 887 GJ. The source of the information on damage is not disclosed; presumably it came from the archives of BASF. The graph, Figure 7 of the reference, suggests a damage pattern similar to the 1948 explosion but rather less widespread damage than at Flixborough (shown in Figure 8 of the reference).

Giesbrecht et al attributes the 1943 explosion to hydraulic rupture following solar heating of an overfilled tank car with a delay to explosion of 10 to 25 seconds. The reference does not disclose its source for this information.

Davenport (1984) quotes a property loss of $53 million (1983) for the disaster.

13.13. LUDWIGSHAFEN - 28th JULY, 1948.

13.13.1. THE GENERAL PICTURE.

This disaster, which occurred almost five years to the day after the 1943 disaster, is considered by some to be almost a carbon copy of the previous disaster in that it has been attributed to the same cause, the hydraulic rupture following exposure to solar radiation of an overfilled car, this time containing dimethyl ether. However, as the present author will demonstrate, this is by no means certain.

There is much more information available on this incident than on the one previously discussed. The following sources were consulted:-

(a) Stahl et al (1949).
(b) BASF (1948).
(c) Giesbrecht et al (1981).
(d) Correspondence from BASF including the provision of photographs.

The principal facts are that a railtank car exploded on 28th July, 1948, in the Ludwigshafen works of BASF. The car was standing in Anilin Strasse at a point between buildings which were then numbered Lu21 and Lu47. The car suffered a catastrophic failure and discharged its entire contents of 30,400 Kg (66,900 lb) of dimethyl ether. There was a devastating VCE which killed 207 and injured 3,818, 500 of them seriously. All those killed were aged between 15 and 65 and only a few were women. There is no evidence that anyone was

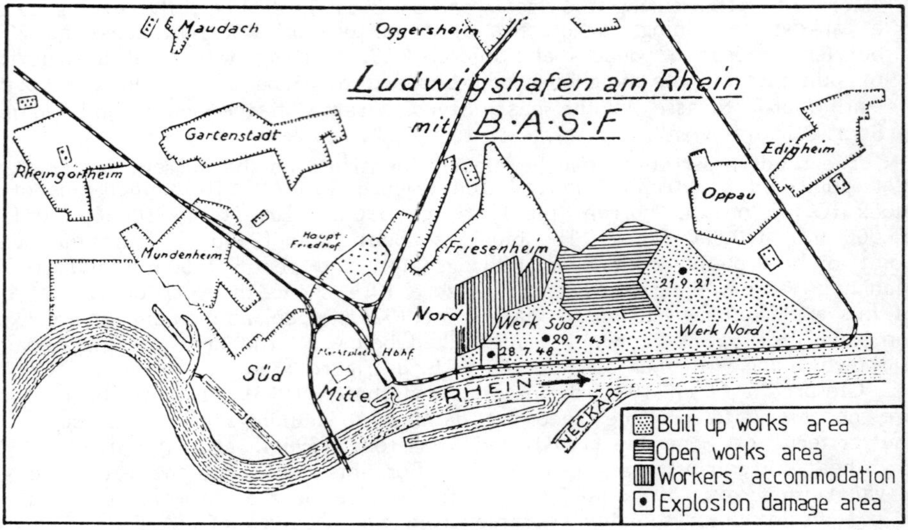

Fig. 13.4. SKETCH PLAN OF BASF SITE.

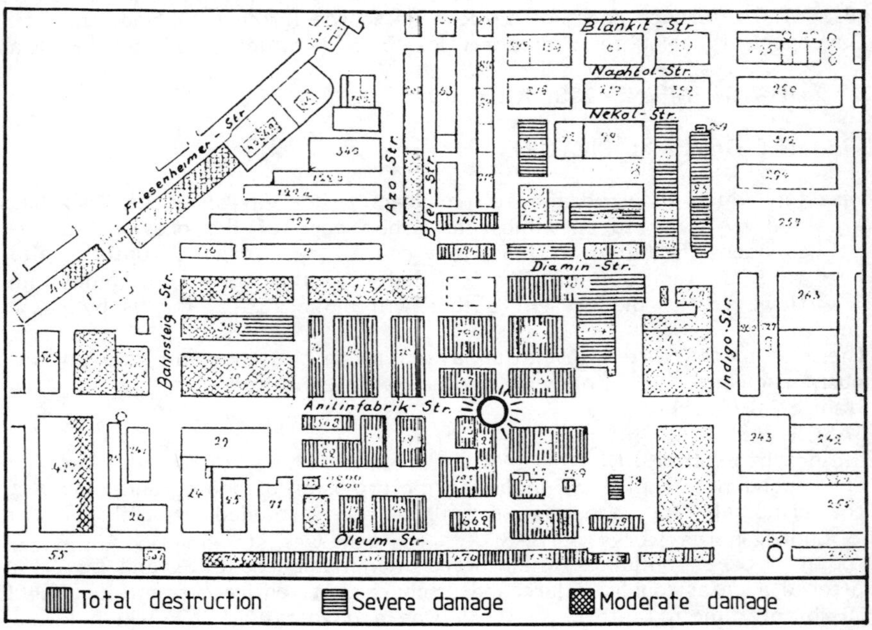

Fig. 13.5. DAMAGE FROM 1948 VCE.

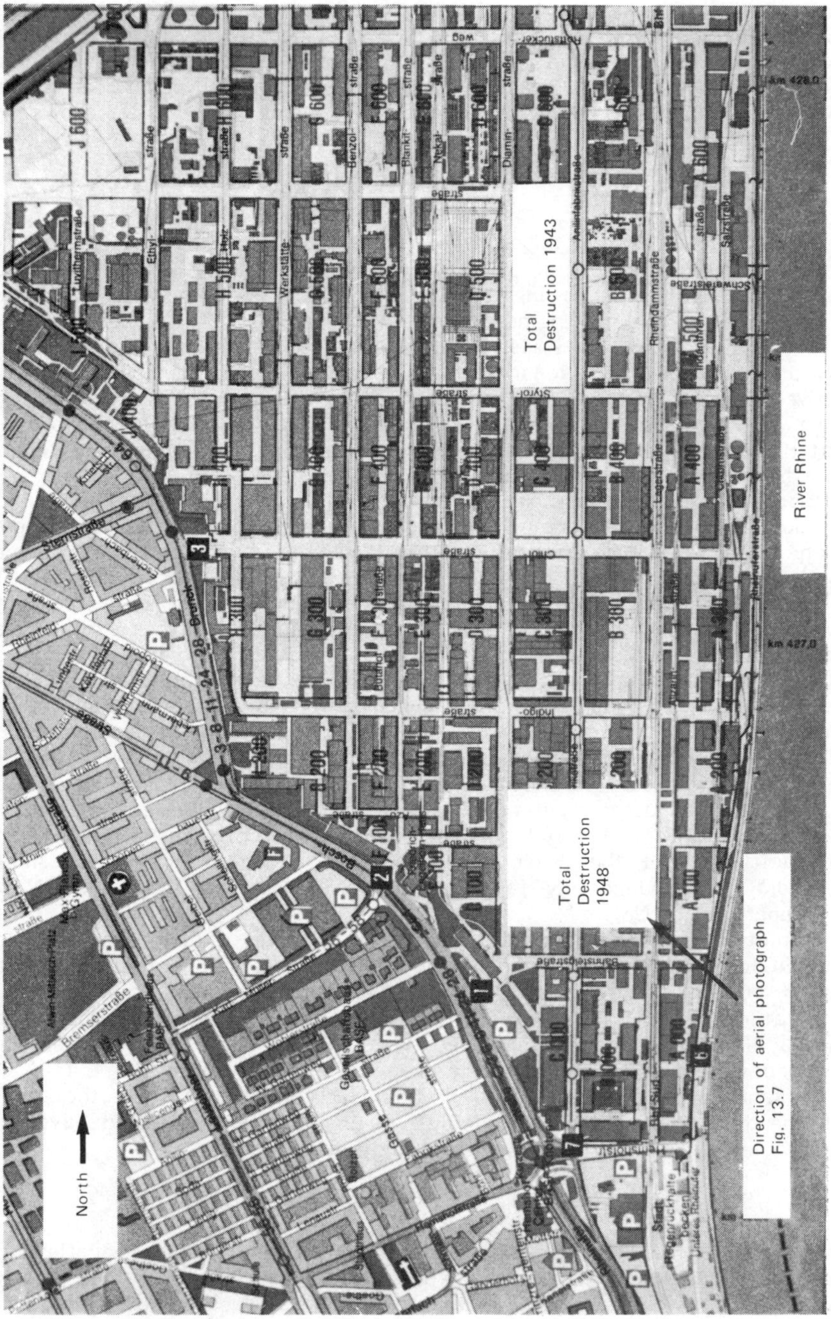

Fig. 13.6. DAMAGE ARES OF 1943 AND 1948 COMPARED.

killed outside the works. The extent of the damage is shown in Figure 13.5., and is contrasted with that of the 1943 explosion in Figure 13.6. where the two areas have been marked on a modern plan provided by BASF. One should not read too much into the relative sizes of the two explosions as shown, but the larger area for the 1948 disaster does accord with its higher death toll. Figure 13.7. is a general view of the disaster area and Figure 13.8. shows the remains of the car.

13.13.2. THE STAHL REPORT.

Stahl et al (1949) is a Report, apparently prepared to serve the ends of justice, by a team of three investigators. At the time of the disaster the area was under military occupation following the Second World War and the team was led by W.R. Stahl, a mining engineer from Washington. The other members of the team were Professor P. Strassmann, a Professor of Chemistry from Mainz University, and Professor G. Richard, a Professor of Pharmaceutical Chemistry of the University of Nancy. The present author has only had access to the main Report and not to its appendices. However, it was assumed by the present author that if any data of crucial significance was missing from the main Report, it would not be found in the appendices.

The Report found, on the basis of the analysis of a routine sample which had been taken from the car and which later was found among the debris, that it had been filled with pure di-methyl ether and that it was free from ether peroxides. The Report, based upon this analysis, also ruled out the possibility that there was any appreciable quantity of non-condensible gas present in the vapour space. It concluded that this left two main alternative explanations for the failure of the car. The first, (a) was that the car had been over-filled and that when it heated up in the sun this resulted in a hydraulic rupture, and the second, (b) was that it had failed under a pressure equal to the vapour pressure of its contents because of a defect in its construction. The team finally settled on (a) and this has now become generally accepted as the cause.

However, it is necessary to understand that the Report, in reaching conclusion (a), was compelled to assume that the actual capacity of the car instead of being its nominal 50,200 l (11,100 gal), had been incorrectly manufactured and its true capacity was of the order of 48,145 l (10,600 gal), or 4% too small. The Report did not produce any factual evidence for this assumption. (Note: gallons are imperial, not US).

The assumption was made because, even after a series of pessimistic calculations, the Report was forced to conclude that, if the car actually had had a capacity of 50.2 m^3, the contents would have had to have been raised to a mean temperature of 55°C (131°F) for the vessel to have been entirely filled with liquid. This, the Report concluded, was an impossibly high temperature and it assumed instead a somewhat more feasible temperature of 40°C (104°F). But, for the tank to be completely full of liquid at this temperature, the actual volume of the tank, based on the calculations in the Report, would have had to be some 4% less than that given on its loading plate.

13.13.3. FLAWS IN THE ARGUMENT ?

Detailed analysis of the Report suggests that there are a number of flaws in its arguments. Of importance to the whole argument is to know the density of the liquid phase of dimethyl ether over the temperature range 20°C to 55°C (68°F to 131°F). Even today this data is not to be found in the standard reference works. Stahl et al relied on a determination of density performed especially for the occasion by the laboratories of BASF. The density temperature relationship so determined is plotted in Figure 13.9. The Report uses this density

Fig. 13.7. AERIAL VIEW AFTER 1948 EXPLOSION.

Fig. 13.8. REMAINS OF DISINTEGRATED TANK CAR.

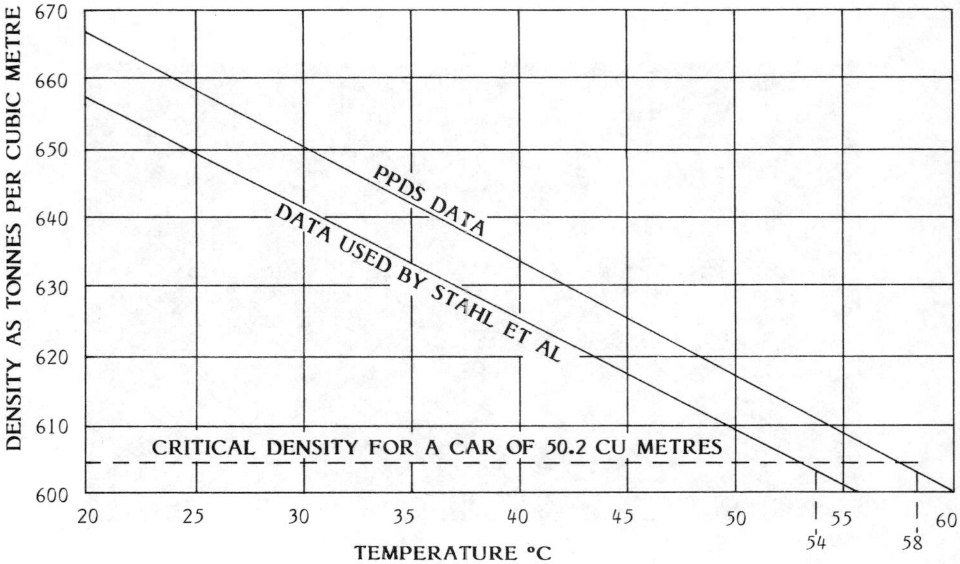

Fig. 13.9. DENSITY/TEMPERATURE RELATIONSHIP FOR DIMETHYL ETHER.

to determine the percentage ullage which would exist at 40°C (104°F) in order to compare it with the German Pressurised Gas Regulations which require an ullage of 5% at that temperature. On the basis of the relationship between density and temperature, the Report calculated the ullage to be only 4.3%. It concluded therefore that, by this criterion, the car was overloaded.

However, a second criterion of the same Regulations required that for 1 kg of dimethyl ether, 1.65 l must be available in a tank car. The weighed quantity of dimethyl ether was 30,380 kg (66,840 lb) maximum which corresponds to a requirement for 50,127 l (11,027 gal). The capacity of the car as disclosed by its weighing plate was 50,200 l (11,100 gal). Thus by the second criterion the car was not overloaded.

However, the question remains as to whether the density/temperature relationship used was correct. Figures for the density of the liquid phase of dimethyl ether have been obtained by the present author from the Physical Properties Data Service of the Institution of Chemical Engineers [Physical Properties Data Service (1984)]. These data are claimed to be the most accurate available. They are also plotted in Figure 13.9. If these data be substituted in the calculation of page 46 of the Report, the ullage then becomes slightly greater than 5%. In other words, the tank was slightly underfilled by either criterion. In all probability the two criteria were just alternative ways of expressing the same rule and were based on a density of 0.636 for dimethyl ether at 40°C, which accords with the PPDS data.

It should be pointed out here that in the present analysis median values are used throughout. The Report is careful throughout to estimate the margins of error for the data used, but is forced to conclude that even if all the errors were to operate in the same direction, this would not significantly affect the conclusions reached.

13.13.4. WHAT WAS THE BURSTING TEMPERATURE ?

To revert to the question of the bursting temperature of the tank, the Report calculates, using its density figures, that the tank would become just full of liquid at 55°C (131°F), i.e. the temperature at which the density would be 30,380/50200 or 0.605. Accepting the PPDS data this density would be achieved at a temperature of 58°C (136°F). It must also be considered that the tank would not burst immediately the dimethyl ether became entirely in the liquid phase, because the walls would stretch before they burst, and this could account for an estimated further 3°C temperature rise. The Report accepted that temperatures of this order were impossible for the bulk contents to achieve and postulated instead that 40°C (104°F) could feasibly be reached. But for it to burst at this pressure the tank must have been smaller than its nominal capacity which had never been directly measured. The Report suggests that the tank would have had to have had a capacity of only 48,842 l for it to burst at 40°C. However, for it actually to have burst at 40°C, it would have had to be full of liquid at ca 37°C for the reason given above. On the PPDS data the density at 37°C would be 0.642 corresponding to a volume of 47,300 l.

However, it is unlikely that even on a July day in Ludwigshafen, the bulk temperature of the 30 tonne tank could have risen even to 40°C. The tank was shielded from direct solar radiation by being encased in wood and might more plausibly be assumed to have had a maximum temperature for the air surrounding it equal to the maximum shade temperature. It has also to be borne in mind that the tanker was orientated north to south which would present the minimum area of exposure to the midday sun. The team did consider whether they ought to do tests on the rate of heat-up of a similar tanker, but did not pursue this line of investigation. Had they done so, they might well have found that the bulk temperature would have been less than the maximum shade temperature because of the need for a temperature gradient if heat transfer is to take place.

Though it might be thought that consideration of the actual meteorological conditions existing at the time was central to the theory that was being put forward, the Report makes no mention of the actual temperatures on that day even though they were available. The present author has been able to obtain the shade temperatures for Mannheim, which is immediately across the river, for July 28th, 1948. The German Daily Weather Report gives for Mannheim: 12.00 hrs 31°C, 18.00 hrs 30°C, with a maximum of 32°C, [German Daily Weather Report (1948)].

This suggests that the figure of 40°C assumed by the Report as being an achievable bulk temperature for the car contents is too high. A more likely figure would be about 29°C (84°F), which would allow for a temperature gradient between the shade temperature and the liquid contents. However, for the reasons given above, the tank would have had to be completely filled at some 3°C less than the bursting temperature. Thus it would have had to be full at about 26°C (79°F). If at that temperature the liquid phase had completely filled the car, based on the PPDS density, the volume of the car would have been only 46,170 l (10,160 gal). The car would thus have had to have had a true volume of only 91.9% of its normal capacity.

13.13.5. METALLURGICAL EVIDENCE.

The team did submit the wreckage to metallurgical examination by Professor Siebel of Stuttgart University of Technology. He noted that the vehicle had been fabricated by watergas welding, a process then becoming obsolete. He noted that failure had principally taken place (80% of its length) along a welded longitudinal seam. Though the weld generally was 90% of the strength of the metal outside of the zone, there were some places where the weld was weaker. To quote from Professor Siebel, "The presence of those weak spots may be one explanation for the rupture of the container walls".

The Report summed up these conclusions as follows: "Obviously, therefore, the rupture of the container could be blamed upon a flaw at some point along this upper longitudinal seam. It could in fact have given way under pressure that was within normal bounds; and the break would have continued along the path of the seam since the seam was slightly weaker than the sheet material itself. The mechanical strength tests . . . do not allow us to say with absolute certainty that this was in fact what happened".

Though absolute certainty was not possible, Professor Siebel's work seems to require assumptions which were more plausible than the assumption that the tanker was too small.

A question which received little attention in the Report was the past history of the tankcar. It was noted, in passing, that it had been damaged in an accident in France the previous year which had knocked its valves off. No enquiry was made into the possible weakening of the tanker which might have ensued from this. Furthermore, the car belonged to an ammonia plant and the plate which gave the tare weight also stated "Highest weight of filling with liquid ammonia 26,800 kg" (Stahl et al, page 57). This would seem to indicate that it could have been used for transporting materials other than dimethyl ether, as, for example, anhydrous ammonia. This substance is known now to be capable of embrittling some steels, [Chemical Industries Association (1975)].

13.13.6. WERE STAHL'S CONCLUSIONS SOUNDLY BASED ?

In the present author's view, the conclusions of the Report are not soundly based. The investigators seemed to have convinced themselves that the tank could not have failed under ordinary service conditions, even though it had previously been involved in an accident. The argument about the true dimensions being at variance with the nominal dimensions could very well have been settled by measuring the wreckage, but no attempt seems to have been made to do this. And it seems difficult to believe that the car could have been in service for 19 years without anyone noticing the discrepancy. At best the view that the tank was up to 8% too small can only be a hypothesis for which factual evidence was not produced.

If an alternative explanation is that the tanker failed through structural flaws this is quite consistent with the view that it took a hot day, on which the vapour pressure in the car was higher than usual, to bring about the disintegration.

The conclusions reached are based on the limited information available. It may be that there were reports other than that of Stahl et al which independently confirmed the conclusions of that Report. The Report may also have given rise to a check of actual tank car capacities which may have shown other examples of undersized cars which would have given weight to the conclusions of the Report. If such evidence exists, it has not come to the attention of the present author.

13.13.7. THE VCE ITSELF.

The description given in Stahl et al of the sequence of events is that numerous witnesses reported a whistling sound like that of a steam locomotive and the appearance of a brownish white cloud before the main explosion. The Report suggests that there were two chemical explosions. The first was a minor external explosion which toppled the car over causing it to disintegrate and that this was followed by the main explosion. Experience has since shown that it is not necessary to postulate that the car was toppled by a chemical explosion. The whistling sound would be associated with the initial cracking and the disintegration would be the result of the ensuing pressure rupture. However, detailed as is the Report of Stahl et al in many areas, it does not attempt to analyse the phenomenon of the explosion. It misleadingly compares it to that of a 'landmine' even though it did not produce a crater.

Giesbrecht et al, as noted above, analyses the well-attested damage pattern of this disaster (Chapter 2 of Stahl et al gives much information on this subject) as shown in Figure 4.7. of this reference, putting forward the view that the heat of combustion involved was 854 GJ. The reference concludes that the maximum overpressure experienced was less than 0.5 bar (7 lbf/in^2).

Davenport (1984) attributes a loss figure of $25 million (1983) which is only half of the 1943 disaster figure though the area of damage was at least as great and there were four times as many fatalities. The TNT equivalence, according to this reference, was 20 to 60 tonnes. The area of total destruction was ca 40,000 m^2 (430,000 ft^2) [which compares with 35,000 m^2 (380,000 ft^2) quoted for the 1943 explosion] and the area of total destruction plus severe damage was ca 300,000 m^2 (3.2M ft^2).

13.13.8. HYDRAULIC RUPTURE AS A HYPOTHESIS.

One general conclusion from the analysis above is that hydraulic rupture is an attractive hypothesis and that it may be more palatable than the hypothesis that, under certain circumstances, pressurised storages may fail under normal operating conditions albeit at the upper end of the normal operating range. Because it is a more convenient hypothesis than the alternative, it is imperative that it be subjected to rigid scrutiny and proof. Such rigid scrutiny and proof was, in the present author's view, lacking in the case of Ludwigshafen 1948.

13.13.9. SOME GENERAL POINTS ON THE EXPLOSIONS.

The very severe effects of this incident accord with the view now generally held that significant overpressure effects after ignition of vapour clouds occur only in circumstances in which there are buildings and other obstructions which create turbulence and which provide conditions of some confinement. The industrial landscape at Ludwigshafen had many such features.

The very high death tolls from these two explosions were no doubt contributed to by high population densities in the vicinity of the points of release. Such high population densities would not be tolerated today.

13.14. PORT HUDSON, MISSOURI, USA. A VCE.

This VCE occurred on 9/12/70 and it followed the rupture of a pipeline carrying liquid propane. It may be the largest VCE to date but it occurred in a sparsely populated area with a warning period which permitted the evacuation of the relatively few inhabitants. There were no fatalities and only slight injuries. Though described at the time as a detonation, it now seems likely that it was a deflagrative explosion initiated by an internal explosion in a building.

As an accident to a pipeline is classified in the US as a transportation accident, the incident was investigated by the National Transportation Safety Board (NTSB 1972) and at the invitation of the latter it was also investigated by the US Bureau of Mines [Burgess and Zabetakis (1972)]. It is the subject of case histories in Gugan (1979) and Wiekema (1984).

Both National Transportation Safety Board (1972) and Burgess and Zabetakis (1973) make use of the term "detonation"; it appears in the title of the latter reference which concludes "The damage at Port Hudson was unmistakably the result of a detonation". Gugan (1979) casts doubt on this, Gugan prefers the interpretation that it was a deflagrative explosion. This doubt was also shared by Advisory Committee on Major Hazards (1979).

However, in the early 70s, the use of the word detonation was perhaps less strictly defined than it is today. The I.Chem.E. Working Party on Nomenclature, [Working Party on Nomenclature (1984)], defines it as follows:-
"DETONATION - An explosion caused by the extremely rapid chemical reaction of a substance in which the reaction front advances into the unreacted substance at greater than sonic velocity".

The mechanical cause of the explosion was a break in an 8" (200mm) pipeline in which propane was being transported at a hydraulic pressure of about 60 bars. After the pipeline fractured there was a twenty-minute delay to ignition which enabled the local population to retreat to a safe distance.

Both of the Reports previously quoted agree that the ignition occurred as the result of vapour penetration into a warehouse built of concrete blocks which was about 300 metres from the point of escape and downwind of it. The building housed deep freeze units and it is believed that a spark from a thermostat ignited it. This building seems to have disintegrated from an internal explosion. No other building in the vicinity was completely wrecked in the way that occurred in other disasters, such as Ludwigshafen 1948 and Flixborough 1974.

The quantity of liquid spilled was estimated by the pipeline operators as about 750 barrels or 60 tonnes. Clearly not all of this could enter into the explosion as part of it would have become diluted with air to such an extent as to be below the lower flammability limit and some would be above the upper flammability limit. In the case of a continuous spillage, eventually a state of equilibrium is achieved in which the rate at which material is becoming too dilute to burn becomes equal to the rate of input of fresh material from the point of spillage. Burgess and Zabetakis estimate that the steady state cloud would have had dimensions of about 500 metres by 16 to 20 metres maximum width and a maximum height of 4 to 7 metres. It would have covered an area of about 6,000 sq metres.

Virtually the entire case that Port Hudson was a detonation rests upon the evidence of the eyewitnesses. National Transportation Safety Board (1972) claims "At 10.44 p.m. the propane air mixture detonated and generated a double boom; the gaseous cloud ignited simultaneously at all points in the valley . . ." In Burgess and Zabetakis the eyewitness evidence is summarised "At about 10.44 the valley reportedly 'lit up'. All witnesses agreed that there was no observable period of heat propagation but a sudden flash as in heat lightning. There was an almost immediate pressure pulse and one of the witnesses, who was walking between automobiles, was knocked down. She was half a mile from the centre of the propane-air cloud".

This description raises a number of difficulties. There just could not be a simultaneous ignition of the cloud, nor could there be an almost immediate pressure pulse at half a mile away. Even at sonic velocities it would have taken 0.5 to 1 second for the explosion to engulf the 500-metre cloud and about 3 seconds for a pressure pulse to travel half a mile. It seems apparent that the eyewitnesses would not be able to discriminate to the extent necessary to establish that there was a detonation. The initiating event, the internal explosion, would have occupied less than half a second perhaps and this no doubt would be what "lit up the valley".

To establish that there was a detonation required forensic evidence of overpressures well in excess of those associated with deflagrations. Such evidence is lacking from the reports. As has been previously mentioned, the only demolition was as the result of an internal explosion; there was a house close to the presumed point of ignition which, though damaged beyond repair, still had its walls intact. There were ornamental shrubs surviving in the garden and fences and a telegraph pole were left standing. All other damaged buildings were rated as class B damage.

Burgess and Zabetakis give a total of 37 structures, presumably houses, within a radius of a mile. The corresponding figure for Flixborough was 156 (excluding structures on the site). The median distance at Port Hudson, i.e. the distance at which 50% of the structures lay, is given as 3.4 miles (5.75 km). In Figure 13.10. the present author has plotted percentage damage at Port Hudson and damage data on houses in the vicinity of the Flixborough explosion against distance. The damage data at Flixborough is derived from Humberside Police (1974). The plot suggests that the damage pattern was little different to that at Flixborough which is universally agreed to have been a deflagrative explosion. Figure 13.11. is a sketch map (based on Fig. 10 of NTSB 1972) which shows the areas affected and which shows that there were a number of areas in which damage was especially concentrated, some of them detached from the main area of damage. The Report refers to these as "caustics" and attributes them to focussing effects produced by atmospheric refraction.

The question of TNT equivalence has been much discussed and there has been a fair measure of agreement in the figures put forward. NTSB gives a figure of 45 tonnes; Burgess and Zabetakis agree broadly with this. Gugan (1979) and Davenport (1984) report similar figures.

Using a technique described in Marshall (1980) which is based on the comparison of areas of light, scattered damage, and which is described further below, the present author has deduced that the TNT equivalence was ca 64 tonnes. This technique is based upon the assumption that the area of light scattered damage is proportional to the TNT equivalence of an explosion raised to the power 2/3. Since the area of light scattered damage at Port Hudson was 1.6 times the area of light scattered damage at Flixborough, then the TNT equivalence at Port Hudson would be equal to $1.6^{3/2}$ times the TNT equivalence at Flixborough. If this equivalence be taken as 32 tonnes, then the ground level TNT equivalence of Port Hudson is 32 x 2.02 = ca 64 tonnes.

The Port Hudson explosion has been assigned some very divergent estimates of yield. Davenport (1984) gives 7.5%. Burgess and Zabetakis (1972), whilst accepting that this figure accorded with a TNT equivalence of 50 tonnes and a spillage of 750 barrels, point out that a much smaller quantity of material than this would have taken part in the actual explosion as by the time it occurred there would be a steady state plume. Based on the quantity of propane in such a steady state plume, the yield would be nearer 20%. Based on the comparison with Flixborough, the yield would be ca 5% (based on the total spillage) or 13% based on the steady state plume.

Port Hudson presents some difficulties for those who have argued that it is the presence of buildings, pipework and vessels which produce the acceleration necessary to develop significant overpressures. However, there seems little doubt that the initiating event was a powerful confined explosion.

To summarise, the total spillage before ignition at Port Hudson was greater than that at Flixborough and the TNT equivalence seems to have been about twice as great. There is no real evidence that it was a detonation.

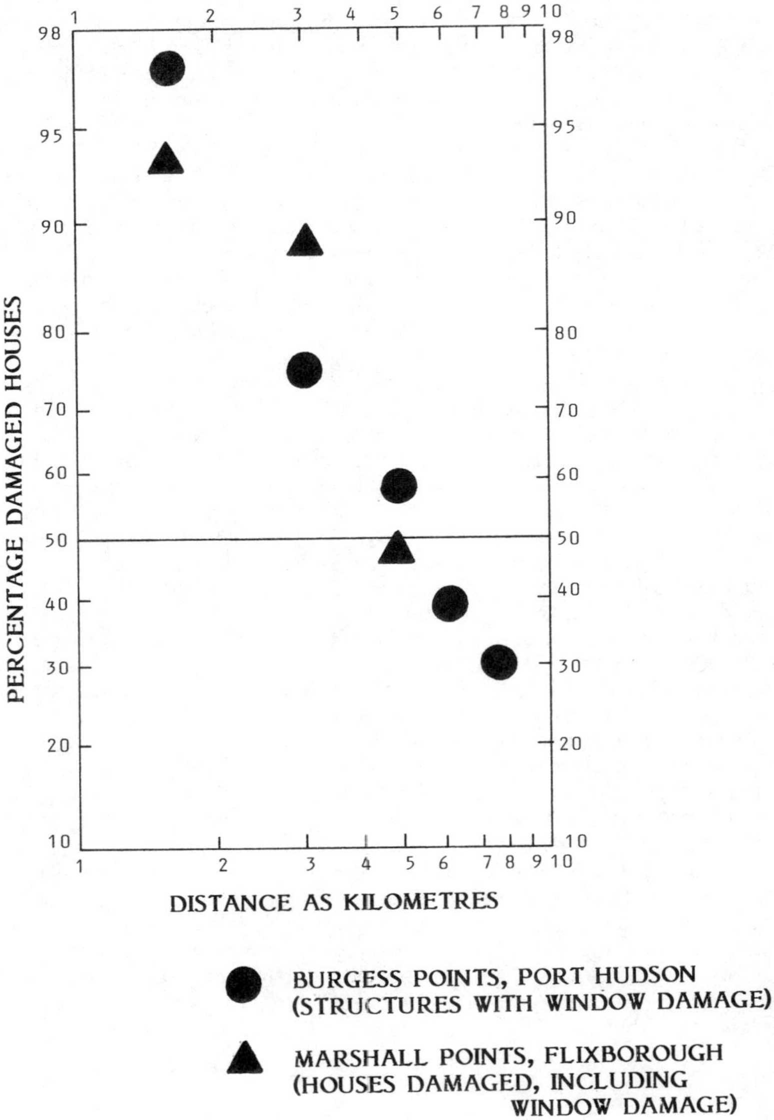

Fig. 13.10. PERCENTAGE OF HOUSES DAMAGED WITHIN A GIVEN RADIUS.

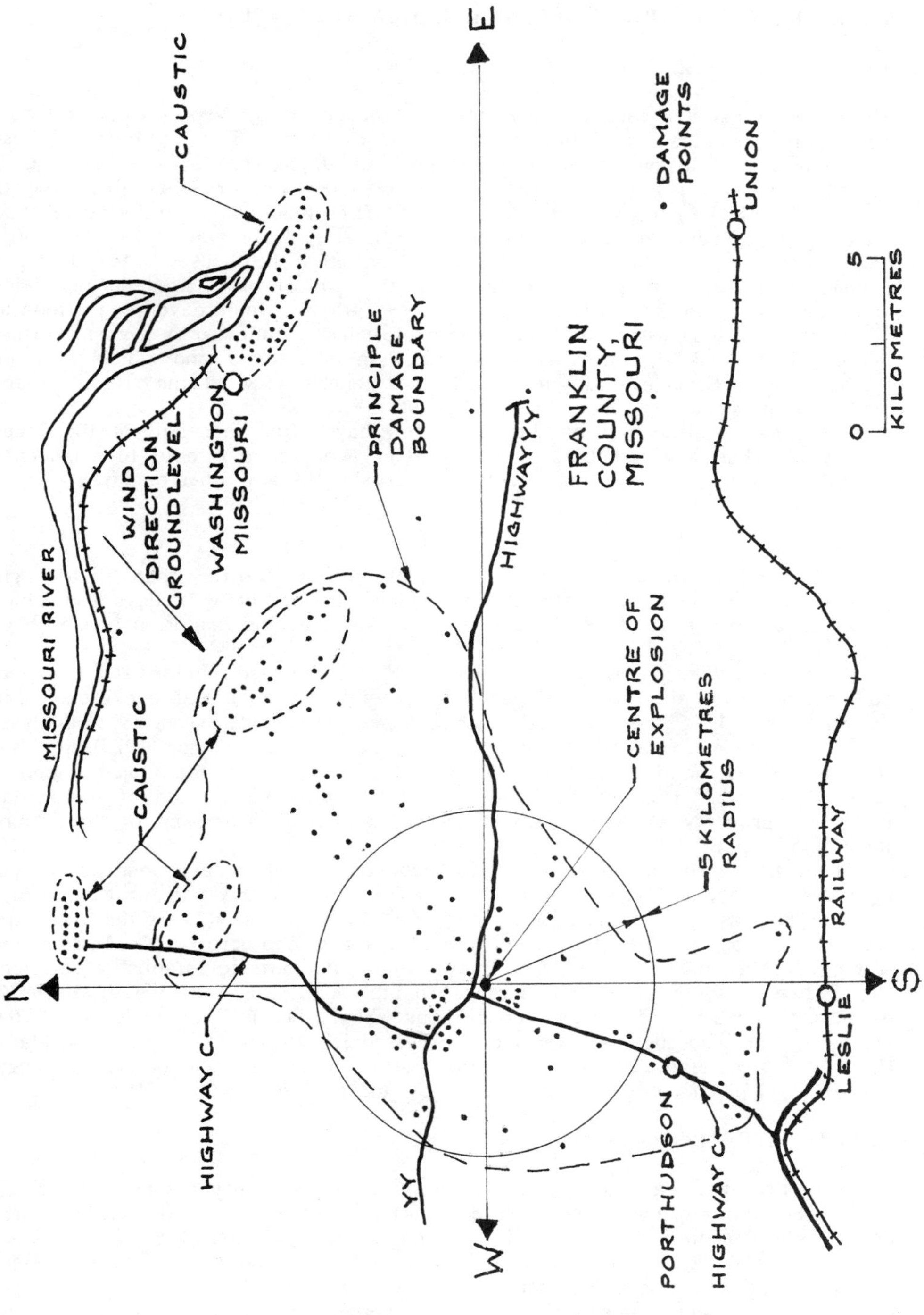

Fig. 13.11. DAMAGE AREA AT PORT HUDSON.

13.15. FLIXBOROUGH, HUMBERSIDE, ENGLAND. A VCE.

13.15.1. GENERAL ACCOUNT OF DISASTER.

There was a massive escape of cyclohexane vapour at the Nypro Works, Flixborough, Humberside, at 16.53 hrs on the 1st June, 1974. This ignited giving rise to a massive explosion which killed 28 men and injured 36 on site and injured at least 53 more off-site. Most buildings on the site were severely damaged, some including the control room and the main office block were totally demolished. There was severe damage to equipment. Off-site, 1,821 houses and 167 shops and commercial buildings were damaged; some of the houses were damaged beyond repair. It may be said that had the explosion occurred during office hours instead of on a Saturday afternoon, the death toll would have been comparable to that of Ludwigshafen, 1948. The explosion was followed by a vast conflagration which has been previously described in Chapter 9 and which inflicted further severe damage. Figure 13.12. is a general view of the site after the explosion and fire, taken from the north west.

The present author visited the site a few days later on behalf of the Trade Unions who had lost members in the disaster and was thus able to experience its full enormity in a way which would be impossible by any other means.

13.15.2. LITERATURE.

No VCE has been made the subject of a more intense scrutiny than Flixborough. It was investigated by a Public Inquiry which sat for some 70 days and which produced a Report which is the definitive account of the explosion [Flixborough (1975)].

The present author's book, Marshall (1979a), in addition to its technical content, analyses the disaster against its social and industrial background and describes how the authorities dealt with the emergency. Grayson (1975) analyses the disaster from the point of view of the Trade Unions. Taylor (1975) discusses its implications for management. Westgate (1975) deals with the human responses to the disaster. Humberside Police (1974) is a detailed account of the human injury and property damage sustained, as well as being an account of the actions taken by the police.

There have been numerous technical publications which have been concerned, in whole or in part, with the disaster. Of these, clearly the most important is the previously mentioned Flixborough (1975). Gugan (1979) discusses the disaster in considerable depth and provides many photographs. Tucker and Webb (1975) is an analysis of the damage, which attempts to establish a relationship between overpressure and distance for the explosion. Sadee et al (1977) extends this work and reaches conclusions about the TNT equivalence of the explosion. Flixborough is extensively discussed in Advisory Committee on Major Hazards (1979) especially in connection with the design of control buildings. There are 11 papers on various aspects of the disaster in I.Chem.E. (1976).

13.15.3. EYEWITNESS ACCOUNTS OF THE DISASTER.

The present author had the opportunity on June 18th to conduct a private enquiry in Scunthorpe interviewing members of the Transport and General Workers Union and of the Association of Scientific, Technical and Managerial Staffs employed at the Nypro Works. From the examination of these eyewitnesses, it was evident that there were four principal sequences of events:-

(1) A rumbling sound, muffled explosion or boom. This was heard by all the witnesses.

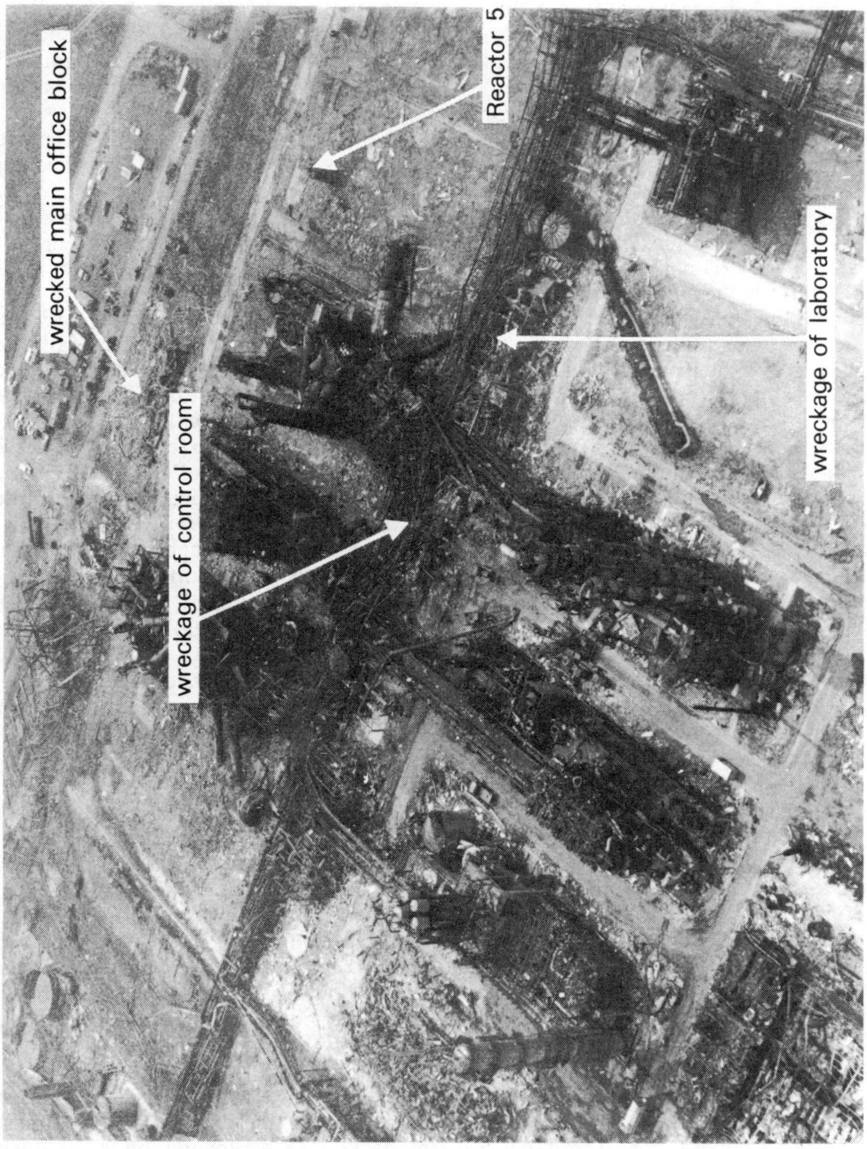

Fig. 13.12. GENERAL VIEW OF NYPRO WORKS AFTER FIRE AND EXPLOSION.

(2) This was followed by a roar, rumbling or whistling. This was heard by six witnesses. The fire alarm went off. This was heard by eleven witnesses. There was a Tannoy announcement of "Fire at 25" (25A was the number of the cyclohexane oxidation section). This was heard by three witnesses. A cloud of vapour developed in the vicinity of 25A. Some described it as "white" and it was seen to rise in the space where Reactor 5 had been. One witness described it as a "mushroom cloud". Other descriptions were "vapour haze", "heat haze", "vapour cloud". Debris was seen to fall.
(3) A witness then saw a "fire ball", another spoke of "fire at 25A", others spoke of "an orange glow then a great flame at 25A", "a glare", "seemed to be a flash from the hydrogen plant", "hydrogen reformer furnace ignited", "flames". One witness described the events after the initial sound as "A vapour cloud followed by falling debris, followed by fire".
(4) There was then an explosion of great severity which was heard by most, but four witnesses only recollect being flung to the ground without hearing an explosion.

There was a conflict of evidence between the witnesses over the time these events took. Some put the interval between (1) and (4) as "seconds" or "20 seconds", but the general consensus was about 1 minute. (Flixborough 1975 concluded that the time lapse was of the order of 2 minutes).

Based upon the evidence presented, it was reported in Marshall (1974):- "The evidence was consistent with the following sequence of events:- (a) The failure of the bridging pipe between Reactors 4 and 6. This would be the muffled explosion or rumble. (b) Vapour then rushed out through the 28-inch branches of Reactors 4 and 6 to produce a roaring sound like a relief valve blowing off. (c) A cloud consisting of tens of tonnes of cyclohexane vapour formed. (d) This ignited, probably at the hydrogen plant. (e) The flame travelled to a region of the vapour cloud in which the cyclohexane vapour/air mixture lay inside the explosive limits. (f) A catastrophic explosion then occurred . . . (g) The blast wave and flying and collapsing debris wrecked the laboratory and control room, office block and gatehouse and caused severe damage to other buildings. (h) Rupture of pipes and vessels in the surroundings then provided fuel for the intense fire which followed". Flixborough (1975) in its conclusions did not contradict any of this.

It may be added that no eyewitness, either inside or outside of the site, obtained a clear view from the front of what happened in the gap in the reactor train which it was subsequently concluded was the point of escape for the cyclohexane vapour.

13.15.4. FORENSIC EVIDENCE.

By forensic evidence is meant non-verbal evidence as to the cause of the explosion such as was compiled from the debris or from film or other recording devices.

The first question to be addressed was, granted that there was an explosion, what type of explosion was it ? A rough assessment of its magnitude was that the damage from it much exceeded that from any single war-time bomb charged with conventional explosive. Comparison could be made with the effects of the numerous Second World War V1 and V2 weapons which had a warhead of the order of 0.8 tonnes, [Collier (1976), Appendix], and it is clear that the damage at Flixborough much exceeded that from a single weapon of this sort. It was possible, therefore, to hazard a guess that the explosion was equivalent to tens of tonnes of TNT. Dense explosives could be ruled out on two grounds, (a) such materials were not stored on site, (b) had dense explosives been involved they would have produced a crater and an area of intense shatter; no evidence was found for this. This left a gas phase explosion either confined or open air. A totally confined explosion (whether vapour or dust) could be ruled out on the grounds that there was no building on the site of sufficient volume to produce an explosion of this magnitude. This left an open air explosion as the only feasible cause. Such an explosion would have to involve tonnes of flammable vapour to result in damage of the magnitude experienced at Flixborough.

It is necessary, therefore, to seek evidence for such a leak, which must have been a rapid leak because otherwise it would have been noticed by the operators. A search quickly identified the likely source, which is shown in Figure 13.13, which views a gap between Reactors 4 and 6 from the west. (The reasons why this gap existed are discussed below). The suspected source was a collapsed by-pass pipe, the disintegration of which left open to atmosphere two 28 inches (0.67 metres) in diameter branches from vessels in a reactor train. This reactor train contained some 140 tonnes of cyclohexane which was being maintained during operation at some 70°C above its atmospheric pressure boiling point. It was obvious that such a disintegration would make it possible for the liquefied gas in the reactor train to flash off vapour at a rate governed only by the size of the apertures. That the explosion was caused by this disintegration was not a source of dispute at the Inquiry and it may, therefore, be accepted that it was this collapse which led to the explosion. Such controversy as existed, and on which the Inquiry spent much of its time, lay in deciding what caused the by-pass pipe to collapse. This question is further discussed below.

Figure 13.14. is a photograph, taken from the west, which shows the by-pass pipe in a shut up or jackknifed condition.

The time at which the explosion took place was identified from a number of sources such as on-site clocks which stopped when their electricity supply failed and from a number of members of the public for whom the sound of the explosion coincided with a particular scene on television. There was also evidence from the atmospheric disturbances recorded by the University of Leicester. The evidence from Leicester, reported in Jones and Spracklen (1974), suggested that relatively small precursor events were recorded in the ionosphere (part of the upper atmosphere) and picked up by a receiver at Leicester. These events took place above Gainsborough at about 17.00 hours B.S.T. and were succeeded 48 secs later by a disturbance which was four times greater than any natural occurrence. The instruments were recording upward movements in the ionosphere using the Doppler principle.

13.15.5. CAUSES.

The root cause of this disaster, in the present author's opinion, lay in the basic design of the plant. The operation which was conducted in the process unit from which the escape originated, (Works No. 25A), was the oxidation of cyclohexane in the liquid phase at a temperature of ca 155°C and a pressure of about 9 bars. Because of the unfavourable reaction kinetics, the oxidation was designed to take place in a train of six reactors each of which overflowed by gravity into the next reactor in the train, the last, Reactor No. 6, overflowing into a horizontal treatment vessel of comparable capacity. The reactor train is shown in course of erection in Figure 13.15. The treatment vessel is not visible as it was installed to the rear of the reactors. The figure clearly shows the stepwise arrangement of the reactor train.

Each reactor had a capacity of ca 20 tonnes and the total inventory of this part of the plant, including treatment units, was circa 140 tonnes. The present author has calculated the theoretical adiabatic flashing fraction for pure cyclohexane, under the conditions stated above, to be ca 40%, (see Sec. 13.15.9). Thus, were a serious loss of containment to occur, up to 56 tonnes of cyclohexane vapour could be discharged. This represented the potential for disaster.

The realisation of this potential was brought about by a chain of circumstances

Fig. 13.13. THE POINT OF RELEASE.

Fig. 13.14. THE JACKKNIFED BY-PASS.

which are described briefly below; for a full account Flixborough (1975) should be consulted.

In the first instance a crack was discovered on March 27th, 1974 on Reactor 5. In common with the other reactors this was of composite construction being fabricated from a 12mm mild steel shell bonded to an internal liner of 3mm stainless steel. Cyclohexane was leaking from the crack so it was apparent that both the shell and the liner were penetrated by the crack which extended for ca 6 metres. It was decided to remove Reactor 5 from the reactor train and to run the system on five reactors by installing a by-pass pipe to bridge the gap left by the removal of Reactor 5. Flixborough (1975) was very critical of the decision to resume production without stripping the lagging from the other reactors to see whether they also were cracked. After the explosion it became clear that Reactor 6 had also been cracked, but less seriously.

Reactor 5 was accordingly removed from the train and parked on spare ground near the reactor train to await repair. It was then realised that the by-pass which would bridge the gap left by the removal of Reactor 5 would have to be cranked to accommodate the difference in level between the outlet of Reactor 4 and the inlet of Reactor 6. Furthermore, it became apparent the stainless steel pipe of the same diameter as the outlet branches (28 inches, 0.67 metres) was not available and that the by-pass would have to be fabricated from 20 inch (0.49 metres) pipe. A sketch of this pipe as it was eventually installed is shown in Figure 13.16.

A crucial feature of the system which completely escaped the attention of those responsible for fabricating and fitting the by-pass was the presence of two stainless steel expansion bellows, one in the outlet branch of Reactor 4, and the other in the inlet branch of Reactor 6. It was the presence of these two bellows which eventually brought about the collapse of the by-pass. British Standard 3351:1971 [as quoted by Flixborough (1975)] requires that "for axial bellows, the piping shall be guided to maintain axiality of the bellows and anchored at adjacent changes in direction to prevent the bellows being subject to axial load due to fluid pressure". (There may be a mistake here and non-axial load is meant, VM). Also, "when expansion joints are used the advice of the manufacturer should be sought with regard to the guiding, anchoring and support of the adjacent pipework". The manufacturers were quite explicit. In Teddington (1974) they state "Use only one bellows between two anchors" and "A pipe between two anchors must be straight". They further recommend "Because axial bellows require positive anchoring and guiding, it is almost invariably impracticable to use them if the pipeline is elevated", and "Only in exceptional circumstances can axial bellows be used out of line between anchors". Finally they state, "You can only have one axial unit between any two anchors". The anchors at Flixborough were the connections to the reactors and this was thus a case in which an elevated pipeline had more than one bellows in a pipe system which was not straight. No attempt was made to contact the manufacturer of the bellows during the period of use of the by-pass.

Teddington (1974) draws attention to the phenomenon of "Squirming". This is a particular case of the failure of a strut which, provided that it is subject to a critical level of stress, will fail by buckling. Figure 13.17. demonstrates the principles involved. These are based upon the proposition that a bellows, if internally pressurised, will give rise to an axial force. If this be restrained then, at a critical level of internal pressure, the bellows will squirm. Where two bellows in a piping system are not on the same axis, the level of critical pressure required to produce squirming will be reduced.

Tests were conducted on site at Flixborough on a replica of the system which was provided with a similar measure of support by scaffolding poles that had been provided for the original assembly. These tests showed that the bellows squirmed at 9.8 kg/cm^2 (9.6 bars) air pressure but that, having squirmed,a pressure

Fig. 13.15. THE REACTOR TRAIN IN COURSE OF ERECTION.

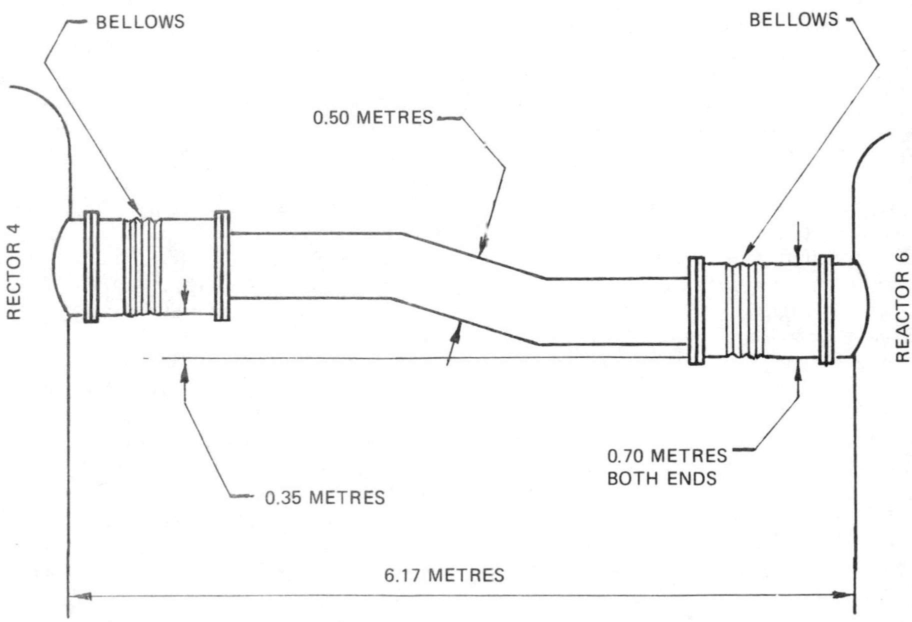

Fig. 13.16. SKETCH OF BY-PASS PIPE.

of 14.6 kg/cm^2 (14.3 bars) had to be applied to cause them to burst. However, only the assembly and not the two vessels were filled with air, as the assembly was blanked off from the adjacent vessels. Therefore when squirming occurred the pressure fell because the squirmed bellows had a greater volume than in their unsquirmed condition. In conditions of operation this would not apply because the liquefied gas then present would provide a constant pressure regardless of the increase in volume. None of the simulations involving bellows ever produced the jackknifing of the cranked pipe; it was considered unsafe to attempt it. Mechanical tests on a replica pipe showed that the pipe could be buckled at about 18 bars internal pressure. Theoretical work by Newlands [Flixborough (1975), Appendix 1] suggested a dynamic mechanism whereby, in a constant pressure system, the vapour in the system, at constant pressure, was theoretically capable of bursting the bellows and following through to jackknife the pipe at internal pressures close to the operating pressure of the system.

Figure 13.18. is a diagram of the forces which would have been likely to have been developed in the system. The forces produced by the two bellows were not in line and, therefore, they produced a couple or turning moment on each bellows. These turning moments were not equal because on the bellows attached to Reactor 4 the force of gravity acting on the by-pass was in the opposite direction to the turning moment and, on the bellows attached to Reactor 6, the gravitational force would reinforce the turning moment. It seems likely, therefore, that the bellows attached to Reactor 6 would have been the first to squirm. Figure 13.14. suggests that in the process of collapse the by-pass spun round. This is because the end which was attached to the Reactor 6 bellows and which had the shorter stub piece, is the end which, as photographed, is near to Reactor 4.

It must be pointed out that detailed analysis of the behaviour under stress of the complete assembly required mechanical engineering knowledge of a high order. This is evidenced by papers such as Newland (1975).

In view of the fact, established by simulation tests, that bellows squirm occurred at 9.6 bars when the safety valve pressure of the unit was set at 11 bars, it is very relevant to ask whether the system was pressure tested before returning the process unit to operation and, if so, with what result ?

Flixborough (1975) was very critical of the procedures adopted by the Nypro management which had been to subject the refitted assembly to a pneumatic test at ca 9 bars, the normal operating pressure, whereas BS 3351 requires hydrostatic testing at a pressure 1.3 times design pressure (in this case, 11.7 bars), i.e. above the safety valve pressure which was set at 11 bars. It expressed the view that, had the assembly been subjected to testing in accordance with the BS, it would have failed under test and the disaster would have been averted.

A further subject for serious criticism was the failure of the Nypro management to provide proper support for the by-pass. What was done was to rely upon a modification to the ordinary builder's scaffolding which had been installed to facilitate the removal of Reactor 5 and the installation of the by-pass. There seems little doubt that this would have been adequate only to withstand the forces due to the weight of the by-pass and its operating contents which it would seem were the only forces in the system of which management had taken cognisance. It might be further commented that had a proper system of support been provided, the disaster might have been averted notwithstanding the other malpractices.

13.15.6. THE 8-INCH HYPOTHESIS.

Flixborough (1975) in its main Report consists of 39 pages of 230 paragraphs. 75 of these paragraphs, or roughly one third of the Report, are taken up by evidence concerned with what has become known as the "8-inch hypothesis".

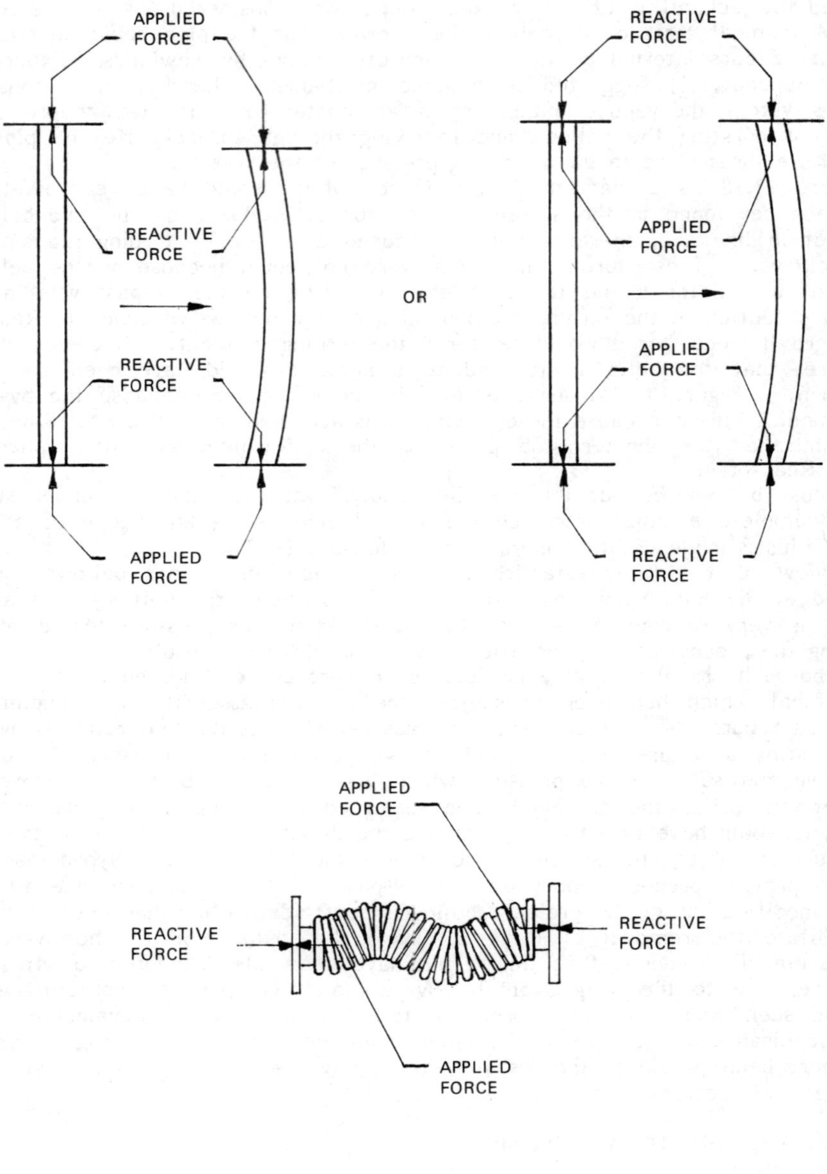

Fig. 13.17. BELLOWS SQUIRM AS AN EXAMPLE OF STRUT INSTABILITY.

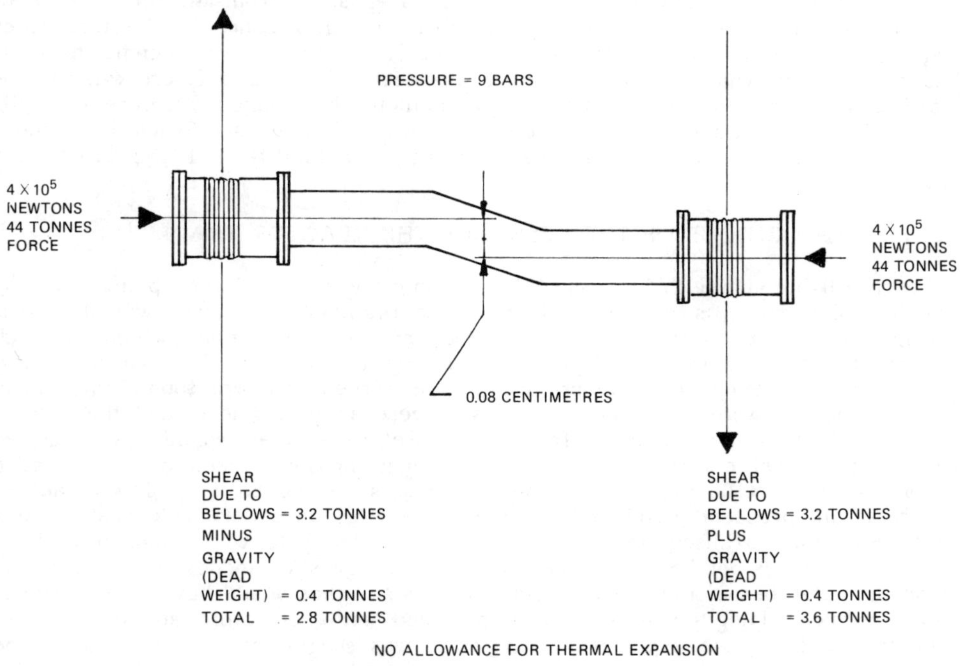

Fig. 13.18. FORCES TENDING TO BUCKLE AND SHEAR THE BY-PASS.

Because of the importance which this assumed at the Inquiry, it seems necessary to comment briefly on it even though the hypothesis was eventually rejected.

When forensic examination was commenced in the vicinity of Section 25A, in addition to the discovery of the jackknifed by-pass pipe and the accompanying open 28-inch branches, it was found that there was a 50-inch (1.25-metre) split in an 8-inch (0.2-metre) diameter stainless steel pipe to the rear of the reactors. The split was a petal fracture on the inside of a bend. This pipe formed part of the treatment system. Were it to be established that the split had developed after the explosion and as a consequence of the fire it could have played no part in its causation. On the other hand, had the split developed before the main explosion it might, and it was contended by some investigators that it did, play a role in bringing about the collapse of the by-pass. It may be stated that there was no disagreement over the proposition that only the collapse of the by-pass could account for the massive explosion. But it was argued that some form of local fire had weakened the walls of this pipe to such an extent that it could no longer withstand its internal pressure of ca 15 bars. A jet of flashing liquid then emerged which gave rise to a small vapour cloud explosion which produced overpressure sufficient to produce instability in the nearby by-pass with the results discussed above and which were generally agreed by all parties.

However, metallurgical evidence was presented by Cottrell and Swann [given in summary in Flixborough (1975) and more fully in Cottrell and Swann (1976)]. These investigators concluded that there was a 3-inch (75mm) crack caused by zinc embrittlement through the pipe being splashed by molten zinc before the surface had become heavily oxidised and that the 50-inch (1.25-metre) crack had started as a creep cavitation crack and had been completed by plastic deformation. Such a split could have been formed in as little as 4 minutes at 950°C and at an internal pressure of 15 bar. It is apparent from this that the small crack must have developed before the split as the pressure in the pipe after the split would have been too low to bring it about. Thus conditions must have existed in which the pipe could be splashed with hot zinc (there was galvanised wire in the vicinity securing the aluminium sheet lagging) before the split developed. The Court accepted the conclusions of Cottrell and Swann in conjunction with what it regarded as the inherent improbabilities of the alternative hypothesis.

13.15.7. A CRITIQUE OF THE DESIGN OF THE REACTOR TRAIN.

Though Flixborough (1975) speaks in its Summary, of the Nypro plant as being "well designed", it is the view of the present author that there was a serious design fault in a system which had 28-inch branches on a plant which had such a potential for disaster. Though the cascade system had the merit of simplicity, it cannot have been cheap to provide the reinforced concrete support structure. An alternative system in which the vessels were at ground level and the process fluid would have been pumped from one vessel to another, would have required pumps and level controllers. But the connecting pipes need only have been about 4 inches (100mm) in diameter. This is on the basis of 150 tonnes of caprolactam produced daily and consuming ca 120 tonnes of cyclohexane a day with 6% being oxidised per pass. The circulation rate would thus have been about 100 tonnes per hour. Perry (1973),[5.31], suggests 5 inches as the economic pipe diameter but in the special circumstances 4 inches would have been permissible. Such piping could have been provided with non-return and excess flow control valves. No explanation apparently was sought by the Court, and none has been provided, to explain the reason for the gross oversizing of the gravity overflow pipes. There was an "off gas" manifold which presumably provided a measure of pressure equalisation and which made this oversizing even less necessary.

13.15.8. TNT EQUIVALENCE AT FLIXBOROUGH.

13.15.8.1. The justification for a detailed analysis.

The question of the TNT equivalence of the Flixborough explosion is dealt with below at some length. The author believes that it is justifiable to deal with this in some detail for the following reasons:-

(a) No other VCE has been subject to the same degree of examination.
(b) The analysis of the Flixborough explosion may be taken as an exemplar of such explosions.
(c) The analysis demonstrates the difficulties of estimating the energy release of a VCE from examination of the damage inflicted.
(d) The analysis demonstrates that highly misleading conclusions may be drawn from viewing particular cases in isolation and in a non-statistical way.

13.15.8.2. Sources of information.

Flixborough (1975) does not report in detail on the TNT equivalence of the explosion. It refers to four sources of information, (1) the ionospheric readings reported in Jones & Spracklen (1974), (2) a report by the Atomic Weapons Research Establishment [Samuels and O'Brien (1974)], (3) an estimate from a barograph chart of a glider in the vicinity, and (4) a report from the Safety in Mines Research Establishment [Roberts (1974)]. From these, Flixborough (1975) drew the conclusion that the explosion lay in the range of equivalence of 15 to 45 tonnes of TNT. The Report does not, however, specify whether this was a groundburst or an airburst equivalence. The difference is important as 1 tonne of airburst is equivalent roughly to 2 tonnes of groundburst, because half of the energy of a groundburst of a dense explosive is dissipated in crater formation and ground shock.

The ionospheric observations (1) gave an upper limit of 55 tonnes of TNT according to Cremer and Warner (1974).

The present author has not had an opportunity to consult (2) the A.W.R.E. blast damage report [Samuels and O'Brien (1974)]. However, the two authors were later associated with Sadee in Sadee et al (1977) on this theme and it may be assumed that this later paper, which is discussed below, represents their mature judgements on the matter. As is shown below, Sadee et al (1974) suggested an equivalence of ca 16 tonnes airburst, (32 tonnes groundburst, VM).

Gugan (1979) refers to (3) and claims that this reference suggested a figure of 18 tonnes of TNT equivalence.

The present author has not been able to consult (4) but Cremer and Warner (1974a) presents a summary of Roberts (1974). This reference discusses the methodology employed and states that damage was referred to airburst models. The combined data give energy in the region of 15 to 45 tonnes.

Cremer and Warner (1974) refers also to signals picked up by the Atomic Weapons Research Establishment (A.W.R.E.), (presumably these were from stations set up to monitor underground nuclear explosions abroad). The disturbance from Flixborough was compared with the disturbance from the accidental explosion of a freighter off the coast of Cornwall on 14/2/72, when 110 tonnes of nitroglycerine were involved. The comparison is the subject of a report as Appendix 2 of Cremer and Warner (1974), which considered that there were considerable difficulties in making direct comparisons and that the Flixborough explosion might have been of the order of 8 to 32 tonnes of nitroglycerine. (This would be about 13.5 to 55 tonnes of TNT, presumably groundburst).

Two papers published since the Flixborough Inquiry have given data on damage and its relationship to distance. These are Tucker and Webb (1975), and Sadee et al (1977).

13.15.8.3. The Tucker and Webb analysis.

Tucker and Webb (1975) presented two sets of data, one set from on-site where the damage was estimated by gross damage to buildings and equipment, and a second from off-site where the damage was mainly estimated from glass damage. These are shown in Table 13.1. Correlations based upon regression analyses of the relationship log overpressure v log distance from epicentre are presented in the paper. The slopes of these regressions (log overpressure/log distance) are roughly 0.875 and 0.94. The slope of the curve in Figure 10.2. (Page 221) which relates log overpressure to log scaled distance approximates to 1.44. The regressions are thus highly discrepant from the standard relationship for dense explosives. The two regressions are also far from consistent with each other, the regression for off-site yielding overpressures for a given distance more than double those for on-site. This is in contradiction to the views of Phillips and Luckritz referred to on Page 253, but these views are not necessarily true. In terms of TNT equivalence, assuming Hopkinson's Scaling Law, this implies a ratio of about 10 to 1.

Tucker and Webb's data points are plotted in Figure 13.19. It is evident that the degree of scatter is very large. The table shows values for equivalent groundburst (M) derived from Figure 10.2., by obtaining the scaled distance

TABLE 13.1. OVERPRESSURE VERSUS DISTANCE, SCALED DISTANCE AND TNT EQUIVALENCE AT FLIXBOROUGH.
Based on Tucker and Webb (1975).

Distance (mean) metres	Overpressure (mean) kPa	Scaled Distance from Kingery & Pannill (1962)	TNT Equivalent Tonnes *
	Inside of works		
70	69	40	5.4
90	28	65	2.62
90	21	80	1.42
105	51	47	11.14
120	34	59	8.40
130	28	65	8.00
130	48	48	20
150	69	39	57
150	17.5	90	6.6
155	24.5	71	10.0
170	21	80	9.6
170	41	52	35
175	21	80	10.46
200	14	105	6.9
200	28	65	29
210	28	65	33
210	48	48	83
220	24.5	71	29
225	24.5	71	32
230	14	105	10
240	15.5	96	15.6
240	18	88	20.3
240	21	80	27
255	34	59	80.7
260	21	80	34
260	17.5	90	24
290	14	105	21
300	9	145	9
350	10	135	17.5
		Mean	22.6
	Outside of works		
475	20	82	194
575	22	76	433
605	14	105	191
700	10	135	140
850	8	160	150
1000	20	82	1813
1200	10	135	702
2500	13	110	11750
2900	2	450	267
3700	2	450	555
3800	2	450	602
5000	3	335	1460
		Mean	1521

* Groundburst

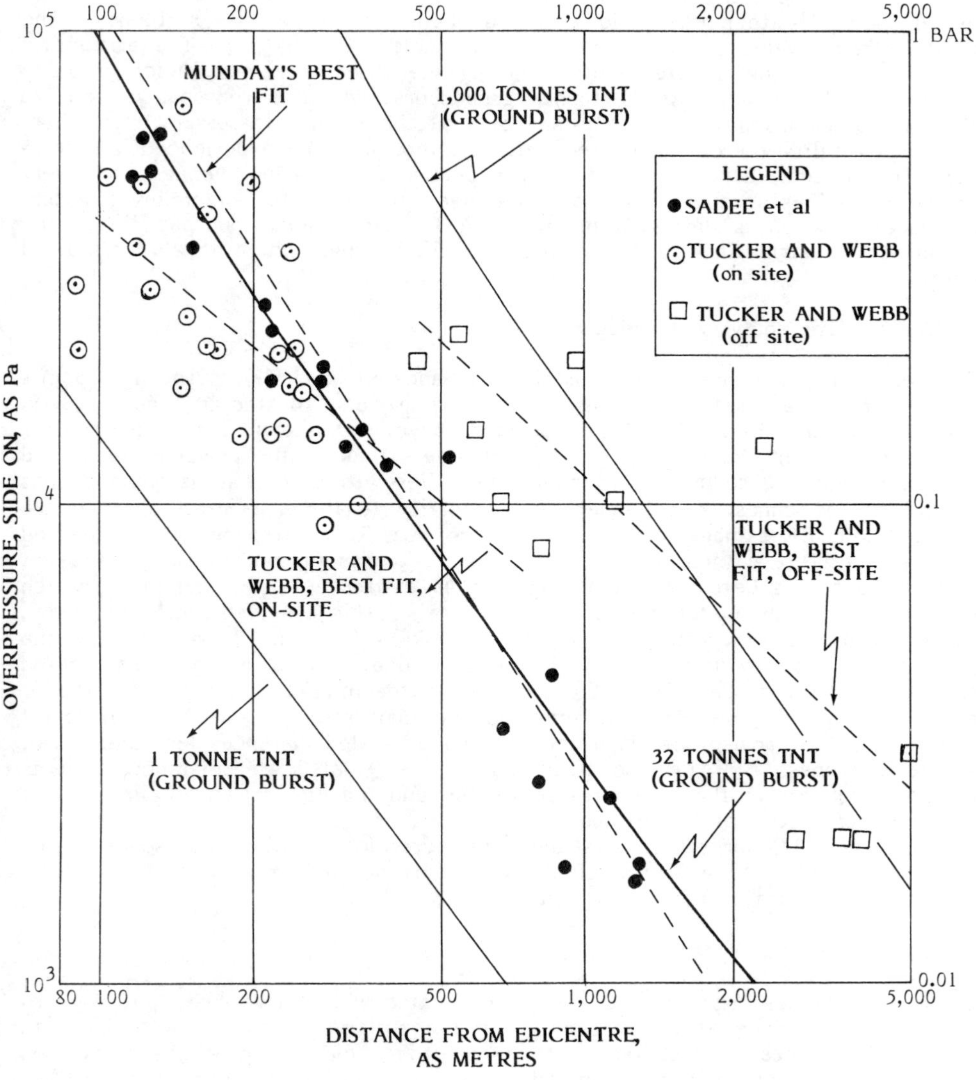

Fig. 13.19. OVERPRESSURE VERSUS DISTANCE AT FLIXBOROUGH.

corresponding to the deduced level of overpressure and finding the value for M which would correspond to the actual distance.

$$M = [\text{Distance}/\text{Scaled Distance}]^3$$

Figure 13.20 is the present author's plot of log TNT equivalence against log distance. He believes this to be a novel approach and that it shows more clearly than a plot of log overpressure versus log distance not only the magnitude of TNT equivalence associated with the individual points, but also their scatter. The figure demonstrates both the wide scatter of Tucker and Webb's data and that their methodology must have been defective in assessing off-site damage as all their off-site points yield TNT equivalence in excess of 100 tonnes. The most extreme value is close to that of a nuclear explosion. The arithmetic mean value for the on-site points, ignoring the three innermost which probably lay within the cloud, was 25 tonnes groundburst which is reasonably close to those of other investigators. It may be that the team of assessors upon whom the responsibility for collecting the data lay, though highly experienced in assessing fire damage, lacked the necessary expertise for relating damage and overpressure. This feature may also have been the case for other investigators at Decatur which is discussed below. A further explanation may lie in the asymmetric distribution of the damage at Flixborough which is also discussed below.

13.15.8.4. The analysis of Sadee et al.

The second paper to be analysed is that of Sadee et al (1977). The data points for that paper are set out in Table 13.2. and are also plotted in Figures 13.19. and 13.20., using the methodology described above. It is obvious that the scatter of the data points of Sadee et al is much less than in the previous paper and that the range of deduced TNT equivalence lies within more reasonable bounds of 6.7 to 78 tonnes. This may simply testify to the high level of expertise in the field of damage assessment of the A.W.R.E. personnel who collected the data. The arithmetic mean value is 32 tonnes (groundburst). Sadee et al give 16 tonnes detonated at a height of 45 metres as a physical model which would most closely match the damage pattern. This use of an airburst model, though it avoids the problem of brisance, has not found much favour. In the present author's view it adds yet a further complication to an already complicated situation and the model does not correspond to the physical realities of a vapour cloud explosion. He would recommend the continued use of the groundburst model with the proviso that it is inapplicable at scaled distances less than about 40 metre tonnes$^{0.333}$, i.e. the cloud radius. At small scaled distances, objects would be subject to the vapour cloud regime and not the brisant regime of the near field of a conventional explosion.

Munday (1975) presented a straight line correlation which is also presented in Figure 13.19. The source of the data was not disclosed, but it may have been the same data as was used by Sadee et al.

13.15.8.5. Combining the data.

Table 13.3. brings together the various estimates which have been put forward for TNT equivalence at Flixborough. In the author's view, the best figure to take is 32 tonnes groundburst. However, there may be problems associated with this on account of the asymmetrical nature of the damage pattern as disclosed in Figure 13.22. which is given in Section 13.16.

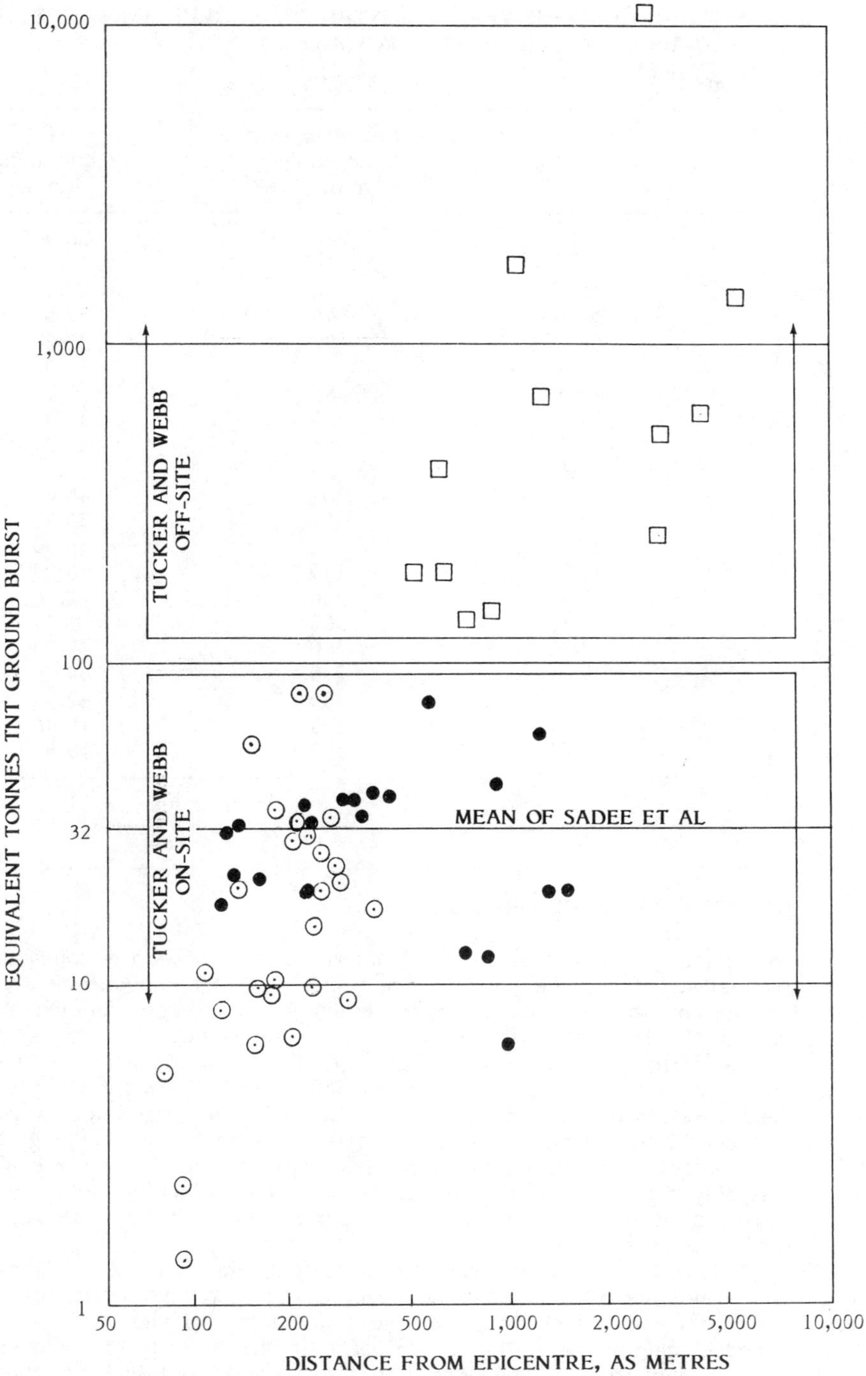

Fig. 13.20. EQUIVALENT TNT VERSUS DISTANCE AT FLIXBOROUGH.

TABLE 13.2. OVERPRESSURE VERSUS DISTANCE, SCALED DISTANCE, AND TNT EQUIVALENCE AT FLIXBOROUGH.
Based on Sadee et al (1977).

Distance (mean) metres	Overpressure (mean) kPa	Scaled distance from Kingery & Pannill (1962)	TNT equivalent tonnes
125	60	42	30.5
120	50	46	17.75
160	35	56	23.32
130	50	46	22.57
135	60	42	33.20
220	27.5	66	37.04
230	18.5	85	19.80
230	24	71	34.00
290	18.5	85	39.71
290	18.5	85	39.7
350	14	102	40.4
335	14	102	35.42
400	12	118	39.00
535	11	125	78.00
700	3.5	300	12.70
945	1.75	500	6.75
825	2.75	360	12.03
885	4.5	250	44.36
1190	2.5	300	62.40
1345	1.75	500	19.46
1340	1.75	500	19.5
		Mean value	31.8
		say	32

13.15.9. THE YIELD AT FLIXBOROUGH.

If there are difficulties in estimating TNT equivalences, these are compounded in estimating yields. Since the yield is based upon the TNT equivalence divided by some function of the quantity of hydrocarbon giving rise to the explosion, even if this quantity be known, there will be as many values of yield as there are differing estimates of TNT equivalence. If there are differing estimates of the quantity involved, then the number of possible values of yield is the product of the number of values of the two variables. There may be differences between "True values" and "Conventional values"; the latter may be based on the ground-burst TNT equivalent and the quantity of hydrocarbon spilled even though it is agreed that only a portion of the spillage actually takes part in the explosion. It certainly seems necessary, when quoting figures for yield, to state clearly the basis on which they are calculated.

Flixborough (1975) made no reference either to the quantity of cyclohexane in the reactor train, nor to how much escaped. There is thus no authoritative figure for the quantity spilled on which to base a calculation of yield.

One method of calculating spillage is to estimate the quantity of cyclohexane originally present and from a knowledge of the theoretical adiabatic flashing fraction, to estimate how much might have vaporised. After the removal of Reactor 5 there were 5 reactors. Based upon Fig. 4 of Cremer and Warner (1974a), the inventory of cyclohexane in each reactor may be estimated at ca 20

TABLE 13.3. ESTIMATES OF TNT EQUIVALENCE AT FLIXBOROUGH.

Source	Groundburst	Airburst	Not specifically stated.
Flixborough (1975)			15 to 45
Ionospheric readings			$\lesssim$ 55
Barograph in glider			18
S.M.R.E.		15 to 45	
A.W.R.E. monitoring stations.	13.5 to 53		
Tucker and Webb (on-site)	25 *		
Sadee et al	32 *	16 = ± 2	

* present author's calculations.

tonnes, though this does not allow for the volume of the agitator and other internal fittings, but no allowance has been made for the increased depth over the lip of the side branch due to hydraulic head. The inventory of the after-reactor, from Fig. 3 of Flixborough (1975), has been taken as ca 25 tonnes assuming it to be 80% full. Thus the total inventory of the 5 reactors and the after-reactor was ca 125 tonnes. It has been assumed that vessels downstream of the after-reactor would have been isolated by a non-return valve on the outlet of the after-reactor.

The theoretical adiabatic flashing fraction (T.A.F.F.) may be estimated from the data in Cremer and Warner (1974a), (Appendix 2), and from the operating temperature of 155°C (Ibid P.43) as follows:-

$$\text{T.A.F.F.} = \frac{\text{Specific heat of the cyclohexane} \times [T_o - T_b]}{\text{Specific latent heat of cyclohexane}}$$

$$= \frac{0.44 \times [155 - 80]}{94} = 0.35$$

It is not claimed that this calculation is an accurate one as there are doubts about the values of the physical constants largely because the contents were not pure cyclohexane, however it is probably accurate enough for the purpose.

Thus, in theory, 44 tonnes could flash off leaving about 80 tonnes of liquid, assuming that the quantity given off as spray may be neglected. This is confirmed by Sadee et al which states that "At Flixborough about 120 tonnes of cyclohexane was available in 5 reactors and one after-reactor . . . After the explosion it was found that the vessels still contained 80 tonnes of cyclohexane so a maximum of 40 tonnes was involved in the explosion". This seems to confirm the calculation of inventory given above and that entrainment was negligible.
It also suggests that 45 tonnes (there is a small discrepancy between the inventory calculated above and that in Sadee et al) can safely be taken as the limit of the release. Thus the yield, based on groundburst equivalency of 32 tonnes and on a 45 tonne release, would represent the minimum value for yield. This minimum is 32/450 or 7.0% if 1 tonne of cyclohexane is taken as theoretically equivalent to 10 tonnes of TNT. The figure would be higher if based on the mass of cyclohexane present in that portion of the cloud which was flammable, or if all the flashing vapour had not been released before the explosion occurred,

or some combination of these.

13.15.10. THE DAMAGE AREA.

The Report, Humberside Police (1974) Appendix A, provides data on damage to property in 17 locations. Excepting for Scunthorpe, the neighbouring town, they were all small in relation to the whole area of damage and could therefore be treated as "cells". The present author has analysed property damage in each of these villages and was able to set them in the rank order which arose from a comparison of the proportion of houses which were damaged. This rank order proved not always to be the same as the rank order of distance from the epicentre. (See Table 13.4). The villages were divided into the categories according to the proportion of houses which had been damaged, (1) whether more than 80% of the houses had been damaged, (2) whether between 40% and 80% of the houses had been damaged, and (3) where damage had occurred but where fewer than 40% of the houses had been damaged. When these were marked on a map of the area, it was possible to draw envelopes round the various locations and to show the marked lack of symmetry in the damage pattern. This damage pattern is used in Section 13.17. to draw a comparison between Flixborough and Decatur. The present author has used this in Marshall (1980) to develop the concept that there should be a relationship between the area enclosed by the outer envelope and the TNT equivalence of an explosion. This area was 117 sq kilometres at Flixborough. This would yield the following:-

$$A_e = \text{ca } 12 \times \text{TNT eq}^{0.666}$$

Where A_e = Area enclosed by envelope of light scattered damage
and TNT eq = groundburst equivalent TNT (tonnes).

The author suggests that in any future case of forensic examination of the scene of a VCE, the territory be divided into cells of equal area based upon a grid and that the percentage of house damage be assessed for each cell. Though the technique of assessing damage for individual villages worked reasonably well for Flixborough, this was because their area was small in relation to the whole area under consideration. Where the area of a location is large, then the principle of averaging becomes unsound because the nearer portion of the area may be very heavily damaged as compared with the further portion. The maximum area chosen should not exceed 1 sq km per cell.

13.15.11. SUMMARY.

The explosion had its primary origin in the presence, in a cyclohexane oxidation unit, of an inventory of a liquid maintained by pressure at 75°C above its atmospheric pressure boiling point and of such size as to yield a vapour cloud containing tens of tonnes of cyclohexane vapour.

Containment was lost because of the collapse of an ill-designed temporary by-pass pipe assembly which was improperly tested and inadequately supported.

That the apertures so presented were of such a size that it was possible for a cloud of such a magnitude to form in less than a minute, does not appear to have a justification in the necessities of design. The vapour cloud ignited, initially to produce a flash fire and then a devastating explosion. Estimates of the magnitude of this explosion vary, but perhaps the best estimate is that it was equivalent to 32 tonnes groundburst TNT.

The Court of Inquiry concentrated its efforts on discovering the causes of the failure of the by-pass and did not address itself to certain wider issues such as those discussed below in Chapter 20. It justified this course by calling attention to the existence of the then rewly established Advisory Committee

TABLE 13.4. PROPERTY DAMAGE AT FLIXBOROUGH.

Name of Village	Distance from Epicentre (miles)	Population	No. of Houses A	No. of Damaged Houses B	No. of Damaged Windows C	B/A	Rank Order B/A	C/8A*	Rank Order C/8A
Amcotts	0.50 (1)	236	77	73	437	0.95	1	0.71	2
Flixborough	0.70 (2)	544	79	72	604	0.91	3	0.95	1
Normanby	1.80 (3)		38	35	173	0.92	2	0.56	3
Burton upon Stather	2.00 (4)	2450	756	644	2803	0.85	4	0.46	4
Gunness	2.40 (5)	1090	333	26	77	0.08	8	0.03	7
Luddington	2.60 (6)	379	131	7	10	0.05	10	0.01	11
Keadby	2.75 (7)	1768	495	5	10	0.01	12	0.0025	10
Thealby	3.00 (8)		63	29	30	0.46	6	0.06	6
Alkboro	4.50 (9)	461	158	62	39	0.39	7	0.03	8
Garthorpe	4.60(10)	295	111	56	210	0.50	5	0.24	5
Winterton	4.90(11)	3997	1503	11	23	0.007	14	0.002	13
West Halton	5.10(12)	249	88	3	10	0.03	11	0.014	9
Whitton	6.00(13)	170	48	3	2	0.06	9	0.005	12
Winteringham	6.25(14)	844	275	3	4	0.01	13	0.002	14
North Bank of Humber	≥6.00			2	2				
	* Assumes 8 windows per house								
Scunthorpe	≥2.00	ca 22000 up to 3 miles.	Not known.	1821	5926	?	?	?	?

on Major Hazards whose Terms of Reference included these wider issues. If criticism can be levelled at the Court it may perhaps be that it devoted too much time to the examination of the "8-inch pipe" hypothesis. Even had this hypothesis been true, it would in no way have detracted from the blame which attached to the management of the works as the by-pass pipe was clearly always on the verge of failure.

13.16. DECATUR, ILLINOIS, USA. A VCE.

The chief significance of this explosion which occurred in a freight marshalling yard on 19/7/74, lies in its apparently anomalously high yield. Gugan (1979) reports the yield as 20-40 tonnes of TNT or 32-65% based upon a 63-tonne release of propane (actually it was isobutane) from a crashed rail car. It is evident from discussion above that these two sets of figures are incompatible. A yield of 20-40 tonnes corresponds to ca 3.2 to 6.5%. It might seem therefore that Gugan (1979) made a mistake in the decimal place. Correspondence with the author, Gugan (1979a), disclosed that the decimal place error existed but that it lay in the yield being understated by a factor of ten, i.e. it should have been 200-400 tonnes of TNT as reported in Davenport (1977). [Davenport (1984) gives a revised estimate of 20-125 tonnes of TNT or 18% max].

The incident is described in National Transportation Safety Board (1975). At about 5.00 a.m. on 19th July 1974, during shunting operations in a rail yard at Decatur, Illinois, a tank car containing 69 tonnes of isobutane was penetrated by the coupler of a box car which it impacted. The impact produced a hole measuring approximately 56 cm by 66 cm, i.e. about a third of a square metre in area. After a delay of 8 to 10 minutes ignition occurred giving rise to a vapour cloud explosion. 7 employees were killed, 33 employees were injured and there were 316 cases of injury, including "anxiety", to members of the general public. There was extensive damage to buildings, the most severe being to a building in the centre of the yard which was used as a hostel by train crews. Most of the fatally injured employees were standing outside this hostel at the time. 283 freight cars were demolished and 312 were damaged. 67 houses were damaged beyond repair and over 600 were damaged in some degree. 11 schools were damaged as were 31 commercial establishments. The total damage was assessed at $18.4 (1974).

The present author examined this incident in Marshall (1980) and concluded, on the basis of National Transportation Safety Board (1975) and of a report of the fire fighting operations, Fire Command (1975), that by no criterion was the explosion at Decatur more severe than the explosion at Flixborough which (see above) was of the order of 32 tonnes of TNT groundburst. The discussion below gives certain details which were omitted from Marshall (1980) which was a summary of a paper presented to the Advisory Committee on Major Hazards.

Figure 13.21. shows a plot of overpressures (as deduced from damage) versus distances for Flixborough and which includes some values of overpressure (as deduced from damage) versus distances for Decatur, which led Davenport to conclude that the TNT equivalence at Decatur lay between 250 and 400 tonnes. It will be seen that there was a wide scatter at Flixborough and that some of the individual points could be used to support estimates for TNT equivalence at Flixborough which are much higher than the ca 32 tonnes now generally accepted. Table 13.4., which is based upon Humberside Police (1974), reinforces this point by showing that damage at Flixborough expressed as a percentage of houses damaged in a particular locality, was by no means strictly correlated with the distance of the locality from the epicentre. Thus even if a group of damaged houses had been used as a basis for estimating TNT equivalence, figures much higher or lower than the mean value could have been obtained according to which group had been chosen.

Table 13.5. is a general comparison of the effects of the two disasters, showing point by point that the Flixborough explosion was the more severe. The only exception lies in the figures for the number of members of the public who were injured. This may be explained, at least in part, by the inclusion of persons treated for shock ("anxiety") at Decatur, there is no record of the numbers of persons suffering from shock but not otherwise injured at Flixborough.

The conclusions reached from Table 13.5. would be invalidated however

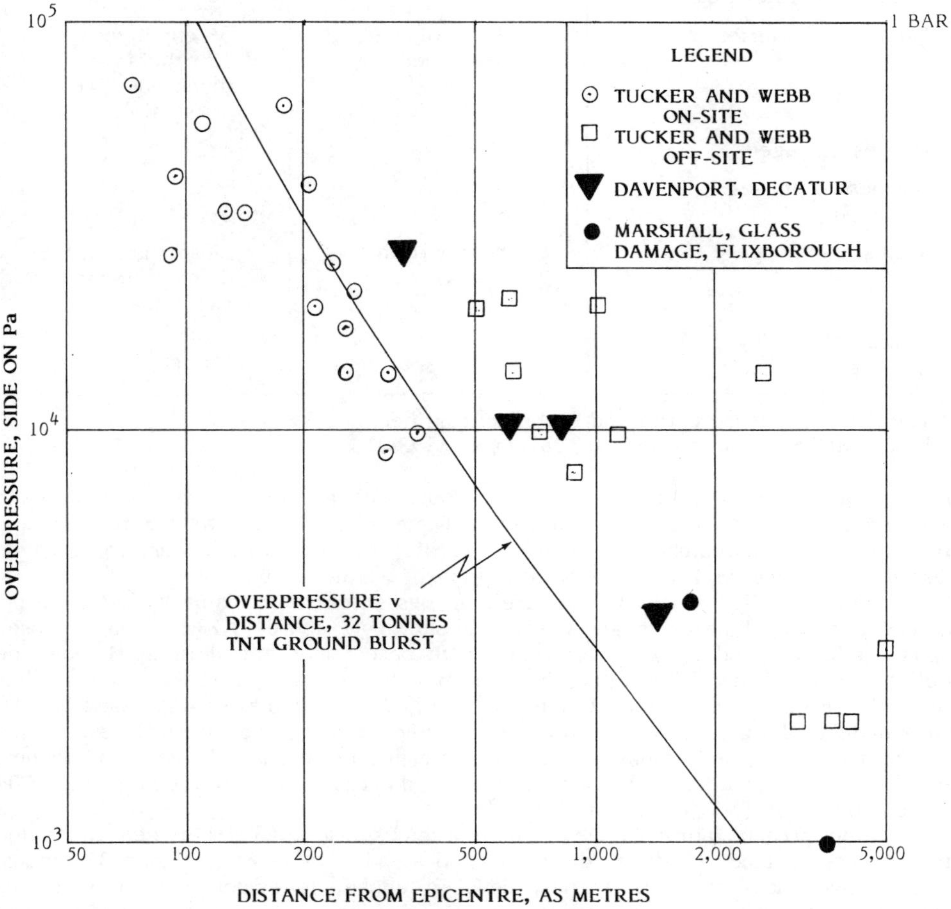

Fig. 13.21. OVERPRESSURE VERSUS DISTANCE, FLIXBOROUGH AND DECATUR COMPARED.

TABLE 13.5. POINT BY POINT COMPARISON.

Subject of Comparison	Decatur	Flixborough
Substance released	Isobutane	Cyclohexane
Pressure	2.5 bars *	8 bars
Quantity spilled	69 tonnes	ca 40 tonnes
Quantity in cloud	ca 40 tonnes *	ca 40 tonnes
Delay to ignition	8 to 10 minutes	2 minutes
Financial loss (1983 dollars)	36.9	140.5
Number of fatalities	7	28
Number of injuries	33 employees 316 others	36 employees 52 others (hundreds given first aid)
Houses damaged	700	1821
Business premises damaged.	31	167
Radius of damage	Most within 1 mile	More than half of damage beyond two miles
Radius of window damage	3 miles	6 miles

* Pressure of release estimated by present author.
Financial loss figures are from Davenport (1984).

if it could be shown that the density of population in the vicinity at Flixborough was higher than at Decatur. However, Table 13.6. shows that the reverse was true and that the number of persons and of properties was higher, for a given distance from the explosion, at Decatur than it was at Flixborough.

Finally, in Figure 13.22., the area of light, scattered damage for Decatur, as deduced from Fire Command (1975) and shown in Figure 13.23., is superimposed upon the corresponding area for Flixborough based upon the data in Humberside Police (1974). It will readily be seen that the Decatur area is much smaller than the Flixborough area. Using the method of calculation discussed in the case history of Port Hudson, the present author has concluded from an estimation of the two areas, and assuming that Flixborough had an equivalence of 32 tonnes, that the TNT equivalence of Decatur was of the order of 2 to 4 tonnes [c f. 200 to 400 tonnes of TNT, Davenport (1977)].

In the present author's view, the enormous discrepancy between the value for TNT equivalence as deduced by Davenport and a value of less than 32 tonnes, indeed much less than 32 tonnes, deduced above, is evidence of the serious pitfalls which can arise in estimating TNT equivalence. This should never be done on the basis of a few examples of conspicuous damage, but must be treated on a statistical basis. Furthermore, as the number of case histories grows, it should be more and more feasible to treat new cases by comparison with past examples in the way the present author has demonstrated. The evidence provided from individual cases of damage must always be viewed against the macrocosmic picture.

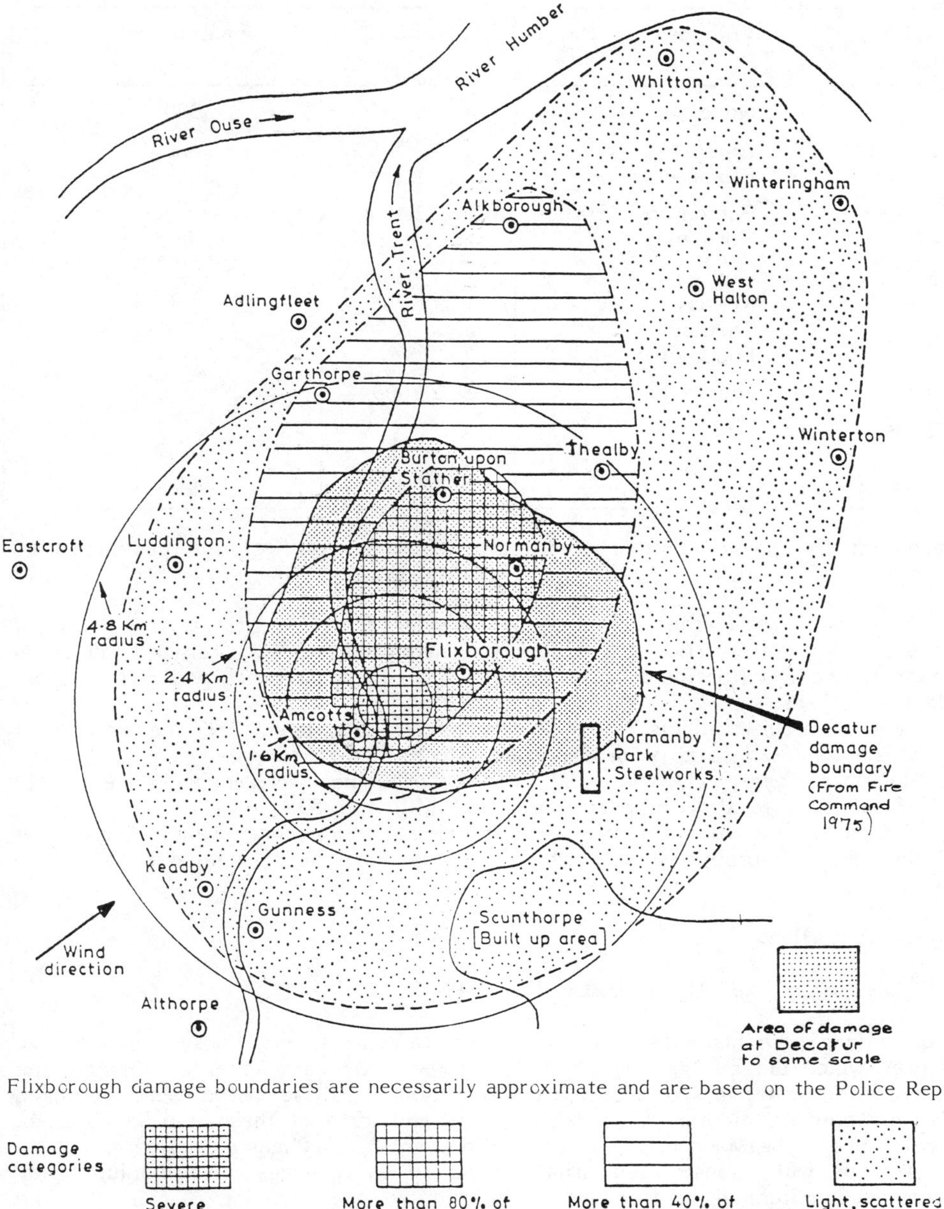

Fig. 13.22. FLIXBOROUGH AND DECATUR DAMAGE AREAS COMPARED.

TABLE 13.6. FLIXBOROUGH AND DECATUR.
POPULATION DENSITIES COMPARED.

	Flixborough		Decatur
Miles Radius	Villages	Inhabitants	Inhabitants
0 - 1/2	None	ca 40	6,000
1/2 - 1	Amcotts	544	
	Flixborough	236	
1 - 2	Burton upon Stather	2,540	
	Normanby	ca 120	
2 - 3	Gunness	1,090	
	Keadby	1,768	
	Thealby	ca 200	
	Luddington	379	
	Scunthorpe	ca 22,000 *	
Total 0 - 3 miles		ca 29,000	52,000

*Based on the present author's estimate that one-third of the population of Scunthorpe lived within a 3-mile radius of the explosion.

There are two further points to be considered. The first is that National Transportation Safety Board (1975) drew attention to two railyard incidents similar to the Decatur incident. These two incidents are recorded in Davenport (1984) as follows:-

(a) East St Louis, Illinois, 22/1/72, 53 tonnes propylene, TNT equivalence 1 to 2.5 tonnes, 1 fatality, property loss $19 million (1983).

(b) Houston, Texas, 21/9/74, 80 tonnes butadiene, TNT equivalence 20 to 57 tonnes, 1 fatality, property loss $26.7 million (1983).

The second is that railway yards provide circumstances which favour the development of turbulence and which provide vertical confinement under the wagons.

13.17. AIRSHIPS.

13.17.1. THE CONSTRUCTION OF AIRSHIPS.

Before giving case histories, it is useful to note some salient features of airships, features which did not greatly alter during the entire airship era. Though the literature does not always specifically mention it, those listed were hydrogen-filled. Some US airships were helium-filled and three of these, the Shenandoah, the Acron, and the Macron, were wrecked in flight by mechanical causes.

Airships were constructed usually of a series of rings of aluminium alloy joined by longitudinal girders and connected by cross-bracing of wire. The outer envelope was of canvas stiffened by cellulose varnish (dope) the function of which primarily was to reduce drag and to provide a weatherproof enclosure for crew and passengers where these were carried.

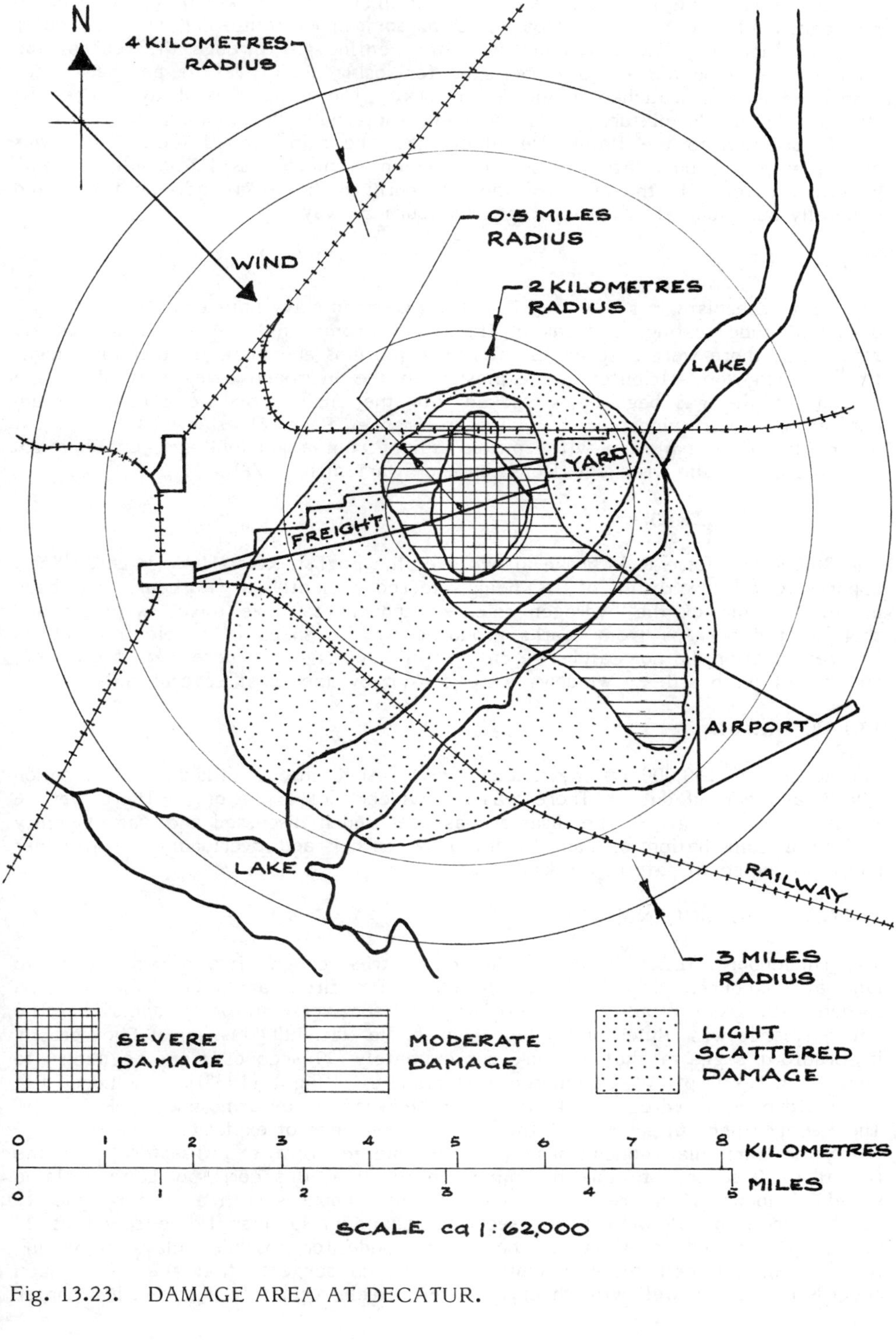

Fig. 13.23. DAMAGE AREA AT DECATUR.

The outer envelope played no part in containing the gas which was carried in gas bags which were of goldbeaters' skin, sometimes reinforced with an outer layer of fabric. There was provision for venting gas through the roof of the airship, this being a necessary operation for losing height. The gas bags were suspended in nets attached to the framework. There was thus always a possibility that flammable mixtures could accumulate internally in an airship.

In addition to the flammable lifting gas, the ship carried fuel. This was originally petrol but, later, diesel fuel, which is much less flammable, became generally used. In the event of the ship burning on the ground, the fuel would generally burn long after the hydrogen had burned away.

13.17.2. LZ.18.

The first case history is that of LZ.18 (L.2, German Navy number). On 17/10/13, during altitude testing, a streak of flame shot from a motor gondola and shortly afterwards there were a series of internal explosions and instantly the ship caught fire. Deighton attributes the explosion to the hydrogen being valved off the bottom of the gas bags being sucked into the engines by the partial vacuum created by their windscreens. This reference, Page 21, shows the ship just before the first explosion with smoke issuing from a gondola and, on Page 23, a photograph of the wreckage.[Deighton and Swartzman (1978)].

13.17.3. R.38. (ZR.2).

The 80,000 cu metre R.38 broke in two over the English River Humber on 24/8/21, apparently as the result of its being steered violently to starboard and back again. Some of the hydrogen released appears to have mixed with air and then ignited perhaps from sparks produced by the snapping of electric cables. The rear section did not catch fire but fell into the river. There were 5 survivors out of 49 on board. Windows were broken by the blast several miles away.

13.17.4. R.101.

The R.101, of 150,000 cu metre capacity, crashed into a hillside in France on 5/9/30 and burned out. There was no evidence of explosion. There were 6 survivors out of 54. The disaster has been much discussed and was evidently due to a combination of bad design, bad weather and overloading. It never had a proper certificate of airworthiness.

13.17.5. THE HINDENBURG.

The Hindenburg, (LZ.129), of 200,000 cu metres, caught fire on coming in to land at Lakehurst, New York on 6/5/37. The fire started near the tail and spread with great rapidity. There were 62 survivors out of 97 aboard. The entire episode was filmed and a tape loop in the National Museum of Photography in Bradford suggests that it took approximately 30 seconds from beginning to end. A US Dept. of Commerce Report, [Ref: Knight (1938)], concluded that there had been a hydrogen leak which had been ignited by atmospheric electricity. The Report ruled out sabotage. There was no evidence of explosion.

Some particular conclusions seem to emerge from this disaster. If the fire took 30 seconds to spread from stern to bows, this corresponds to a flame speed of about 8.5 metres per second. This compares with a burning velocity for hydrogen of 3.5 metres per sec and maximum laminar flame speed of 28 metres per second. Allowance has to be made for possible delays in passing from an ignited cell to an unignited cell which suggests that the time taken accords reasonably well with theory. Alternatively, using the equation in Chapter

8.4. for fire ball duration, this would give a duration of 10 seconds. However this calculation does not take account of the time delays in travelling from one cell to another. In any case, there are differences between the situation under discussion and the classical fire ball situation.

A final point of significance is that so many survived. This seems to argue that the effects of thermal radiation from burning hydrogen are less than from, say, the burning propylene of the San Carlos disaster. The survivors must have been very close to the burning hydrogen. However, hydrogen burns with a markedly non-luminous flame.

13.17.6. SOME CONCLUSIONS ON AIRSHIP ACCIDENTS.

Taking account also of the discussion in Section 12.4.7., the most general conclusions are these:-

(1) In the event of a crash, the hydrogen is by no means bound to ignite.
(2) The most likely result of ignition is the deflagration of the hydrogen.
(3) An explosion is not likely to ensue but, when it does, only a portion of the hydrogen is likely to be involved.
(4) The circumstance most likely to give rise to an explosion is mechanical break-up with delay to ignition.

Chapter 14

TOXIC RELEASES

14.1. INTRODUCTION.

14.1.1. NOMENCLATURE.

For the purposes of this chapter "Toxic" will be defined as in I.Chem.E. (1984) as:-
TOXIC - A property of substances which, when introduced into or absorbed by a living organism, destroy life or injure health.

Following Wingate (1976), "Toxicology" may be defined as follows:-
TOXICOLOGY - Study of poisons, an extension of pharmacology.

Other terms will be defined as they arise.

14.1.2. SCOPE OF CHAPTER.

The chapter is in no sense a treatise on toxicology in the medical sense. It is concerned with the problems of containment of toxic substances and with the consequences of loss of containment. The discussion of toxic effects is, however, inevitable as is also a discussion of hazard in terms of data on toxic effects. The limitations imposed by conflicting evidence will be discussed at appropriate points in the chapter.

It will be concerned with acute events, i.e. those of short duration, typically an hour or less, and not with the effects of persistent releases which give rise to occupational ill-health or to similar illness to those living in the vicinity. Such chronic problems are very important in their own right, and are studied under such headings as occupational hygiene or atmospheric pollution or water-borne pollution. They do not form part of the subject matter of this book.

Unless specially indicated, the discussion is concerned with potential effects on human beings and not on animals or vegetation.

14.2. THE NATURE OF THE PROBLEM.

14.2.1. THE EFFECTS OF TOXIC SUBSTANCES.

There are three main routes of entry into the human system, ingestion, that is through eating or drinking, through the skin, or by inhalation. As will appear later, this latter route is far and away the most significant route in the acute events with which this chapter deals.

The literature on toxic effects is heavily biased towards the chronic effects produced by inhalation or skin contact especially as the result of occupational exposure and the acute effects consequent upon ingestion whether by accident, malice or in the course of attempted suicide. Such ingestion is referred to as "poisoning". For some of the gases encountered in industry, there is perhaps more information to be obtained from works on chemical warfare than from other sources.

It should, however, be noted that whereas flammable gases constitute a hazard which is largely independent of their chemical composition in that the effect considered is that of burns and the hazard of all explosives lies in their common ability to create overpressure, the effects of any toxic substance is peculiar to that substance or to substances closely related to it chemically. Thus the symptoms produced by acute exposure to chlorine, carbon monoxide or hydrogen sulphide are completely different.

Chlorine is an irritant of the lungs and the principal cause of death arises from the victim drowning in the fluid secreted to counteract the irritation. Carbon monoxide combines with the haemoglobin of the bloodstream and thus interferes with the vital exchange of oxygen with the tissues. Hydrogen sulphide paralyses the central nervous system.

For this reason, after considering general matters, there will be a discussion of a number of commercially significant toxic substances treated individually. These will be illustrated by case histories in Chapter 15 where any are available.

14.2.2. THE MAGNITUDE OF THE PROBLEM.

As with incidents arising from the release of flammable gases, the statistics relating to major releases of toxic substances are neither very abundant nor very reliable. In Marshall (1977) is listed 25 incidents occurring worldwide, and involving chlorine in quantities of 1 tonne or more, over the previous sixty years, and 6 incidents occurring worldwide, and involving ammonia in quantities of 15 tonnes or more, over the previous ten years. Advisory Committee on Major Hazards (1979) lists 17 chlorine incidents which overlap the incidents in Marshall (1977), and 11 incidents involving ammonia. Both references quote a single incident involving phosgene.

In Marshall (1979) it was reported that the present author had abstracted, from a number of compilations on releases of liquefied gases, a total of 200 separate incidents, and of these incidents 30, or 15%, were toxic releases. The remainder were releases of flammable gases. The reference commented that this suggested that toxic releases were not the principal major hazard problem. Though, as will be demonstrated, the hazard is the cause of considerable public apprehension.

Confirmation of this has come from Kletz and Turner (1979). The data in this reference suggests that over a fifteen-year period, out of a worldwide total of 1307 deaths from major incidents involving fires, explosions and toxic releases in both fixed installations and transport, there were 104 due to toxic releases or 8%. Of the injuries, there was a total of 4285, of which 1343 or 32%. This indicates that the ratio injury to death for toxic releases is markedly different to that in fires and explosions. However, it may be said that if these statistics were to be updated to the time of writing, they would be seriously upset. This is because a single toxic release in India in December 1984, has killed more than 2500 people. Here the lethal agent was methyl isocyanate. It appears that 45 tonnes of it were stored in an underground tank which had become pressurised by becoming heated as the result of chemical action. But this does not upset the conclusion that the majority of Major Hazard Accidents are concerned with fire and explosion.

The toxic properties of substances are much more difficult to measure than are their physical properties and these difficulties are discussed below. However, this being admitted, there is general agreement that some substances are non-toxic, as, for example, nitrogen at ordinary pressures, others are moderately toxic as are ammonia and carbon monoxide, others are highly toxic as is chlorine, and others are hyper-toxic, as are the nerve gases. This, taken in conjunction with the varied modes of attack on the human body, makes a systematic description of acute toxic hazards very difficult.

14.2.3. THE LITERATURE.

There are three broad classes of literature in the field. However, actual works on the field of toxicology, though they may be principally concerned with one or other of these classes, may overlap in their subject matter with the other classes.

Firstly there is main stream toxicology which is concerned, in the main, with ingestion and which includes a great deal that may be classified as under the heading of consumer protection. This includes investigation of the potentially harmful effects from, for example, cosmetics, food additives or drugs. The literature of this subject is voluminous and has little direct bearing on the topic under discussion. A recent book concerned with poisoning is Vale and Meredith (1981). This gives statistics of poisoning in the UK and includes statistics on inhaled toxics.

Secondly there is occupational toxicology. This is much more concerned with chronic than with acute effects. An early reference is Legge (1934) which describes the scene in the UK in the early thirties and which gives data on occurrences of "gassing". An important modern reference is Patty (1978). A key reference work is International Labour Organisation (1983) which gives data on the toxic properties of many substances in common industrial use. The NIOSH Registry of Toxic Effects of Chemical Substances is an annually updated compendium of information on toxic properties. The present author possesses the 1978 edition and it is this edition, NIOSH (1978), to which reference is made below. Figure 14.1. is an example of a typical substance entry from the Registry. The Registry contains a foreword describing in detail the significance of the various data categories and how they were arrived at. A useful pocket guide is NIOSH/OSHA (1978). This gives measures of acute toxicity as well as occupational exposure limits. There are, in addition to the general works on occupational toxicology, numerous monographs on particular areas or on specific substances such as lead.

Thirdly there are works concerned at least in part with acute events. Reference has been made to the relevance of works on the military use of toxic agents. Only a select list can be made of such works; those cited are works which the present author has consulted. They are Fries and West (1921), Prentiss (1937), Haldane (1938), and Sartori (1939). Home Office (1938) describes many aspects of gas warfare which are relevant to the industrial scene and Royal Army Medical Corps. (1925) gives information on the treatment of victims of chlorine, phosgene and other agents. Later works on chemical warfare do not have much to say about chlorine or phosgene as these are regarded as out-moded as weapons, however important they may be as industrial gases at the present time. Modern works include Rose (1968) and Robinson (1968). Recently published is the definitive Haber (1986). A comprehensive bibliography of works on chemical warfare is obtainable on request from the Imperial War Museum, Lambeth Road, London. Books and papers of individual case histories will be referred to in the appropriate paragraph of Chapter 15. Certain papers on likely fatalities from chlorine will be referred to in Chapter 18.

(A) (1) (3) (4) (5)

AA1111111 RTECSIUM

(2)	UPDT: 8101	CAS: 9999–99–9 MW: 357.88 MOLFM: C-Re-S-Te	
(6)	SYN:	NIOSHRTECSIUM · RTECS (DOT) · RTECSIAM (French) · TSL (OBS.)	
(7)	MTDS:	mma-sat 50 ug/plate	(10) PNASA6 72,5135,75
		cyt-hmn:leu 100 umol/L/8H	MUREAV 4,53,67
		dlt-mus-orl 35 mg/kg	TXAPA9 23,288,72
(8)	IRDS:	skn-hmn 5%/1M nse MOD	JAMAAP 135,1920,68
		skn-rbt 10 mg/24H open MLD	AMIHBC 10,61,54
		eye-rbt 20 mg	AMIHBC 10,61,54
(9)	TXDS:	ihl-hmn TClo:4000 mg/m3/7M TFX:CNS	AHBAAM 116,131,36
		orl-rat LD50:800 mg/kg	MarJV# 26 Apr 76
		orl-rat TDLo:70 gm/kg/78W-1 TFX:NEO	NTIS** AD900-000
		ihl-rat TCLo:100 ppm/7H/(6-15D preg) TFX:TER	TXAPA9 28,442,74
		ivn-mus LDLo:1800 mg/kg	JNCIAM 5,251,45
(11)	AQTX:	AQUATIC TOXICITY RATING: TLm96:100-10 ppm	WQCHM* 2,-,74,
(12)	REVIEW:	CARCINOGENIC DETERMINATION: ANIMAL SUSPECTED	IARC** 1,61,72
	REVIEW:	THRESHOLD LIMIT VALUE-air:50 ppm	DTLVS* 4,358,80
	REVIEW:	TOXICOLOGY REVIEW	PLMJAP 6(1), 160,75
(13)	REGS:	OSHA STANDARD-air:TWA 50 ppm (SCP-J)	FEREAC 39,23540,74
		DOT-ORM-A, LABEL-NONE	FEREAC 41,15988,76
(14)	CRIT DOC:	OCCUPATIONAL EXPOSURE TO RTECSIUM recm std-air:TWA 10 ppm	NTIS**
(15)	STATUS:	SELECTED BY NTP FOR CARCINOGENESIS BIOASSAY AS OF SEPT 1980	
	STATUS:	NIOSH MANUAL OF ANALYTICAL METHODS, VOL 3 S255	
	STATUS:	NIOSH CURRENT INTELLIGENCE BULLETIN 41, 1980	
	STATUS:	REPORTED IN EPA TSCA INVENTORY, 1980	

A. RTECS Accession Number, a sequence number assigned to each substance. See Format.
1. Substance Prime Name. See 1 in text.
2. Date when substance entry was last revised. See 2 in text.
3. Chemical Abstracts Service Registry Number. See 3 in text.
4. Molecular Weight of this substance. See 4 in text.
5. Molecular or Elemental Formula of this substance. See 5 in text.
6. Synonyms, common names, trade names, and other chemical names. See 6 in text.
7. Mutation Data. See 7 in text; also see Figure 2.
8. Skin and Eye Irritation Data. See 8 in text; also see Figure 3.
9. Toxicity Data. See 9 in text; also see Figure 4.
10. CODEN, an acronym for the references from which the data and other citations were abstracted. See 10 in text; also see Figures 2, 3, and 4.
11. Aquatic Toxicity Rating. See 11 in text.
12. Review of this substance. See 12 in text.
13. Standards and Regulations for this substance promulgated by a Federal agency. See 13 in text.
14. A Criteria Document supporting a recommended standard has been published by NIOSH. See 14 in text.
15. Status information about this substance. See 15 in text.

Fig. 14.1. A TYPICAL SUBSTANCE ENTRY IN THE NIOSH REGISTRY.

14.3. CAUSES OF RELEASE.

14.3.1. RELEASE FROM PRESSURISED STORAGE.

A number of toxic agents, of which chlorine and ammonia are important examples, are stored as liquefied gases under pressure usually less than 10 bars. If containment be lost a portion of the liquid flashes; the size of this fraction depends upon the substance and the temperature. The size of this fraction for chlorine and ammonia is shown in Figure 5.5. The formation and subsequent behaviour of the ensuing vapour cloud have been discussed in Chapter 7. Possible causes of pressure vessel failure have been discussed in Chapter 6.

14.3.2. ESCAPE FROM NON-PRESSURISED STORAGE.

Some toxic agents which are liquids at ordinary temperatures are stored in non-pressurised storages fitted with breathing vents and with suitable equipment, such as activated charcoal absorption columns, to prevent escape to atmosphere.

One possible cause of loss of containment could be overpressurisation by inert gas such as nitrogen as, for example, by failure of a reducing valve where inadequate provision exists for automatic correction of this fault. A further cause could be floating off the toxic layer by admission of water.

Unanticipated levels of heat input could represent a further cause. This could arise through solar radiation or a nearby fire.

The admission of substances which react with the contents is a cause which has given rise to toxic emissions even where the contents of a storage have been of relatively low toxicity. There have been examples in the UK in which hydrochloric acid and bleach liquor (sodium hypochlorite) have been inadvertently mixed in this way and have given rise to an emission of chlorine.

Admission of a substance capable of catalysing polymerisation or decomposition could liberate heat capable of boiling off part of the contents.

14.3.3. ESCAPE FROM REACTORS.

The general question of runaway reactions has already been discussed in Chapter 6. There have been a number of case histories of toxic releases from runaway reactions. These are discussed, with special reference to dioxin, in Chapter 15.

14.4. ESTIMATION OF TOXICITY.

14.4.1. INTRODUCTION.

For gas/vapour releases, as has been demonstrated in Chapter 9, it is possible by a combination of the laws of the physical sciences, of engineering and of meteorology, to make approximate predictions, on a statistical basis, of the time/concentration integral of toxic material at a given point following a loss of containment. What this prediction does not do, of itself, is to predict the level of harm which will ensue to a human being at that point. To do this requires, among other factors, an input of toxicological data.

14.4.2. HUMAN EXPERIMENTATION.

If the data which are used to calculate the time/concentration integral are recognised as inexact, this applies with even greater force to the toxicological data. To predict human casualties requires a measure of the level of concentration of a toxic substance at which a given level of harm is suffered by human victims. It has to be accepted that there never can be a single index for this. This is because there is a considerable variation between persons on account of such factors as age, physique, state of health, etc. To simplify the discussion, let us assume that someone in good health and aged between 30 and 40 could be a typical person. There would also have to be a standard level of activity as the total quantity taken in (the dose) during the duration of the exposure to the toxic atmosphere would be highly dependent on the respiration rate. How then could this be predicted ? The most obvious method of finding out would be to administer the toxic gas to the typical person at a typical level of physical activity and see what level of dose would produce a given level of harm. This would constitute human experimentation.

There are a number of ways in which human experimentation may be carried out. One is by the use of volunteers. Where only the lower harm levels are explored, this has proved feasible to conduct under safeguards. Haldane (1927) describes experiments in which the investigator was his own guinea pig. Human experimentation in which lethal doses are administered to unwilling subjects is socially taboo and is outlawed by civilised countries.

But there is evidence for the level of lethal doses for certain toxic substances which arises from voluntary acts. These are acts of suicide both where this is successful and where it is unsuccessful.

Also, especially from poisoning by ingestion, there is evidence from accidental administration or from murder, though criminal poisoning is rarer today than it used to be. The quantity ingested may be known either from physical evidence such as the quantity taken from a container, or by post mortem examination. Over the course of time, sufficient data may accumulate to enable a minimum fatal dose to be established for common toxics which will be expressed, in accordance with universal practice, as mg/kg of body weight. Information on this subject appears in works on forensic medicine as, for example, Smith (1945). This latter also gives information on the effects of toxic gases including war gases.

However, the acts described in the last two paragraphs cannot properly be described as "experimentation". The persons performing the acts are not doing them to seek scientific truth and they lack all the normal disciplines associated with scientific research. This is not to say that the knowledge cannot be gained in absence of other evidence by scientific analysis of these cases.

14.4.3. EVIDENCE FROM WARFARE.

Under conditions of war, things are done which constitute a form of human experimentation. The present author is prepared to draw lessons from these, however abhorrent the acts may have been, if the knowledge helps to avoid chemical disasters.

There is evidence that there was a considerable element of scientific planning in the initial gas attacks of the First World War and of follow-up of its results. This is evidenced by the role played by the distinguished German scientist Fritz Haber, in the planning and supervision of the chlorine gas attack at Ypres in April, 1915. This role is described in Haber (1986) and in Goran (1967).

The implications of the scientific information to be derived from the study of its use in war will be further discussed in the case histories below and in Chapter 18.

14.4.4. LIMITATIONS OF HUMAN EXPERIMENTATION.

The limitations of the methods of gaining knowledge of toxic effects on humans are considerable. Were there no other methods, we only have a very limited amount of information about the effects of poisons in general, and would know even less about the toxic effects of many substances which are common in industry. This is because few of them have been used for murdering people, as suicidal agents or in warfare. There have been many substances introduced into industry on which, fortunately, we have no historical experience of fatality among humans, but for which there is a powerful case that they should be controlled.

This state of affairs can, at least partially, be remedied by animal experimentation.

14.4.5. ANIMAL EXPERIMENTATION.

The advantage of animal experimentation is that it is scientific. All the disciplines of scientific investigation can be applied such as the use of control groups, the use of pure strains, especially among smaller mammals, the establishment of groups closely similar in age and physique, the quantities of substance administered can be accurately measured and, perhaps above all, results can be achieved which can be demonstrated to have statistical significance. Without animal experimentation the data in NIOSH (1978) would be only a small fraction of what is currently presented. Information is presented in NIOSH (1978) on standard techniques of animal experimentation.

There are however disadvantages. The data vary between one species and another, sometimes by many orders of magnitude. Hay (1982), P.33, claims that for dioxin (2,3,7,8, tetrachlorodibenzo-p-dioxin) the toxicity, for guinea pigs, expressed as LD_{50}, has been reported as being as low as 0.6 micrograms/ kg of body weight, whereas for golden hamsters, the same index is reported as being 3000 micrograms/kg. (For definition of LD_{50}, see below). The ratio between the toxicity in the two animal species is thus 5000 to 1. If animal data for this case is to be translated into human data the question is, which of these data is the appropriate measure ?

From the point of view of predicting human fatalities in a situation in which there is a sudden release of toxic substances, a factor which the present author has claimed is almost universally ignored [see Marshall (1983)], is that the mortality rate is for untreated animals whereas in the human situation the number of actual fatalities would be appreciably reduced below the potential number by medical treatment of the victims.

One considerable difficulty associated with animal experiments is their cost. Another is the growing level of public opposition to animal experimentation in principle. However the public pressures for safeguards from toxic hazards is producing a growing demand for toxicological data which, at present, cannot be obtained except from animal experiments.

14.5. THE PRESENTATION OF TOXICOLOGICAL DATA.

14.5.1. NOMENCLATURE.

There are a number of ways of expressing toxicity. Data for ingestion is always expressed as mass of toxic agent per kg of body weight. (For an average human adult, the body weight is ca 70 kg). However, even assuming that the data from animal experiments are transferable to humans, there is no unique figure for an animal of a given species. This is because animals vary in susceptibility. According to Bridges (1984), variations in LD_{50} can be two or threefold for a given substance depending upon the sex, age or strain of the animal.

Furthermore, experiments on animals in which all the animals die, do not indicate the threshold level at which death occurs. For this reason, toxicity data are often expressed as LD_{50} which is the dose which will kill 50% of an exposed population of designated animals. According to Bridges (1984), ". . the

LD_{50} is best viewed as an approximate indication of lethal potency and it must never be regarded as an absolute figure".

Values of the subscript differing from 50 may be reported in the literature. LD_0 for example denotes the lowest recorded lethal dose. The slope of the line which relates the value of LD_x to x varies considerably between one toxic agent and another. According to Bridges (1984), the ratio between LD_{50} and LD_{01} for a number of pesticides has shown a 70-fold variation.

As has been pointed out above, the ingestion route is not usually the most important route for acute chemical releases. The most important route here is by inhalation. For this reason, data which disclose the LC_{50}, or LC_x, are more relevant. But such figures are meaningless unless related to the time of exposure. For fairly high levels of concentration, the following relationship derived by Haber and quoted in Prentiss (1937), holds:-

$$C_T \times T = K \quad \text{where}$$

C_T = concentration of toxic agent
T = time of exposure
K = a constant

However, were this formula to be valid for all levels of concentration, the concept of a threshold limit would be meaningless. There is considerable experimental evidence that Haber's formula is not correct at low concentrations. Prentiss presents a graph for the effect of phosgene on dogs. According to Haber this should be a rectangular hyperbola, but experimental results demonstrate that at lower concentrations, the time taken to produce death is higher than would be deduced from Haber's formula. To give an example, at 200 mg/cu metre concentration, experiment shows, according to Prentiss (1937), that instead of the 25 minutes predicted by Haber, 45 minutes would be required to produce death.

Just how exact the figures in Prentiss are cannot be ascertained, but there is little doubt of the general truth of these observations. For example, NIOSH (1978) gives for hydrocyanic acid a value of LC_0, for human inhalation, of 200 mg/cu metre with a 10-minute exposure. From the Haber formula, it would follow that the corresponding time for 10 mg/cu metre should be 200 min. In fact 10 mg/cu metre is very close to the threshold limit, i.e. infinite exposure time without harm. The reason for this, in general, is that the body has a capacity for absorbing low concentrations of toxic substances and either metabolising them or excreting them or some combination of the two.

It follows from this that the concept of relating directly the time concentration integral, referred to above in section 14.4.1., to human mortality is not valid. This integral is a form of product of concentration and time which takes account of fluctuating concentration, which would be the rule in an actual release situation. However, the value of this integral on its own is not sufficient to predict mortality as two equal integrals, one which relates high concentration with low values of time will produce a higher level of mortality than the integral of low concentration with high level of time.

Lethal doses expressed as the integral of concentration with time are not readily translatable to lethal doses as mg/kg of body weight. To be able to do this would require knowledge of the fraction of the toxic agent retained in the lungs and the rate of respiration. In mammals it is generally assumed that there is a constant ratio between rate of respiration, volume/time, and body weight. The standard figure for humans is taken as ca 7 litres per minute when at rest, but it can exceed 30 litres per minute during exercise, [Encyclopedia Britannica (1951), Vol.19, P.217].

A report, Withers (1985), the findings of which will be discussed below under the heading of chlorine, suggests a relationship:-

$$LD_{50} \quad \propto \quad ct^{0.5} \quad \text{where}$$

c	=	concentration of chlorine
t	=	time of exposure

The reference states that the exponent (0.5) is an approximation only and that the data are insufficient to warrant any more precise value. In the present author's view, it will be best not to assume too readily that the exponent may not vary from one toxic agent to another. Clearly the equation cannot hold for very low concentrations as this would imply that there could not be a threshold limit below which harmful effects, let alone fatality, could not occur.

14.6. THE CLASSIFICATION OF TOXIC AGENTS.

14.6.1. PHYSICAL CHARACTERISTICS.

14.6.1.1. Dispersive energy.

By "dispersive energy" is meant the energy which is available within the agent which enables it to be brought into contact with the victims. The quantity of energy will vary with the total pressure under which the agent is confined and with its vapour pressure under the conditions of confinement.

The agents which possess the highest level of dispersive energy are the liquefied gases such as chlorine and ammonia. Phosgene, which is also a liquefied gas, has a lower dispersive energy than chlorine at the same temperature because of its lower vapour pressure. Agents maintained as liquids above their atmospheric pressure boiling point in process units will have a dispersive energy which is related to their vapour pressures.

Highly volatile liquids which, though inherently less dangerous than liquefied gases, are liable to have a high rate of evaporation, especially in wind.

Liquids of low volatility and solids may be assumed not to possess inherent dispersive energy. To disperse them requires a "vector". Some form of explosion, not necessarily a violent explosion, or admixture with an effluent stream such as flue gas, is needed.

14.6.1.2. Persistence.

War gases may be classified as "persistent" or "non-persistent". The former are liquids of moderate to low volatility of which the classic example is mustard gas. Others are released as gases and their persistence varies with atmospheric conditions and terrain. Whereas mustard gas might have a persistence measured in days in summer and weeks in winter, chlorine had a persistence measured in minutes or hours.

In industrial practice the differences may not be so clear cut. This is because tonnage spillages of liquefied gases such as chlorine or ammonia give rise to an immediate flash-off of, say, 15%. But the bulk of the spillage will evaporate only as fast as heat can transfer into the residual pool. Therefore a large spillage may take days to evaporate. (See chlorine case histories below). On the other hand dioxin, which is also discussed as a case history below, is a solid with a half-life in soil measured in years.

A factor which is important in the industrial context is how easily an area over which a toxic liquid or solid of low volatility has been deposited may be decontaminated. Some dangerous substances may be readily neutralised; an example of a substance of this kind is mustard gas. This could be neutralised by hypochlorites which are chemicals found in most kitchens in the form of household bleaches. [see Home Office (1938)]. An extreme case of the other sort is dioxin. No satisfactory decontaminant has been found for this in spite of extensive research.

However it may be said, in general terms, that the intrinsic persistence of toxics is least for the loss of containment of buoyant toxic permanent gases. Of these, carbon monoxide and hydrogen cyanide are mong the few examples. These are followed by ammonia, which is a liquefied gas of relatively high buoyancy (but see case histories below). Then follows chlorine, a liquefied gas much denser than air, then come highly volatile liquids such as methyl isocyanate, then liquids of low volatility such as mustard gas and finally solid toxics such as dioxin.

14.6.2. BIOLOGICAL CHARACTERISTICS.

14.6.2.1. Mode of attack.

As has been pointed out above, toxic agents vary greatly in their mode of attack on the human system. The author has not been able to find a classification of toxics, based on their mode of attack, which is appropriate to the chemical and process industries. There have been several classifications of war gases, some of which are quoted in Prentiss (1937). One of these, the British between-the-wars classification, the present author has modified to take account of present day circumstances. These are, with examples of each,:-

(1) Lung irritants. Chlorine, Phosgene, Chloropicrin, [Trichloronitro methane].
(2) Vesicants. Mustard gas [bis (2-chloroethyl sulphide)].
(3) Sensory irritants. Diphenyl chloroarsine.
(4) Lachrymators. Bromobenzyl cyanide.
(5) "Blood gases". Carbon monoxide.
(6) Incapacitating agents. BZ Agent.
(7) Nerve gases. Tabun, Sarin, Soman.

However, such lists do not include many of the toxic agents to be encountered in industry such as ammonia and methyl isocyanate. On the other hand, many of the substances in the war gas list have no current industrial importance. Thus, though both sensory irritants and lachrymators may be chemical hazards, in the author's view they do not rank as **major** chemical hazards. Incapacitating agents are, by intention, non-lethal. The nerve gases listed appear to be manufactured solely for their end use which is to be stock-piled for military purposes and do not function as intermediates in organic synthesis. Thus they do not have industrial significance. They will, however, be briefly discussed below along with those pesticides to which they are chemically related.

14.6.2.2. Inherent toxicity.

It would be useful to be able to rank toxics into broad categories of severity of the hazard presented. Figure 14.2., part of which is derived from table 1 in Bridges (1984), shows however that there are no internationally agreed standards of classification. Moreover the categories are heavily biased towards ingestion; they are LD_{50}s from oral administration to rats (except for Col.9). The most likely industrial hazards, those arising through respiration or through skin contact, are not covered.

The columns are as follows:-

(1) As in EEC Directive on Major Accident Hazards of 24/6/82 Annex IV. Bridges (1984) claims that Col.2 of table 1 of that reference is from this Annex, but the classifications in Bridges do not correspond with Annex IV.

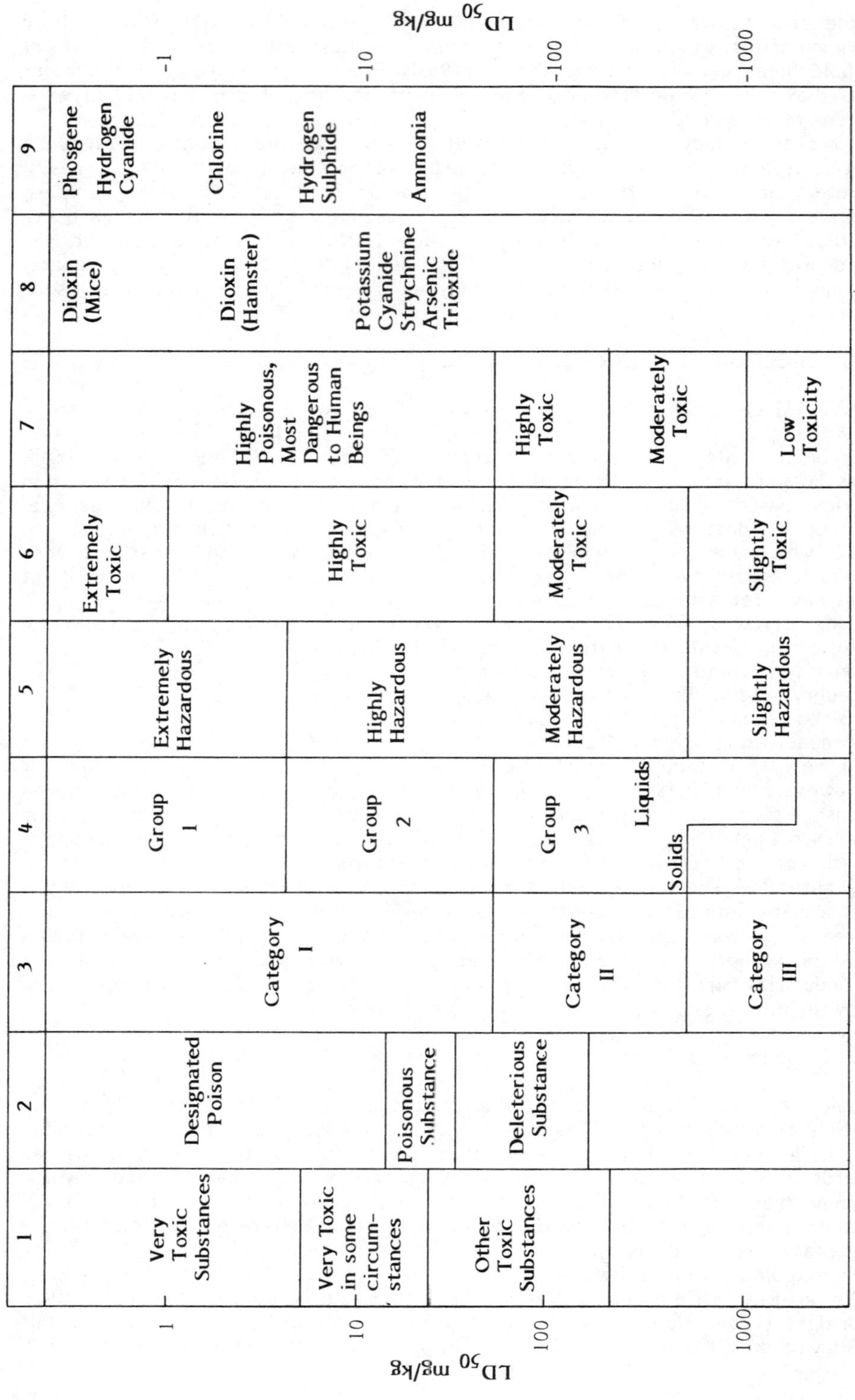

Fig. 14.2. INTERNATIONAL RANKING OF TOXICITY.

(2) Poisons and Deleterious Substances Control Law (Japan).
(3) Federal Insecticide, Fungicide and Rodenticide Act (USA).
(4) Transportation of Dangerous Goods (United Nations).
(5) Classification of Pesticides (World Health Organisation).
(6) Hodge and Sterner Classification quoted in Bridges (1984).
(7) Classification adopted in USSR, derived from International Labour Organisation (1983).
(8) Examples of acute toxicity by ingestion taken from NIOSH (1978).
(9) Indicative levels of LD_{50} derived from inhalation data in CW (1971) and NIOSH (1978). These are nominal equivalents and are based upon the following assumptions:-
   (a) All the toxic materials inhaled enters the system.
   (b) At high concentrations the product of time and lethal concentration is a constant.
   (c) The index, respiration rate per kg of body weight, is as in human subjects, i.e. 10^{-4} cu metres per minute per kg of body weight based on a respiration rate of 7 litres per minute (resting) and an average weight of 70 kg. Thus:-

$$\text{Dose (mg/kg)} = \frac{\text{Toxic concentration (mg/cu metre) x time (mins)}}{10^{4}\text{ (kg mins/cu metres)}}$$

Where the lethal concentration is given in parts per million, this may be converted to mg/cu metre by multiplying by (mol wt of toxic x 0.0423).

14.6.2.3. Injury to fatality ratio.

The use of the word "casualty" in the military context means any loss of personnel which reduces the fighting strength of a unit. Thus the term includes battle deaths, non-battle deaths such as deaths from natural causes or in transport accidents, battle injuries severe enough to cause permanent disability (permanent casualties), battle injuries leading to temporary disability (temporary casualties), non-battle injuries, taken prisoner, deserted, etc. When used in a civil context the term "casualties", if used in predicting the likely outcome of the realisation of a major hazard, unless broken down into categories such as dead, permanently disabled, temporarily disabled, etc, is liable to confuse and alarm the public who may believe that a "casualty" is someone killed.

Different toxic agents differ in their ability to injure as well as to kill. In the First World War, according to Prentiss (1937), Table X, for all belligerents, and for all causes of injury or death, there were approximately 3 non-fatally wounded per battle death. But for gas casualties the ratio was 13 to 1. Thus poison gas brings in its train more injuries per death than does high explosive. It does not follow that all toxic agents display the same injury to death ratio. There seems general agreement that mustard gas gave a particularly high injury to death ratio as compared, say, with chlorine. From the military point of view, especially when on the defensive, mustard gas was very effective. But the question as to whether chlorine or mustard gas is the more dangerous in an industrial situation is more difficult to answer. If, for a given mass of the agent, A produces W fatalities and X injuries, and B produces Y fatalities and Z injuries, and other things are equal, provided $W \geqslant Y$ and $X \geqslant Z$, it is clear that A is more dangerous than B. But if $W \geqslant Y$ and $X \leqslant Z$, a more difficult decision arises. This problem will be further discussed in Chapter 18.

14.6.2.4. Long-term effects.

The extent to which the long-term effects of exposure to toxic agents is known with certainty depends upon the availability of the subsequent medical records of those subjected to acute exposure. Trevethick (1973), of which the author was a medical practitioner in industry, states "It is very doubtful that chronic effects occur from inhaling chlorine. After each acute episode - recovery is complete . . . Chronic poisoning (by phosgene) is unknown in industrial practice but repeated acute exposure can lead to chronic chest disease . . . Where recovery takes place after a severe episode of cyanide poisoning there may be permanent effects due to malfunction of parts of the central nervous system".

There seems little doubt that a number of agents, of which mustard gas is a prime example, can cause permanent blindness, disfiguring scars and other disabilities.

The author has not been able to find evidence of long-term effects from acute sub-lethal exposure to nerve gases. Heath (1961) implies that there are none.

There has been considerable controversy over possible long-term effects from dioxin. It is well known that exposure to dioxin can cause the disfiguring disease of chloracne in some exposed subjects. This does clear up eventually though in severe cases this may take about two years. It has also been alleged, based upon animal experiments, that teratogenic effects (damage to an unborn child) may occur. Medical examination of 34 foetuses, 30 from legal abortions, 3 from miscarriages, from women in the Seveso area, suggested that one may have been abnormal possibly exhibiting Down's syndrome. Hay (1982) quotes the medical team as saying "the results of the embryo-morphological studies do not have any hint of damage brought about by exogenous agents (i.e. dioxin, VM)". The team added a rider that this did not necessarily indicate that dioxin does not bear a risk for mother and child. Analysis of birth defects subsequent to the disaster has given no clear picture. A certain number always occur in any population; Hay (1982) quotes a figure for Western Europe of 2.5 to 3 per hundred. The rate for the post-disaster period at Seveso, according to Hay (1982), was somewhat less than for Europe or for Italy as a whole. Other links have been alleged such as with cancer, but there seems little evidence of an indisputable character for any of them. However it may be as well, for the present, to leave a question mark over the long-term industrial importance effects, other than chloracne, which may be caused by dioxin.

14.6.2.5. Industrial significance.

It would be very useful to be able to draw up a table showing the quantities of the various toxic substances by inventory and by the numbers of installations on which any given inventory is exceeded. Unfortunately this is impossible because the information is not published. There is now a published list of UK sites which fall under the terms of the CIMAH Regulations (see Chapter 19), but these do not differentiate the hazards.

But it can be said that there are some things which are well known. One is that both chlorine and ammonia are stored in tanks with capacities of the order of hundreds, if not thousands, of tonnes at both points of production and points of usage. On the other hand sulphur dioxide, which is produced on a considerably greater scale than chlorine, is hardly ever stored in quantity. This is because the greatest quantity produced is in the manufacture of sulphuric acid where it goes straight from a sulphur burner into a catalytic reactor where it is converted immediately into sulphur trioxide. The greatest part of this latter substance is then immediately converted into sulphuric acid in an absorption tower. Neither sulphur dioxide nor sulphur trioxide are stored in quantities which reflect the scale on which they are produced.

In general, large-scale storage is associated with centralised production and a large number of dispersed users. On the other hand, it is clear that some toxic substances which attract considerable attention from the mass media, of which dioxin is an example, have very little industrial importance.

14.6.3. THE RANKING OF TOXIC AGENTS.

Based on the classification given above it is possible, in principle, to give a ranking to toxic agents. This could be done by giving a weighting to each of the seven factors discussed above and then grading each agent on a scale of, say, three or five points against each of the factors. Table 14.1. exemplifies this approach. In it, inherent toxicity is given a maximum weighting of 8, industrial significance a weighting of 4, and the rest of the factors 2. Within each factor the ranking is in the ratio of 2:1.

The application of this system of weighting suggests that dioxin in the industrial context ranks only marginally higher than chlorine.

TABLE 14.1. RANKING OF TOXIC AGENTS.

Factor	Chlorine	Ammonia	Mustard Gas	Dioxin
Dispersive energy	2	2	0	0
Persistence	1	1	2	2
Industrial significance	4	4	0	0
Mode of attack *	2	2	1	1
Inherent toxicity	4	0	8	8
Injury to fatality ratio	1	1	2	2
Long-term effects	0	0	2	2
Totals	14	10	15	15

* Based on assumption that attack through the lungs is more likely to kill before medical treatment can be brought to bear than would ingestion.

Chapter 15

CASE HISTORIES OF TOXIC RELEASES

15.1. INTRODUCTION.

Because the toxic properties of the substances discussed are unique to each substance, the author has decided to preface each case history or set of case histories with an account of the toxic and other properties of the substance.

Dioxin is dealt with at considerable length because of the attention it has attracted even to the extent that the EEC Directive on Major Accident Hazards [European Community Directive (1982)] is familiarly known as the "Seveso" Directive on account of the incident at Seveso which is discussed as a case history. Similar considerations have been applied to Bhopal.

References to TLV and IDLH, which are defined in Appendix 1, are taken from NIOSH (1978). Because the time function attached to them is indeterminate they do not represent a "dose". Their ability to indicate rank order of inherent toxicity is limited. Physical data, unless otherwise indicated, are drawn from Science Data Book (1971) or from Weast (1981) and Weast (1985).

15.2. CHLORINE.

15.2.1. PROPERTIES.

Chlorine (Cl_2) has a molecular weight of 71. It is a greenish yellow gas which is 2.5 times as dense as air. It has a critical temperature of 144°C and an atmospheric boiling point of -35°C. It is handled as a liquefied gas. Its adiabatic flashing fraction at various temperatures of storage is given in Figure 5.5. of this book and is about 17% at 15°C.

It is an intensely irritating gas. There has been, even recently, considerable controversy surrounding the quantification of its toxicity. This is in spite of its historical importance as a war gas and its present-day industrial importance. A recent report, I.Chem.E. (1985a), concludes, on the basis of a critical analysis of animal experiments, that the LC_{50} for humans is 300 to 400 ppm for 30 minutes exposure. Using the formula given on P.323 of this book, this gives an LD_{50} of ca 3.5 mg/kg which compares with the value of 19,000 mg min/cu metre given in SIPRI (1971), which is equivalent to 1.9 mg/kg. The TLV for chlorine is 3 mg cu metre and it is regarded as having a IDLH level of 75 mg per cu metre.

Chlorine attacks the lungs, producing oedema, i.e. a copious secretion of fluid. This fluid is abstracted from the bloodstream and leads to thickening of the blood. The most immediate treatment that can be given, and it can be given to persons without medical qualifications, is to administer oxygen.

15.2.2. INDUSTRIAL SIGNIFICANCE.

Chlorine is widely used in industry and has important uses outside of the chemical and process industries as, for example, in water treatment. It is produced almost exclusively by the electrolysis of brine together with the concomitant production of a molecular equivalent to caustic soda. Storages with inventories exceeding 50 tonnes are quite common.

Legge (1934), in a review of accidents reported to the factory inspectors in the UK between 1908 and 1931, records that there were 177 cases of gassing by chlorine but only 1 fatality. He attributed the low fatality rate to the fear that chlorine aroused in the work force which made them evacuate an area as soon as they could smell any of the gas escaping.

Detailed evidence on accidents involving chlorine was provided at the Pheasant Wood, Thornton Cleveleys, Lancashire, Public Inquiry of September 1981. The author was the Assessor at this Inquiry which was concerned with permissions for housing in the vicinity of a major chlorine producing and using installation. The evidence is summed up in Harris (1981). Harris stated that UK production of chlorine was currently running at 1 million tonnes p.a. and 7 million tonnes p.a. for the remainder of Western Europe, out of a world total of 30 million tonnes p.a. During the last 25 years in Western Europe, there have been 8 fatalities out of 550 million man hours worked. This gives a F.A.R. (Fatal Accident Rate) of 1.5 which compares with 2 for British industry as a whole and 3.5 for the British chemical industry as a whole. An appendix to Harris (1981) gives the distribution of fatalities country by country. This shows that of the 8 fatalities, 5 occurred in Spain and only 1 in the UK. Put in another form, and very approximately, this works out at about 1 fatality for every 12 million tonnes of chlorine produced in Western Europe. This is much better than the record of the coal industry expressed as fatalities per million tonnes produced which is, for the UK today, ca 1 fatality by death or disease per million tonnes. Harris (1981) included case histories of chlorine releases, some of which will be discussed below.

15.2.3. USES IN WAR.

There is general agreement among historians that the first significant gas attack in the First World War occurred on the Eastern Front at the Battle of Bolimov in Poland. This was on January 19th, 1915. [See Dupuy and Dupuy (1970) P.950, and Haber (1986) P.25]. It appears that the Russians attached so little importance to it that they even failed to inform their allies of the new weapon. There is also general agreement that the attack, which was by shells, was largely nullified by the bitter weather which reduced the volatility of the gas used. [see Wren (1972)].

The next gas attack was at Ypres on April 22nd, 1915. The details which follow the author has drawn from the study of a number of works, some of which, being standard works of military history, will not be individually referenced. The paragraphs which follow in quotes are reproduced from Marshall (1977). Comments in square brackets [. . .] in these paragraphs are current insertions. References quoted are listed along with the other references in this book.

"Let us look first at the chlorine gas attack on 22nd April, 1915 in the Ypres sector. Here some 168 tonnes of chlorine were released by the Germans in a light steady wind on a four-mile (7-km) front. This resulted in 5000 deaths according to some estimates. The French colonial troops defending the line were taken completely by surprise and had nothing but improvised protection. If the figures are to be believed (and they have been questioned) in terms of fatalities per unit mass of gas, this was the most punishing of any attack in the war, yielding a mortality index of about 30. [The concept of the mortality index is discussed in Chapter 18]. This index appears to be much higher than that recorded for any large-scale non-military release. [It has probably now been exceeded by the Bhopal, 2/12/84, disaster].

As compared with non-military releases, there were many factors which would have led to higher casualties at Ypres.

(1) Military discipline which ordered troops to stand fast.
(2) The geometry of the release was along a line at 90° to the wind direction and 7 km long. Escape to the edge of the cloud was thus ruled out for the vast majority of the victims.
(3) Meteorology. Release was ordered only when the weather was favourable for inflicting the maximum casualties.
(4) Terrain. Trenches and dugouts formed pockets which filled up with gas.
(5) Rapidity of release. This took only 5 minutes.
(6) Presence of other harassments of war, shell fire, machine gun fire, which would deter the victims from seeking escape.

Against these may be set the circumstance that those affected were all fit and aged between 15 and 45.

The incident is surprisingly poorly documented. [The author has been unable to obtain any further information on it either from the Imperial War Museum or from the Chemical Defence Establishment, Porton Down. This may be partly explained by the fact that no British troops were stationed in the sector in which the attack took place]. It might be inferred from what evidence (or lack of it there is) that the cloud did not reach [the town of] Ypres which was about 5 km away." [This appears to be confirmed by the UK Chemical Defence Establishment (1976)].

"We can only roughly estimate how many men there were in the affected zone. The Germans . . . estimated 5000 dead and 15,000 gassed. British sources [History of the Great War (1923)] record that 350 died in hospital out of 7000 treated. [Adding these together] would give a total of about 27,000 troops affected of whom 5350 (20%) died. This squares reasonably well with the usually accepted figure of two divisions in the sector. The population density, assuming that the troops were caught in a zone 4 miles wide by 1 mile deep, would be about 7,000 per square mile which is somewhat lower than in a British suburb." But Haber (1986) considers that the figure of 5,000 deaths quoted by Hanslian and Prentiss is a serious exaggeration. A figure of 1,000 killed is quoted below table 10.2. of the reference.

Figure 15.1. is taken from Wren (1972) and shows an aerial view of a cylinder gas attack. Figure 15.2. is from Hanslian (1938) and is a plan of the Ypres sector at the time of the April 22nd, 1915 chlorine gas attack.

It may be stated in general terms, and this agrees with the conclusion in Haber (1986), that the use of chlorine produced results which disappointed the protagonists of chemical warfare in World War 1.

15.2.4. INDUSTRIAL RELEASES.

Table 15.1., which is taken from Advisory Committee on Major Hazards (1979), which in its turn was based on Marshall (1977), lists 18 releases of tonnage quantities of chlorine which have been recorded between 1939 and 1976. Brief reference will be given below to a further two incidents which have occurred since this table was compiled.

Some of the 18 releases are the subject of more detailed comment below. Unless separately referenced, the information is derived from Harris (1981).

15.2.5. BATON ROUGE, LOUISIANA, 10/12/76.

Information on this incident also appears in Eilers (1976). A spillage of 90

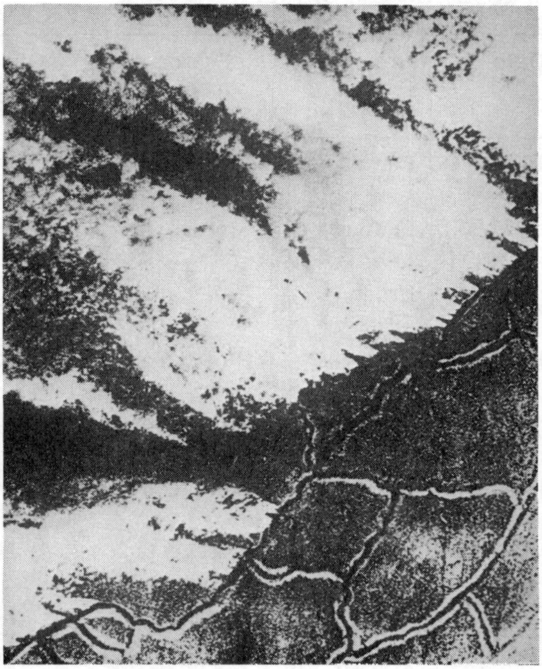

Fig. 15.1. AERIAL VIEW OF GAS ATTACK.

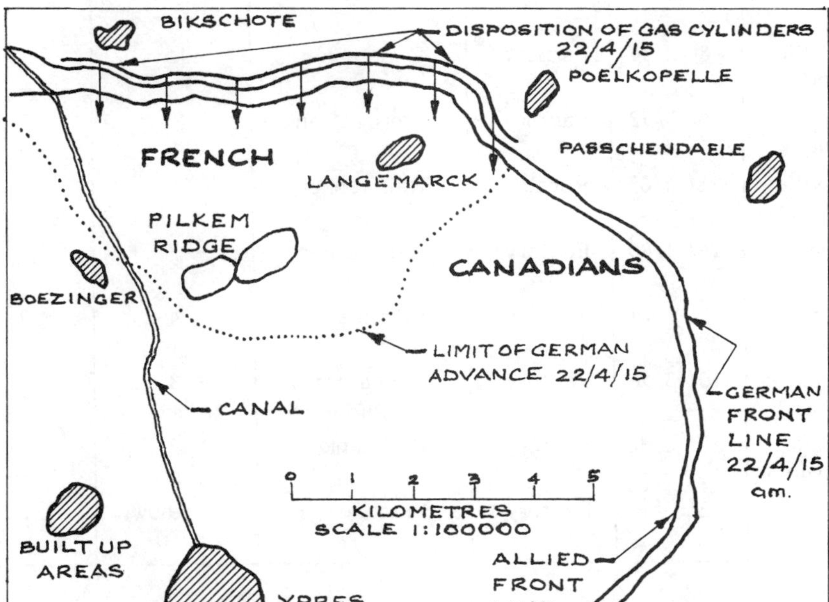

Fig. 15.2. PLAN OF YPRES SECTOR, APRIL 1915.

TABLE 15.1. RELEASES OF CHLORINE.

Location	Date	Area/site	Source of leakage	Quantity released (tonnes)	No. of fatalities
Baton Rouge Louisiana	10/12/76	Factory	Storage tank	90	0
Rauma Finland	5/11/47	Factory	Storage tank	30	19
Cornwall Ontario	30/12/62	Urban/ Rural	Rail tanker	28	0
Griffith Indiana	13/ 3/65	-	Rail tanker	27	0
La Barre Louisiana	31/ 1/61	-	Rail tanker	27	1
St. Auban France	13/12/26	Factory	Storage tank	24	19
Syracuse New York	10/ 5/29	Factory	Storage tank	24	1
Zarnesti Rumania	24/12/39	Factory	Storage tank	24	60
Wyandotte Michigan	1917	Urban	Storage tank	17	1
Chicago Illinois	4/ 2/47	Urban	Rail tanker	16	0
Niagara Falls New York	8/ 2/34	Factory	Rail tanker	15	1
Walsum W.Germany	4/ 4/52	Factory	Storage tank	15	7
Brandtsville Pennsylvania	28/ 4/63	Rural	Rail tanker	8	0
Mjodolen Norway	26/ 1/40	Factory	Rail tanker	7	3
Freeport Texas	1/ 9/49	Factory	Pipeline	4	0
Lake Charles Louisiana	10/ 3/56	Factory	Connecting pipework	3	0
Johnsonburg Pennsylvania	12/11/36	Factory	Rail tanker	2	0
Mobile Alabama	12/ 7/64	Factory	Pipeline	Unknown	1

tonnes was lost at the rate of 18 tonnes per hour. There were no fatalities even though the incident occurred in the vicinity of a major city. This was because the wind blew the gas cloud across the Mississippi river which is ca 1 kilometre wide at this point. The territory on the far bank of the river, over which the gas cloud dispersed, was largely uninhabited.

Some 10,000 university students and 500 members of the general public were evacuated as a precaution. Some process workers had to receive medical attention.

The release was caused by a relatively minor explosion of a mixture of natural gas and air in part of the system resulting in a chlorine storage tank falling from its supports onto a protruding object which punctured it. The size of the puncture has not been reported.

That more serious effects did not occur seems to be due to these factors:-

(1) Low population density downwind.
(2) Slow release through small aperture.
(3) Atmospheric temperature appropriate to December.

15.2.6. RAUMA, FINLAND, 5/11/47.

This was a catastrophic failure which is alleged to have been caused by overfilling and subsequent hydraulic rupture of a tank not fitted with a relief device.

30 tonnes were released causing 19 deaths; most of the victims were within 150 metres of the tank. People in houses 200 metres away survived.

15.2.7. LA BARRE, LOUISIANA, 31/1/61.

This is not discussed in Harris (1981). There is an account of it in Joyner and Durel (1962) which is written from a medical standpoint.

6000 US gallons, which the present author calculates to be ca 35 tonnes or rather more than is given in Table 15.1., were spilled after a rail car crashed due to derailment. The cloud appears to have been partially confined on its northern side by the levee of the Mississippi river. Many animals were killed including 49 pigs, 4 mules, 2 cows and 1 horse. 320 chickens also died. The location of these animal deaths enabled the potentially lethal area, for persons caught in the open, to be estimated at ca 15 km^2. The cloud was evidently rapidly lethal to animals up to 2 km from the site of the derailment. A sketch provided in the reference shows the area to lie entirely downwind and in the direction of the wind.

Air analyses were carried out within three hours of the spillage. Levels of ca 400 ppm (1200 mg/cu metre) were recorded at 75 metres from the wreck up to 7 hours after the spillage. It was not until 9 hours after the release that the level had dropped to 8 ppm (24 mg/cu metre) and that only in the immediate vicinity of the wreck. It has to be borne in mind that the bulk of the chlorine spilled would form a pool which would evaporate rather slowly.

100 persons were affected. Only 15 were detained in hospital of whom one, a child of 11 months, died some 5 hours after exposure. 10 of those hospitalised developed pulmonary oedema; 5 were unconscious on admission. The reference gives details of the medical treatment given. This was primarily the administration of oxygen. Atropine was found to be beneficial.

The child who died was in a house situated about 50 metres from the spillage. The child's father had tried to relieve the infant's choking and gasping by taking him into the open air but conditions there were worse than inside the house. A 2-month-old infant who remained in the house, and the child's father, survived.

15.2.8. ST. AUBAN, FRANCE, 13/12/26.

This disaster, in which a spillage of 24 tonnes killed 19, is notable for having been caused by an internal explosion resulting from the reaction between chlorine and hydrogen which is a co-product of the electrolytic process. Most of the deaths occurred within 50 metres of the tank.

15.2.9. ZARNESTI, ROUMANIA, 24/12/39.

This disaster, which it is believed was caused by a hydraulic rupture, spilled 24 tonnes and killed 60 people. This is the highest death toll of any industrial chlorine disaster. Most of the dead were close to the tank but some were killed at a railway station 250 metres away. One person was killed half a mile away, the most distant human fatality from chlorine in peace time.

There was only a light wind and dispersal was slow. Many escaped through moving to higher ground.

15.2.10. MISSISSAUGA, TORONTO, CANADA, 10/11/79.

This incident, which is not discussed in Harris (1981), involved a derailment of 25 or 26 cars of a train made up of 106 cars. Of these, 1 contained chlorine, 11 propane, 3 toluene, and 3 contained caustic soda. The incident has not been well documented in the technical press though there is an extensive press report in Sunday Star (1979).

The immediate cause of the accident was an overheated axle box which seized up and caused the derailment.

According to Sunday Star (1979), the chlorine car which ruptured originally contained 90 tons (it is not clear whether this means long tons or short tons), and at the end of the incident 30 tons still remained in it. The tank was eventually patched and towed away. The area was a mass of tangled wreckage and several of the propane cars caught fire. One exploded, propelling wreckage for a considerable distance. Some have taken the view that the thermal updraught from the fires helped to prevent the dispersion of the chlorine. 100 firemen fought the blaze for 6 days.

Sunday Star (1979) does not record anyone being gassed and certainly no-one was killed. The police evacuated over 200,000 people over an area of about 125 sq kilometres. Some were away from their homes for nearly a week.

15.2.11. MONTANA, MEXICO, 1/8/81.

This was also initiated by a railway accident in which the brakes failed on a train of 32 chlorine cars. The train ran free for 30 miles and then derailed in a narrow valley. A total of about 300 tonnes was spilled. Montana, a village of some 20 houses, was engulfed by the cloud and 17 were killed. It is believed that up to 500 people were exposed, many of whom had to be treated for chlorine gassing. Thousands fled from the nearby town of Cerritos.

15.2.12. OVERVIEW OF CHLORINE CASE HISTORIES.

I.Chem.E. (1985a) which independently reviewed the case histories of chlorine releases notes that, with one exception, the fatalities occurred within 400 metres of release and mostly within 250 metres.

15.3. AMMONIA.

15.3.1. PROPERTIES.

Ammonia (NH_3) has a molecular weight of 17 and has a density which is equal to 0.6 times that of air at the same temperature. This should not be taken to mean, in the event of loss of containment of ammonia stored as a liquefied gas, that the released gas will necessarily be buoyant. There is evidence that some ammonia/air clouds formed in this way have been denser than the surrounding air. It can be shown, mixing ammonia vapour at the temperature of its atmospheric pressure boiling point of -33°C with air at an ambient temperature of, say, 20°C, that whatever the proportions in which they are mixed, it always yields a mixture lighter than air at the ambient temperature. In order to yield higher densities, it is necessary to postulate adiabatic saturation of the air by evaporation either of ammonia droplets entrained in air or by wind chilling of the residual pool below -33°C. Ball (1970), and Shaw and Briscoe (1978), seem to show that the rate of conduction of heat from the surroundings will always maintain the temperature of an ammonia pool sufficiently close to its atmospheric pressure boiling point as to rule out the latter possibility. However, the former possibility is no doubt feasible during the flashing stage. Beveridge (1981) reviews this problem but without coming to definite conclusions.

Ammonia, as previously noted, has an atmospheric pressure boiling point of -33°C. It has a critical temperature of 132°C. In these properties it closely resembles chlorine. Like chlorine it may conveniently be handled as a liquefied gas. Its vapour pressure/temperature and adiabatic flashing fraction/temperature relationships are also very close to those of chlorine, (see Fig. 5.5). Unlike chlorine, however, substantial quantities of ammonia are shipped as a refrigerated liquid. At the Marsa el Brega installation in Libya, to which the present author was at one time Safety Adviser, the production rate was 1000 tonnes per day, all of which was exported by ocean-going tanker. The quayside storages had to hold tens of thousands of tonnes for convenience of shipping. This was all held in refrigerated storage at slightly above atmospheric pressure. Ammonia is also handled by cross-country pipeline in the USA if not elsewhere. Table 15.2. gives examples of 4 spillages from US pipelines.

Ammonia is much less toxic than chlorine. It has a TLV of 50 ppm (35 mg/cu metre) and an IDLH of 500 ppm (350 mg/cu metre). NIOSH (1978) gives only one reference to LC_{50} which is 4837/60 mins for mice. Using the formula given on Page 323, this gives an LD_{50} of 21 mg/kg (cf. 3.5 for chlorine quoted in section 15.2.1. above). Too much significance should not be attached to the figure of 21 mg/kg for ammonia because of paucity of data, but the fact that ammonia is less toxic than chlorine is universally agreed. It is significant that NIOSH (1978) has so little data on ammonia as is also the case with chlorine. It reflects the fact that the quantity of data in the literature on the toxicity of a chemical seldom reflects the industrial importance of the chemical.

Trevethick (1973) regards pulmonary oedema as the most dangerous consequence of breathing ammonia and recommends the administration of atropine sulphate. Splashes on skin and eyes can cause severe burns.

15.3.2. INDUSTRIAL SIGNIFICANCE.

Ammonia is a very important chemical. It is used primarily for the manufacture of fertilisers but is the starting material for the manufacture of a great variety of chemicals. It is today virtually the sole source of nitric acid. Ammonia is generally manufactured in a few large plants with stream capacities of 1000 tonnes per day and upwards with the product being sent to the point of usage by road, rail, ship or pipeline. As noted above, it may be transported as a liquefied gas or under refrigeration. Transport under refrigeration is more safe than as a liquefied gas. The scale of production worldwide is roughly the same as chlorine.

Legge (1934) gives 105 cases of gassing in the UK with 9 deaths, all deaths occurring singly, between 1907 and 1931. The author has not been able to obtain comprehensive statistics for fatalities involving ammonia such as are given for chlorine in Harris (1981). Table 15.2., which is taken from Advisory Committee on Major Hazards (1979), lists 11 major spillages of ammonia which caused 41 fatalities. This contrasts with Table 15.1., which lists 18 major spillages of chlorine which killed 113. An important source of information

TABLE 15.2. RELEASES OF AMMONIA.

Location	Date	Area/site	Source of leakage	Quantity released (tonnes)	No. of fatalities
Floral Arkansas	5/ 6/71	Rural	Pipeline	600	0
Enid Oklahoma	7/ 5/76	Urban	Pipeline	500	0
Conway Kansas	6/12/73	Rural	Pipeline	277	0
Landskrona Sweden	16/ 1/76	Port	Ship-storage connection	180	2
Blair Nebraska	16/11/70	Rural	Storage tank	160	0
Crete Nebraska	18/ 2/69	Urban	Rail tanker	90	9
Belle West Virginia	21/ 1/70	Urban	Rail tanker	75	0
Texas City Texas	3/ 9/75	Urban	Pipeline	50	0
Potchefstroom S. Africa	13/ 7/73	Urban	Storage tank	38	18
Houston Texas	11/ 5/76	Urban	Road tanker	19	6
Lievin France	21/ 8/68	Urban	Road tanker	19	6

on ammonia incidents is the CEP series on ammonia plant safety which comes out at the rate of 1-2 volumes each year. This series, which is published by the American Institute of Chemical Engineers, has appeared under a number of variants of title and sometimes discusses topics which seem to have little relevance to ammonia plant safety.

15.3.3. USES IN WAR.

There is no record of ammonia being used as a toxic gas in war.

15.3.4. INDUSTRIAL RELEASES.

15.3.4.1. Floral, Arkansa, USA, 5/6/71.

This appears to be the largest spillage recorded in the literature. A pipeline burst releasing ca 600 tonnes. There were no human fatalities. The ammonia reached a watercourse and killed thousands of fish. The incident is noted in Chemical Engineering (1971).

15.3.4.2. Potchefstroom, South Africa, 13/7/73.

This incident, which occurred in a fertiliser plant, has been described in detail in Lonsdale (1975). In it the dished end of a 50 tonne storage failed. A piece of the dished end, which was 2.9 metres in diameter, blew out and the piece, which represented about 25% of the metal plate making up the end, travelled for 40 metres before coming to rest. 38 tonnes of ammonia, at a storage temperature of ca 15°C, were released killing 18 people, 6 of them outside the works fence. The reference is very informative, giving a sketch map showing the positions of those who were found dead, those who subsequently died, and those who escaped. This is reproduced in Figure 15.3.

Of those working within 50 metres of the escape, 2 were found dead and 5 were rescued but subsequently died. Of those working between 50 metres and 100 metres, none died immediately but 5 died after rescue. Between 100 and 150 metres, 1 died after rescue. 5 were rescued between 150 and 200 metres from the escape but died subsequently. 65 cases of non-lethal gassing were reported.

Of some significance for the problem as to whether ammonia/air clouds are buoyant was the observation by some of the victims that the air was more breathable near the ground.

The reference gives details of the metallurgical examination of the tank. The dished end was found to be very brittle and a report by an independent metallurgist criticised the failure of the company to have the vessel stress relieved by heat treatment after it had been repaired some four years previously.

15.3.4.3. Houston, Texas, 11/5/76.

In this incident, a tank truck crashed through a crash barrier on a freeway and fell onto the road 10 metres below. 19 tonnes of ammonia were released immediately. There were 6 deaths; though not stated this total must have included the driver who would probably have died regardless of the load carried. 4 of the deaths were immediate, more than 100 were injured, [see McMullen (1976)]. Figure 15.4. is a photo reproduced from Fryer and Kaiser (1979). This was taken 1 minute after the crash.

There is evidence from this incident that the cloud initially was less buoyant than air as it produced a considerable area of brown scorched grass. The author has estimated that the area was approximately 1 sq kilometre from comparisons between photographs in Fryer and Kaiser and maps in McMullen (1976) [see Figures 15.5. and 15.6].

Samples of the air in the vicinity were taken by technicians of the City Health Department. However, sampling did not begin until 2.5 hours after the incident. Levels by then showed concentrations of the order of micrograms per cu metre and were indistinguishable in level from analyses for the area taken some days prior to the incident.

15.4. PHOSGENE.

15.4.1. PROPERTIES.

Phosgene or carbonyl chloride, $COCl_2$, has a molecular weight of 99. The gas has a density of 3.5 times that of air and the liquid a density of 1.4 tonnes per cu metre. It is a volatile liquid or colourless gas with an atmospheric pressure boiling point of 8°C. It has a critical temperature of 182°C. It may be handled as a liquefied gas at all temperatures ordinarily encountered in the atmosphere. Its theoretical adiabatic flashing fraction at 20°C is ca 5%.

Phosgene is highly toxic. It has a TLV of 0.1 ppm (0.4 mg/cu metre) and

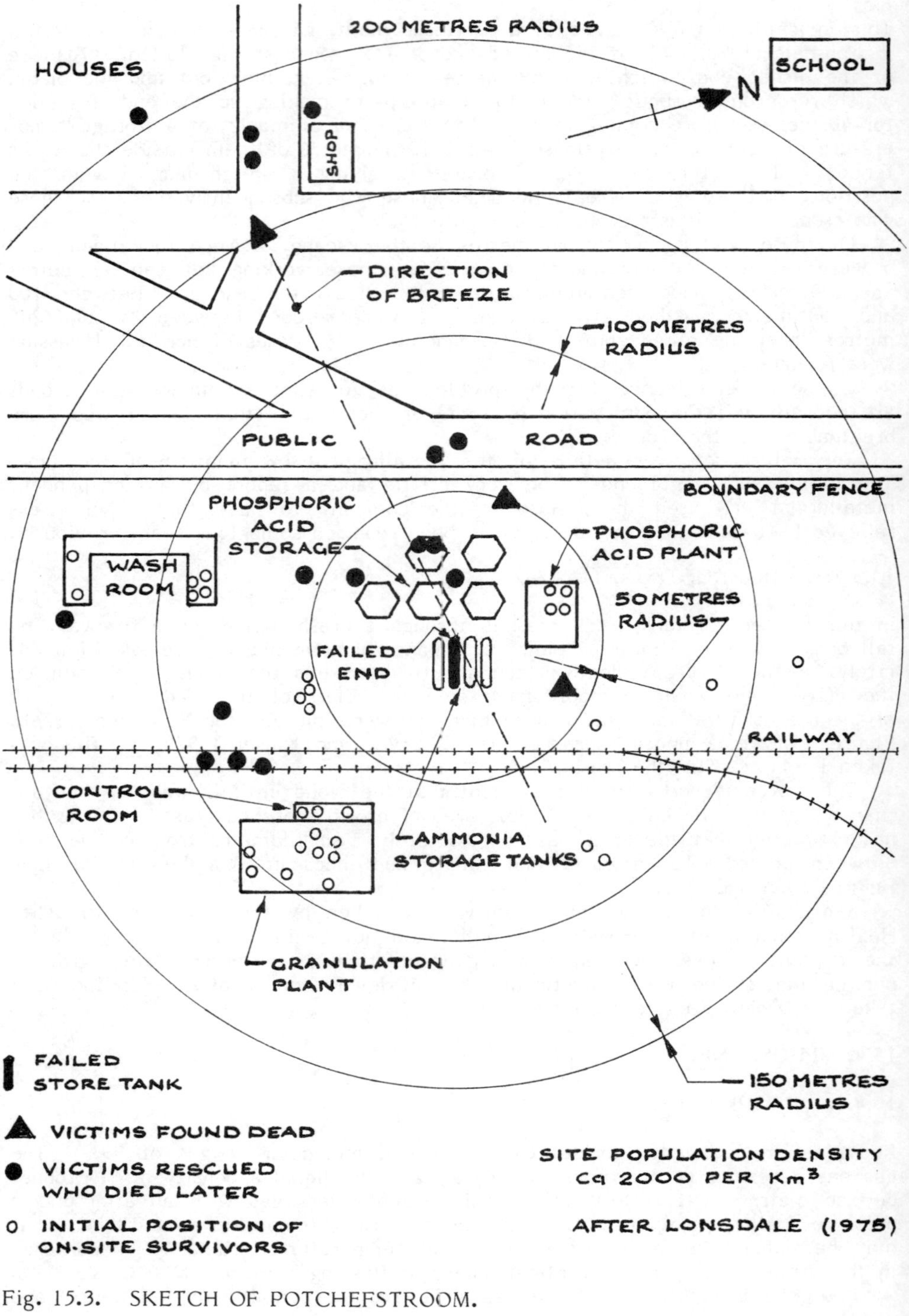

Fig. 15.3. SKETCH OF POTCHEFSTROOM.

an IDLH of 2 ppm (8 mg/cu metre). In low concentrations it smells like new-mown hay. Unlike chlorine, it does not have an immediately adverse effect. As Home Office (1938) puts it, "There is frequently a period of 'well being' between the initial effects produced by phosgene and the appearance of the more serious symptoms. This is the meaning of the so-called 'delayed action of phosgene' ". Trevethick (1973) suggests that there may be a delay in the onset of symptoms of up to 48 hours. The principal symptom is severe pulmonary oedema; administration of oxygen is recommended.

SIPRI (1971) suggests that LD_{50} for phosgene is ca 0.3 mg/kg based on the figure quoted of 3,200 mg min/cu metre as the LC_{50} value. NIOSH (1978) quotes for human inhalation an LC_{50} of 3200 mg/cu metre but without quoting the duration of the exposure which must clearly have been very short. Thus the data for this important toxic is sparse. Prentiss (1937), which is not a very reliable source, quotes US investigators as rating phosgene as being 11 times as lethal as chlorine and their German counterparts as rating it as 16 times as lethal. The ratio between LD_{50} for chlorine as given in Withers (1985) and quoted above, and the SIPRI value for phosgene, leads to a value of toxicity ratio, phosgene to chlorine, of 12 to 1. However, relating the values for phosgene and chlorine given by SIPRI yields a ratio of only 6 to 1.

15.4.2. INDUSTRIAL SIGNIFICANCE.

Phosgene is not produced on anything like the scale of ammonia or chlorine; US production is ca 0.6 million tonnes p.a. compared with about 16 million tonnes of ammonia and of chlorine. It is nevertheless an important industrial chemical. Over recent years it has become a source of isocyanates which in their turn are transformed into polyurethanes. 85% of US production of phosgene is taken up in this way.

Legge (1934) records 106 cases of gassing between 1907 and 1931, most of these in the war years 1917 and 1918. There were only two fatalities, both in the 1920's, in the UK.

There is only one incident with multiple fatalities on record. This occurred at Hamburg in 1928 and is described below.

15.4.3. USES IN WAR.

Phosgene was extensively used in the First World War. Prentiss records 23 major cloud attacks in which it was used mixed with chlorine and 43 gas shell attacks, 29 of them where it was used on its own. It does not appear to have been as effective relative to chlorine as its toxicity would suggest. Possibly its relatively low volatility would reduce its effectiveness and this would be why it was mixed with chlorine for cloud attacks.

The subject of phosgene in war is extensively discussed in Haber (1986). It may be noted that a substance called "diphosgene" by the Germans in World War 1 is not a dimer of phosgene. It was in fact trichloromethyl chloroformate

15.4.4. RELEASE IN INDUSTRY - HAMBURG, 1928.

Only one release which led to multiple fatalities is on record. This took place in Hamburg, Germany, on May 28th, 1928. The present author has two sources for this, one is JRUSI (1928) and the other is Harris (1981). According to the first reference, the incident took place near the Hofe Kanal at the upper end of the harbour district. It killed 11 persons. A changing wind blew the gas in a number of directions. As it was Sunday, and an election Sunday at that, there were relatively few people about and this reduced the death toll. The spillage was "300 or more cubic feet of gas". The author believes this to be

Fig. 15.4. PHOTOGRAPH TAKEN 1 MINUTE AFTER HOUSTON TANKER CRASH.

a gross underestimate and thinks it might refer to the volume of liquid. If it were 300 cu ft of liquid this would be about 12 tonnes, Harris (1981) quotes 11 tonnes. There were 171 people treated in hospital. There is reference to delayed effects, in one case it was claimed of one week. People were affected up to eleven miles away. 350 people were evacuated for a day. Harris claims that 2 persons in a rowing boat 100 metres away lost their lives and that 2 died 300 metres away.

15.5. CARBON MONOXIDE.

15.5.1. INTRODUCTION.

Carbon monoxide has been included in the case histories because it causes more deaths in the UK than any other toxic gas. Nevertheless, the author does not rate it as a major chemical hazard within the meaning attached to that term in this book. This is because, with very few exceptions, deaths from carbon monoxide occur under conditions of confinement and they generally involve only one person.

15.5.2. PROPERTIES.

Carbon monoxide, CO, is a colourless, odourless, tasteless non-irritating gas. It has a molecular weight of 28 and thus is only slightly less dense than air. It has a critical temperature of -139°C and is thus a permanent gas. It is one of the few permanent gases which are toxic, fluorine is another.

It is only moderately toxic. It has a TLV of 50 ppm (55 mg/cu metre) and an IDLH of 1500 ppm (1650 mg/cu metre). NIOSH (1978) gives three values for LC_{50} for three different mammals, the mean value being 3323 ppm or, using the formula on Page 323, a dose of 94 mg/kg. They are all for 4 hours' exposure and the apparent dose may be much lower for higher concentrations and shorter times. Only a portion of the carbon monoxide inhaled is retained so that the actual retained dose is much lower than the apparent dose. It should, in principle, be possible to obtain the retained LD_{50} for humans with considerable statistical reliability from post mortem examinations of victims. This is because carbon monoxide combines with the haemoglobin (Ha) of the blood to form carboxyhaemoglobin (COHa) which then prevents the blood from transporting oxygen. The COHa can be analytically determined. According to Health & Safety Executive (1984), death can occur when the Ha in the blood is 80% saturated. Using figures given in Bell et al (1965), the present author calculates the dose at 80% saturation to be 18 mg/kg. Bell et al claim that the affinity of CO for Ha is 250 times greater than is the affinity of oxygen.

Appreciable quantities of COHa are to be found in the blood of cigarette smokers. According to Health & Safety Executive (1984), non-smokers show .1 to 2.5% (based on total Ha) of COHa from the ordinary metabolism, whereas smokers can show 5 to 15%. The same reference suggests that death can occur with 70-80% saturation of the Ha level and, with 80-90%, death will occur in minutes.

15.5.3. NON-OCCUPATIONAL SIGNIFICANCE.

According to Vale and Meredith (1981), approximately 1000 persons died from carbon monoxide poisoning in the UK in 1979 out of a total of 4000 deaths from poisoning. There are a number of ways in which people can die from carbon monoxide poisoning. A proportion commit suicide by inhaling motor exhaust fumes which, according to Health & Safety Executive (1984), contain 10,000 to 100,000 ppm. (Compare LC_{50} of ca 3,000 over 4 hours). Others

Fig. 15.5. PHOTOGRAPH OF DAMAGE TO GRASS AT HOUSTON, 4 DAYS AFTER CRASH.

Fig. 15.6. PHOTOGRAPH OF DAMAGE TO GRASS AT HOUSTON, 4 DAYS AFTER CRASH.

die inadvertently from carbon monoxide generated by gas-fired heating appliances. This is because the burning of carbonaceous fuels always has a tendency to generate carbon monoxide, especially under conditions of oxygen deficiency. The worse circumstances are those in which there is an inadequate flue, or inadequate admission of fresh air, or some combination of the two. In such cases death may be due to the absence of oxygen as much as to the presence of carbon monoxide.

The figures quoted above from Vale and Meredith seem discrepant with the statistics of the UK Office of Population Censuses and Surveys [OPCS (1979)]. These give, for accidental poisoning, i.e. not suicide, and for England and Wales only, in 1978:-

Motor vehicle fumes	24 fatalities	(21 domestic)
Incomplete combustion of domestic fuels.	33 "	
Other carbon monoxide	9 "	(8 domestic)

British Gas (1983) have provided the following figures for accidental poisoning from CO arising from piped gas:-

	1978	1979	1980	1981	1982
Fatalities	63	95	84	102	70

15.5.4. INDUSTRIAL SIGNIFICANCE OF CARBON MONOXIDE.

Carbon monoxide is a constituent of various synthesis gases used in the manufacture of methanol and other alcohols. It is a constituent of blast furnace gas and coke oven gas. It is used in the carbonyl process for the purification of nickel. It is the starting point, along with chlorine, in the manufacture of phosgene.

As has been noted above, the number of fatalities in industry arising from carbon monoxide has fallen considerably.

Deaths from the gas were formerly frequent, but many of the processes which caused these deaths have been abandoned as technologically obsolescent. Legge (1934) records over the period 1907-1931, 1899 gassings by carbon monoxide, 224 of them fatal. This is an average of more than 10 fatalities per year as there were no records for two of the years. Today deaths from carbon monoxide are not separately recorded in the Reports of the UK Factory Inspectorate. In 1974, 329 cases of gassing in industry by all causes were reported, 6 of them being fatal. The corresponding totals in 1977 were 395 and 3. [See Health & Safety Executive (1977) and (1979)]. However, OPCS (1979) gives for 1978 zero deaths from carbon monoxide poisoning in industry.

15.5.5. MOSELLE RIVER, FRANCE/LUXEMBOURG, 17/1/82.

In this incident a river barge collided with the support for an overhead pipeline carrying carbon monoxide. The pipeline was severed and there were 5 fatalities no doubt caused by the escaping gas giving no warning signs.

15.5.6. SIGNIFICANCE IN WAR.

In spite of many properties which would favour its use such as cheapness, ease of manufacture and lack of warning symptoms, carbon monoxide has never been used as a war gas. This is doubtless because of the difficulty of handling a compressed permanent gas as opposed to a liquefied gas. Its buoyancy would also be against its use. Nevertheless, it caused injuries and possibly fatalities in the First World War. As Royal Army Medical Corps.(1925) put it ". . . the bursting of H.E. shell, the firing of machine guns in confined spaces and the explosion of mines, all produce noxious gases such as . . . carbon monoxide which may cause serious casualties among those exposed to them".

15.6. MUSTARD GAS.

15.6.1. INTRODUCTION.

Mustard gas is discussed because though it has no industrial importance, it is listed in Annex III, [European Community Directive (1982)], where it is given an inventory of 1 kg which is the inventory appropriate to the most dangerous category of toxic substances.

Mustard gas was used to a considerable extent in the First World War, it was used in the Abyssinian War of 1935, and there are reports of its use in the Gulf War between Iran and Iraq. There was an incident in the Second World War in which the gas was not being used offensively which is described below.

15.6.2. PROPERTIES.

According to a table given in Clarke (1968), mustard gas, [bis (2 chloroethyl) sulphide], in the purified form known as "distilled mustard", is a colourless to amber coloured oily liquid with a faint garlic odour. It is a solid below 14°C and has a boiling point above 200°C. It has a volatility of 960 mg per cu metre at 25°C. It has a delayed action effect varying from 1 to 48 hours to the onset of symptoms, no pain is experienced at the time of first contact. Symptoms include severe blisters to the skin, injuries to the eyes, leading in some cases to temporary or permanent blindness, and injuries to the lungs as the concentrations in the vapour phase may be lethal. As used in the First World War, it was at a lower level of purity than that available today. It is one of the types of gas known as a vesicant.

NIOSH (1978) gives for a human subject a lowest recorded lethal dose through the skin of 64 mg/kg and an LC_{50} by inhalation of 1500 mg/cu metre for 1 min. The former transforms to a lethal dose of ca 4.5 g/kg and the latter to a lethal does of 10.5 mg/kg. There is thus a considerable discrepancy between the skin and inhalation doses. However, the NIOSH concentration is very much higher than anything likely to be encountered in the open air (cf. the volatility of 960 mg/cu metre at 25°C).

According to Hersh (1968), mustard gas, (known under the code name of HD), is currently classified by the US military authorities as a "non-lethal blister gas though at certain exposures it can be lethal".

Mustard gas is thus not a gas at all, but a volatile liquid which has a vapour pressure which is high enough to cause injury in confined spaces. It has no inherent disperse energy; this had to be provided in war by a small explosive charge.

Mustard, if promptly treated by commonly available bleaches containing hypochlorites, can be neutralised though any which has already penetrated the skin will cause injury. These common reagents are effective at decontaminating clothing or buildings, [Home Office (1938)].

15.6.3. MUSTARD GAS IN THE FIRST WORLD WAR.

Prentiss (1937) argues very strongly that mustard gas was very effective at producing casualties in the First World War, quoting a figure of 60 lbs (27 kg) per casualty. For war gases as a whole, there were 20 casualties per fatality according to Prentiss. However, according to Table XVII in Prentiss, which gives data on the principal gas attacks, for mustard used on its own, ca 12,000 tonnes produced 61,500 casualties out of which 1,130 were fatalities.

Haber (1986) quotes a total production during the war and suggests that the German production of mustard during the war was of the order of 7,500 tonnes (including 20% of solvent) as compared with 87,000 tonnes of chlorine and that the Germans produced about 38 million gas shells. Haber expresses the view that it is virtually impossible to quantify the effects of gas warfare and it is not possible from this reference to obtain clear figures on, for example, the ratio of fatality to injury.

15.6.4. THE SINKING OF THE S.S. JOHN HARVEY.

Hersh (1968) quotes an example from the Second World War. A ship, the S.S. John Harvey, which was carrying about 100 tonnes of mustard gas, was subject to a severe air attack in an Italian harbour. There were 600 casualties, 83 of them fatal, when the sailors from this and other ships attacked in the harbour at the same time, tried to swim through water covered with a layer of mustard gas. It was not realised till too late what had happened and this delay may have been responsible for a substantial number of fatalities which might otherwise not have occurred.

15.7. ORGANOPHOSPHORUS COMPOUNDS.

15.7.1. GENERAL.

The objective of this subsection is to discuss (a) a group of compounds containing phosphorus which have considerable importance as insecticides and which, in varying degrees, are toxic to man, and (b) to discuss a number of closely related compounds which, though not of commercial importance, are toxic in the extreme and have been contemplated as possible war gases. These latter are the so-called "nerve gases".

The question as to whether these constitute major chemical hazards is an open one. There is no doubt whatsoever that many of the compounds in this class are highly toxic. By oral administration the LD_{50} of parathion, an insecticide, at 5 mg/kg, [Heath (1961)], is lower than that of strychnine at ca 23 mg/kg, [NIOSH (1978)]. Sarin, a nerve gas, has an oral LD_{50} of 0.28 mg/kg, [Heath (1961)]. It is not certain that some of these compounds, for example Tabun, a nerve gas, are manufactured today at all. Even for those that are manufactured a vector has to be postulated which, in the event of an accident, will disseminate the substance in a form sufficiently concentrated as to do harm. Parathion, for example, is sprayed on crops at a concentration of 0.01 - 0.10%. It may be noted that European Community Directive (1982) lists a number of insecticides in its Annex III.

15.7.2. PROPERTIES.

The compounds in question have the general formula which is given in Figure 15.7., which also gives the structural formulae for some insecticides and nerve gases. These are derived from Heath (1961), Rose (1968), and SIPRI (1971), from which further information set out below is also derived. According to Kirk-Othmer (1981), there are some 25 million theoretically possible toxic compounds with this general formula, of which perhaps 100,000 have been synthesised. Production levels in the mid-seventies in the US were of the order of thousands of tonnes per annum.

The mode of action of these compounds is similar to that given for the nerve

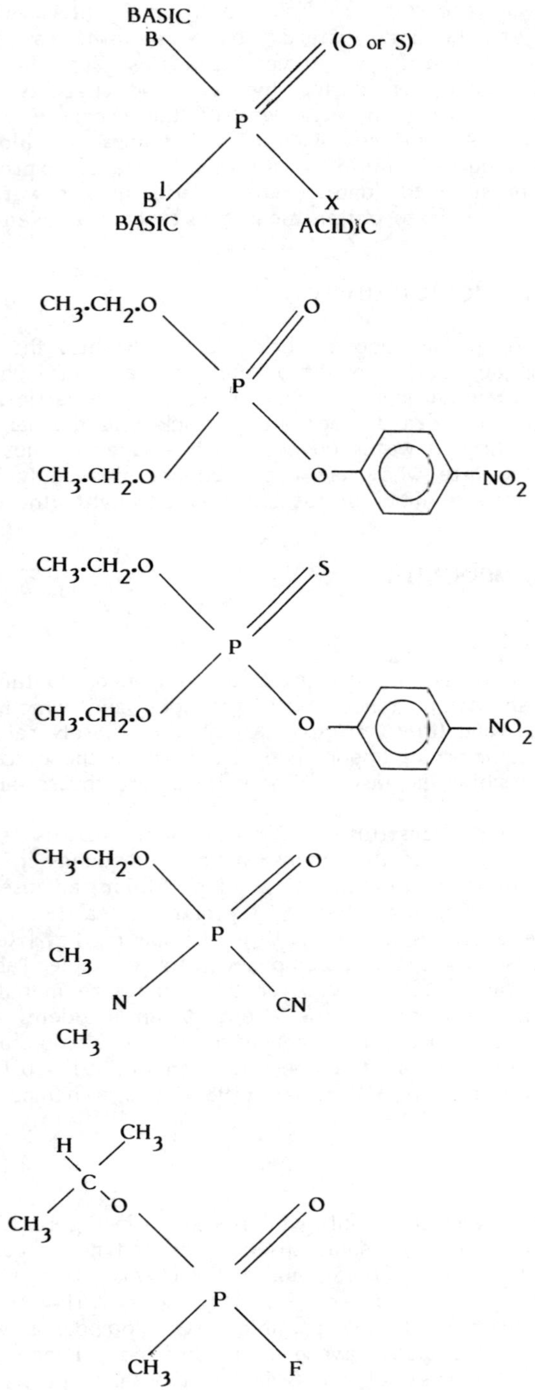

GENERAL FORMULA

PARAOXON
(INSECTICIDE)

PARATHION
(INSECTICIDE)

TABUN
(NERVE GAS)

SARIN
(NERVE GAS)

Fig. 15.7. STRUCTURES OF ORGANOPHOSPHORUS COMPOUNDS.

gases described in Rose (1968) as follows:- "The nerve gases are anti-cholinesterase agents, working by blocking the enzyme which the body uses to destroy one of its chemical nerve signal transmitters after it has done its job. This has two effects. One is that control is lost over the affected part of the nervous system. The other is that a large concentration of the chemical transmitter rapidly builds up within the body, and that chemical is, itself, a powerful poison. The body is first incapacitated, and then forced to poison itself".

The action of the insecticides follows this route but is presumably designed specifically to interfere with the nervous system of insects without being unduly toxic to man. The nerve gases on the other hand are specifically designed to be extremely toxic to humans.

Human symptoms, at low dosage, include anorexia, nausea, sweating and tightness in the chest. At higher doses there is vomiting, loss of control of the bladder and bowels, and, eventually, paralysis of the respiratory muscles. Some of the compounds strongly affect the eyes, [Heath (1961)].

15.7.3. PARATHION.

Insects, according to Kirk-Othmer (1981), are responsible for crop losses equal to about 10% of the potential agricultural output of the USA. The production of insecticides is thus of great commercial importance.

Parathion is one of the most widely used insecticides. It is a "wide spectrum" insecticide being effective against the Colorado potato beetle and a variety of pests which attack cotton.

Its LD_{50} value is given by Heath (1961) as 3 mg/kg when administered intravenously to rats; according to Kirk-Othmer (1981) it is 6-15 mg/kg for rats. It is thus a highly toxic substance and has been used in many cases of suicide. Accidental poisoning has often taken place through the skin because the victims have failed to understand that lethal doses can enter the body in this way. According to Heath (1961) there were 121 cases of phosphorus insecticide poisoning in California in 1954 and, in Japan, there were 6000 cases of parathion poisoning in the years 1953-59. It is not known how many were fatal or whether the exposure was acute or chronic.

15.7.4. THE NERVE GASES.

The term nerve "gas" is, strictly speaking, a misnomer as they are all liquids with boiling points higher than those of water. They are, however, sufficiently volatile as to be highly dangerous in the vapour phase. Because of this fact, and that the term is well established, they will be referred to below simply as nerve gases.

The boiling points and volatilities (expressed as grams per cu metre of air when saturated at 20°C) of four nerve gases are given below; this data comes from Kirk-Othmer (1979). The LD_{50} values are for inhalation and come from SIPRI (1971), as do the percutaneous values. These latter are not expressed statistically in the reference.

The first three of these compounds were developed in Germany in the Second World War. Tabun was stockpiled but never used. According to Borkin (1979), this was because Hitler was unsure about whether his enemies possessed it and feared retaliation. Rose (1968) claims that the process for making Sarin was not perfected by Germany during the war because of corrosion problems but that these problems were overcome by the Americans after the war. Since that time, it is claimed that the gas has been stockpiled in the US.

An account of the symptoms arising from the sub-lethal doses of Sarin administered to human volunteers is summarised in Heath (1961), together with accounts of the more severe symptoms arising from higher doses resulting from accidental

TABLE 15.3. PROPERTIES OF NERVE GASES.

Gas	Boiling point °C	Volatility g /cu metre	LD_{50} (inhalation) mg/kg	Percutaneous toxicity mg/kg
Tabun	256	610	.04	14
Sarin	147	21,900	.01	35
Soman	167	3,060	.005	1.4
VX *	298	10	.001	0.085

* Formula not published in open literature.

exposure. According to this reference, the course of symptoms is more or less the same with all the organophosphorus compounds; it is the dose required to produce them which varies.

Therapy includes the administration of oxygen, artificial respiration (presumably not mouth to mouth) and massive doses (i.e. 50 mg per patient) of atropine. Therapy is claimed to have been successful in 50% of cases of severe parathion poisoning.

It is impossible to say whether these compounds have any commercial significance as intermediates in the manufacture of insecticides. Cox (1976) claims that Sarin is so used but the present author has not been able to confirm this.

A recent military development has been the "Binary Weapon" which is described in general terms in Kirk-Othmer (1979). In this weapon, two reactants are charged separately into a bomb where they are separated by a membrane. Neither reactant is particularly toxic, but their reaction product is a nerve gas. The bomb is primed by puncturing the membrane. This arrangement, it is claimed, considerably reduces the hazards of manufacturing and storing nerve gas weapons.

15.8. DIOXIN.

15.8.1. GENERAL.

Though there had been a number of previous industrial releases, dioxin first came into international prominence as the result of an incident at Seveso, North Italy, on 10th July 1976, which contaminated several square kilometres of agricultural land for periods, in some cases, measured in years. There were no fatalities but there were, resulting from this incident, appreciable numbers of sufferers from the chronic disease of chloracne, the nature of which will be described below.

The significance of the Seveso incident as an example of a major chemical hazard will be discussed further in Chapter 18. The discussion below will be concerned (a) with the physical properties of dioxin, (b) the biological effects of dioxin, (c) the industrial significance of dioxin, (d) with the causes and circumstances of the Seveso incident, and (e) with its aftermath.

A further reason for the current prominence of dioxin as a public issue has arisen from the use of herbicides by the US armed forces in Vietnam. A particular herbicide, Agent Orange, has been shown to have contained dioxin as an impurity and it has been alleged that US Army personnel suffered injury as a result of its use. As a result, there has been extensive litigation in the US courts.

Moreover, the herbicide 2,4,5-T, (trichlorophenoxyacetic acid), which is in common use, formed 50% by weight of Agent Orange. This has led to fears, especially by the trade union movement, that chronic exposure to 2,4,5-T arising, for example, through crop spraying, may lead to dioxin poisoning through its presence as an impurity.

Dioxin is a trivial name for 2,3,7,8-tetrachlorodibenzo-p-dioxin, the structural formula of which is given in Figure 15.8. The name "dioxin" was also the trade name, now abandoned, of a totally different compound, the acetic ester of 2,6-dimethyl-m-dioxan-4-ol with which it should not be confused. Nor should it be confused with dioxane (diethylene dioxide) which is a dimer of ethylene oxide and is sometimes called p-dioxin, tetrahydro. Neither of these latter contain chlorine which appears to be essential to the harmful properties of dioxin and its related compounds.

Dioxin is also used as a generic name for a group of compounds, the polychlorinated dibenzo-p-dioxins or PCDDs. To avoid confusion, the word dioxin, when used below, will refer solely to 2,3,7,8-tetrachloro-dibenzo-p-dioxin; the generic group which includes it will be referred to as the PCDDs.

In the discussion below, though it will be necessary to deal with chronic effects, these will, except for passing references, be related only to those arising from acute exposures. The incidents discussed will therefore be acute incidents which disseminated dioxin and gave rise to chronic harm as a consequence of this.

15.8.2. THE LITERATURE.

The literature on the subject of dioxin is voluminous. In the treatment below the author has endeavoured to restrict the discussion to a selection from those works which he has personally consulted. To facilitate the review of this literature, 14 such references are tabulated in Table 15.4., roughly in chronological order, and are analysed as to their principal areas of subject matter.

The first three references were published prior to Seveso. The first two relate to an incident at the Coalite Works, Bolsover, Derbyshire on April 24th, 1968. This was a precursor of Seveso. There was one fatality in this incident, it was due to falling masonry. Milnes (1971) is a relatively short note of the incident with some discussion of the mode of formation of dioxin and its toxicity and biological effects. May (1973) discusses this incident from the point of view of a medical practitioner concerned with the treatment of the victims. These numbered 79 and all suffered from chloracne. Carter et al (1975) relates to an outbreak of dioxin poisoning at a horse breeding farm in Missouri, USA, over a period from June 1971 to January 1974. The cause was the use of an oil residue, which was contaminated with dioxin, to spray an arena in which horses were exercised. 85 horses were exercised on the arena, 62 became sick and 48 died. There were numerous deaths of birds, cats, dogs and rodents. The reference details symptoms displayed by animals and gives details of the analytical techniques used to detect dioxin.

Cattabeni et al (1978) is a multi-author work which is concerned principally with analytical techniques for detecting and measuring dioxin, with its biological effects and with techniques of decontamination. The contributions on these subjects are more detailed than in any other works cited except perhaps Homberger et al (1979).

Crow (1978), though written after the Seveso incident, has very little to say about that incident. It is a medical treatise which gives information on the disease of chloracne and also on substances other than dioxin which can give rise to the disease. The reference demonstrates that all the "chloracnegens" discussed by Crow have a structural resemblance. They are symmetrical molecules with two benzene rings linked through nitrogen or oxygen, or condensed as in naphthalene derivatives, and typically having four chlorine atoms in the molecule as in Figure 15.8.

Seveso (1978) is the Official Report of the Parliamentary Commission set

(a) Structural formula of dioxin.

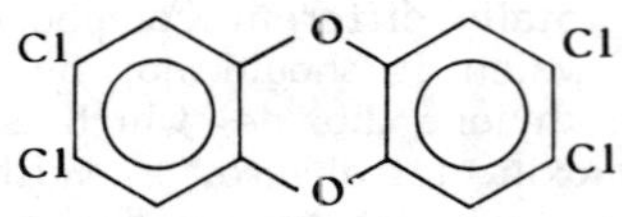

(b) Numbering of derivatives of dibenzo para dioxin.

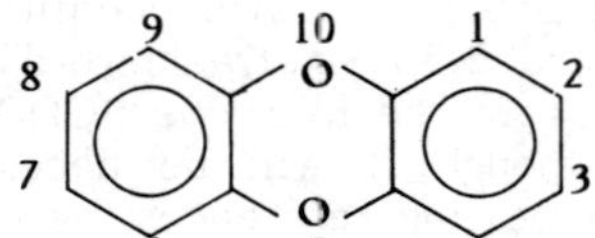

(c) Formation of dioxin.

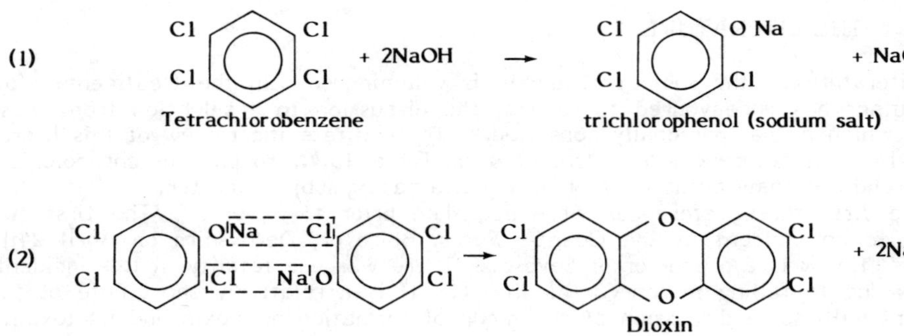

(d) Other chloracnegenic compounds.

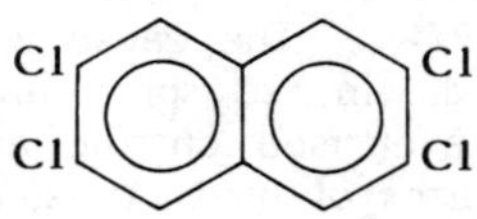

Tetrachloronaphthalene

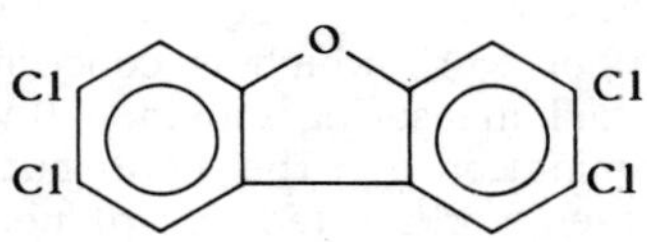

Tetrachlorodibenzofuran

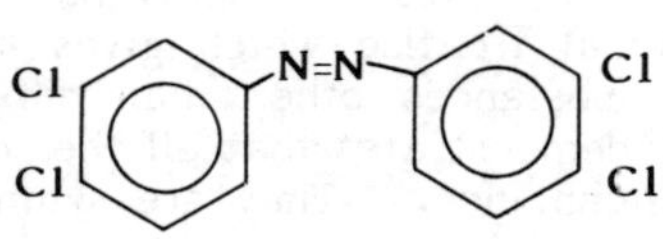

Tetrachlorodiazobenzene

Fig. 15.8. DIOXIN AND RELATED COMPOUNDS.

TABLE 15.4. ANALYSIS OF THE SUBJECT MATTER OF A SELECTION OF REFERENCES TO DIOXIN, TO THE SEVESO INCIDENT, AND TO RELATED TOPICS.

P = Paper
B = Book

REFERENCE			Chemical and Physical Properties of Dioxins.	Analytical Detection of Dioxin.	Modes of Formation of Dioxin.	Biological Properties of Dioxin.	Biological Properties of Related Compounds.	Chloracne as a Disease.	Incidents Involving Dioxin other than Seveso.	Causes and Circumstances of Seveso Incident.	Nature and Extent of Injuries at Seveso.	Medical Follow-up of Persons Exposed to Dioxin.	Environmental Damage at Seveso.	Problems of Decontamination.
Milnes	1971	P	M		X				X					
May	1973	P				X	X	M	X			X		X
Carter et al	1975	P		X		X			X					
Cattabeni et al	1978	B	X	M	X	M				X				M
Crow	1978	P				X	X	M	X					
Seveso	1978	B			X	M			X	M	M	M	M	X
Whiteside	1978	B		X		X	X	M	M	X	X		X	X
Homberger et al	1979	P			X	M				X	M	M	M	X
Holmstedt	1980	P				X			M	X	X	M		
Theophanous	1981	P			M					X				
Hay	1982	B	M		X	M	M	M	M	M	X	X	X	X
Rice	1982	P		X	X					X		X	M	M
Coulston & Pocchiari	1983	B				M			X	X	X	M	M	M
Cardillo et al	1984	P			M					X				

M = Major Treatment.

Blanks do not necessarily imply a total lack of reference to the subject.

up to investigate the causes and circumstances of the Seveso disaster. There is some doubt as to which was its year of publication, as this does not appear to be stated in Health & Safety Executive (1980) which is the English translation. Hay (1982) gives it as June 1977, but this cannot be true as, according to Seveso (1978), the Commission's members were not appointed until July 1977. An Act of the Italian Parliament of 24th January, 1978, extended its life for 6 months so it may be reasonable to assume that it was presented in the summer of 1978. Hence it has been referenced as Seveso (1978). Marshall (1980a) presented an analysis of the findings of the Commission. In it, attention was drawn to the unfortunate omission of a negative in translation in Health & Safety Executive (1980). This was in the passage in which the Commission concluded that the initial injuries were most probably caused by the non-dioxin components of the toxic cloud. A notable feature of Seveso (1978) is that it has a bibliography on dioxin and related compounds with 674 entries.

Whiteside (1978) reviews adverse effects associated with the use of the herbicide 2,4,5-T which contained dioxin as an impurity with especial reference to its use in the Vietnam War. There is a brief account of the circumstances of the Seveso disaster which tends to play down the role of the major constituents of the reactor as causing immediate injury to people and damage to vegetation. However, the reference contains much information on the trauma induced by the disaster in the affected population in the form of anxiety. There is an appendix which gives a detailed account of chloracne resulting from long-term exposure to PCDDs and related compounds.

Homberger et al (1979) is a very detailed account of the toxicity of dioxin, and the effect of the incident on the health of the local population including their medical follow-up after the event. The reference discusses the effects on animals, crops and the land, including the problem of decontamination. The authors of this reference all worked either for Givaudan or for Hoffmann-La Roche, i.e. the owners of the Seveso plant. This circumstance has militated against the universal acceptance of its conclusions of which perhaps the most important were (a) that only relatively few cases of chloracne made their appearance, most of them mild, and (b) that there were no systemic effects. However there is little in the most recent publications, as, for example, in Coulston and Pocchiari (1983), which presents the views of independent specialists, which contradicts these conclusions.

Holmstedt (1980) is particularly notable for describing a number of dioxin incidents, other than Seveso, and their effects.

Theophanous (1981) deals with the mechanism by which the runaway reaction occurred at Seveso. The conclusions reached are accepted in Coulston and Pocchiari (1983) and are discussed in detail below.

Hay (1982) is probably the most comprehensive work to date on dioxin and its related compounds. Like Whiteside, it covers a wide field taking in the events of the war in Vietnam and problems of landfill as well as the Seveso and other industrial incidents. It is particularly detailed on the biological effects of dioxin and the other PCDDs.

Rice (1982) is chiefly concerned with the environmental effects of Seveso and with the problems of decontamination.

Coulston and Pocchiari is a multi-author work which is concerned chiefly with the effects of dioxin on human health and upon the environment and with problems of decontamination.

Cardillo et al (1984) accepts the conclusions of Theophanous (1981) and develops them to recommend measures for improving the safety of processes in which 2,4,5-trichlorophenol is manufactured.

15.8.3. CHEMICAL AND PHYSICAL PROPERTIES OF DIOXINS.

The key references quoted above are chiefly concerned with biological effects and do not contain much information on the chemical and physical properties of dioxin and other PCDDs.

According to Rappe, in Cattabeni et al (1978), the first example of a PCDD, the octachlorine derivative, was synthesised by Merz and Weith in 1872 and dioxin itself was synthesised by Sandermann in a 25 g quantity in 1957. [Sandermann et al (1957); this reference has not been consulted by the present author]. The laboratory technician who had performed the work had to be hospitalised with chloracne. According to a letter from Sandermann, reproduced by Holmstedt in Cattabeni et al, the same hospital had previously treated patients who had contracted chloracne in a local factory in 1953.

Milnes (1971) describes the isolation of a product of an exothermic runaway reaction of a mixture containing 1,2,4,5-tetrachlorobenzene with caustic soda and ethylene glycol. This was a white crystalline solid with a melting point of ca 300°C which proved identical with synthesised dioxin when subjected to mass spectrographic and gas liquid chromatographic analysis. This, in an indirect way, gives a melting point of ca 300°C to dioxin.

Hay (1982) gives an extremely low value to the solubility of dioxin in water, 2×10^{-4} ppm. It is more soluble in fats (44 ppm in lard oil), in hydrocarbons (570 ppm in benzene) and most soluble in chlorinated organic solvents (1400 ppm in dichlorobenzene). Even this latter value would only correspond to the category "very slightly soluble" as used in reference works.

Hay describes dioxin as being stable to heat, acids and alkalis. The reference claims that an exposure of 21 secs to a temperature of 700°C decomposes only 50% of dioxin. Rice (1982) discusses the thermal stability of dioxin and quotes Stehl et al (1973) for the statement that "dioxin is completely decomposed by pyrolysis at 800°C with a residence time of 21 seconds". Rice considered this statement not to be sufficiently informative and gives further information which suggests a half life of ca 10^{-4} secs at 1200°C.

In short, dioxin is extremely inert chemically, and virtually insoluble in most solvents. It remains an unanswered question as to how it is that a substance so inert chemically and physically can be so active biologically.

There is a considerable body of discussion in the literature cited in Table 15.4. on the structure of compounds similar to dioxin, including the PCDDs. The compound dibenzodioxin, the formula for which is given in Fig. 15.8., is capable of being substituted by up to eight chlorine atoms. These are numbered according to the principles used by the journal, Chemical Abstracts, as shown in Fig. 15.8. The number of theoretically possible substituted compounds is given as 75 by Rappe in Cattabeni et al (1978); not all of these have been synthesised. The phenomenon of isomerism plays a role here and it does not follow that two compounds with the same number of chlorine atoms necessarily have the same degree of biological activity. To give an example, there are two isomers of monochlorodibenzodioxin, (MCDD). These are 1-MCDD and 2-MCDD. (All the other positions 3,4,6,7,8, are indistinguishable chemically either from 1 or from 2 MCDD). However for dichloroDD there are 10 possible combinations which are non-identical, for trichloroDD there are 14, and for tetrachloroDD there are 22. For higher numbers of substituted chlorine atoms the numbers diminish, being one only for the octo substituted compound. Any statement regarding toxicity must specify not only the numbers of chlorine atoms but also their positions as isomers do not have identical properties.

Hay (1982) also discusses the properties of the chlorinated dibenzo-furans of which the analogue to dioxin is shown in Fig. 15.8. He claims that this compound is only less toxic than dioxin by a factor of 2-4.

15.8.4. ANALYTICAL DETECTION OF DIOXIN.

The extremely toxic properties of dioxin necessitate highly sensitive techniques of chemical analysis. Rice (1982) points out that any zoning of areas of hazard at Seveso would have been impossible without analytical monitoring. The degree of sensitivity claimed by Rice for the techniques then available was the detection of as little as 5 pg (5 picograms or 5×10^{-12} grams). Hay is critical of the UK's Office of the Government Chemist for using a mass spectrometer which can only measure dioxin in 2,4,5-T to levels of parts per million whereas the latest model can measure in parts per million million. The techniques of measuring are described by Buser and by Holmstedt in Cattabeni et al. Buser describes the use of gas chromatography incorporating glass capillary columns in which the chromatograph is coupled to a mass spectrometer. The operation of the mass spectrometer by a technique known as mass fragmentography is described in Holmstedt (1980).

15.8.5. MODES OF FORMATION OF DIOXIN.

The primary question here is to discuss how dioxin may be formed as an unwanted by-product in the synthesis of 2,4,5-trichlorophenol. This is because it was this process which was being used at Seveso in 1976 and also at the Coalite Works in 1968 which has been previously referred to. The basic process in both cases was the same and is that shown in Fig. 15.8. In it 1,2,4,5-tetrachlorobenzene was reacted with sodium hydroxide in either methanol or ethylene glycol, to yield sodium trichlorophenate. This was then acidified to yield 2,4,5-trichlorophenol, the desired product, as shown in Fig. 15.8. However a side reaction can give rise to dioxin whereby two molecules of sodium trichlorophenate can condense to give rise to one molecule of dioxin and two molecules of sodium chloride. This is shown in Fig. 15.8. These equations say nothing about the relative proportions of sodium trichlorophenate and of dioxin which are formed. As they stand, they suggest that all the product could be dioxin. In practice, even under the worst conditions, the proportion which is dioxin is very low; but then only tiny quantities of this can do a great deal of harm. It seems to be generally agreed that, at the reaction temperatures of ca 180°C which were generally adopted, the formation of dioxin is very small indeed but that it increases markedly with increasing reaction temperatures. It does not follow from this that all the dioxin formed in the reaction appeared in the product as sold; at Seveso there was an incinerator designed to operate at ca 1000°C to dispose of residues which were known to contain dioxin, [Seveso (1978)].

The discussion of the mechanism whereby the reaction products at Seveso became the subject of a runaway reaction will be deferred to a later paragraph.

Alternative mechanisms for generating dioxin are discussed in Cattabeni et al (1978) and in Hay (1982). These include the pyrolysis of a number of chlorinated hydrocarbons such as occurs in municipal incinerators. Such effects properly fall under the heading of chronic events, specifically in this case, atmospheric pollution. An acute event giving rise to dioxin contamination is reported by Kim in Coulston and Pocchiari. Here a fire occurred which involved a transformer which was charged with about 5 tonnes of a mixture of poly-chlorinated biphenyls and chlorinated benzenes. The soot from this fire, which started in the basement of a building, contained about 3 ppm of dioxin.

15.8.6. BIOLOGICAL PROPERTIES OF DIOXIN.

The biological properties of dioxin have been extensively discussed. Out of the 13 works cited in Table 15.4., 5 give it a major treatment. One of these, Cattabeni et al, devotes one-third of its space to this topic and another, Coulston and Pocchiari, is devoted almost entirely to the subject. Clearly only a brief review of the topic is feasible here.

There does not appear to be on record any clear account of human fatalities from dioxin. Hay (1982) gives a table in which it is noted that there was 1 fatality associated with a case of occupational exposure in the US in 1960 and 2 fatalities associated with a case of occupational exposure in Czechoslovakia over the period 1964-69. Hay does not give a case history for the US example but does give one for the Czech incident without mentioning fatalities. He gives reference to a fatality in Germany in 1959 but does not list this in the table even though the legal finding was that death was due to dioxin, a finding on which Hay casts doubt.

There are a number of reports of animal experiments. Seveso (1978) gives a table of LD_{50} values for a number of animal species. These range from .006 mg/kg (oral?) for a guinea pig to .114 mg/kg (oral?) for a mouse. The first of these values has attracted much attention and led to claims that dioxin is the most lethal synthetic substance known to man. However it is apparent that for different animal species there is a variation of at least two orders of magnitude. To quote Bridges (1984) "Ingested TCDD (dioxin) is about 100 times less acutely toxic to mice than to guinea pigs, and the Syrian hamster is about 600 times less susceptible than the guinea pig. Differences of this magnitude between three rodents species make extrapolation from rodent to man highly problematical". Brown and Bridges (1983) however warn "We still do not know with any certainty how acutely toxic TCDD is to man, but it would be a surprisingly incautious person who did not feel that it would be wise to assume that TCDD is a highly toxic substance, even in the absence of definitive, as distinct from circumstantial, human data". Further data on animal experiments is given in considerable detail in a number of chapters of Cattabeni et al, in Olsen et al in Coulston and Pocchiari, and in Hay.

It is of considerable interest that the most characteristic sub-lethal symptom in man, chloracne, cannot be produced in the most common laboratory rodents. According to Crow (1978) it appears only in the rhesus monkey, the hairless mouse and in the rabbit. In the case of the latter, it appears only on the inner surface of the ears, even when it is ingested. This property has made a rabbit a very useful test animal for suspected chloracnegens.

The sub-lethal symptoms in man are well established. They include skin damage such as chloracne, darkening of the skin and excessive growth of hair, damage to the liver, heart, pancreas and lungs, damage to the nervous system with weakness of the lower limbs and loss of sensory perception, and neurasthenia and depression.

Animal experiments have indicated the possibility of teratogenic effects (damage to the unborn foetus) and carcinogenic effects. Neither of these have been confirmed in humans.

15.8.7. BIOLOGICAL PROPERTIES OF RELATED COMPOUNDS.

The relative toxicity of dioxin and other PCDDs has been studied by McConnell and Moore and is reported in Cattabeni et al as well as elsewhere. Their results are displayed in the form of a chart, Figure 15.9. This shows strikingly that 2,3,7,8-substitution is the most toxic PCDD both for guinea pigs and for mice but that the ratio between LD_{50} for the two species is at its highest for this derivative. Hay also discusses this question.

Crow (1978) presents the results of a study of those compounds which display chloracnegenic properties. The formulae for these are presented in Fig. 15.8. They display structural resemblances such as symmetry of the molecule and the presence either of two benzene rings or two condensed benzene rings. Crow comments that all have similar arrangements of electrons on the outside of

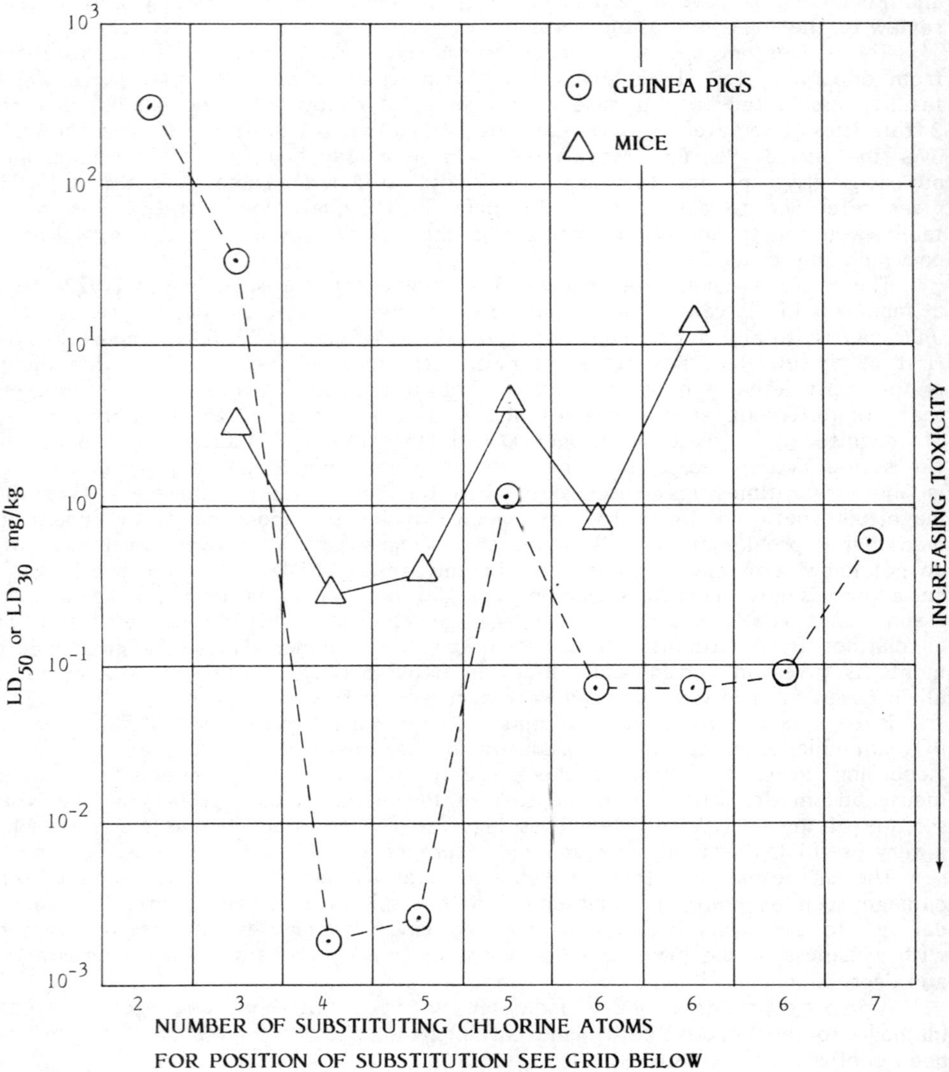

9								X	
8	X		X	X	X	X	X	X	X
7		X	X	X	X	X	X	X	X
6							X		X
4					X	X			X
3		X	X	X		X	X	X	X
2	X	X	X	X	X	X	X	X	X
1				X	X	X	X	X	X

Fig. 15.9. RELATIVE TOXICITIES OF CHLORINE SUBSTITUTED p DIOXINS.

the molecule. The present author has been concerned with advising a major company over chloracne associated with the azo and the azoxy compounds set out in Fig. 15.8.

15.8.8. CHLORACNE AS A DISEASE.

Though it is also discussed in May (1973) and in Hay (1982), the most detailed account of the disease comes in Crow (1978).

Chloracne has been recognised as an industrial disease since the turn of the century. According to Whiteside and to Hay it was first recognised in the chlorine industry in Germany where the symptoms were attributed to contact with chlorine itself. It was later recognised that this was not so and that the cause was chlorinated (or brominated) aromatic compounds. The present author believes that in the chlorine industry the disease was caused by contact with the tars which were used in the past to seal the electrodes. In the UK chloracne was legally prescribed as an Industrial Disease in the Regulations of 1948.

As described by Crow, chloracne is a symptom of toxic attack and may indicate that systemic poisoning is occurring. It is a disease of the skin which may be very mild and practically symptomless or it may lead to serious disfigurement and permanent scarring. It attacks the body at specific sites especially around the eyes and behind the ears. The remainder of the face, neck and shoulders, genitalia, chest and lower trunk may be involved but only very rarely the hands, feet or legs. In severe cases the entire body may be covered with horny spines like a nutmeg grater or the pigmentation may be so altered that a white man may become hard to distinguish from a negro. Chloracne manifests itself in the form of blackheads and resembles adolescent acne except that it has the peculiar distribution described above.

The time taken to recover depends upon the severity of the attack. Mild cases clear up in months but severe cases have been known in which symptoms persisted 15 years after the victim was removed from contact with the chloracnegen. Severe cases of this sort are liable to give rise to permanent scarring.

Crow also discusses the use of an enzyme extracted from the liver, arylhydrocarbon hydroxylase (AHH), to give a measure for assessing suspected chloracnegens. The degree and duration of enzyme induction is claimed to correlate well with the acnegenic and systemic toxicity of the compound being tested.

15.8.9. INCIDENTS OTHER THAN SEVESO INVOLVING DIOXIN.

Seveso (1978), Whiteside (1978), Hay (1982), and Holmstedt (1980), include tables or case histories of examples of industrial exposure to dioxin. These naturally overlap. Hay gives the most comprehensive table with 24 cases listed. Of these only 6 involved explosions and these are tabulated below in Table 15.5. Seveso (1978) claims that an incident at Dow Chemicals USA involved an explosion but Hay does not confirm this.

The Monsanto incident is treated as a case history in both Holmstedt and Hay. It has been the subject of intense medical follow-up described in these two references which has claimed that a study of 121 workers exposed over a period of 30 years showed a lower than anticipated mortality. There was no evidence of excess mortality from cancer.

The 1953 BASF incident is also treated as a case history by Holmstedt and by Hay. A cohort study of the 75 persons involved was made 25 years later. Mortality did not appear to be notably different from control groups. Hay draws attention to the incidence of 6 gastrointestinal cancers where 3 would have been expected. It was at this factory that there was subsequently a death attributed to dioxin which was certified as "necrotic pancreatitis". Neither

TABLE 15.5. EXPLOSIONS INVOLVING ACUTE EXPOSURE TO DIOXIN.

Date	Location	Product	Solvent	No. affected
1949	Monsanto, West Virginia.	TCP	Methanol	228
1953	BASF, Ludwigshafen.	TCP	Methanol	75
1956	Rhone/Poulenc, Grenoble.	TCP	Methanol?	17
1963	Philips-Duphar,Amsterdam.	TCP	Methanol?	106
1968	Coalite, Bolsover, UK.	TCP	Glycol	90
1976	ICMESA, Seveso, Italy.	TCP	Glycol	ca 1000

TCP = 2,4,5-trichlorophenol, Glycol = ethylene glycol.

Holmstedt nor Hay seem convinced that the death was due to dioxin.

The 1963 Philips-Duphar explosion is also treated as a case history in both references. Medical follow-up of this case did not include control groups but mortality was not considered excessive. Philips intend, in the future, to monitor the health of 27 exposed persons remaining in their employ and to match it against a control group.

The 1968 Coalite explosion has been described as a case history in Milnes (1971), May (1973), Seveso (1978), Holmstedt (1980), and Hay (1982). This latter case history is particularly detailed. The Coalite accident may have been the first TCP manufacturing explosion in which the solvent was ethylene glycol. No really satisfactory follow-up of the exposed cohort has been made and, according to Hay, the attitude of the Coalite Company has left much to be desired. Hay also offers criticisms of the conduct of the UK Health and Safety Executive.

The Seveso explosion case history will be discussed in the next paragraph.

15.8.10. CAUSES AND CIRCUMSTANCES OF THE SEVESO INCIDENT.

15.8.10.1. General.

The causes and circumstances of the Seveso incident have been discussed extensively. There appears to have been at least six books published on it and on issues related to it; the number of papers which discuss it probably runs into hundreds. It has been the subject of great attention from the mass media with innumerable newspaper articles and television and radio programmes. Its name has been applied to the EEC Council Directive 82/501/EEC, which has become popularly known as "The Seveso Directive".

To discuss it fully could form the subject matter of an entire book and clearly the accounts given in the paragraphs which follow cannot present the whole truth. The present author has aimed however to present "nothing but the truth" in the condensed summary which follows without being able to tell the whole truth. In the discussion which follows, unless otherwise stated, the source of information is Seveso (1978), which may be referred to as "the Report".

The basic facts concerning this incident are as follows. A batch of 2,4,5-trichlorophenol (TCP) was being prepared in a reactor. The reaction between the initially charged tetrachlorobenzene (TCB) and caustic soda in the presence of ethylene glycol and zylene had been completed. There had been partial removal of the xylene and glycol by vacuum distillation when the shift ended. Heating and agitation were turned off. Some 7.5 hours later the bursting disc on the vessel blew as the consequence of a runaway reaction and a plume containing highly caustic material was formed to the south east of the plant. The deposition of this plume led to human injury and serious environmental damage.

15.8.10.2. The Seveso factory and its process.

The factory at Seveso was owned by a company called Industrie Chimiche Meda S.A. which will be referred to below by its initials ICMESA. The company was a subsidiary of Givaudan of Switzerland, itself a subsidiary of Hoffmann-La Roche. The factory occupied 52,000 sq metres and employed about 150 people. It produced a variety of chemicals, mainly aromatic compounds. The production of 2,4,5-trichlorophenol (TCP) was of the order of 150 tonnes per annum and it was all sold to Givaudan who used it for making hexachlorophene, an antibacterial compound.

The TCP was manufactured from 1,2,4,5-tetrachlorobenzene by reaction with caustic soda in the presence of ethylene glycol and xylene (see Fig. 15.8). Though the ethylene glycol is described more or less universally as a "solvent", experience of organic synthesis would suggest that it plays a chemical role through the formation of alkoxides which bring the sodium into the reaction as an organic compound (see Fig. 15.10). This is made more likely as the xylene was added to enable water to be distilled azeotropically from the mixture. This would shift the equilibrium in the equation given in Fig. 15.8. to the right. This had some bearing on the composition of the resulting cloud where the unreacted sodium would be present as an organic compound rather than as an aqueous layer of sodium hydroxide solution.

15.8.10.3. The reaction vessel.

The Report is uninformative on a number of major issues as was commented in Marshall (1980a). The Report does not include any information about the reactor except that it was fitted with an agitator, that it had provision for steam heating and water cooling, and that it was protected by a bursting disc set to rupture at ca 3.5 bar. The purpose of the bursting disc is stated in the Report. It was there to protect the vessel against overpressure during operations in which the contents were transferred elsewhere by nitrogen pressure. The provision of a vent pipe leading out onto the roof, in the present author's view, was an entirely reasonable concomitant of this. The question as to whether ICMESA should have provided a catch system to take care of the entirely different circumstance of a runaway reaction will be discussed later.

The size of the vessel is nowhere stated nor is there any drawing of the vessel, still less of the system of which it formed part. The Report points out that it lacked any form of automatic control. There is no information in the Report on the quantities charged to form the batch which became the subject of the runaway reaction.

According to Stevens (1980), which apparently bases its information on a reference which the present author has not personally consulted, [Ferraiola (1979)], the vessel was of stainless steel and was heated by limpet coils which could carry either steam or water. According to Wilson (1982), which also quotes Ferraiolo (1979), the vessel capacity was 10 cu metres. The vessel drawing, which appears in both Stevens and Wilson, is reproduced in Figure 15.10., but this is little more than a sketch and lacks detail. For example, water for cooling would almost certainly have been admitted to the bottom of the coil rather than the top. Systems such as these are to be found throughout the world where chemicals are made by batchwise processes. They differ in their detailed arrangements and in their materials of construction but there is a strong family resemblance between them.

$$\underset{\text{Ethylene Glycol}}{HOCH_2.CH_2OH} + NaOH \rightleftharpoons \underset{\text{Monosodium Glycoxide}}{HOCH_2CH_2ONa} + H_2O$$

$$\underset{\text{Monosodium Glycoxide}}{HOCH_2.CH_2ONa} + NaOH \rightleftharpoons \underset{\text{Disodium Glycoxide}}{NaOCH_2CH_2ONa} + H_2O$$

$$\underset{\text{Ethylene Glycol}}{2\ HOCH_2.CH_2OH} \longrightarrow \underset{\text{Diethylene Glycol}}{HOCH_2CH_2.O.CH_2CH_2OH} + H_2O$$

$$\underset{\text{Monosodium Glycoxide}}{HOCH_2CH_2ONa} \longrightarrow \underset{\text{Sodium Glycolate}}{HOCH_2\overset{\overset{O}{\|}}{C}ONa} + H_2$$

$$\underset{\text{Disodium Glycoxide}}{NaOCH_2.CH_2ONa} \longrightarrow \underset{\text{Sodium oxalate}}{NaO\overset{\overset{O}{\|}}{C} - \overset{\overset{O}{\|}}{C}ONa} + 2H_2$$

Fig. 15.10. (a) SOME REACTIONS OF ETHYLENE GLYCOL AND ITS DERIVATIVES.

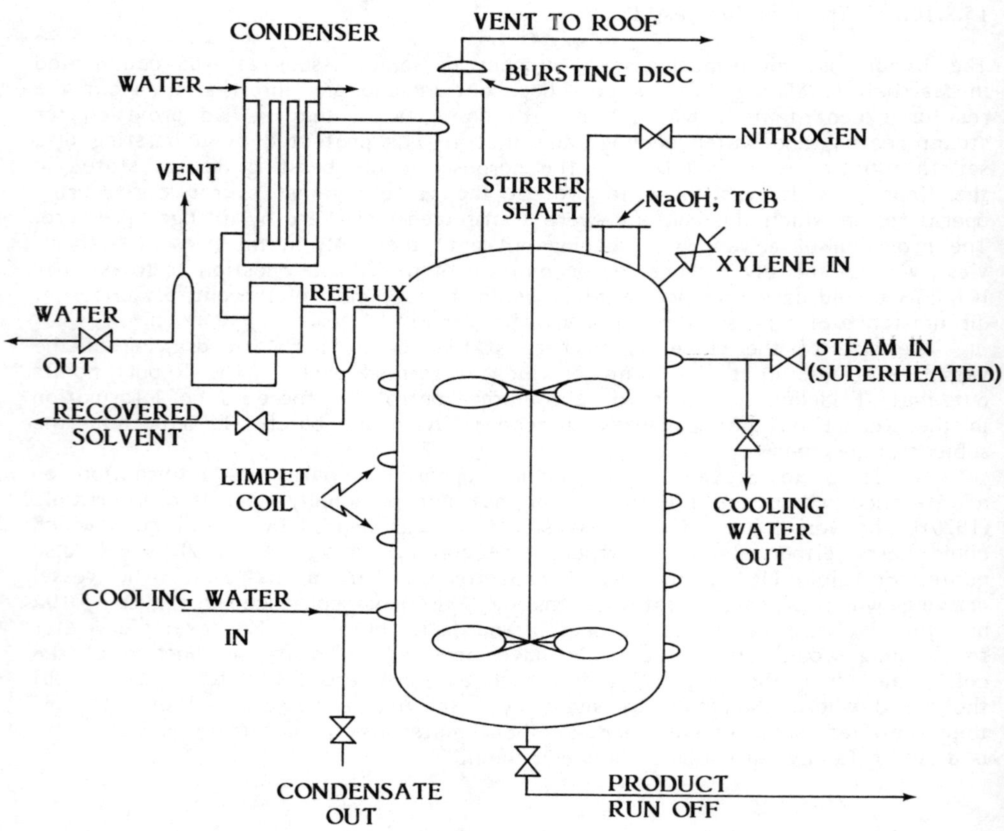

Fig. 15.10. (b) SCHEMATIC DRAWING OF SEVESO REACTION SYSTEM.

15.8.10.4. Quantities charged.

Stevens (1980), based upon Ferraiolo (1979), gives information on the quantities charged to the vessel. These are incorporated in Table 15.6. which includes the present author's calculation of the molar equivalents and the quantities at the end of the reaction. The right-hand column shows the estimate in Rice (1982) of the quantities in the vessel after it was shut down. Quantities are in kgs; mols are kilogram moles.

TABLE 15.6. QUANTITIES OF RAW MATERIALS AND PRODUCTS.

Material.	Quantity Charged (Stevens)	Mol Wt.	Mols	Mol Ratio	Product @ 100% Yield	Product (Rice)
Tetra-chlorobenzene.	2000	216	9.3	1.0	nil	
NaOH	1000	40	25.0	2.7	256*	562
Ethylene Glycol	3235	62	52.2	5.6	3235**	2800
Xylene	609				? ?	
Trichlorophenol (Sodium Salt)		209	9.3	1.0	1940	2030
NaCl		59	9.3	1.0	544	542

* As equivalent NaOH but present as ethoxylates
** Some of this probably converted to diethylene glycol

The quantities given in Rice (1982) agree fairly well except for the figure for NaOH. Overall there is reasonable agreement with the concept that the total quantity charged was ca 7000 kg. Without going into detailed calculations, but assuming that the density of the charge was ca 1,200 kg per cu metre, this suggests that the vessel was rather more than half full.

The Report makes much of the argument that the molecular ratios charged were at variance with those recommended by the parent company, Givaudan, and that this contributed to the accident. It cites the Givaudan Patent as giving mol ratios for TCB : NaOH : Glycol of 1 : 2 : 11.5, as compared with 1 : 2.7 : 5.5 above. This argument shows a very naive approach to chemical patents which are usually at pains not to pinpoint the conditions which lead to optimal results. In fact the ratio TCB : NaOH of 1 : 2 is impractical as it is absolutely necessary to provide an excess of NaOH, otherwise the reaction would not go to completion. The ICMESA practice of adding ca one-third excess NaOH therefore cannot reasonably be criticised. Nor is there the slightest evidence that reducing the quantity of glycol in a charge in any way contributed to the accident. There was however a strong economic justification for doing this as it made possible a more intensive use of vessel capacity. In general, to criticise a manufacturer for seeking to optimise a process by adjustment of process variables, is to ignore the realities of chemical manufacture.

15.8.10.5. Variations to the batch programme.

The Report is on stronger grounds in criticising ICMESA for varying the programme by which the batch was processed. The Givaudan patent provides for acidulation of the product to liberate the TCP as phenol before distilling off the xylene and ethylene glycol. This order was apparently reversed in ICMESA standard practice. As the Report argues, if the batch had been acidulated before it was left at 5.00 a.m. on Saturday, 10th July, 1976, the disaster could not have occurred. In the present author's view this reversal was not done as a mere whim on the part of the management, it was done because it was thereby possible to distil off the xylene and glycol without their being contaminated by TCP as the sodium salt may be regarded as involatile whereas the liberated phenol is volatile. It is also true that the management had permitted the batch to be started at such a time that it could not be finished during the normal working cycle. Had this not been done it may reasonably be argued that the accident would not have occurred either. However this begs the question as to whether the management had reasonable grounds for believing that a reaction, such as led to the accident, could occur if the batch, without acidulation, was left at a temperature, unstated in the Report, but which may be inferred to be below 185°C, [Pages 42 & 43 of Health & Safety Executive (1980)]. Wilson (1982), source unstated, gives it as 158°C. This question will be discussed below.

15.8.10.6. Nature and composition of material discharged.

The nature and composition of the material which was released, apart from its dioxin content, has received little comment. If the safety plate had ruptured at its set pressure of 3.5 bars (gauge ?) and it be assumed (a) that the cause of the pressure rise had been the generation of permanent gases, and (b) that the free volume in the vessel was half its volume, i.e. 5 cu metres, it would follow that the initial discharge was a puff of ca 16 cu metres. Thereafter the rate of discharge would be approximately equal to the rate of generation of the permanent gas. It seems reasonable to believe that this permanent gas would be accompanied by the vapours from the other constituents and by entrained liquid. On reaching the outside atmosphere cooling would occur and the deposit might well have superficially resembled damp snow. Wilson (1982), without quoting a source for the information, says "Some children actually ran into the cloud as it descended dropping like snow flakes". These snowflakes would have inflicted severe injury on anyone on whom they fell. Only Rice of the references consulted gives information on the plume which was generated. This will be discussed in Para 15.8.13.

The argument in the Report that "if the safety plate (bursting disc) had been set to burst at lower values, this would have occurred at a less advanced stage in the reaction, with less harmful effects", seems to be without any obvious justification and it does not attempt any. As will be seen later the reaction which produces this overpressure is very violent and its buildup is very rapid indeed.

15.8.10.7. Could the management have foreseen the accident ?

The question with which this paragraph concludes is the one concerned with the extent to which ICMESA or its parent company Givaudan might reasonably have foreseen the circumstances of the accident. The Report has no hesitation in claiming that "the risks involved in TCP production at the ICMESA establishment were well known to the responsible officers of the two companies". The Report clearly did not accept the view of the management that such an incident was impossible and their claim that experiments carried out subsequently, under varied conditions, had failed to reproduce the incident in the laboratory. The Director General of Givaudan had claimed that "no possibility of an exothermic reaction with a system of this type had been reported in the literature".

The Report attached considerable importance to the occurrence at Coalite which is referred to several times and obviously took the view that what had happened at Coalite could happen at Seveso and that this was the heart of the matter. But was it ?

Milnes (1971) said "Small scale experiments in which TCB hydrolysates were allowed to heat up to higher temperatures than the normal 180°C showed a significant exothermic reaction after residual ethylene glycol had distilled off. Hydrolysates of all isomers of TCB and mixtures of caustic soda and ethylene glycol all had the exothermic reaction in common starting in the 230°C region and proceeding rapidly and uncontrollably to around 410°C. This exotherm, which we attributed to the decomposition of sodium 2-hydroxy ethoxide ($NaOCH_2CH_2OH$) was invariably accompanied by a huge volume of white vapours". May (1973), describing the same incident says "At midnight on 23rd April 1968, the reaction temperature reached 175°C and thereafter it rose continuously for 50 minutes. When it reached a point somewhere in excess of 250°C an explosion of considerable violence occurred and the supervising chemist was killed by falling masonry. Subsequent analysis indicates that the explosion was of a double nature. When the temperature had reached 225°C the process began to generate its own heat and the initial rupturing of the vessel was due to the advance of this exothermic reaction. Ethylene glycol and dichlorobenzene (used at Coalite where xylene was not used) vapours were released . . . these constituted an explosive mixture which was detonated by an electric lamp . . ."

May (1973) was written by a medical practitioner and cannot carry the weight in technical matters that would be carried by a paper written by a chemist or a chemical engineer. However it seems apparent that the technical information in this reference was provided by others. May seems to infer that the rise in temperature over the 50 minutes was due in the beginning to some source other than self-heating until 225°C was reached. Stevens (1980) attributes this temperature rise to a malfunctioning of the oil heating system but refers to Milnes (1971) and May (1973) as the source of this information though it is not explicitly stated in either.

There was thus a thermal barrier to surmount. Below about 230°C any reaction within the mass has been shown (see below) to be so slow as to be undetectable by any but the most sensitive calorimeters. As the reaction mass was left at about 160°C this meant that it had to be heated through 70°C before a runaway reaction could start. This, it would seem, would have required the input into the charge, assuming a specific heat for the mass of ca 2.5 kJ per kg (0.6 cals per gram), of ca 10^9 J or ca 300 kW hrs. If the shell of the reactor had weighed 1 tonne and had had a temperature at close down of, say, 250°C this would have provided only 40 x 10^6 or about 4% of the heat required.

Matters stood at this point for about 5 years until the publication of Theophanous (1981). This paper first of all drew together all the evidence which showed that, even under adiabatic conditions, self-heating below ca 230 was too slow to account for this temperature rise. In practice conditions could not be adiabatic as the vessel would be losing heat to its surroundings. The rate of heat loss for the vessel cannot be calculated from the literature as information on the thermal insulation of the vessel is totally lacking. The reference also accumulated evidence to show that around 220°C self-heating could become sufficiently rapid to initiate a runaway reaction. Finally the paper put forward convincing evidence derived from experiment which showed that if agitation had been stopped the surface layer of the reaction mass could be heated by radiation from the dry walls of the reactor to the initiation temperature for the runaway reaction. The paper drew attention to a fact not referred to in Seveso (1978) that the steam though only at 12 bars gauge, corresponding to a saturation temperature of 190°C, was admitted to the reactor in a highly superheated condition according to Theophanous, i.e. at 300°C. Thus the dry walls would have a temperature at shut down of ca 300°C. Thus though ICMESA thought that they were avoiding the hazard of hot oil they unwittingly substituted the hazard of hot steam. There is no information given in this, or any other reference, as to why steam, with so high a degree of superheat, was used. The superheat could not have arisen, in such a degree, from passing through a reducing valve, i.e. throttling. It is well known that no advantage attaches to the use of superheated as opposed to wet steam.

It remains to discuss the nature of the highly exothermal reaction. Milnes (1971) attributed this to a reaction which dehydrogenated the monosodium alkoxide of ethylene glycol to sodium glycolate and the disodium derivative to oxalic acid. Both reactions are oxidations and hence are exothermic. There seems no reason to disagree with this conclusion. Stevens (1980) seems to doubt it by claiming that the Coalite accident occurred after the glycol had been removed. This does not seem to be true. Even if it were it is not the glycol which is involved but its involatile sodium derivatives which would be left behind after any distillation process. Stevens confirms that the sodium salt of TCP is stable up to 300°C and hence it must be the glycol derivatives which are responsible for the exotherm.

It seems impossible to sustain a charge of lack of foresight when it took six years to demonstrate a mechanism whereby the reaction mass at ca 160°C could have its temperature raised by 70°C into the temperature range at which the exotherm could occur. If this accusation of lack of foresight cannot be sustained then the charge that there was a failure to provide a catch system must fall also.

In Marshall (1980a), the present author made a number of points which he would no longer wish to defend. But one which he believes remains entirely valid is his comment "the predominant view in the Report is that of the chemist and the toxicologist rather than that of the chemical engineer". As shown above the solution to the problem could not be arrived at without detailed consideration of the hardware, a topic which received only the most perfunctory attention in the Report where it seems to have been dismissed as irrelevant.

15.8.10.8. Safety recommendations.

Cardillo et al (1984) analyse Theophanous (1981) and they reach conclusions regarding future safety in processes for the production of TCP.

But the reference makes a claim which is open to question. It supports the view of Stevens (1980) that the decomposition reactions cannot be ascribed to the decomposition of the sodium derivative of glycol on the grounds that explosions had been reported in Milnes (1971) "after ethylene glycol removal". However, as has been pointed out above, neither the Milnes paper nor the May paper claim that the ethylene glycol had been removed before the Coalite explosion and that even if it had, this would still leave the sodium derivative behind. What Milnes reported was that when reaction mixture was heated in the laboratory to ca 230°C some glycol was evolved which is not surprising as the atmospheric boiling point of ethylene glycol is ca 190°C. The reference confirms that there is no evidence in the literature of thermal instability of the sodium salt of TCB below 250-300°C. It accepts that Stevens' unconfirmed assertion that the Coalite accident was caused by over-heating due to a fault in the oil heating system.

The reference gives a number of results of research which prove that self-heating does not occur below 200-300°C. It analyses also the experiments reported in Theophanous (1981) in which an electrically heated steel slab partly immersed at the surface of water only heated up the top layer. The reference comments that this was "in contrast with what would be expected". This remark is surprising since it is common knowledge that it is natural convection which plays the predominant role in heat transfer in unagitated liquids and this operates only when the liquid is heated in its lower layers.

Cardillo et al recommend that the maximum temperature which is to be permitted in the manufacture of TCP should be 180°C below which temperature the reaction mass is thermally stable and dioxin formation is low. Attention is drawn to the dangers of lines clogging due to sublimation of the TCB, a problem which has led some manufacturers to add materials in which TCB is soluble but which do not enter into the reaction and which will reflux round the system and keep the TCB in solution. The reference also drew attention to the need to avoid highly superheated steam as a heating medium. Why this was used at Seveso has not been explained as it possesses no advantages over saturated steam when used to heat reactors, in fact the contrary is true. Its use was the Achilles heel of the Seveso process.

Cardillo et al nowhere discuss agitation as a means of prevention though this has a considerable influence on the avoidance of hotspots and hot layers. It seems to the present author not open to doubt that the creation of a hot layer at Seveso could not have occurred in the presence of even gentle agitation. But there seem to be no literature references to the viscosity of the reactant mass which is a physical property very relevant to the assessment of the efficiency of agitation. This may be a further indication of the scant attention which has been given to the hardware aspects of this accident.

The question of operating procedures is not discussed by Cardillo et al. The present author is of the view that had the Seveso reactor had its coils flushed through with cooling water before it was left to stand over the weekend, the accident would not have occurred.

15.8.10.9. The question of a trapping system.

The Report says "The consequences of the incident, and in particular those affecting the external environment in the vicinity of the factory, could have been avoided, or at least contained, if the plant had been equipped with a system for the collection and destruction of any compounds which might possibly escape - as in fact occurred in this concrete case . . . The management of ICMESA have stated that no destruction system was provided, since they were certain that an exothermic reaction could not occur in their establishment . . . However, the incident which occurred on 10th July 1976 gives the lie to this statement". This is taken from Health & Safety Executive (1980) and the original may have suffered in translation. But, taking it at its face value, the question arises, gives the lie to what ? Is the lie given to the assertion that the management were certain in their own minds, or is it given to the impossibility of the occurrence ? All managements have to exercise judgement as to whether trapping devices are necessary or not as these can be very expensive items of equipment which may, by their presence, introduce hazards into the process which were never there before. The present author's view is that, after weighing up the evidence, the management exercised reasonable judgement in not fitting a catch system. After all, even the hindsight at the disposal of the Investigatory Commission failed to provide a satisfactory explanation of why the runaway reaction occurred.

15.8.10.10. Some general conclusions.

Confining the discussion at this stage to the duty of the ICMESA to avoid the accident and to the procedures adopted by the Commission a number of points emerge.

(a) The Commission failed to conduct a proper investigation of the engineering aspects of the equipment used to process the batch. Instead they appeared to see the solution to the problem purely in terms of chemistry. Had full particulars of the vessel and its system including details of the services to it been published as was done in Flixborough (1975), a solution might have been arrived at earlier.

(b) As with many disasters in the chemical and process industries the accident would not have occurred without the simultaneous presence of a number of adverse factors. These included the suspension of processing before acidulation, the use of highly superheated steam in the coils, the failure to cool them down and the cessation of agitation before the charge had cooled beyond the possibility of adverse reactions.

(c) The strictures upon ICMESA for failing to foresee the runaway reaction and to provide a catch system were unjustified.

15.8.11. NATURE AND EXTENT OF INJURIES AT SEVESO.

15.8.11.1. Human injuries.

The human injuries suffered at Seveso fell into three classes:-

(a) Burns due to skin contact with highly caustic substances.

(b) Chloracne.

(c) Other systemic effects.

The Report gives the following statistics:-

Population of Zone A (the most heavily contaminated)		733
Population of Zone B		4,800
Population of Zone R (6 neighbouring communes)		22,000
Population of additional areas subject to epidemiological study, 5 communes.	ca	190,000

Population denotes resident population, not the numbers present at the time of the release.

During the period 23rd July to 31st August, 477 persons were identified by the dermatological outpatient department at Seveso as having skin injuries which could be identified as arising from the Seveso release. Of these, 113 were resident in Zone A (24% of population) and 278 were resident in Zones B & R (1% of population). This left 56 cases not resident in Zones A, B, R. It has to be remembered that though residence denotes a high degree of probability that the person was in a particular zone at the relevant times, it does not follow that a person resident in Zone R for example was not in Zone A at the time of the release.

The nature of these initial manifestations was that of contact with caustic substances. Certainly the Report notes that:-

(a) There was rapid regression of the symptoms in the first 15 days of August.

(b) Only 34 of the 447 patients were subsequently found to be suffering from chloracne.

This latter is a very significant point as it strongly suggests that acute contact with material discharged from the vessel, even though it injured the skin, in more than 90% of the cases did not give rise to chloracne. It was only after mid-August, according to the Report, that symptoms of chloracne presented themselves.

Unfortunately the world's press presented a confusing picture by showing the patients who had been bandaged to treat caustic burns and representing them as victims of chloracne.

During September and October, 44 cases of chloracne were identified, all of these persons were present in Zone A when the accident occurred. A further 6 cases from this Zone were diagnosed in November and December. This brought the total from Zone A to 50 or 7% of the resident population. Of the 44 cases none were aged above 21. Further screening of 7,077 patients at the dermatological outpatients department with some 44,000 school children brought to light a further 79 cases. There were a further 6 cases identified elsewhere making a total of 135. Later still in 1978, further screening identified 52 further cases making 179 in all. It might be inferred that the cases brought to light by screening represented only cases so mild that medical attention had not been sought.

The Report in a table suggests that in early 1977, only 9 patients could be classified as serious and that by the beginning of 1978, there were no cases classified in that category.

The Report also provides information on systemic effects other than chloracne which suggests that some of the inhabitants of Zones A and B suffered neurological damage. There seemed to be no correlation between diagnosis of neurological damage and incidence of chloracne. Homberger el al (1979) suggests that the converse is also true; that those diagnosed as suffering from chloracne were free of systemic symptoms.

15.8.11.2. Animal injuries.

There were large numbers of deaths among the animal population. Among those which died there were many which had suffered chemical burns attributed to sodium trichlorophenate by Homberger et al. Some were burned on the extremities and others were burned in the throat. Homberger et al seem not to have considered that some of the caustic effects could have been caused by sodium substituted glycol. The reference records that small animals succumbed to the effects more rapidly than large ones. Rabbits died first and horses last. All animals which had been in Zone A were slaughtered (except for a few kept for research purposes) as a precaution against dioxin entering the food chain. Some 78,000 small animals, mainly chickens, were slaughtered as were about 700 large animals. The numbers of those which died from toxic effects were ca 3000 small animals and 12 large animals. Homberger et al remark that a plot of the positions of dead animals correlated well with the dioxin content of the vegetation. (This does not necessarily mean that the animals died of dioxin poisoning; the dioxin would play the role of a tracer which would denote the presence, or previous presence, of material discharged from the reactor. Dioxin would persist on the vegetation for example after rain had washed away the soluble sodium compounds).

15.8.12. MEDICAL FOLLOW-UP OF PERSONS EXPOSED TO DIOXIN.

The general subject is one for which the present author does not feel competent to attempt a critical appraisal. Nevertheless, attention is drawn to some of the reference sources available and to some of the general conclusions which they put forward.

May (1973) comments on treatment of 79 cases of chloracne arising from the Coalite accident. The great majority of cases made almost complete recovery in four to six months. Ten men recovered only after steam bathing followed by ultraviolet treatment. No other illness attributable to dioxin made its appearance. In March 1969, there were two further cases of chloracne among contractors men, one of whom had a son who also showed symptoms. Hay reports that there was no evidence of chromosome damage but some evidence of damage to the immune system. Another worker claimed that there was evidence of impaired liver function in some subjects.

Seveso (1978) deals extensively with medical follow-up which included analyses of death rates, birth rates and abortion rates. No clear picture emerges however. Homberger et al claim that, broadly speaking, nothing abnormal was disclosed. "So far the pathology related to these parameters has remained within the range which is current for the population of this region and could not be correlated with the exposure to dioxin".

A very detailed study by Bruzzi which forms part of Coulston and Pocchiari (1983), gives an account of the difficulties involved in setting up a sound epidemiological study in the area of Seveso. This study agrees with the conclusion that there is nothing in the mortality pattern which correlates with exposure. With regard to abortions, matters are different. In the period immediately following the incident, the rate of induced abortions was higher by a statistically significant amount. It is believed that these were caused by fears of foetal damage arising from exposure to dioxin. The rate of induced abortions fell after 1968. There seems no clear indication of correlation of birth defects with exposure though there is some suggestion that this may be the case in a few instances. As to cancer incidence, the latency period is generally too long for study to reach any conclusions yet. Bruzzi agrees with the finding that there is a statistically significant correlation between some neurological symptoms and exposure to dioxin.

15.8.13. DAMAGE TO THE ENVIRONMENT.

15.8.13.1. The area affected.

Seveso (1978) gives little or nothing in the way of eyewitness accounts of the disaster and says nothing about the duration of the discharge. It notes, in passing, that a foreman heard a whistling sound coming from the reactor and he therefore entered the plant and applied cooling water to the reactor. Rice (1982) suggests that the emission lasted 2 or 3 minutes and formed an inverted cone shaped cloud which drifted for a very short time to the north east and then, after a change in the wind, it drifted for most of the time to the southeast. The plume was variously estimated to have a height of 20 to 50 metres. Whether this cloud had the same composition throughout the period of emission has not been discussed in any of the literature the present author has consulted. It might be thought that the last material to be emitted would be richer in dioxin than the initial puff because the vessel was then at the higher temperature which favoured dioxin formation.

Seveso (1978) provides a sketch of the areas affected; Figure 15.11. is based on this sketch. There were essentially three zones determined by chemical analyses of vegetation and soil. Zone A was the zone of heaviest contamination with a mean level of contamination of 240×10^{-6} grams per sq metre and an area of 1.08 sq kilometres. Zone B had a mean level of contamination of 3×10^{-6} grams per sq metre and an area of 2.7 sq kilometres. Zone R, the zone of minor concern, had levels of less than 5×10^{-6} grams per sq metre and an area of 14.3 sq kilometres. The total area of concern was thus 17.1 sq kilometres.

Rice (1982) calls attention to the existence of a "window" in the urban landscape to the south east of the factory and that it was in this "window" that the greatest part of the fall-out was deposited.

Pocchiari et al in Coulston and Pocchiari (1983) give a series of detailed maps of the area which show such data as the extent of the built-up area, the zones of contamination and the distribution of sampling points.

15.8.13.2. Evacuation.

There was no prompt evacuation of even the most severely affected areas. Though the accident happened on July 10th, it was not until July 24th that Zone A was designated on the basis of analytical results; it covered originally only 0.15 sq kilometres. It was decided then to evacuate the area and to prohibit further access. By July 26th, 225 people had been evacuated. By the end of August, it had been extended a number of times and covered 1.08 sq kilometres. By then a total of more than 730 people had been evacuated. Zone B was designated on the basis of analysis and subjected to controls but not evacuation except for young children in daylight hours and women in early stages of pregnancy.

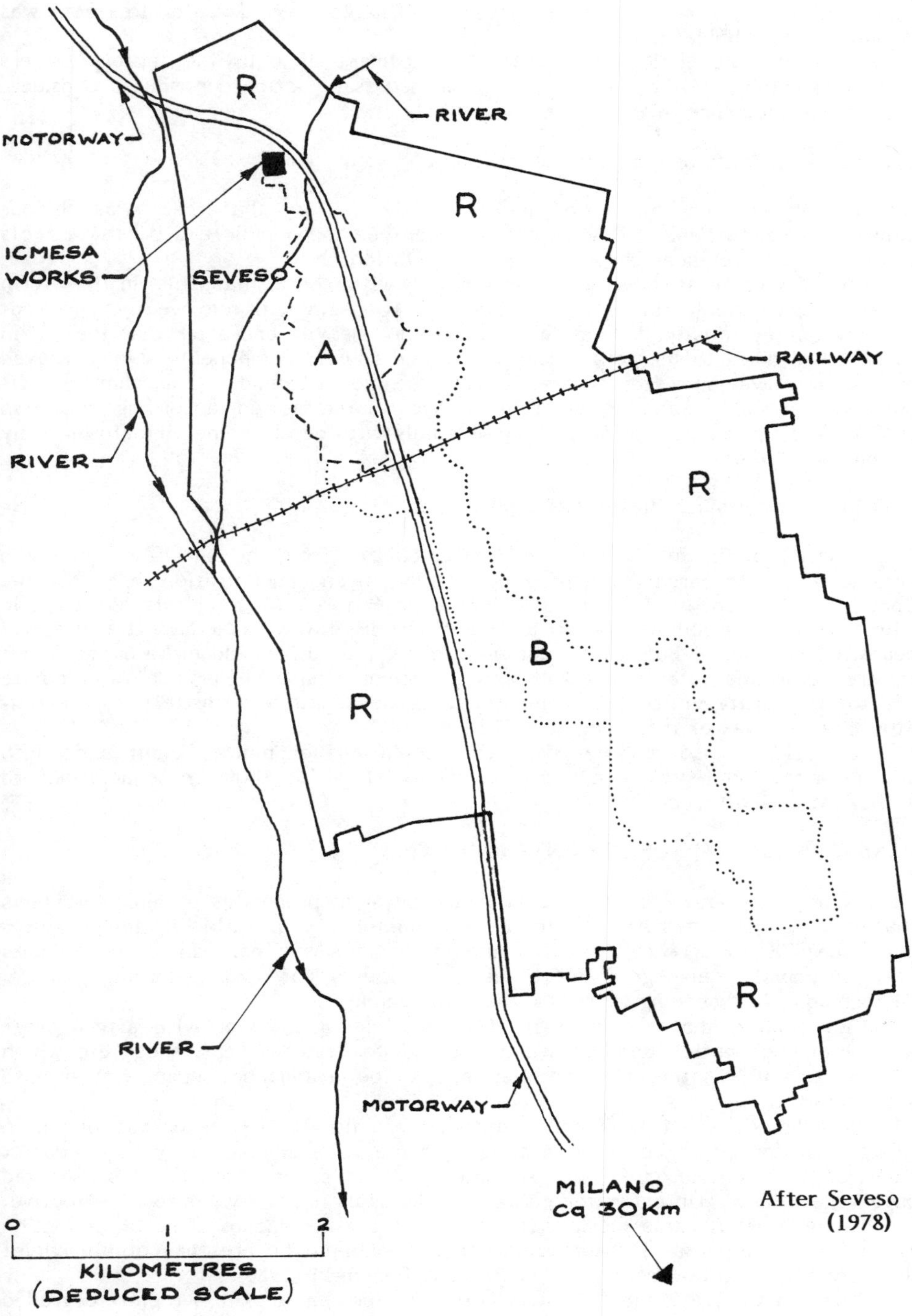

Fig. 15.11. SKETCH MAP OF CONTAMINATED AREAS AT SEVESO.

Food and water were closely monitored. The Zone R "Zone of Respect" was designated as a buffer zone.

There can be little doubt that the tardiness of both the factory owners and the municipal authorities in taking the necessary action contributed considerably to the incidence of chloracne.

15.8.13.3. Effects of dioxin on vegetation and soil.

It was only in the immediate vicinity of the factory that there was obvious damage to vegetation. This must have been due to phenolic and caustic effects as dioxin has never been shown to injure vegetation.

The dioxin must have been associated with these materials in the form of an aerosol which clung to vegetation. To begin with the weather was hot and dry but eventually, in mid-August, it rained heavily and after that the dioxin was washed down to the soil. The phenol and caustic components were dissolved and carried away in water courses leaving the dioxin behind. This clung tenaciously to the soil. Samples taken up to a year later showed, according to Seveso (1978), that the concentration fell away rapidly with depth, being virtually entirely in the top 20 cm.

15.8.13.4. How much dioxin was deposited ?

The quantity of dioxin deposited may be calculated from the sum of the products obtained from the areas of each zone and the mean concentration in each zone. Zone A thus accounts for 240 g and Zone B for 8 g. It is not possible from the information at the present author's disposal to calculate the quantity deposited on Zone R but it does not seem likely that this would make a significant difference to the total of ca 250 g. Attempts have been made to calculate the total quantity emitted as compared with the quantity deposited; Rice estimates that this was of the order of 1.75 kg.

Rice (1982) also provides data on calculations of plume height and width and of isopleths (equal concentration contours) which show reasonable accord with what actually occurred.

15.8.14. PROBLEMS OF DECONTAMINATION.

Comment given above on the chemical and physical properties of dioxin suggests that it is a difficult substance to decontaminate. It is stable at temperatures below ca 700°C, unreactive with common reagents and only difficultly soluble. The microbiological degradation of dioxin in soil is slow and, according to Rice, no technique for speeding it up has yet been found.

Rice reports that it was planned to build an incinerator with after-burner with a planned outlet temperature of 1200°C to destroy crops, sheds etc which had been contaminated, but the plan had to be abandoned because of intense local opposition.

A technique which was believed to be promising was to degrade dioxin in sunlight in the presence of a material capable of acting as a hydrogen donor. Olive oil was considered to be a suitable hydrogen donor which could be sprayed onto vegetation. Unfortunately, though field trials had shown it to be effective, a decision to use it came too late for it to be a feasible means of decontamination. The scientific aspects of the degradation of dioxin in the presence of ultraviolet light are discussed by Crosby in Coulston and Pocchiari (1983).

Ayres in Coulston and Pocchiari (1983) claims that dioxin is rapidly degraded by refluxing it in a carbon tetrachloride solution of ruthenium tetroxide. The author in discussion stated that this would not be a feasible technique for destroying it in soil.

15.8.15. GENERAL CONCLUSIONS.

Though the incident at Seveso has achieved international notoriety, it was an accident in which no one died and the injured all recovered apparently with no permanent ill-effects.

The dioxin released proved to be capable of sterilising for agricultural use for a period of years approximately 4 sq kilometres (1 1/2 sq miles). No satisfactory means of decontamination could be evolved. The incident has been connected by a number of authors with the manufacture and use of the herbicide 2,4,5-T but the factory responsible for the emission was not engaged in making this product.

Though the management was heavily censured for failing to anticipate the possibility of a runaway reaction, it took many years to discover a mechanism by which this could take place. However, in seeking to avoid the use of hot oil, the management substituted super-heated steam which, by a mechanism unknown at the time, was capable of raising the temperature of the top surface of the liquid to the requisite temperature. Had one of a number of simple measures been taken, the incident would not have occurred.

The Official Inquiry concentrated its attentions on the commercial, chemical and toxicological aspects and dismissed the engineering, particularly the chemical engineering, aspects. By so doing they failed to discover the true cause of the runaway reaction.

The Official Inquiry was on sound ground in censuring both the management and the local authorities for the long delay before evacuation was ordered, a delay which must have increased the incidence both of caustic and phenolic burns and of chloracne.

15.9. METHYL ISOCYANATE.

15.9.1. GENERAL.

Methyl isocyanate, which is referred to below as "MIC", was responsible for a disaster at Bhopal, Central India, on December 3rd, 1984, in which an escape of its vapour killed around 3,000 people and may have injured 200,000 more. This disaster, the circumstances of which are set out below as a case history, was the worst industrial disaster in the world's history. Because of the magnitude of the disaster, and the widespread media coverage which it engendered, the author has judged it appropriate to subject it to extended coverage.

In regard to the disaster itself it has to be said that, at the time of writing, the law suits arising from it have yet to be settled, and this factor may be inhibiting candid discussion of all the causes and circumstances.

15.9.2. INDUSTRIAL SIGNIFICANCE.

MIC is manufactured from two readily available substances, phosgene and monomethylamine, by a two-stage process.

(1) CH_3NH_2 (methylamine) + $COCl_2$ (phosgene) → $CH_3NHCOCl$ (methylcarbamoyl chloride) + HCl + heat

(2) $CH_3NHCOCl$ →(heat) CH_3NCO (methyl isocyanate) + HCl

The principal use of MIC is in the manufacture of the insecticide Sevin (1-naphthyl-N-methyl-carbamate). Sevin is much less toxic to humans and animals than is, for example, parathion and it is much less persistent than DDT. The annual production of MIC in the US in the mid-seventies was 12 to 14,000 tonnes and of Sevin, ca 24,000 tonnes. It was Sevin which was the end product of the Bhopal process where the scale of production, according to Worthy (1985), was about one-tenth of that of the US plant of the parent company.

15.9.3. PHYSICAL PROPERTIES.

MIC has the formula CH_3NCO and it has a molecular weight of 57. Is is thus twice as dense as air at the same temperature. Its atmospheric pressure boiling point has been variously quoted in the past but it is generally agreed today to be ca 39°C. It may be noted that, until its most recent edition, the usually authoritative Weast has erroneously reported it as 59.6°C. Weast (1985) now gives it as 39.1 to 40.1°C. NIOSH/OSHA (1978) gives it as 20°C at 375 mm Hg and 38.8°C at 760 mm. It is therefore a highly volatile liquid at room temperature and is not a liquefied gas at atmospheric temperature.

MIC reacts with water in the presence of acids and alkalis and also in the presence of iron, tin, copper and their salts. It is capable of polymerising to give dimers, trimers and higher polymers, [Kirk-Othmer (1981)].

15.9.4. TOXIC PROPERTIES.

There was little data in the open literature on the acute toxic properties of MIC prior to Bhopal which may reflect the fact that up till that time there does not seem to have been a fatal accident involving the substance. According to Dagani (1985), one of the reasons for a lack of animal experiments was the intense discomfort caused to the experimenters by even tiny leaks. However little information has been published since. For example, Union Carbide (1985), which is the Company's official report on the Bhopal disaster, provides no information on the subject. Nor is any provided in Browning (1985) which is referred to below. However, such data as did exist prior to Bhopal ought to have provided sufficient warning that MIC is a very dangerous substance to inhale and to have jolted the principal manufacturers and users into conducting the necessary research into its acute toxicity.

Some of the data on the toxicity of MIC which were known prior to Bhopal are given in Table 15.7. where, for comparison, data are also given for three other gases whose highly toxic properties are well recognised. These data indicate that MIC had a much lower control limit for chronic exposure than the other three (a value which has stood for at least 10 years), that it was rated as more immediately dangerous to life and health than two out of the three, and that it had the lowest value for the lowest recorded lethal concentration and the lowest LC_{50} of the four. It was only in its oral lethality that methyl isocyanate appeared to be only moderately toxic. Even the concentration for "immediately dangerous to life and health" for MIC may be set too high. Dagani (1985) quotes German experiments on human volunteers in which irritation appeared in 1 to 5 mins at 2 ppm and an exposure for a similar length of time at 21 ppm proved unbearable.

Though the criteria for chronic exposure and for acute exposure for toxic substances do not always correlate well it still seems very significant that out of 480 substances for which threshold values are given in Health & Safety Executive (1984), only 10 were more severely rated than MIC and they were mainly highly toxic cumulative metallic poisons.

MIC is now known to cause prompt deaths by pulmonary oedema and to injure the eyes, the stomach, the liver and the skin. Chronic effects, some irreversible, are recognised though there is general agreement that little is known about these long-term effects. Dagani (1985) claims, however, that there is no evidence for carcinogenic or mutagenic effects.

TABLE 15.7. TOXIC PROPERTIES OF MIC COMPARED WITH OTHER TOXIC GASES.

Substance	Control or recommended limit (1) ppm	IDLH (2) ppm	LD_{50} Oral (2) mg/kg	LCL_o Inhalation (3) ppm/time exposed	LC_{50} Inhalation (3) ppm/time exposed
Methyl Isocyanate.	0.013 (1)	20	71 (rat) 120 (mouse)	17.5/60 mins (mouse)	5.0/240 mins (rat)
Phosgene.	0.10 (4)	2		10/20 mins (guin.pig) 50/30 mins (rat) 80/30 mins (dog)	?
Chlorine.	1.0 (4)	30		500/5 mins (mammal) 800/30 mins (dog) 873/30 mins (human)	137/60 mins (mouse) 293/60 mins (rat) 400/30 mins (human) (5)
Hydrogen cyanide.	10 (4)	50	3.7	200/10 mins (human) 120/60 mins (human) 200/5 mins (mammal)	323/60 mins (mouse) 484/60 mins (rat)

Notes.
(1) Control limit from HSE (1984), converted to equivalent MIC.
(2) From NIOSH/OSHA (1978).
(3) From NIOSH (1982).
(4) Recommended limits from HSE (1984).
(5) From I.Chem.E. (1985a).
Where values have been given in the original reference as mg/cu metre, these have been converted to ppm by the present author.

Definitions:-
IDLH Immediately dangerous to life and health; the maximum concentration for escape within 30 mins without irreversible effects on health.
LD_{50} Lethal dose, 50% kill.
LCL_o Lowest published lethal concentration.
LC_{50} Lethal concentration, 50% kill.

Browning, who has been the Vice President of Union Carbide responsible for Health, Safety and Environmental Affairs since 1976, stated, during discussion of his paper, Browning (1985), "The long term health effects of that gas (i.e. MIC) have not been studied in great detail . . . The experience with MIC up until the Bhopal tragedy was from people who were exposed to very small quantities, it was a very irritating gas and they removed themselves from the exposure. People had not anticipated massive exposures of the kind that were experienced at Bhopal . . ." But New York Times (1985) claimed on 30/1/85 that the Union Carbide Corporation technical manual which was distributed to the management at Bhopal stated that MIC "may cause fatal pulmonary oedema".

Waite et al (1985) consider that MIC should be rated at 2 to 5 times the toxicity of phosgene. The present author considers that until more extensive data are available it should be rated at 25 to 30 times the toxicity of chlorine. Waite et al draw attention to the ability of MIC to react with water vapour in the air but give no indication of the reaction kinetics on which there appears to be no information in the literature. Note may be taken that, in the liquid phase, the reaction between MIC and water is rather slow in the absence of catalysts and it would seem unlikely that the gas phase reaction at the same temperature, where the concentration of the reactants is far lower, would be very much higher. It may be that in the past excessive reliance had been placed upon reaction with atmospheric moisture to nullify or mitigate the effects of MIC vapour. The mean residence time of a molecule of MIC in the vapour cloud at Bhopal would have been of the order of two minutes. Browning's remarks quoted above hardly seem to indicate that water vapour in the air prevented the gas from giving rise to intense irritation. However, Dagani (1985) claims that reaction with moisture in the air has created difficulties for those conducting animal experiments.

15.9.5. THE BHOPAL DISASTER.

15.9.5.1. General description of the disaster.

The disaster occurred on December 3rd 1985 at Bhopal, Central India, a city of some 800,000 population. The factory from which the release of MIC originated was owned by Union Carbide India Ltd, a company in which the principal shareholder was the Union Carbide Corporation of Danbury, Connecticut, USA. UCC had a 50.9% holding and was represented on the Board of Directors of UCI. The plant, which produced the insecticide Sevin, appears to have been entirely under Indian management so far as day-to-day matters were concerned.

According to Union Carbide (1985), the first untoward event in the sequence which led to the disaster was noticed at about 2300 hrs on Sunday, December 2nd 1984, when the pressure of MIC storage tank No. 610, which contained some 41 tonnes of MIC, was noted as being 10 psig (0.66 bar) though, apparently some 40 mins earlier, it had been only 2 psig (.13 bar). At 00.15 hrs there was a report that there was a leak of MIC and the tank pressure had risen by then to 30 psig (2 bars). It then rose very rapidly and within moments was off scale, i.e. more than 55 psig (3.75 bars). The safety valve blew and the concrete casing of the tank was heard to crack. The operator turned on the caustic vent scrubber but apparently there was no indication from a flowmeter that caustic soda was being circulated over the scrubber. MIC fumes were noted around the plant and the safety valve continued to blow off. Union Carbide (1985) gives a range of time, between 01.30 and 02.30 hrs, when the safety valve ceased to blow. The temperatures of the contents of the MIC storages were not logged; indeed it has been claimed in New York Times (1985) on 30/1/85 that there was no column to record it in the log book. At about 00.45 hrs, according to the same reference, the temperature of tank 610 had gone off scale, i.e. $\geqslant 25^{\circ}C$. A member of the management estimated that the wall temperature of tank 610 at 05.30 hrs was hot to the touch, i.e. 45-60°C.

The fire brigade were summoned and attempted to produce a water curtain at the top of the scrubber outlet stack which was about 30 metres high and from which the issuing plume rose a further 3 metres higher. Some claimed that the jets from their hoses could not reach high enough but Union Carbide (1985) claims that subsequent tests showed that the jets were capable of reaching to this height. Considerations of mass transfer rates suggest to the present author that such an exercise would be largely cosmetic.

For an account of what happened outside of the plant, it is necessary to refer to sources other than Union Carbide (1985) which makes no mention of the effects on the local population, indeed that report makes no reference at all to the toxic properties of MIC. Such rather scanty information as has been available to the present author is mainly based upon reports in the New York Times, [New York Times (1985)], in Chemical & Engineering News (1985), and upon his recollections of UK television programmes on the subject. Examination of the various reports suggests that they include a fair amount of repetition.

Because of the time at which the disaster occurred, the vast majority of the local population were in bed where it is believed that many died. Large numbers, and this is especially true of those living close to the factory, were living in very primitive conditions in closely packed shanties which could afford virtually no protection from the ingress of gas as some were little more than tents. The population density has been estimated in Slater (1985) as 25,000 per sq kilometre, which is about 6 times the density in a UK town. It was almost certainly higher still in the vicinity of the factory. According to Dagani (1985), those who lived in well-constructed houses, especially those with more than one floor, escaped the worst effects. Breathing through wet cloths proved helpful.

The local hospitals were soon overwhelmed by those affected and the staff apparently had no idea what was the causative agent or how to treat the victims. The State Director of Public Health has claimed, [New York Times (1985)], that a factory doctor told him, as people were dying at the hospital, that the gas was not lethal and only caused eye and lung irritation, "it's not so toxic as to create any problem". New York Times (1985) is critical of the failure of the management to alert the local people to the dangers and to organise liaison with the local authorities responsible for dealing with emergencies. It may be said that at the time of Bhopal such actions were not required by law in the countries of the EEC though at the time of writing such requirements are coming into force.

Treatment of some of the victims has been complicated by a view taken by some of the local medical profession that high cyanide levels have been found in the blood of some of the victims. In their view, the appropriate treatment is to give sodium thiosulphate which is used in first aid treatment of victims of hydrogen cyanide gas. Dagani (1985) commented on this in February 1985 and, according to a UK television film, this treatment was still being given a year after the accident. Dagani claims that US medical opinion is very sceptical of the diagnosis.

The exact number of dead and injured is not known and may never be known. New York Times reported the death toll as 2,000 with 200,000 injured, but others since have put the fatalities at a higher figure. Slater (1985) has suggested a range of 2,000 to 10,000, Kharbanda (1985) has suggested 2,000 to 5,000. Worthy (1985) estimates that 5,000 were killed and 10,000 permanently injured. Figure 15.12. is a sketch map of the affected area with an indication of the likely extent of the cloud.

15.9.5.2. The process.

There were five principal process areas on the site. Apart from the MIC plant there was a phosgene manufacturing plant, a plant for producing Sevin from MIC, a plant for manufacturing alpha naphthol, which was one of the reactants in the Sevin process, and which had been closed down as it could not produce a product of the necessary specification, and a pesticide formulation plant. None of these four have been implicated in the disaster by any investigation

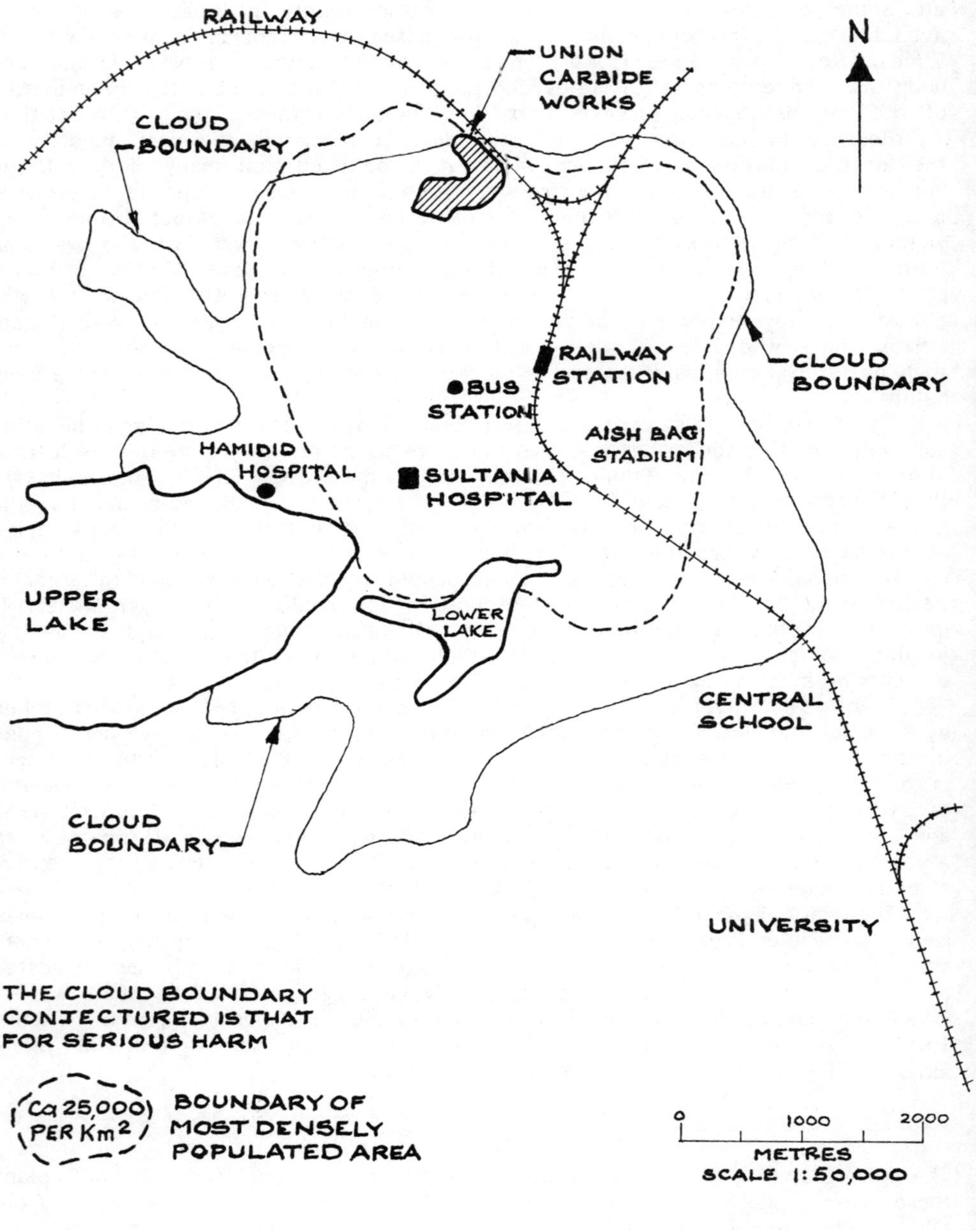

Fig. 15.12. SKETCH MAP OF AFFECTED AREAS AT BHOPAL.

team and they will not be further discussed.

As indicated in section 15.9.2. above, the manufacture of MIC is a two-stage process and the two stages are set out in a block flow diagram, Figure 15.13. According to Union Carbide (1985), monomethyl amine was brought into the factory in tank cars and was reacted with excess phosgene in the vapour phase to produce methylcarbamoyl chloride (MCC) and hydrogen chloride. After quenching with chloroform the product was stripped of excess phosgene which was then recycled. The stripped MCC was then pyrolysed to yield crude MIC which was distilled to separate it from chloroform, MCC and residues. The distilled MIC then passed to one of three stainless steel type 304 storage tanks. It was in one of these, tank 610, that the runaway reaction occurred.

This tank is sketched in Figure 15.14. It had a nominal capacity of 15,000 US galls or ca 57 cu metres. It measured 40 ft (13 metres) in length by 8 ft (2.43 metres) in diameter. It had a design pressure of 40 psig (2.72 bars) and a hydraulic test pressure of 60 psig (4 bars). The tank was mounded in earth and the mound was covered in concrete. It was fitted with a relief valve with a rupture disc on the upstream side and was provided with a pressure gauge intermediate between the rupture disc and the relief valve. Dry, high purity, nitrogen was connected to the tank.

Provision was made for there to be continuous circulation of the tank contents through a heat exchanger in which it was cooled by chloroform which had been chilled in a refrigeration unit. Chloroform was used to prevent the possibility of water getting into the tank from the coolant system.

The relief valve exhausted into a scrubber where vented MIC could be saponified with caustic soda. Residual gas, with any undissolved MIC, could then pass to atmosphere through a 100 ft (30 metre) stack. Alternatively, the treated gas from the scrubber could be diverted into a flare stack of similar height. It appears that there was also a direct route from the vents to the flare stack.

It would thus appear that there were three safeguards against the discharge of MIC to atmosphere. In the first place the storage tanks were refrigerated which would slow down any adverse exothermal reactions and thus give time for remedial action. Secondly, discharges could be treated with caustic soda which would saponify the methyl ester to yield involatile sodium isocyanate and relatively innocuous methanol. Thirdly, discharges could be flared off to yield harmless gaseous products. That none of these provisions were effective will be demonstrated below.

15.9.5.3. The search for the immediate cause.

Following the disaster, the Union Carbide Corporation sent a team, which included seven engineers and scientists, to India. They arrived on December 6th and remained for 24 days. It is apparent that, because the Indian Government was conducting its own investigation, considerable constraints were imposed upon the team who were, among other things, not permitted to interview everyone they would have wished. Their report, Union Carbide (1985), which was presented at a press conference in March 1985, deals very largely with their examination of tank 610 especially their estimate of the quantity of material which remained and its chemical composition. The results of these investigations, together with their hypothesis, based upon these data, as to the cause of the runaway reaction which, it is universally agreed, was responsible for the boiling off of the MIC, are given in the Report. The team additionally had the task of disposing of the remaining stocks of MIC which they accomplished.

The team estimated that somewhere between about 5 tonnes and 10 tonnes of material remained in tank 610. Thus out of 41 tonnes originally in the tank between 30 and 35 tonnes of MIC plus gaseous decomposition products, and possibly some liquids and solids, escaped from it through the relief valve.

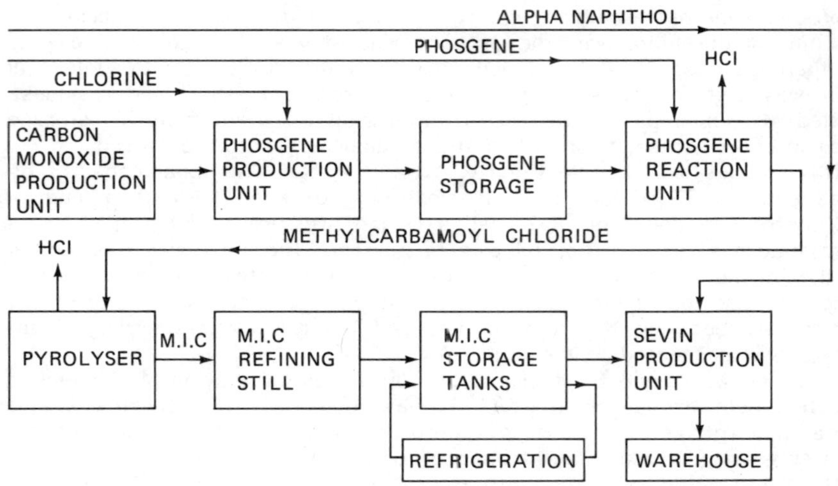

Fig. 15.13. BLOCK FLOW DIAGRAM OF MANUFACTURE OF MIC.

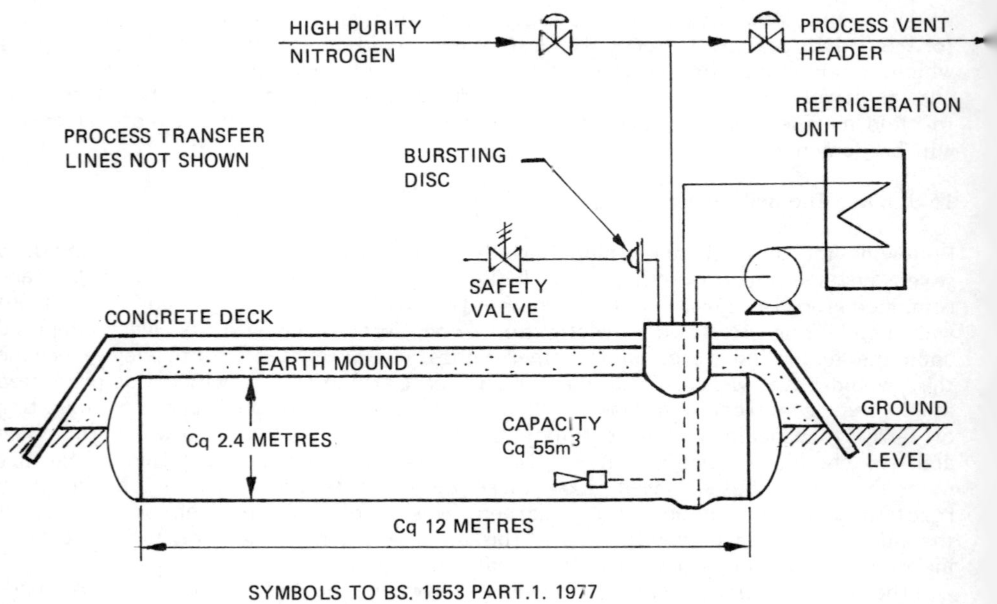

Fig. 15.14. SKETCH OF TANK 610.

Two core samples were taken from the residue which was solid and these were found to contain only minor quantities of either methyl isocyanate or water. There were substantial (ca 5%) quantities of hydrolysable chlorine (i.e. chlorine not bound to carbon atoms, VM) which was mainly in the form of hydrochlorides of amines. There were traces of iron, chromium and nickel whose ratio was roughly that of the stainless steel of which the tank was constructed. No sodium was found.

The main constituent was the trimer of methyl isocyanate which formed about half of the residues. There were also significant quantities of dimethyl isocyanurate which is formed when methyl isocyanate reacts with free isocyanic acid.

Highly significant was the presence of a number of compounds such as methylamine hydrochlorides, dimethyl urea and trimethylbiuret amounting to more than 10%, whose formation can only be accounted for by a reaction between MIC and water. The replication of the reactions in the laboratory led the team to conclude that between 500 kg and 1000 kg of water must have entered the tank. The formation of these compounds in which water is a constituent of the reaction requires quantities of water which can be calculated from considerations of stoichiometry. This is unlike the case of reactions where water is merely a catalyst and may need only be present in trace quantities.

The hypothesis advanced for the event is that a quantity of water got into the tank, either by accident or as the result of sabotage, and it began to react with the MIC to give, initially, monomethyl amine and CO_2 which latter would pressurise the tank. The heat evolved promoted corrosion of the stainless steel by chloroform which was a constituent of the MIC because of previous misoperation of the distillation column. The team concluded that there may have been between 750 and 1500 kg of chloroform present. The corrosion of the iron then catalysed the exothermal polymerisation of the MIC and the runaway reaction then ensued. The heats of reaction, according to Worthy (1985), are between 540 and 585 BTUs/lb (1250 to 1360 kJ/kg) of MIC hydrolysed or polymerised.

15.9.5.4. Why did the precautions fail ?

That the use of refrigeration had been discontinued six months previously was admitted in Union Carbide (1985a) by R.Van Mynen, the chairman of the investigation team which went to India, but he offered no explanation for why this had been done. However, an Indian Trade Union Research Group in URG (1985) has alleged that it had been closed down simply to reduce the running costs of the factory. This closure, however occasioned, had taken place against a background of the stated view of the parent company, as expressed in their safety documents, that refrigeration was essential and that MIC should be maintained in storage at $0^{\circ}C$. Without refrigeration the temperature of MIC in storage would tend to equilibriate to the mean diurnal temperature of Bhopal which can reach ca $30^{\circ}C$ in July. It has been stated in New York Times (1985) that the high temperature alarm which was provided to monitor the effectiveness of the refrigeration had simply been taken out of commission when refrigeration was abandoned and that during the previous summer the temperature of the tank contents had sometimes gone off scale, i.e. $\geqslant 25^{\circ}C$. Thus the prime system of protection was inoperative.

There has been controversy as to whether the scrubber was operative. It appears that, when tested the following day, the pump circulated liquor quite normally and it is claimed that the flow meter which showed no flow when the operator started the pump was merely blocked. The wall of the reservoir under the scrubber was found to be hot the following morning indicating that absorption had taken place. However, no record exists of the caustic soda content either before or after the disaster. But serious doubts exist as to whether the scrubber was adequately sized to cope with, in very rough terms, 15 tonnes of MIC per hour. It maybe that it was only intended to cope with such venting of MIC as accompanied routine operation of the plant. The release rate above is equivalent to 4.0 kg per second. At atmospheric pressure and, say, 50°C, this is 1.85 cu metres per second. The diameter of the scrubber was quoted at 5ft 6ins (1.7 metres) in Union Carbide (1985) and its cross-sectional area was thus 2.26 sq metres. This would give an empty tower velocity of 0.8 metres per second which, though high, does not appear inordinately so, but this could depend on the size of the Berl saddles with which it was packed. However, what may be more germane are the mass transfer coefficients in a system which involves chemical reaction and where the reactions are relatively slow saponification reactions. However, there can be no doubt that the scrubber failed to operate efficiently, whether this was because caustic was not circulated or because the scrubber was inadequately sized to cope with emergency conditions. This latter can only be checked with full knowledge of the mass transfer coefficients and the design of the scrubber system. It may be that the facility for discharging the effluent gas from the scrubber outlet to the flare stack was a recognition that the scrubber on its own could not cope with emergency loads.

The final safeguard was the flare stack which was out of commission and, according to some reports, at the crucial time there was a length of pipe missing from its inlet.

15.9.5.5. How did water get into the tank ?

Though everyone agrees that water got into the tank there has been controversy as to how it got there. Union Carbide (1985) concluded that there are only two possibilities, either because water used for washing a header had leaked past a number of allegedly closed valves, which they considered improbable, or because it had been introduced as a deliberate act of sabotage. Slater (1985) has provided a scenario in which he claims that an attempt by the operators to transfer material from tank 610, at a time when there was about 1 tonne of water in a header, opened a vent valve and allowed water to get in. An argument against the water getting in through passing valves lies in the fact that after the relief valve had ceased to blow and the tank cooled down, a partial vacuum developed on the tank.

15.9.5.6. The state of labour relations at Bhopal.

Union Research Group (1985) was severely critical of the state of labour relations at the factory and attributed the main blame for the disaster on the cuts which had been made in manning levels. The process manning levels on the MIC plant had been halved and there had been equally serious cuts in the maintenance crews. Morale was very low and this set up a vicious circle whereby the most experienced people had left thus making for heavier burdens on those remaining. The report alleged that there was no time to take safety precautions and that this had applied particularly to the abandonment of the practice of plating off stop valves. The report paints a picture of the factory being run on a shoe string. Generally speaking, what the report suggests is that a factory where a great deal of reliance was placed upon human intervention, the manning levels were those more appropriate to a fully automated process. Confirmation for the point regarding the turnover of experienced men seems to be provided in Worthy (1985) which claims that this was commented on in the report of a safety audit conducted by the parent company in 1982.

15.9.5.7. The question of inventory.

In the present author's view, the principal cause of a disaster on the scale which occurred at Bhopal lay in the original design concept which provided for the storage of up to 120 tonnes of what it may be claimed is one of the most highly toxic chemicals handled on an industrial scale. (The EEC inventory for MIC which triggers off the special controls appropriate to a major accident hazard installation, and which was formulated before Bhopal, was 1 tonne). But was this large inventory essential ? The present author would cast doubt upon this. It seems clear from Browning (1985) that the Union Carbide Corporation did not attach importance to the level of inventory as it did not consider that anything of the sort which happened at Bhopal could occur. There seems no evidence that Union Carbide ever stressed the importance of maintaining a minimal inventory, indeed the provision of tanks with a total volume of 171 cu metres was an encouragement to keep large stocks. It would appear from Worthy (1985) that the original process for making Sevin did not use MIC. The same reference speaks of a process developed by the Bayerwerk in which the MIC is used as fast as it is made and of a process developed by LaPorte which has a maximum inventory of 10 kg of MIC.

15.9.6. GENERAL CONCLUSIONS.

Methyl isocyanate is one of the most highly toxic substances in common use yet there were few data on its toxicity prior to Bhopal. However such data as existed indicated that MIC was more toxic than chlorine, phosgene or hydrogen cyanide. The Bhopal disaster was the worst industrial disaster in world history because of a design which tolerated a very large inventory of this highly lethal chemical. Moreover the factory in which it was used had a management which seemed unaware of the crucial necessity of maintaining certain safeguards which had been built into the process. The economic problems of the factory encouraged the abandonment of these safeguards.

The scale of the disaster was undoubtedly contributed to by the time of day at which it occurred, by overcrowding, by totally inadequate housing, by a shortage of medical facilities and by a failure to alert the medical authorities to the problems they would have to face in the event of a serious leak of MIC. The death toll at Bhopal will be further discussed in Chapter 18 "The Quantification of Major Chemical Hazards".

Chapter 16

OTHER MAJOR CHEMICAL HAZARDS

16.1. INTRODUCTION.

In paragraph 4.6.4.1. there was a discussion on the general effect on the human body of the effects of changes in its environment. The effects of thermal radiation, of blast waves which raise the ambient pressure, and which may produce localised pressures as a secondary effect, and of excessive concentrations of toxic substances, have been given extended treatment in the chapters dealing with fires, explosions and toxic releases. It remains to discuss other environmental changes to see whether they present hazards which, if realised in an acute situation, may give rise to large-scale fatality or injury. Certain environmental changes will not be discussed; ionising radiation and electric potential lie outside the scope of this book, and noise, though it may give rise to chronic problems of ill health, does not constitute an acute hazard.

This leaves the following subjects for discussion:- (1) high temperature effects other than those arising from thermal radiation or flame impingement, (2) cryogenic hazards, (3) hazards arising from low or high oxygen concentrations, (4) corrosive hazards, (5) hazards from pathogenic organisms used in chemical processes, and (6) the possibility that there may arise, in the future, chemical hazards as yet unanticipated.

It will be seen that the hazards discussed lie at the margins of the subject. However, the author considers it important to discuss them in order that the treatment of major chemical hazards given in this book shall be as comprehensive as possible. He also believes that they should be discussed against the background that the major hazards described in the book have nearly all come into existence during the recent past and that, correspondingly, new hazards may arise in the immediate future.

16.2. INJURIES FROM HOT LIQUIDS.

The treatment of death and injury from high temperatures in Chapter 8 was concerned with the effects of fires which injure through heat radiation and through flame impingement. But there exist dangers from hot liquids and their vapours which can give rise to scalds. From the medical point of view there is little difference between burns, which may be defined as being caused by dry heat in excess of 60°C, and scalds, which may be defined as being caused by moist heat in excess of 50°C. The medical treatment is the same.

There can be little doubt that many have been killed or injured by scalds arising from the rupture of steam boilers, steam pipes and steam heating appliances. Though the author has excluded steam boilers etc from the scope of this book, in principle there is no difference between injuries from steam and hot water and the injuries produced by other hot liquids and their vapours as, for example, petroleum products undergoing distillation.

However, the injuries produced have usually occurred within a short range of the point of leakage and have usually affected only a small number of people at a time. Sometimes, where the issuing liquids have later caught fire, the effects of the resulting conflagration have obscured the injuries produced by the initial event.

However, for the category of really high temperature liquids, especially those hot enough to cause charring of flesh, the distinction between a burn and a scald has no meaning. Molten metals fall into this category.

16.3. THE SCUNTHORPE DISASTER.

16.3.1. GENERAL DESCRIPTION.

On November 4th 1975, at around 02.47 hrs, at a steelworks in Scunthorpe, Humberside, England, there was an explosion in a rail-mounted ladle (a "torpedo"), which showered nearby personnel with molten iron. Of the 23 persons in the vicinity, 4 were killed immediately, 7 died in hospital and 8 others required hospital treatment for the burns they sustained. This explosion is described in Health & Safety Executive (1976) which is the report of the Factory Inspectorate into the incident.

16.3.2. DETAILED DESCRIPTION.

Shortly before the explosion, at around 02.15 hrs, a fire had broken out due to a burst at one of the blow pipes at the base of a blast furnace. Such a fire was not an unusual event and was only an indirect cause of the explosion. But it did influence the explosion because it concealed for a time a water leak which followed the blowing out of a steel plug from the system used to water-cool the tuyeres. Not only did it conceal the leak for a time but the flames prevented anyone getting near to it to stop it. As a consequence water flowed over the cast house floor and down the channels provided for the molten iron. It then ran into the torpedo which contained about 170 tonnes of molten iron at about 1450°C. It has been calculated that the quantity of water which entered was about 2 tonnes. There was no apparent immediate reaction, but this was in accord with previous experience which suggested that water flowing onto molten iron did not create a serious hazard.

The blast furnace shift manager decided that it was necessary to pour the contents of the torpedo as soon as possible as he considered that the cooling effect of the water could lead to serious operating difficulties by causing solidification at the top outlet. He accordingly gave orders to move it and it was during this operation, after a locomotive had been coupled up and had moved off, that the explosion occurred showering the surroundings with ca 85 tonnes of molten iron.

The opinion of a team of technical investigators was that the most probable cause was that the shunting operations caused the contents of the torpedo, which was a horizontal cylinder about 12 metres long by about 2.25 metres diameter (both internal) with a top opening for pouring, to surge backwards and forwards. This then trapped water at one end and resulted in a violent steam explosion which ejected about half of the molten iron. There was some evidence of consequent external blast equivalent to 0.5 kg of high explosive.

Subsequent to the explosion the stump of a screwed plug made of free-cutting steel was found near the furnace. It was largely corroded away and though it could not be definitely identified as the actual steel plug which had been fitted in the copper pipe on the tuyeres' cooling system and which was missing, its condition was a clear indication that steel plugs could become badly corroded in such service.

16.3.3. THE POSITIONS OF THE DEAD AND INJURED.

Appendix 11 of Health & Safety Executive (1976) gives a plan of the immediate vicinity of the blast furnace and on it are marked the positions of 20 of the 23 men stated to be present at the time. It marks the position of only 4 fatally injured persons and so, presumably, it does not mark the positions of those injured who later died in hospital. All those marked were within 45 metres of the explosion and the two most distant of those marked as fatally injured were 40 metres away.

16.3.4. FINDINGS.

Those findings which were of relevance to the chemical and process industries seem to be:-
(1) Though it was entirely reasonable to expect that the cooling system at some time would leak water, no attempt was made in designing the casting floor to ensure that such leaks could not contact molten iron.
(2) There were no emergency plans worked out in advance to deal with the situation which occurred or any similar situation.
(3) Though there were undoubted hazards attached to running off iron, at least four persons were present whose presence was not essential to the operation.
(4) Some of those present were not provided with the appropriate protective clothing.

16.4. LOW TEMPERATURE HAZARDS.

16.4.1. CRYOGENIC LIQUIDS.

There is some controversy over what constitutes a cryogenic liquid. Perry (1973) implies that it is a liquid with a boiling point -200°F (-129°C). For convenience in this discussion it will be taken to include liquid ethylene and liquid ethane as well as oxygen, nitrogen and methane which latter are handled as liquids on a large scale. All these liquids, except for ethane, boil at temperatures which are lower than those found naturally on earth; the boiling point of ethane is slightly higher than the minimum temperature recorded at the "pole of cold" in the Antarctic.

The injuries caused by contact with such liquids may be described as "cold burns". There is little in the literature written for engineers on the subject of cold burns. These are sometimes equated with "frostbite" of which the symptoms and treatment are described in an article with that heading in International Labour Organisation (1983). The present author would draw a distinction between cold burns which arise from being splashed by cryogenic liquids, and frostbite which arises through relatively prolonged exposure to very cold air. However the distinction between the two may not be sharp, especially if the frostbite arises from exposure to a very cold, high velocity air stream.

These liquids may all do severe harm if spilled. Wading through even shallow pools, unless in special footwear, may in a short time do severe harm to the feet and if this led to the person falling into the pool, this could quickly cause death. Inhaling the cold vapours above the pool may seriously damage the lungs and respiratory passages. Cold burns may occur and if much of the body is affected, death may ensue. Localised severe cold burns may lead to gangrene if skilled medical attention cannot be brought to bear. The effects of cold burns, and the treatment recommended for them, are discussed in British Cryogenics Council (1970) and in Du Pont (1969). Unlike the case with severe heat burns, severe cold burns do not destroy the nerves and thus the rewarming of affected tissue is likely to be accompanied by severe pain and require the administration of morphia or other pain killers.

There are general recommendations for safety precautions and for protective clothing in British Cryogenics Council (1970).

Spilled cryogenic fluids may cause death or injury through causes other than cold as, for example, through asphyxiation or through the subsequent ignition of the vapour cloud. Death by asphyxia will be discussed below, the effects of ignition have already been discussed. It may be that, in some cases, these other effects may have been recorded as the cause of death when in reality extreme cold may have been the cause. For example, in Cleveland 1944 which is discussed as a case history above, the sudden spillage of over 1000 tonnes of LNG, may have killed by cold, before the fire reached them, a number who were engulfed by the spillage.

There appear to be no case histories of multiple deaths arising from the spillage of cryogenic fluids causing fatal cold burns on their own, uncomplicated by, for example, fire. However it cannot be concluded that such incidents may not occur in the future.

16.4.2. COLD INJURIES FROM THE SPILLAGE OF LIQUEFIED GASES.

As has been argued in detail in earlier chapters, liquefied gases possess considerable dispersive energy and their effects may thus have a much wider range than is likely to occur with the spillage of cryogenic liquids. It is therefore more feasible that multiple injuries, or even deaths, may occur, from causes other than fire, explosion or toxicity, in the event of loss of containment of liquefied gases. Reference has been made in the case history of San Carlos to the possibility that some people may have been seriously injured or even killed by the arctic blast of cold vapour and liquid droplets arising from the rupture of a propylene tanker.

A group of refrigerants have been developed in which the hazards of flammability and toxicity are low, or non-existent. These, of their nature, fall into the category of flashing liquids and some have atmospheric pressure boiling points well below 0°C.

Examples of these refrigerants are the Freons ("Freon" is a trade name of E.I. Du Pont). Safety aspects of handling Freons are given in Du Pont (1969). Generally speaking these are non-flammable (some can extinguish fires) and non-toxic though some can give rise to anaesthesia, Freon 12 is an example. Some of the lower boiling members of this class may present hazards of cold burns or frostbite uncomplicated by flammable, corrosive or toxic effects.

Spillages of such liquefied gases as chlorine and ammonia may produce cold burns but the corrosive and toxic effects of these substances will much outweigh the effects due to cold.

16.5. ASPHYXIA.

16.5.1. THE NATURE OF ASPHYXIA.

Asphyxia is caused either by having the supply of air to the lungs cut off as, for example, in drowning or by being buried in particulate matter, or by breathing an atmosphere which is deficient in oxygen. The vast majority of cases of asphyxia occur in confined spaces.

16.5.2. FATALITIES FROM DROWNING.

To deal with the example of drowning, this can occur as the result of the bursting of storage tanks especially if these are not bunded. Water is sometimes stored in tanks on chemical or process plants, especially treated water such as boiler feed water or desalinated water. If the view be taken that water is non-hazardous and of relatively low value, there may be a temptation to omit bunding with consequent dangers of drowning. Failure of such store tanks by initial sucking in through pumping from an inadequately vented tank is not uncommon. In any case, many liquids are of lower density than water and persons engulfed cannot swim in them even if they are otherwise relatively innocuous.

A somewhat bizarre example of a drowning accident, leading to multiple fatalities, is that which arose from a tank containing 9,000 tonnes of molasses. This burst in Boston USA on January 15th, 1919. The rupture killed 21 and injured 40. Many horses were also drowned. Considerable property damage occurred and the elevated railway partly collapsed because of hydraulic forces acting on supporting girders. The cause was eventually attributed to inadequate design. It is described in ENR (1919) and ENR (1920).

16.5.3. ASPHYXIA FROM PARTICULATE MATTER.

Many have died in the chemical and process industries through becoming buried in powdered material, though this has rarely taken the form of a multiple accident.

An extremely serious accident occurred at Aberfan, South Wales on October 21st, 1966, in which a heap 250 metres high and containing millions of tonnes of coal waste, slid down a hill side and killed 147 people, 116 of them children, and mostly from asphyxia. The slide engulfed two schools and a row of houses. It was initiated by an accumulation of water at the base of the heap. It is described in Aberfan (1967).

The Aberfan disaster was not a process industry disaster in the terms of this book, and the scale of it was far greater than could be anticipated to occur in these industries; it does, nevertheless, provide a warning that heaps of powdered material may shift, especially if not on level ground, nor in adequately designed bunkers, and with possible serious consequences.

16.5.4. ASPHYXIATION THROUGH OXYGEN DEFICIENCY.

Asphyxiation through oxygen deficiency occurs in circumstances either in which air has been depleted of oxygen ("vitiated air"), as for example by a redox reaction which produces a solid oxidation product or relatively non-toxic carbon dioxide, or in which air has been displaced in whole or in part by other gases. The principal effect of the latter is the failure to support life rather than inherent toxicity. Both forms of asphyxiation are far more likely to cause fatalities in confined spaces than in the open air.

The problem of vitiated air has been noted in old mine workings where the oxygen has been consumed by slow oxidation processes and has been replaced by carbon dioxide. In one case known to the author which occurred in Lancashire, England, in 1976, there were multiple fatalities at the entrance to an abandoned drift mine. These arose as a consequence of a sudden fall in barometric pressure which drew vitiated air into the entrance tunnel. Similar problems could arise in sewerage or drainage systems, especially in the aftermath of a sudden fall in barometric pressure.

Table 16.1., which is taken from Du Pont (1969), illustrates the effects of breathing air containing diminishing concentrations of oxygen. The effects are for persons at rest, they may be more severe where exertion is present.

Asphyxiation as the result of air being displaced wholly or partially by non-respirable gases has long been recognised as a problem associated with work in confined spaces. Deaths from carbon dioxide have long been noted. This gas is of course appreciably heavier than air and can form low level layers.

TABLE 16.1. SYMPTOMS OF OXYGEN DEFICIENCY.

Oxygen % by Vol.	At rest, signs and symptoms of oxygen deficiency.
12 - 14	Respirations deeper, pulse up, coordination poor.
10 - 12	Cheyne-Stokes respiration, giddiness, poor judgement, lips blue.
8 - 10	Nausea, vomiting, unconsciousness, ashen face.
6 - 8	8 min, 100% of animals die; 6 min, 50% die and 50% recover with treatment; 4-5 min, all recover with treatment.
4	Coma in 40 seconds, convulsions, respiration ceases, death.

16.6. THE EFFECTS OF EXCESSIVE OXYGEN CONCENTRATION.

16.6.1. THE INDUSTRIAL SIGNIFICANCE OF OXYGEN.

The production of liquefied oxygen (LOX) is an important branch of the process industries. The current production of tonnage oxygen in the US is of the order of 13 million tonnes per annum. Of the US production, 70% goes into steelwork and a further 10% into other metal-working operations. In 1982, oxygen production in the US ranked 4th after sulphuric acid, nitrogen and ammonia [see Malpas et al (1984)]. Much of this is produced in "captive" or "on-site" plants, i.e. plants whose output is dedicated to the requirements of a nearby installation, typically a steelworks. Some steelworks consume well in excess of 1,000 tonnes per day of liquid oxygen. Such captive plants may also distribute oxygen as a gas by pipeline. One captive installation in the UK has a network of pipelines exceeding 20 km. Liquid oxygen is also transported by road or rail in specially designed tank cars and there are many storages at users' premises. Booth (1973), which describes the structure of the industry, gives evaporation losses from such storages as 0.1 - 0.2% per day. In addition, innumerable cylinders are employed in providing small-scale users with gaseous oxygen.

16.6.2. TOXIC EFFECTS.

Though, in high concentrations, oxygen may be regarded as toxic, this problem lies outside the scope of this book. This is because oxygen toxicity can only occur either through short exposure at pressures of 3 bars or more, or through chronic exposure at lower pressures.

16.6.3. ENHANCED COMBUSTION.

The major problem with high oxygen concentration lies in enhanced combustion. This is a problem which received world attention following the deaths of three American astronauts in a space capsule on the ground on January 27th, 1967.

Increased oxygen concentration has a number of effects:-

(1) It increases the ease of initiating combustion. Threshold spark energies are lowered; the sparks released on earthing a human body after friction with carpets etc may be sufficient. An apparently extinguished cigarette or pipe may reignite and burn fiercely.

(2) Rates of burning are increased. According to Brown (1978) they increase roughly in the ratio of the oxygen concentration, i.e. at 1 bar the rates are five times higher than at ca 0.21 bar, the normal concentration. The effects of this increased rate will manifest themselves in extremely rapid spread of fire in clothing, denim overalls may be consumed in 20 seconds resulting, according to Brown, in a total body burn. Body hair will also burn very rapidly.

It could be, though the present author has not found supporting evidence for it in the literature, that under conditions of high oxygen concentration, flesh may be capable of self-sustaining combustion in spite of its high water content, i.e. behave like wood or paper.

The effects described above are enhanced still further if the oxygen is above atmospheric pressure and/or temperature. These conditions can only occur in confined spaces and are not applicable to open air releases.

(3) Materials which are considered, in normal circumstances, to be of low flammability may burn readily in an oxygen-rich atmosphere. It is well known, for example, that oils and greases must not be used in piped oxygen systems, but there are many other substances, some of which are listed in British Cryogenics Council (1970), which may be handled relatively safely in air but which are dangerous in oxygen-rich atmospheres. Brown notes that conventional "flame proof" fabrics fall into this category.

(4) The effect of higher oxygen concentration on gas appliances will be, on account of the higher burning velocity, for flames to tend to "strike back", i.e. to burn at the inlet jet. There is no reason to assume that flames will be hotter. Even in a 100% oxygen atmosphere, the appliance will admit excess oxygen as it formerly admitted nitrogen. This will cool the flame just as nitrogen did.

(5) Fire extinguishing becomes more difficult as smothering is not effective. Fire extinction must be prompt and only water sprays which spray the whole body are likely to be effective. Brown suggests that a delay of longer than 20 seconds may lead to fatality.

Certain fire-extinguishing agents may give off toxic fumes in oxygen fires. The author knows of an incident in the past in which carbon tetrachloride was used to attempt to extinguish an oxygen fire. This resulted in a worker having to be hospitalised.

The Health & Safety Executive has published a guidance note on fires and explosions due to the misuse of oxygen,[Health & Safety Executive (1985c)].

16.6.4. HOW FAR CAN ENHANCED OXYGEN LEVELS REACH ?

Dilution of oxygen-rich clouds with air is mixing them with a gas which, self-evidently, also contains oxygen. Thus a 50/50 mixture of oxygen and air contains more than 50% oxygen. The formula for computing the concentration after mixing air with pure oxygen is:-

O = $1 + 0.21\ R\ /\ R + 1$ where
O^C = oxygen fraction in the mixture
R^C = ratio, air : oxygen.

If the maximum concentration considered to be safe be taken as 24% then R, from the equation, = 25. This compares with R = 20 to dilute methane to its lower flammable limit, R = 25 to dilute hydrogen to its lower flammable limit, and R = ca 40 to dilute acetylene and propylene to their lower flammable limits. On this basis, the distance of travel of oxygen before mixing with air will render it harmless, is comparable with that of the mixing distances for the dilution of flammable gases with air.

16.6.5. ARE THERE CIRCUMSTANCES IN WHICH OXYGEN MAY BE DANGEROUS IN THE OPEN AIR ?

There appear to be no case histories of serious incidents following the spillage of liquid oxygen or the escape of gaseous oxygen from a pipeline. But, based upon the above considerations, fatalities cannot be ruled out from oxygen spillages. The CIMAH Regulations now being enforced in the UK provide that an inventory in storage of 2000 tonnes or more of liquid oxygen may render it subject to major hazards controls; (pipelines are not subject to the CIMAH Regulations).

16.6.6. LIQUID OXYGEN HAZARDS PRODUCED BY LIQUID NITROGEN AND LIQUID AIR.

The respective boiling points of liquid oxygen and liquid nitrogen are -183°C and -196°C. One of the consequences of this is that if liquid nitrogen is in contact with air, mass transfer takes place. Thus oxygen liquefies and nitrogen vapourises. So, what starts as liquid nitrogen may end up as being largely liquid oxygen. Air contacting a surface on the other side of which is liquid nitrogen preferentially condenses out liquid oxygen. As liquid air has a boiling point much closer to liquid nitrogen than it has to liquid oxygen, similar effects can take place with liquid air.

Many organic substances, if they soak up liquid oxygen, can become highly explosive. Indeed explosives manufactured by soaking charcoal, for example, with liquid oxygen, were at one time in common use. However such explosives have been found to be unstable and to explode without having to be detonated. But they are so cheap to make that, in spite of the dangers, they are still being used in some parts of the world, [see Fordham (1980)]. It is possible, therefore, for the mass transfer effects described above to lead to explosions in the chemical and process industries; one is described as a case history below.

16.6.7. TWO CASE HISTORIES.

A serious accident involving oxygen-enriched air occurred on 23/9/76 at Wallsend, on the River Tyne, England, where a guided missile cruiser, HMS Glasgow, was being fitted out. Shortly after the commencement of work in the morning, a fire of exceptional severity broke out below decks and 8 men lost their lives. A subsequent Inquiry established the cause as leaks on oxygen lines which caused an over-night build-up of a high concentration of oxygen, [see Health & Safety Executive (1977)].

There was an explosion in a UK factory for processing potato crisps in 1979 which killed two men and injured another. The incident is reported in Health & Safety Executive (1981), though the reference does not give a precise date for the occurrence. The explosion occurred after a quantity of roast pork had been soaked in liquid nitrogen to embrittle it in preparation for grinding it to be used as a coating for the crisps. It was being handled mechanically when an explosion occurred. An inquiry established the cause as mass transfer between the liquid nitrogen and air which had condensed out oxygen.

16.7. CORROSIVE SUBSTANCES AS MAJOR CHEMICAL HAZARDS.

16.7.1. THE NATURE OF CORROSIVE SUBSTANCES.

The substances under discussion may also be described as "toxic" as, if ingested, they are undoubtedly able to cause death and, in some cases, they give off fumes which, if inhaled, may also be lethal. However, the effect under discussion is the ability of corrosive substances to cause chemical burns to the skin which give rise to serious injury or to death. Corrosive substances are particularly dangerous to the eyes.

They may be classified as acids, alkalis or other corrosives.

16.7.2. THE LITERATURE OF CORROSIVE HAZARDS.

Unlike some of the toxic gases discussed above, because corrosive substances have so long been handled in laboratories, and there have been so many small-scale accidents involving them, there is a considerable literature on their effects and on appropriate treatment. The main difference between laboratory accidents and major hazard accidents is one of scale both of the size of possible releases and the numbers affected. Medical treatment is likely to be the same for both.

It is possible to cite here only a few major works which deal with the effects of corrosives; other works not cited may be equally important. Standard works include Sax (1984), Steere (1977), NIOSH/OSHA (1978) and Bretherick (1981). Brown (1978) discusses medical treatment and the pathology of corrosive injuries; Trevethick (1973) also discusses medical treatment of corrosive injuries.

16.7.3. SOME INDIVIDUAL CORROSIVES.

16.7.3.1. Mineral acids.

The acids, nitric, sulphuric and hydrochloric, when in concentrated form, are all highly dangerous corrosives. Nitric and hydrochloric acids give off toxic fumes. Sulphuric acid, in the form of oleum which is essentially a solution of sulphur trioxide in sulphuric acid, also gives off toxic fumes. Sulphuric acid and nitric acid are extensively used as a mixture known as "nitrating acid". The burns which arise from these acids, especially sulphuric and nitric acids, have a considerable thermal component as their reaction with flesh is highly exothermic. This is one of the reasons for treating such burns by copious irrigation with water.

These acids are of great industrial significance and hence each are manufactured in the UK on a scale of millions of tonnes per annum. The quantities stored are correspondingly large. In consequence multiple accidents may occur by loss of containment. The acids are all very reactive, sulphuric acid and hydrochloric acid generate hydrogen in contact with metals and on this account may present an explosive hazard which could lead to dispersion in the form of droplets. Nitric acid is a powerful oxidising agent which can give rise to fires and explosions. A handbook of reactive hazards, Bretherick (1979) devotes several pages to the hazards posed by the reactions of concentrated nitric acid.

16.7.3.2. Hydrogen fluoride.

This substance, either as pure hydrogen fluoride (BPt 19.5^{o}C), or as hydrofluoric acid (40%, 42% and 48% with water) is particularly harmful to the flesh. Special treatment is required and this is described, for example, in Bretherick (1981). The Health and Safety Executive have produced a wall poster (F2250) on the treatment of HF burns and information is also available from the UK Chemical Industries Association, [Chemical Industries Association (1978)]. Even dilute solutions are dangerous, one reason being that with such solutions there may be no immediate effect but the delayed effects may be serious.

Any company manufacturing or handling hydrogen fluoride in any of its forms is strongly advised to have full liaison with the local medical authorities; fore-warning is essential to ensure that the proper treatment is available. This requires calcium gluconate gel which is a specialist pharmaceutical which is available from only one source in the United Kingdom.

Failure to provide prompt and adequate medical attention may lead to gangrene and amputation of a limb.

16.7.3.3. Organic acids.

The two fatty acids most likely to cause serious injury by corrosive effects are formic acid and acetic acid. The former is found in the sting of the ant, from which it derives its name, and in nettle stings. It can cause severe blisters. Acetic acid is less dangerous but extensive splashing, unless promptly treated, may lead to painful burns. Both formic and acetic acid are manufactured and stored in sufficient quantity to be capable of giving rise to multiple fatalities.

Monochloracetic acid and acetic anhydride are also dangerous corrosives.

16.7.3.4. Corrosive and caustic substances.

Into this category falls sodium hydroxide, which is of great commercial importance, and potassium hydroxide, which is much less significant commercially. Both have similar properties. They are handled as concentrated solutions (e.g. 40% caustic soda), or as solids (e.g. flake caustic). The solutions, on loss of containment, are more likely to give rise to multiple injuries than the solids. Injuries are due to chemical reaction with tissue which includes saponification of body fat. The reaction is exothermic and will, therefore, also give rise to thermal burns. As with acid burns, washing with copious quantities of water is imperative. It is advisable to wash all parts of the body, including the hair, as soon as possible after being splashed with these alkalis. The eyes need to be given protracted washing if it is suspected that any caustic liquor has entered.

In the Seveso incident, discussed in section 15.8. above, 477 people were injured by burns from the highly caustic contents of the reactor.

16.7.3.5. Ammonia.

The information in the literature is mainly concerned, so far as splashing is concerned, with the effects of ammonia as a 35% solution in water (SG = 0.88). Even as an aqueous solution ammonia causes chemical burns. When the liquid arises from loss of containment of ammonia as a liquefied gas, cold burns rather than thermal burns are likely to ensue. It seems doubtful whether, in such circumstances, chemical burns would be the most serious problem as the inhaled vapours could cause greater injury.

16.7.3.6. Phenol (carbolic acid).

Phenol is an important organic chemical with a production rate in the US of circa 1.5 million tonnes per annum. Though a solid at ordinary temperatures, its low melting point ($43^{o}C$) enables it to be shipped and processed in the form of a liquid.

In spite of its being known as "carbolic acid", it is difficult to classify phenol with the mineral acids, even though it reacts with alkalis to give phenates.

Its behaviour towards tissue is very different. It is a very dangerous substance on the skin. It is rapidly absorbed and can attack the kidneys. Deaths have resulted from a hand being covered with phenol. One problem is that it acts, initially, as a local anaesthetic which delays the onset of pain. This may deceive the victim into a false sense of security.

In the Seveso incident the caustic contents of the reaction which caused burns to 477 people, also contained trichlorophenol which may have contributed to the injuries sustained.

16.7.3.7. Phosphorus.

Multiple injuries from phosphorus have occurred through the incorporating of the white (otherwise known as yellow) allotrope of phosphorus into incendiary weapons such as napalm. This is because white phosphorus will ignite spontaneously in air. Though a solid, it has a low melting point ($44^{o}C$) and it is usually shipped in liquid form. Unburned white phosphorus, lodged in burn injuries may severely aggravate the effect of burns, [Brown (1978)].

A phosphorus railcar exploded at Brownson, Nebraska, US, on February 4th, 1978. The incident is described in Hymes (1985). There was a train crash in which 31 cars of a mixed freight train were derailed. The car in question overturned and was canted at 15° to the horizontal. Three other cars were lain on top of it. A leak of liquid phosphorus took place and the flow was stemmed by the fire service building a dyke around it. The phosphorus ignited producing dense white, moderately toxic, fumes of P_2O_5. After 7 hours the tanker exploded scattering burning blobs of phosphorus over about 0.15 sq kilometres. Several persons were burned by it some sustaining third degree burns. Others received mechanical injuries from the explosion. Hymes sets out several theories as to why the car exploded but states that the favoured theory was that water trapped above the phosphorus (the substance was being transported under a layer of water) heated up to ca $250^{o}C$, at which its pressure would be ca 38 bars, sufficient to burst the tank.

16.8. MAJOR ENVIRONMENTAL HAZARDS.

16.8.1. ENVIRONMENTAL RATHER THAN HUMAN HAZARDS.

The question as to whether there are major hazards to the environment which are not at the same time hazards to human life, is an open one. Perhaps there are some pesticides which are so selective that massive releases of them are capable of injuring the environment without damaging humans. Hay (1982) appears to consider the herbicide 2,4,5-T to be harmless to humans. This is because he seems to be concerned to show that the harm that was done to the people of South Vietnam by the dropping of this herbicide as a defoliant was entirely caused by the presence in it of dioxin as an impurity. 2,4,5-T is 2,4,5-trichloro-phenoxyacetic acid; Hay quotes an LD_{50} for it of 300 mg/kg for oral administration in rats. NIOSH (1978) gives 7 values for the LD_{50} for this by oral administration, they vary from 100 to 500 and their mean is 354 mg/kg. Comparing this with dioxin, where the values quoted by Hay are generally of the order of around 0.1 mg/kg, suggests that 2,4,5-T in an acute event would do far more harm to the environment than to the human population. 2,4,5-T is not listed in Schedule III of the European Directive on major accident hazards, [European Community Directive (1982)].

A possible route for a pesticide to do harm to the environment rather than to inflict direct injury to humans, would be for it to be discharged into a watercourse. A case history of such an event is given below.

16.8.2. CASE HISTORY OF A DISCHARGE OF PARAQUAT.

On Saturday 13th February, 1982, a fire occurred at Morley, Leeds, England, in a warehouse storing paraquat and diquat-based herbicides contained in drums, plastic bottles and cardboard cartons, and also octylphenol (MPt 40°C) in paper sacks. Though the fire brigade responded promptly, an intense fire developed which consumed the octylphenol and much of the wooden pallet storage system. The fire resulted in a quarter of the herbicide stock being liberated, which drained, together with fire fighting water, into a water course in close proximity to housing. Widespread environmental pollution resulted from the incident when the River Calder became contaminated. The herbicide-saturated warehouse foundations continued as a source of contamination during rain storms long after the main consequences of the incident had been realised. No one was injured. [See Advisory Committee on Major Hazards (1984)].

It maybe noted that neither paraquat nor diquat are listed in Schedule III of the EEC Directive on major accident hazards, [European Community Directive (1982)].

16.9. POSSIBLE PATHOGENS FROM CHEMICAL PROCESSING.

16.9.1. GENERAL.

There is not, at the present time, any apprehension that the use of what is now termed "biotechnology" could lead, in the immediate future, to acute hazards from pathogens released from chemical processes. There are many fermentation processes in use; though some of these are of very long standing as, for example, the fermentation of sugars to give ethanol, there has been a rapid growth in recent years in the number of substances made by fermentation processes. It may therefore be said that for some of these processes the operating experience is of short duration.

But a more important factor may lie in the growth of genetic engineering, otherwise known as recombinant DNA or rDNA technology. Though the potential importance of rDNA lies mainly in other fields, it may in the future be applied to novel processes in the transformation of chemicals and chemical wastes. In which case some of the apprehensions currently being expressed about laboratory-scale work may apply with greater force to such processes especially having regard to their greater scale.

16.9.2. FERMENTATION.

The fermentation of sugars to yield alcohol is probably as old as human history. However it was not till the end of the nineteenth century that it was applied to the production of other organic chemicals such as lactic acid. Originally these were anaerobic processes which broke down complex molecules into simpler ones. Today, according to Kirk-Othmer (1985), the modern definition is "any microbial action controlled by man to make useful products". Aerobic fermentations make possible the synthesis of large molecules and have been applied to the production of antibiotics, organic acids, enzymes and vitamins. The reference gives the view that the potential of the application of genetic engineering is such that there may be a complete revolution in the fermentation industries over the next ten years.

It may be noted that chronic occupational illnesses connected with fermentation processes are already recognised but there are no case histories to the author's knowledge concerned with acute events.

16.9.3. GENETIC ENGINEERING.

There will obviously be great interest in the possibility of producing micro-organisms by rDNA techniques which will be custom-built for existing or novel processes. These techniques are biological and are generally conducted today in the laboratory on a scale of less than 10 litres.

However, both in the UK and abroad, a very prudent attitude is taken towards even laboratory-scale work because of the apprehension which attaches to the impact of previously unencountered organisms upon humans, animals or plants. The UK has set up a Genetic Manipulaticn Advisory Group (GMAG) and Regulations have been made requiring all those wishing to commence work in the field to inform both the HSE and GMAG. Regulation in the US is described in volume 4 of Moo-Young (1985), though this reference does not attempt to analyse the hazards which may arise. It recognises that in the US, such technology has already moved out of the laboratory into the field of industry. It claims that early fears have been laid to rest and that it is entirely possible that large-scale applications of rDNA will be so risk-free that little if any regulatory controls will be called for.

This, however, remains to be seen. It is only prudent, for the present, to put up a marker in this area.

16.10. ARE THERE ANY OTHER KINDS OF MAJOR CHEMICAL HAZARD ?

If the taxonomy of changes to the environment propounded in Chapter 4 is fully comprehensive then the analysis in this chapter will have covered the whole range of types of major hazard, other than fires, explosions and toxic releases, which can arise from the chemical and process industries.

However, even if this be so, it is impossible to catalogue the whole range of harmful substances and their effects let alone to try to predict the properties of new substances which may achieve technological importance in the future. Nevertheless such properties are not likely to lie outside of the broad categories of hazard which have been discussed in this and preceding chapters.

Chapter 17

METHODS OF CONTROL OF MAJOR CHEMICAL HAZARDS

17.1. SCOPE OF CHAPTER.

Chapter 17 is the first chapter in the part of the book which is concerned with the control of major chemical hazards and it may be helpful at this point to review the scope of this, and the following three, chapters.

Chapter 17 will be concerned first with the general principles underlying the control of major chemical hazards and which are aimed at the prevention of their realisation and with the limitation of their consequences.

It has to be recognised that if a hazard exists, the likelihood of its realisation can never be reduced to zero. Control, if it cannot eliminate the hazard entirely, must aim to reduce the individual or societal risk associated with the hazard to a level which it is "tolerable", "acceptable", or "justifiable" to society as a whole. The question as to what constitutes "tolerable", "acceptable", or "justifiable" is difficult and to some extent unresolved and depends to a considerable degree upon the public perception of the risks. The problems of public perception are discussed.

Quantification is a vital part of the process of assessment of hazard and risk and are today the subject of detailed analysis. Some of the techniques used in the area are discussed in the chapter. However, quantification is discussed in more detail in Chapter 18.

The interaction between legislative and managerial control is complex. Generally speaking, though there are some exceptions such as that of nuclear power in the UK, legislative control only appears after managerial control has been found wanting. But, for reasons of convenience, the author has decided to treat legislative control in the EEC and the UK in Chapter 19, before discussion of the principles of managerial control in Chapter 20. This chapter will, among other topics, consider the need to restrict the numbers of persons in hazardous areas and the related problem of the siting and construction of control buildings.

17.2. GENERAL PRINCIPLES OF HAZARD CONTROL.

17.2.1. HAZARD AND LAW.

17.2.1.1. Legal relationships.

The law can only concern itself with hazards which are under the control of human beings. For example, being bitten by a wild animal can only be a subject of legal redress if it can be shown that the animal has an owner. If it can be established that there is a person responsible for an injury, then it may be possible to obtain redress. This can take two forms, civil redress or criminal redress.

In Britain, civil redress takes the form of suing for damages in a civil court. In Britain the person suing, the plaintiff, has to prove negligence, i.e. that the defendant had failed in the duty of care towards the plaintiff. Such a failure constitutes a "tort" or wrong. Any damages awarded are paid over to the plaintiff. It is not essential to prove that the activity complained of was "against the law", only that the defendant was negligent in allowing it to happen. However, many torts are also crimes.

The second form of redress is a criminal prosecution. Though private individuals may "lay information", private prosecutions are now practically unknown. Most prosecutions in England are, at the time of writing, in the hands of the police. Soon this system is to be changed and the responsibility for prosecutions is to be put into the hands of a Public Prosecutor. However, prosecutions for breaches of the occupational health and safety legislation in the United Kingdom remain in the hands of the Factory Inspectorate and the police have no jurisdiction in this area.

A criminal prosecution may result in a fine or imprisonment or both. Conviction does not entail, of itself, the handing over of damages to the victim.

Civil cases are concerned with wrongs as between one individual and another, whereas criminal cases have within them the element of an injury to society. As pointed out above, an act may constitute a tort as well as a crime.

17.2.2. STATE CONTROL POLICY IN BRITAIN.

Chicken (1975) analyses the general principles whereby hazards are legally controlled in Britain. Though this work does not discuss the chemical and process industries specifically, the general conclusions arrived at seem applicable to these industries. Chicken surveys a number of areas, road transport, air transport, factories, nuclear reactors and atmospheric pollution and looks at the policies of control including the role of "interest groups" and the influence which they exert.

The terminology in Chicken differs somewhat from that of this book and in the comments which follow the present author has, to some degree, translated Chicken's terms. Chicken concludes that hazards may be realised through lack of recognition that they exist or lack of knowledge on how to prevent them from being realised. Some are realised because of failure to consider the consequences of actions. Individuals or governments may consider certain levels "acceptable" even though these levels may not be quantified. The reasons for acceptance may be many and varied.

When significant hazards are identified, a legal framework is established to control them and a specialist inspectorate is established to monitor them. Government generally precedes the establishment of controls by taking advice from interested parties. When controls are established, research is instituted to solve the problems of control and to improve safety.

Older technologies are, in general, less tightly controlled than those that are newly established. The object of control policy is to keep the risks within acceptable bounds but within an allocation of resources for the enforcement of the policy which is considered to be in harmony with other demands for resources.

Policy making is primarily a civil service rather than a governmental activity. Cabinet involvement is usually small and there is little scrutiny by Parliament or interest by the political parties. Hazard control is a specialist-based sub-system of the general policy making system. Chicken forecasts that future developments would include the requirement for those designing plant to have to evaluate whether their associated hazards and risks fall within existing public standards.

Chicken uses the system of classifying interest groups developed in Wooton (1970), (not consulted by the present author). This recognises three types of group, economic, integrative, and cultural. By way of example, in the light of this classification, the Confederation of British Industries (CBI) is economic, the Institution of Chemical Engineers is integrative, and Friends of the Earth cultural. They are recognised as operating at three levels: local groups, regional and national groups, and the highest level groups. Examples may be residents' associations at the lowest level, an individual trade union at the middle level, and the CBI, the Trades Union Congress (TUC), and major professional institutions at the highest.

Interest groups play a role in making hazard control policy. Though Chicken (1975) was written before some of the current institutions came into being, it might be noted that in the field of occupational health and safety the policy making body, the Health and Safety Commission, is made up of three groups of an integrative type operating at the third level. These are the CBI, the TUC, and representatives of the local authorities. This pattern is followed in the Advisory Committees which may also have independent experts in membership. Professional institutions are not usually directly represented but may have a number of their senior members sitting on such committees. This was so with the Advisory Committee on Major Hazards which had four senior members of the Institution of Chemical Engineers among its members.

Applying the analysis in Chicken (1975) to the problem of major chemical hazards, the need for control was recognised in Robens (1972) and then by the public at large after the Flixborough disaster of 1974. Policy was then shaped by the HSC and its executive arm, the HSE, which is a civil service-type body advised by a committee, the Advisory Committee on Major Hazards. This was made up of independent experts in membership of major professional institutions together with representatives of the CBI, the TUC and of the Local Authorities. Its work was barely influenced by Government or by political parties. Legislation was proposed, as will be discussed below, based on the advice of this Committee. The enforcement of any legislation would be carried out by a specialist section of the Factory Inspectorate.

17.2.3. SELF-REGULATION AND STATE REGULATION.

The analysis in Chicken (1975) does not devote any attention to the operation of "self-regulation". There is clearly no such thing as pure self-regulation, in that all activities which contain within them an element of hazard are constrained by the possibility of legal redress for negligence. However many hazardous activities are not made subject to statute as, for example, motor racing on a privately owned circuit, and it is convenient to regard any such activities as subject to self regulation within the rules and codes of conduct of a voluntary association. The principle of self-regulation as it was envisaged in Robens (1972) had a rather different meaning. Robens envisaged self-regulation as a system whereby, within a framework of a general statutory duty, employers would be free to make their own rules and safety codes and would not rely upon organs of state to formulate them.

Though the Health and Safety at Work Act [HASAWA (1974)] incorporates the spirit of the principle of self-regulation, this is to some degree a counsel of perfection and the subsidiary legislation in the form of Regulations made under the Act, and many other previously enacted Regulations incorporated into the Act, contain much detailed prescription of measures to be taken.

Some of these are absolute requirements such as the requirement to fence machinery. However such requirements, which arose from earlier factory legislation, were based on the concept that all factories were generically similar as indeed they were to a large degree at the time the original Factories Acts were passed. Today the gulf which separates, for example, the operation of a catalytic cracker from the operation of a cotton mill is enormous. Figure 17.1. shows a typical 19th century cotton mill. It is, among other factors, the impossibility of reducing chemical processes to a common pattern which makes self-regulation the only practicable basis for control in the chemical and process industries. However certain areas of these industries, such as, for example, those concerned with the fencing of machinery or the operation of steam boilers, or the electric supply, remain under prescriptive control.

An evolution may be seen whereby hazardous activities, when in their infancy, are self-regulated under the civil law. Later, when there is public concern about the growth of hazard, a demand grows for regulation. In the past such regulation was primarily prescriptive but today it may consist in reproducing civil obligations in the criminal law and thus making certain acts both tort and crime.

The most recent development is that concerned with demands for the quantitative assessment of the risks posed to the public by hazards, a subject discussed below. One point which may be noted is that one effect of statutory control may be the imposition of more uniform standards of what constitutes the threshold value of individual and societal risk. Threshold values are discussed later.

17.3. PUBLIC PERCEPTION OF RISK.

17.3.1. THE PROBLEM.

A problem which has arisen is that, in some areas of control of hazards, there is a wide discrepancy between the public appraisal of the level of risk of injury to individuals or society associated with a given hazard, and the appraisal of this level by specialists in the field of risk analysis.

There is now a considerable literature on this subject of which the author has found himself able only to consult a small proportion. Nor is he able, for considerations of space, to be able to analyse all that he has consulted. He believes that the most significant input has been made by the Royal Society in Royal Society (1981) and Royal Society (1983) and it is from contributors to Royal Society (1981) that he has abstracted what seem to him to be the most cogent contributions to the discussion.

17.3.2. THE APPROACH OF THE PSYCHOLOGISTS.

Lee (1981) comments, in relation to the assessment of risks to the public, "It is assumed that accurate scientific measurement provides the best estimate and that public perceptions, if discrepant, are "irrational". The present author would enter a criticism of this immediately. "It is assumed" says the reference, but it does not say by whom. It does not seem to follow that because there is a discrepancy between estimates of, say, the height of a mountain, by surveyors and by members of the public, that there is any general view among surveyors that such discrepancies arise from "irrationality" on the part of the public. It seems more likely that they would put it down to lack of instruments, or of lack of knowledge of trigonometry or failure to take account of the altitude at which the estimate is made etc. The discrepancy in the estimation of risks is quantitative, not qualitative, as the specialists agree with the public that there is some level of risk.

However, this being said there are many points in Lee (1981) which seem

This is a photograph of a part of the premises of Messrs. Howarths of Salford, Lancashire. As far as can be ascertained the photo was not in any sense intended to highlight the dangerous conditions in such mills. It shows many features common to factories of the time. The looms are driven by belts from overhead line shafts which in turn would be driven by a steam engine. There is a great deal of unguarded machinery. The women appear to have bound up their hair but are not wearing caps and their dress is unsuitable for work in such circumstances. The factory appears to be lit by electric light. The potential for personal accidents was obviously considerable and there was no doubt a considerable fire risk. The illustration is from Engineering of July 1894.

Fig. 17.1. A TYPICAL 19th CENTURY WEAVING SHED.

more relevant. For example that fear, or its lesser form, anxiety, has been with mankind since the beginning of evolution. Or that the public is far less fatalistic than it used to be. Lee points out that a hazard has meaning only in human terms and that the public may also take account of suffering as well as death. Lee treats death as a "Simple binary event, you are either dead or you are not - whereas suffering lies on a continuous scale of severity". This seems not to take cognisance that the statistics of deaths in accidents subsume the suffering associated with them. Whereas the majority of people have had some experience of long-drawn-out illness, only a minority in their lifetime witness violent death. Even at second hand they are spared the details as, for various reasons, the media avoids giving a full account of what happens when, for example, people are burned to death through being trapped in crashed cars. Lee also argues that "The awfulness of a catastrophe may lie not merely in the loss of life and suffering involved but in the violation of people's sense of justice or moral rightness".

Lee acknowledges that criticisms of the shortcomings of objective assessment have led to a much greater degree of sophistication in the practice of risk assessment including the development of cost-benefit analysis as a component. Cost-benefit, Lee contends, is not evenly spread.

The reference also discusses research techniques for the comparison of perceptions of risk drawing attention to the difficulty of portraying them in two, let alone three dimensions. The public's evaluation will differ in their assessment according to whether it is personal safety or public safety which is at stake. Familiarity or unfamiliarity will also alter perception.

Lee perceives at least some element of irrationality in public perceptions. "There is plenty of evidence that people select and distort what they attend to, to reduce conflict, in some cases suppressing entirely any consideration of the unpalatable aspects of a hazard and in others reconstructing unpleasant aspects so that they are seen in a way that is more congenial to the main thrust of their personal evaluation".

Slovic et al (1981) claims that all forms of risk assessment contain a large component of subjective judgement. "Someone, relying on educated intuition, must determine the structure of the problem, decide the consequences to be considered, and select the relevant data and interpret it". In the present author's view this says very little as it could be said about any scientific analysis. It is useful therefore to remember that no scientific analysis is totally objective nor is public opinion totally subjective. The truth might be better put that scientific analysis is predominantly objective whereas public opinion contains a large measure of subject judgement. Papers such as Slovic et al presumably claim to be predominantly objective and have, as their aim, the introduction of greater objectivity into the study of risk.

The reference goes on to criticise the lay evaluation of risks giving examples of the bias introduced by recent events or mirrors of such events in fiction (for example in novels, cinema or TV). It quotes research which showed that when people were asked to estimate the numbers of deaths per annum in the US from various causes, having been told that there were 40,000 deaths per annum in motor accidents, the responses showed a bias in over-estimating accidental deaths, murder, food poisoning and violent death in general, and an under-estimation of deaths from cancer and diabetes. Deaths from disease were under-estimated by a factor of ten. Put another way, the public overestimate deaths from rare events and underestimate deaths from common events. The reference attributes some of the responsibility for this to media bias in that the media over-report sensational events.

Slovic et al is also critical of judgements both by experts and by lay persons considering that these are frequently over-confident. It reports research in which people were asked to give a judgement on which of two lethal events occurred the more frequently and to estimate the odds that they were correct. One in eight of those giving odds of 100 to 1 or more were wrong. Of those offering 10,000 to 1, one in ten were wrong ! An example of a judgement on which heavy odds were given was the incorrect assertion that murder is more frequent than suicide in the US. But the paper also criticises certain expert judgements for over-confidence.

Research has also reported on reactions to a number of characteristics associated with hazards which are set out below. The present author has set out his own judgement of the chemical and process industries in relation to these factors and his estimate as to how the general public would respond.

TABLE 17.1. FACTORS IN RISK PERCEPTION.

Factor.	Public perception of risk.	Author's perception of risk.
Involuntary in nature	High	Low
Unfamiliar	High	Low
Frequency of death	Med	Low
Unknown to science	High	Low
Catastrophic	High	Med
Causing dread	High	Low

The author believes that on none of these factors he would, as someone familiar with the problem, see things the way the public would.

17.3.3. SOCIAL FACTORS.

The two papers considered above represent the point of view of psychologists. Though they both freely acknowledge the existence of social interaction, this aspect was not considered in detail. For example, when considering the objectivity of scientific investigation, Lee drew attention to some psychological factors which tend to limit objectivity. But there are others which may be more serious. Lee assumes that the investigator is striving to be honest and that the impediments to objectivity are those common to all scientists. (In using the word "scientist" here, the present author intends it to refer to anyone, whether or not they would describe themselves as a scientist, who is governed by the disciplines of science when conducting an analysis of a problem). There may be impediments to objectivity which arise from the relationship between a scientist and the source of the funds which make the analysis possible. A risk analysis conducted by an employee of the XYZ Chemical Company on their behalf may be objective but public suspicion would undoubtedly attach to it. On the other hand, a risk analysis conducted by an academic might be regarded far less open to suspicion. However the funds for this would have to come from somewhere and if they came from the XYZ Chemical Company this might still leave the analysis open to suspicion.

So-called independent investigators are open to social pressures. In Universities there is the hope of promotion or simply of academic prestige. Appearances on television, for example, may be seen as helping to achieve these.

The Official Report of the Flixborough Inquiry, [Flixborough (1975), Paras 155, 171, 172 & 193], contained criticisms of a number of expert witnesses for lack of objectivity in the pursuit of a particular theory. There have been numerous allegations that scientific papers attributing high levels of risk to consumer products have been instigated by commercial rivals but without this fact being made plain. Such allegations have included claims that risks associated with the birth control pill arose from research covertly financed by the manufacturers of an alternative method. There have been claims that a sweetening agent, which was banned for three years in the US, was put off the market by covertly financed research which was also fraudulent. Scientific fraud is a subject which cannot be covered here; suffice to say that it exists and there are well authenticated cases of blatant fraud such as the Piltdown Man affair.

The present author has found no evidence which leads him to suppose that non-objectivity among experts is significantly responsible for widening the perception gap by a bias towards understating risk. It seems to him that experts hired by, or acting on behalf of, interest groups are as likely to exaggerate risk, as experts hired by the owners of hazardous installations are to represent the risks as being lower than they really are.

17.3.4. THE PUBLIC PERCEPTION OF CHEMICAL HAZARDS.

One factor not considered by Slovic et al was perceived necessity. Cars are considered necessities, nuclear power probably is not. Because cars represent so high a level of convenience to their owners the mortality, disablement or non-disabling injury associated with them is ignored or played down. On the other hand, people generally believe, both in the USA and in the UK, that electric power could be generated from alternative sources just as it has been for more than a century. They tend to reason that the problem of risk associated with nuclear energy could be removed by doing away with the hazard. That there are hazards associated with, say, the mining of coal, is not disputed but the activity of mining coal is seen as a "voluntary" activity whereas nuclear power is perceived as being "imposed" upon the surrounding population.

Does the public see the chemical and process industries as necessities ? Personal experience suggests that there is little consciousness of the indispensable role of the chemical industry in providing the basis of present-day living. As an example of this, when the author was engaged to report on an incident in Derbyshire in which a cloud of oleum and sulphur trioxide had been released and was discussing this with an indignant householder, she expressed the opinion that "such places ought to be closed down". In her kitchen in which the conversation took place, there were at least two household products which were made from oleum and sulphur trioxide manufactured in that very works. This problem is discussed in Marshall (1980b) and in Marshall (1984).

Table 17.2. "Hazardous chemicals make household products", is taken from this latter reference. It demonstrates that many commonly used products involve the use of dangerous chemicals, most of them, when present at certain levels of inventory, being classifiable as major chemical hazards.

17.4. THRESHOLD LEVELS FOR RISK.

17.4.1. THE HISTORICAL PERSPECTIVE.

It is impossible to say when questions of safety were first introduced into questions of public policy regarding new technological developments. The Old Testament in Deuteronomy 22, verse 8, provides "When thou buildest a new house, then shalt thou make a battlement for thy roof, that thou bring not blood upon thine house, if any man fall from thence". (Deuteronomy was probably written in the 7th century B.C.). This may well indicate the sequence that when flat roofed houses first appeared there was "self-regulation" and their builders (presumably also their owners) suited themselves about providing battlements (parapets). Later, after a number of persons had been killed or injured, the public authorities intervened and prescribed a solution which had already been adopted by some builders.

TABLE 17.2. HAZARDOUS CHEMICALS MAKE HOUSEHOLD PRODUCTS.

	ACIDS									ALKALIS		FLAMMABLES			TOXIC ORGANIC CHEMICALS
	HYDROFLUORIC ACID	HYDROCHLORIC ACID	HYDROCYANIC ACID	NITRIC ACID	SULPHURIC ACID	SULPHUR DIOXIDE	OLEUM, SULPHUR TRIOXIDE	ACETIC ACID	CHLORINE	AMMONIA	CAUSTIC SODA	OLEFINES	OTHER HYDRO-CARBONS	HYDROGEN	
GENERAL															
Paper						X					X				
Paint					X				X						
Petrol	X											X	X		
CLEANING AGENTS															
Soap											X				
Bleaches									X		X				
Detergents							X						X		
Floor Cleaners										X					
Dry Cleaners									X			X			X
HARDWARE															
Refrigerants	X	X							X		X		X		
Non-stick Coatings	X	X							X		X		X		
Floor Tiles		X										X			
Cigarette Lighters													X		
FOOD															
Drinking Water									X						
Margarine														X	
Sugar						X									
Grain Whisky					X										
CLOTHING															
Nylon										X		X	X	X	
Rayon					X			X			X				
Acrylics			X									X			
Dyestuffs				X	X								X		X
GARDENING															
Fertilisers				X	X					X					
Pesticides									X		X				X

In more recent times, questions of safety were certainly raised during the consideration of Parliamentary Bills for the construction of railways. These proceedings were the equivalent of present-day public inquiries. Hunter (1977) describes the hearings in 1825 of a Bill to gain approval of the Liverpool to Manchester railway. There were 150 separate objections and the Committee considered the Bill during 37 sessions. George Stephenson, the engineer designate, was warned not to quote a speed of 20 mph for his locomotives. One politician warned him that he must restrict his claim to lower speeds "or he would be regarded as a maniac . . . and damn the whole thing". One author claimed that "even the bulk of intelligent opinion in 1825 could not believe that any machine could travel at over ten miles an hour without disintegrating and killing everyone in sight". That particular Bill was thrown out. The author has read some of the arguments of the time and there seems evidence that even as far back as a century and a half ago there were some who used the argument of safety because it seemed more high-minded than argument based on self-interest.

Examples could be multiplied. There was the UK Locomotives Act which was passed in 1865. This imposed a general speed limit of 4 mph (2 in towns) and required the vehicle to be preceded by a man carrying a red flag. This lasted, in one form or another, for 31 years. It may have saved lives but it is now generally agreed to have done immense harm to the infant British motor industry.

One thing seems quite clear and that is that there was nothing quantitative about the balance that was then struck between the advantages of technical progress and the price which had to be paid in human lives.

It may also be argued that there was then a great gulf fixed between the public tolerance of those hazards which could kill or maim members of the public, and those which could only kill or injure industrial workers. Coal mining was a case in point where it was difficult to conceive of anyone being injured at the surface by even the most devastating explosion underground. Figure 12.1. has shown that there were several years in the British coal industry in which between 400 and 500 were killed in disasters alone. In Bryan (1975), his Table 9.1. gives for 1853 to 1862 an annual average of 1000 miners killed out of a labour force of ca 250,000 or 4 persons p.a. for every 1000 employed. This is an F.A.R. of circa 133 compared with the figure of 3.5 for the UK chemical industry today. Even up to 1930, the total deaths exceeded 1000 p.a. for a workforce of ca 1 million.

Unfortunately few present-day works concerned with risks to the public display any historical perspective. In consequence their readers see it all as a newly arisen, and growing, problem. Kates (1977) put it as follows, "By almost every type of societal indicator, except one, hazardous events have been increasing . . . The one exception is the statistical record of hazard consequences . . . There frequently appears an enormous divergence between this record and the perception of hazards by scientists, the public and officialdom".

17.4.2. BALANCING ECONOMIC ADVANTAGE AGAINST RISK.

It has been a feature of the last 20 years that a new feature has emerged, the quantification of individual and societal risk. It might perhaps be argued that this quantification has outpaced the quantification of benefit. Whereas a century ago, with some exceptions, technological and economic advantage thrust aside considerations of risk to the public, let alone the workers, there is some evidence that, to a limited degree, the pendulum has now swung the other way. Today, in some instances, it may be considerations of risk which thrust aside technological and economic advantage especially where risk to the public is involved.

The comparison of risk and cost/benefit is fraught with many difficulties.

The most fundamental of these problems is the lack of a common denominator. There are in principle three possible solutions to this. One is to put a financial value on human life. Another is to express economic advantage in terms of lengthening of human life. A third is to find an alternative entity in which both might be expressed which is neither length of life nor money.

The concept of evaluating human life in financial terms is one which arouses vehement opposition in some quarters. It is held that human life is sacred, which puts it outside the scope of financial evaluation. But when the workings of society are examined there is little or no evidence that this is the view of governments or, for that matter, the electorates on which they rely for support. Even where lives of identifiable, as opposed to statistical, persons are at stake, governments are prepared in certain instances to put them at jeopardy rather than hand over cash. This applies to the cases of hi-jacked aircraft or other vehicles where ransom demands are made. Governments have, in general, refused such demands even where this refusal has posed a significant threat to the lives of passengers. This has been because it has been considered that higher issues have been at stake and, in taking this view, governments have received widespread public support.

At a more general level, evaluations of human life in financial terms are everyday matters. These take the form of court judgements of compensation for loss of life. They also take the form of decisions on financial expenditure which is intended to save life. For example, it is well known that lighting of roads results in a reduction in the number of motor accidents, but the question remains "How much to spend to save a life ?". This question is discussed in Fortune (1986). In it different methods of analysis lead to conclusions which range from \$650,000 to \$7,000,000 (1986), though several methods of analysis favour the lower figure. It seems to the present author that different levels of skill, wealth or social standing are likely to be reflected in different values placed upon an individual life.

It would be going beyond the scope of this book to attempt any detailed discussion of the problem, but attention may be called to Holdgate (1981) where some aspects of this evaluation are discussed.

The alternative approach seems on the face of it to be without validity. Figure 17.2., which is derived from Preston et al (1972) and I.I.S.S. (1981), shows that, except for some very poor countries, over a span of per capita national income of about ten to one, there is no correlation between this income and the expectation of life at birth. Thus the possession of motor cars or television sets or the taking of holidays abroad makes no apparent contribution to longevity. What they do contribute to is to the quality of life.

This is even more difficult to quantify than the financial value of human life.

The third alternative, to find a common denominator which is measured neither in money nor in length of life, does not seem to have been attempted.

It would appear, therefore, that the attempt to equate cost-benefit with the disbenefit, risk, must be abandoned as far as this book is concerned.

17.4.3. DETERMINING THE THRESHOLD OF RISK BY EQUIVALENCE.

It has been demonstrated above that it is not possible, for the present at any rate, to treat the benefit/disbenefit problem by an algebraic sum which has to yield a positive answer. Faced by this an alternative solution has evolved. This is to examine long-existing benefit/disbenefit situations where it is judged that the public have reached a decision by subjective means that the disbenefit is outweighed by the benefit. If the level of disbenefit, i.e. the risk, can be then quantified, this would provide a threshold value for risk.

Rothschild (1978) put forward such a solution to the problem. Rothschild

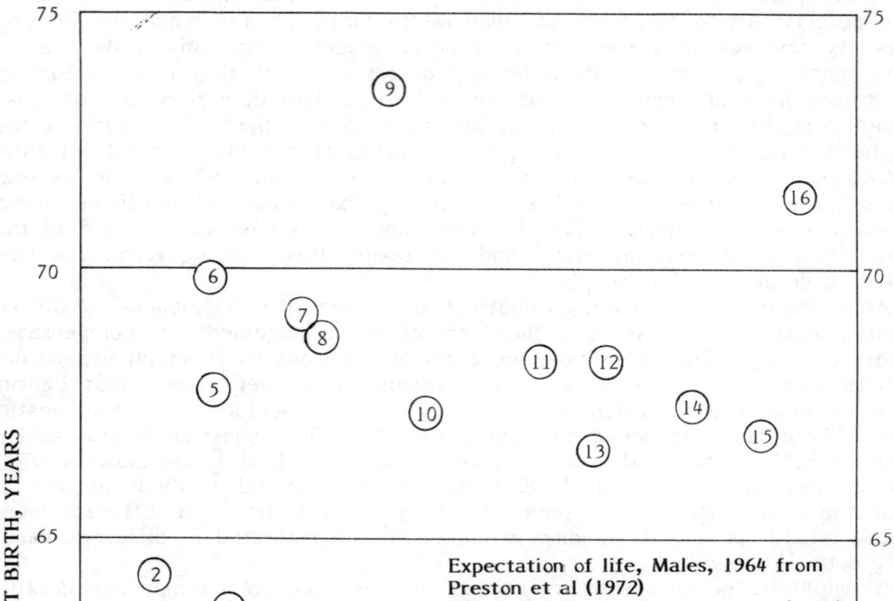

Fig. 17.2. EXPECTATION OF LIFE v PER CAPITA GNP.

proposed that any beneficial activity which posed a lower level of risk than arises from using the roads should be permitted on the grounds that the public shows acceptance of the toll of the roads.

In essence this substitutes the subjective balance made by the public (or at least that section of it which owns and drives motor cars), between the benefit and the risk of motoring for some, as yet unrealisable, technique of quantitative assessment of other benefit/risk situations.

This, in the present author's view, is a simplistic approach. For reasons discussed above the subjective balance struck by the public is governed by a variety of factors. In the particular balance chosen by Rothschild, there is widespread agreement that the public places an exceptionally high value on the benefits attached to car ownership and that by many objective tests, the benefits may be shown to be both high and real. Rothschild appears to regard all benefits as equal as they can all be matched to the same level of risk. But this is manifestly unsound. In some cases the public may well regard as acceptable much higher levels of risk for what they regard as much greater benefits as, for example, in heart transplant operations. In others, and nuclear energy is cited above, rightly or wrongly, the benefits seem to the public to be slight or dubious and there is a demand for much lower levels of risk than Rothschild would regard as a suitable threshold.

A further factor which Rothschild does not take account of is that in the case of motor cars, benefit and disbenefit is spread very evenly today. This is not true of, say, the problem of living near a chemical works. The benefits arising therefore accrue to the public as a whole, whereas the disbenefits are concentrated within a radius of about a kilometre. The local population may regard their share of the risk as disproportionate.

The implications of the Rothschild standard may be calculated. If it be assumed that in the UK, 1 person in 500 lives in the vicinity of a chemical or process plant, and taking the figure given in Preston et al (1972) for risk for death from motor vehicles as the mean of that given for males and for females at age 30 in 1964, i.e. circa 0.0001, this would yield ca 10 public deaths per year from the chemical industry. At present the figure for the UK is less than 0.1 per annum as there have been no such fatalities in the last twenty years. The adoption of the Rothschild standard by the chemical and process industries would thus lead to a considerable deterioration in public safety. (It should be noted that if the risk for dying from traffic accidents is related to the length of time of exposure, then it is higher than the figure quoted above by a factor of more than 10. It is not clear which level of risk was intended in Rothschild but the present author has assumed it to be the figure for a year, in which there may have been, say, 400 hours driving).

An alternative approach, which shares with the Rothschild approach the concept of a standard of comparison, but which uses a different standard, is discussed below. It is based upon using "life tables" from which it is possible to deduce the risk of dying in the next twelve months, from all causes or from a specified cause, at any chosen age for either males or females. Such tables are set out in the previously cited Preston et al (1972). Tables are provided for many countries and at various dates, some tables going back to the 19th century. It would be a simple exercise to derive corresponding figures for other European countries or for North America from this source.

Figure 17.3. has been derived by the present author for males in England and Wales for 1964. It has the form of a "bath tub" curve and demonstrates that the minimum level of risk of dying either by natural causes or by all causes lies at age 10. (It may be noted in passing that the bulge in the "all causes" curve is mainly caused by young men killing themselves on the roads). These levels seem to the author to form a useful yardstick for assessing imposed risk. Plotting the Rothschild standard on Fig. 17.3. shows that it lies very close to

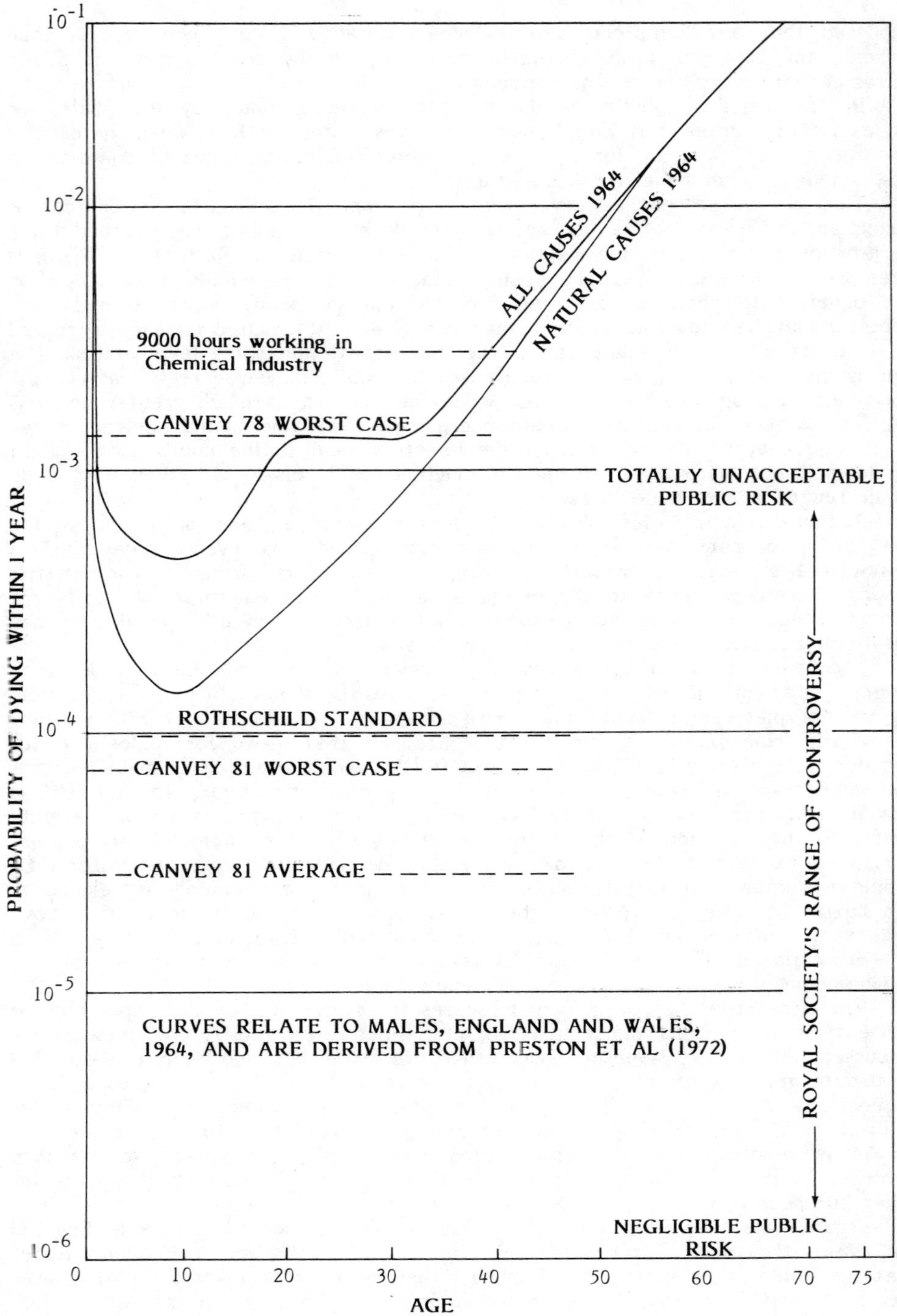

Fig. 17.3. CERTAIN LEVELS OF RISK COMPARED WITH AVERAGE RISK OF DYING WITHIN ONE YEAR OF AGE N.

the level of risk of dying from natural causes at age 10 and would increase the risk of dying from all causes at that age by 25%.

The level of risk for people living at Canvey Island, Essex, England, was assessed in 1976 after there was public agitation over a proposal to build an oil refinery at this locality at which there was already a methane terminal, major oil storages and nearby chemical plants. The report of this assessment was published in Canvey (1978). It concluded that the worst levels of risk were 1.3×10^{-3} or 13 times the Rothschild standard. Such a level would increase the all causes level at age 10 by a factor of 4. The publication of these figures did not produce any marked outcry, presumably few realised what they meant. However a reassessment after certain recommended improvements had been carried out, produced a drastic reduction in the levels as was disclosed in Canvey (1981). The new levels were 6.5×10^{-5} for the worst case and 3.5×10^{-5} for the mean. In view of the high concentration of hazardous industry in Canvey, such values may be regarded as "worst case" levels.

There is no generally agreed threshold for risk in the UK. Royal Society (1983) takes the view that, depending on the level of benefit to be anticipated, risks in the range 10^{-3} to 10^{-6} represent the range of controversy. The present author is of the view that a level of risk of 5×10^{-5} which is 1/10 of the risk of dying from all causes for a child of 10, should be the maximum to be permitted. This corresponds to the mean level in Canvey (1981). However the actual record disclosed by the argument given earlier that ca 500,000 members of the public at risk from UK chemical and process plants have not suffered a fatality in at least 20 years, would suggest that the actual level of risk is more of the order of 10^{-7}.

17.5. HOW RISK MAY BE ASSESSED.

17.5.1. INTRODUCTION.

The discussion above assumed that it is technically feasible to assess risk. It is necessary now to give brief consideration as to how this is done. It must be said at the outset that predictive assessments of this character are necessarily inexact. Farmer (1984), however, concludes that though the estimates of probability may be inexact by a factor of 3, up or down, this does not alter the essential message they give.

The purposes of risk analysis are neither passive nor contemplative. Risk analysis is a guide to action where the risk arising from an activity is judged to be excessive in relation to the likely benefits. This action may take such forms as design changes or "add on" safety features. The estimation of the value and purposes of risk analysis given in works such as Kunreuther and Linnerooth (1983), which regard risk analysis as a purely contemplative science, is mistaken in the present author's view.

17.5.2. AN ILLUSTRATIVE EXAMPLE.

The steam boiler is a useful illustrative example of how risk may be calculated. Examination of this problem throws up many issues which are of general significance.

Firstly it is necessary to distinguish between different types of boiler as these will have different levels of risk associated with them. UK legislation includes under the heading of "boilers", equipment designed to raise steam and also any closed vessel into which steam is admitted. The discussion will be concerned with vessels for raising steam. There is then a broad distinction between boilers in which the heating arises from combustion within the boiler structure, and waste heat boilers in which the heating medium is a hot process gas. Boilers in which combustion takes place may be shell boilers, of which the Lancashire boiler is an example, or water tube boilers. The example chosen is the Lancashire boiler which is the prototype of today's "packaged" boilers.

The Lancashire boiler has had a long history and this may seem at first sight to offer the advantage of a considerable historical experience. This is especially so as there has been a requirement in the UK since 1882 to conduct an enquiry into all boiler explosions; some 3,400 Reports of these enquiries were in existence in 1977 [see Warwick (1977)]. Unfortunately this is true to only a limited extent because so many changes have been made to the design and construction of such boilers and their auxiliaries. Some of these changes have been made in the interests of safety, others have been made in the interests of more economical operation.

As has been pointed out in Chapter 6, failure may lie on a spectrum from a pinhole leak to a catastrophic failure and, in principle, it is possible to predict the risk of any of them. For illustrative purposes, only the risk of catastrophic failure will be discussed. It may be useful to note here that this exemplifies the principle that hazard analysis must precede risk analysis; in this instance one mode of realisation of the hazard has been selected out of a number of potential modes and risk analysis then is applied to predict the risk of this particular realisation occurring.

But catastrophic failure can occur through more than one route of which two immediately present themselves. These are shell failure and fire tube failure. In principle there are two modes of shell failure, (a) rupture due to excess pressure, and (b) rupture at or below design pressure. The former implies failure of the pressure relief system as it is implicit in boiler design that occasional overpressure due to demand fluctuation is inevitable. Such a risk can be determined as a function of the risk of safety valves failing to function per se. On this there are considerable data as such equipment is widely used in industry and the manufacturers have usually researched this area. Since steam boilers, by law, have to be fitted with two safety valves, the probability of this happening is, on the face of it, F^2 where F is the probability of a safety valve failing to operate. However this may be understating the probability as one, or both, may have been tampered with. Such could occur with faulty installation which placed one or both downstream of a stop valve; Warwick (1977) gives an example of this happening. Safety valves have been known to have been deliberately blanked off. Or one or both may have been supplied to a wrong specification or have been wrongly set.

It will be noted that the two situations are being considered. One is the failure of boilers, per se, the other is failure of equipment which maybe in much more general use than are boilers. Such equipment includes various kinds of valves, pumps, piping and control instruments as well as the electrical installation.

A second point to note is that the risk will not be, in general, a single-valued expression but will represent a range of probability which will require statistical treatment.

Failure of the shell under operating conditions may occur for a number of reasons which the author will not attempt to deal with exhaustively. They include faulty design, the choice of incorrect or faulty material of construction, incorrect fabrication or assembly; such faults have previously been discussed in Chapter 6. They usually show up early in the life of the boiler; they are sometimes referred to as "teething troubles". A plot of risk of failure versus age for boilers, as indeed for many other items of equipment, resembles the shape of Figure 17.3. It has a high initial value analogous with infantile mortality, a minimum value in middle life and a rising value analogous to senility in humans, and when failure may arise from creep, fatigue or from corrosion, which latter may be caused by inadequate control of feed water quality.

In many of the late life failures a significant contribution may be attributed to human failures in operation.

Failure of the fire tubes may be sketched briefly. This has mainly arisen in the past through allowing the water level to fall so that the fire tubes, instead of being wholly submerged have part of their area in the steam space. This is a phenomenon identical with part of the "BLEVE" scenario but is of far greater antiquity. Originally entire reliance was placed on the boiler attendant maintaining this level by manual control but today the level is automatically controlled and backed up by low level/high level alarms. Risk in this instance may be calculated from the historical record of such automatic control equipment.

The illustrative example shows that the evaluation of risk for a steam boiler is fairly complex; many of the issues have only been treated sketchily. When the exercise has to be performed for, say, an entire petrochemicals plant, the task becomes exceedingly complicated.

17.5.3. RISK AND RELIABILITY.

As has been pointed out in Chapter 4, risk can be expressed in two ways. It may be expressed as reciprocal time, e.g. as $X \times 10^{-v}$ years; that is the measure of the probability of the event occurring during the next unit of time, usually a year. When risk is a measure of the probability of one event following another, for example the probability that if A fails then B, which is a back-up device, will also fail, risk is then expressed as a number between 1.0 and zero. If the probability of failure is 1 then the presence of B will not alter the risk associated with A as the risk associated with $A \rightarrow B$ is the product of the risk of A and the risk of B. If the risk associated with B = 0.1, for example, then the combined risk is 0.1 x risk of A.

There is a vast amount of data on the risk of failure of standard components which is collected in the literature under the subject heading of "reliability". This is the inverse of risk; when risk is expressed as reciprocal time, reliability is expressed as time. When reliability is expressed as probability this is equal to 1 minus risk expressed as probability.

17.5.4. FAULT TREE ANALYSIS.

The computation of risk in complex plants is considerably assisted by fault tree analysis. This is not, in itself, a measure of risk; neither is it a measure of hazard. It is an algorithm or formal technique for setting out strings of sequential events which can lead to a "top event". Though almost universally referred to as a "fault" tree it is a neutral device which equally can set out the paths which can lead to a desired goal. Used as a fault tree, as was pointed out in Marshall (1980c), it is equally applicable as a tool for analysing the sequence of events which lead to the failure of a wrist watch or to the melt-down of a nuclear reactor. The present author described the technique in that reference more or less as follows:-

> "The methodology generally used for predicting the risk of failure of systems involves the construction of an algorithm based upon the principles of Boolean algebra and in which it is possible to assign probabilities to each of the events . . . Such a "tree" . . . identifies the event sequence which could lead to a specified failure event. The use of fault tree analysis in the chemical industry is described in Powers and Tompkins (1973).
> Such fault tree analysis, though providing an algorithm for risk assessment also has an equally important function, that of the qualitative description of failure modes, i.e. of the paths which can lead to any sub-event or top event. A sub-event featured as a "component failure" is capable of featuring as a top event in a subsidiary fault tree analysis. For example an electric motor failure can appear as "component failure" which leads

to a sub-event which leads to a top event. But an electric motor failure can feature as a top event in a fault tree analysis of its own. The boundaries of a fault tree analysis are, in principle, unlimited but, in practice, they are drawn as the result of an analyst's judgement.
It is necessary to emphasise that there is no such thing as "pure risk". Fault tree analysis presupposes firstly the existence of a system, whether actual or projected, and secondly that the system can be represented in the form of a circuit diagram. The events which lead to the top event and the top event itself are the realisation of hazards. Hazard analysis must precede risk analysis".

One drawback to risk analysis using fault trees is its neutrality. It contains no value judgements regarding the relative importance of the events which it describes. However it is possible to take short cuts by looking at the top events first as, for example in the prediction of the risk of steam boiler failure, by confining the analysis to the tree which leads to catastrophic failure. This would be justified if what is being looked at is risk to the public.

A second draw-back is its limited applicability to the chemical and process industries. Fault tree analysis would seem to have its greatest applicability to mechanical/electrical systems especially where the events are of a binary nature, i.e. on/off as for example the blowing of a fuse. Process malfunctioning certainly includes elements of this particularly in its automatic control systems. However much of process malfunctioning lies not in simple on/off situations but departure from norms. Though it is possible to treat departure from norm as a binary situation in one sense, the sense that a variable is either inside or outside of a predetermined range of values, the setting of this range may require a fair amount of selective judgement and the consequences to which it could lead may be difficult to predict.

It would be interesting to discuss whether the application of fault tree analysis could have prevented either Flixborough or Seveso; the author has failed to discover any literature which investigates this general question.

17.5.5. HAZARD AND OPERABILITY STUDIES.

A system of formal analysis which is more specific to the chemical and process industries than fault tree analysis is that known as "Hazard and Operability Studies". It was devised by ICI Ltd and was described in Lawley (1974). Though it is of value in predicting circumstances which can lead to the realisation of major hazards, this is not its principal function which is to maximise day-to-day operability and to avoid personal accidents. It was defined in Chemical Industries Association (1977) as:-
"The application of a formal critical examination to the process and engineering intentions of the new facilities to assess the hazard potential of mal-operation or mal-function of individual items of equipment and the consequential effects on the facility as a whole".

This appears defective as a definition not least in that it says nothing about the technique. I.Chem.E. (1984) defines it as:-
"A study carried by application of guidewords to identify all deviations from design intent (and) with undesirable effects for safety and operability". "Guidewords" are defined as a "list of words applied to system items or functions in a hazard study to identify undesired deviations".

Chemical Industries Association (1977) and Kletz (1984) describe the application of HAZOP (as it is usually known). HAZOP is based upon an examination of individual process items on a flow sheet. The design intention of each item is considered in turn this usually being included in its specification. A series of guidewords are then applied to these process intentions and the implications are considered. There are seven basic words, NO (or NOT), MORE, LESS,

AS WELL AS, PART OF, REVERSE and OTHER THAN, but other appropriate words such as SOONER, LATER, or HIGHER, LOWER are also permissible. These are applied in turn, and where appropriate, to such intentions as flow, temperature, pressure, concentration etc. The consequences which would arise are then assessed in relation to safety and plant operability. Where these are judged to be significant appropriate remedial measures are devised which, if they entail modifications to the system, must be the subject of a further HAZOP. Such an analysis is thus diagnostic and remedial.

Examination of the case histories in the book suggests that only a few would have been predictable through the use of HAZOP.

There are, however, other drawbacks to the operation of HAZOP of which the considerable demands it makes upon manpower may be the most serious.

Chapter 18

QUANTIFICATION OF HAZARDS

18.1. INTRODUCTION.

The problem of establishing that level of inventory of an identified hazardous substance at which the hazard it presents may be judged to constitute a **major** hazard, is a difficult one.

Today, arising out of the previously referred to European Directive, twelve European countries are in the process of passing and implementing laws to control major chemical hazards all of which are based upon mandatory threshold levels for more than 180 substances or class of substances as laid down in the Directive. Such laws, if they are to be enforced, require the expenditure of substantial resources both by the state and by private industry. But if the laws are too stringent they consume resources which might be better deployed elsewhere. If they are too lenient they fail to give proper protection to the public.

Yet, in spite of the very large sums of money at stake, there has been little published theoretical analysis of the basis upon which such threshold values have been established, and virtually all of the papers published on the subject are from the pen of the present author. On the one hand there have been arguments from the mass media and from public pressure groups that the threshold levels are set too high and that the law is in consequence too lenient. Such arguments were put forward in the British House of Commons when the CIMAH Regulations (see Chapter 19) were laid before it. On the other hand some sections of industry in the EEC are arguing that the thresholds, in certain areas, appear to be far too stringent.

This chapter will be devoted to an analysis of the problem and will attempt to establish a set of principles upon which threshold values may be determined.

18.2. THE PROBLEM RESTATED.

Those public bodies which have been concerned with the problem have all stated it as a problem of **major** hazards. All have agreed that such hazards present a risk of injury to the public as well as to those working in the installation. The Terms of Reference of the ACMH and Article 1 2c of the Directive have also considered that such a hazard may present a risk of damage to the environment as a concomitant of personal injury. They have also conceived the possibility that damage to the environment may be produced without producing personal injury. The Directive puts it ". . . leading to a serious danger to man, either immediate or delayed, inside or outside of the establishment, and/or to the environment". The question as to whether a hazard may be realised which presents a major risk of environmental damage without, at the same time, presenting a major risk of injury to people, has been discussed in Chapter 16.

It seems beyond dispute that accidents such as the Oppau explosion in 1921, or the Ludwigshafen VCE of 1948, the Flixborough VCE of 1974, or the Bhopal toxic cloud of 1984, were of such a magnitude as to constitute, in the minds both of experts and of the public, the realisation of major hazards. They represent the extreme end of the spectrum of chemical hazards and the justification of the need for exceptional measures to control hazards of exceptional potential.

On the other hand, there seems to be general agreement that hazards which normally present a risk of killing one or two persons do not constitute major hazards. To take an example, there seems little doubt that the cylinders of LPG which are common articles of commerce and which hold perhaps 10 kg of gas, possess the potential to cause one or two fatalities. But to seek to control such cylinders under the legislative category of "major hazards" would bring the concept into disrepute.

The threshold level for **major** hazard must therefore lie somewhere between the potential for causing scores, if not hundreds of deaths, and the potential for killing one or two people. The question is where does it lie ?

18.3. THE MEASUREMENT OF POTENTIAL.

18.3.1. POSSIBLE METHODS OF MEASUREMENT.

In the section above it was assumed that the potential for causing fatalities is a valid measure of the potential of a hazard. It seems inescapable that any realisation which leads to major loss of life will be regarded by the public as a matter for serious concern and that potential for fatality must be one measure of a hazard.

A second measure, in principle, is the potential for producing widespread injury. This potential will be examined below.

A third possible measure is the potential to cause widespread damage, which may be interpreted as damage to the environment. Such damage may be measured, at least partially, though not necessarily exclusively, in financial terms.

18.3.2. MEASUREMENT BY FATALITIES.

One characteristic of measurement by fatalities is that it is relatively unambiguous. Lee (1981) commented on death being "a simple binary event, you are either dead or you are not . . .". However, though there is some truth in Lee's observation, there is not the total simplicity that Lee claims. If a body is dragged from a fire then the conclusion may be obvious. But if the victim is still alive and dies later, the question as to whether the fatality has resulted from the fire may be attended by forensic difficulties, especially if several months elapses before death. But, even so, the statement "relatively unambiguous" still holds as such cases of long-delayed deaths are relatively few, especially in relation to fires and explosions, though perhaps less so in relation to toxic injury. It is well known that this problem presents many difficulties where nuclear accidents are concerned but such accidents are not the concern of this book.

A further factor lies in the reporting of such events. Fatalities provide an index of sensation for the media and events involving fatalities are more likely to be reported than those which do not. This problem of under-reporting of incidents not involving fatalities has been discussed in Section 3.2. above.

18.3.3. MEASUREMENT BY INJURY.

There can be no doubt that the realisation of major chemical hazards can give rise to non-fatal injury and that it is not necessary to cite evidence to this effect. It is one of the characteristics of such realisations that they lead to multiple deaths, to serious injury and to minor injuries and that, in general, the severity of the injuries sustained falls off with distance. It is also generally true that the numbers affected in some categories of realisation increase as the severity decreases. This is analogous with the "Heinrich Triangle" described in Heinrich (1950). The Heinrich Triangle shows that, for industrial accidents, there are far more minor injuries than there are serious injuries and far more serious injuries than fatalities. However it is only an analogy as Heinrich was examining the chance that a personal accident could give rise to any given level of injury. In personal accidents many of the hazards which give rise to them are most unlikely to give rise to anything more serious than minor injury. Major chemical hazards have the capability of giving rise to multiple accidents and it would be expected, in general, that they would have the capability of giving rise to the whole spectrum of injury.

The question which then arises is whether a particular type of hazard may be expected, in a typical realisation, of giving rise to multiple cases of serious injury without a corresponding, and possibly lesser, number of fatalities. Were there to be, for example, scores of such injuries in an incident of which a substantial number led to permanent disability, would this be regarded by the public as the realisation of a major hazard ? The answer may well be yes. But the question then becomes, would it be reasonable to predict, for a given class of hazard, that this would be the most likely outcome and that for this class fatalities would be rare ?

The historical experience of major fires and explosions is strongly against this view. In the case of severe fires, fatalities are common, and it cannot be said that in these instances the typical realisation is one which results in large numbers of injury and that cases of death are untypical.

However the experiences of toxic releases may be different, at least for some substances. Seveso is widely regarded as a disaster but it was characterised by hundreds of cases of injury, some of them long-lasting, but without fatalities. The historical experience shows that the causative agent, dioxin, had been involved in a number of previous incidents in which it had given rise to multiple injury but seldom, if ever, to death. Thus if society is of the view that the hazard presented by certain toxic substances constitutes a major hazard, even though the historical experience shows that it gives rise to large numbers of injuries, but not to fatalities then, in cases of this sort, it may be reasonable to measure the potential consequences in numbers injured rather than numbers killed. Even so, because of the relative rarity of the circumstance, this would constitute, not the norm, but a departure from the norm.

A problem which arises from counting injuries is that, unlike death, there may be many degrees of injury or of permanent disablement. Statistics may differ from one country to another because of differences as to what constitutes serious injury.

A further question is whether the injuries associated with certain hazards may be of a specially unpleasant character. For example, in the minds of many people injuries from nuclear radiation appear to arouse more aversion than, say, those from fire. This particular question lies outside the scope of this book, but it is raised to illustrate the problem. It would seem that both fires and explosions can give rise to crippling and agonising injuries. It is an open question whether the sub-lethal toxic effects of certain substances may fall into a special category of unpleasantness which makes the injuries worse than those sustained from fires or explosions. It would seem, even if such special effects can be demonstrated, very difficult to quantify them.

18.3.4. MEASUREMENT BY FINANCIAL DAMAGE.

Where there is a realisation which gives rise to widespread death and/or injury financial loss is an inevitable accompaniment as the victims have to be compensated. But there may be widespread property damage in addition which could amount to sums in excess of $100,000,000. Where there are fires or explosions, financial loss through property damage may be regarded as an inevitable accompaniment.

In the case of toxic releases, animal deaths are likely to accompany human deaths and this clearly constitutes a source of financial loss but it seems unlikely that this could be the major index of a disaster. It is not so obvious that damage to plant and equipment and to domestic premises are a necessary accompaniment of toxic releases. Some gases may damage control equipment and others may bleach furnishings but the financial loss from this source is likely to feature less than that in the aftermath of an explosion.

18.3.5. SUMMARY.

From the above it may be concluded that, as a norm, the potential of major chemical hazards may most effectively be quantified as potential fatalities. In exceptional circumstances, and where a case can be made, the potential may be expressed alternatively as potential injuries or financial loss.

18.4. THE ISOTRAUMIC APPROACH.

It follows from the argument above that if all fatalities be regarded as equivalent, then one hazard, the potential of which is measured in fatalities, would be equal to another if its potential in fatalities is equal. Thus disparate hazards such as explosives and toxic gases may now be compared with each other. This is the view put forward in Marshall (1977).

It follows also that the threshold levels for the control of different hazards should be those which are assessed as being likely to give rise to an equal number of fatalities. The author coined the description "isotraumic" (Gr = equal wound) for this approach in Marshall (1982) and in Marshall (1983a).

It remains now to determine the isotraumic threshold level and then to consider how, for each hazard, or class of hazards, the relationship between inventory and potential fatalities may be determined.

18.5. DETERMINING THE ISOTRAUMIC THRESHOLD LEVEL.

18.5.1. THE CRITERION OF NASH.

There is no absolute criterion which can be used to establish the number of fatalities, if such be chosen as the measure of what constitutes the point of transition from an "accident hazard" to a "major accident hazard". Nash (1976) declared that his criterion for inclusion of an incident in his book was that it had caused twenty or more fatalities. The rule is adhered to in large measure by this author, out of about 350 major fires and explosions he records between 1910 and 1975, only 6 have fewer than 20 fatalities and these mostly had exceptional levels of property damage. Thus 20 fatalities could constitute the criterion of a major accident hazard. However it could be argued that, as Nash was concerned with all kinds of fire and explosion, the numbers involved, had he chosen, say, ten as the threshold limit, would have been too large to give details in the space available. The plot of F v $\geqslant$N for fires in the UK, given in Figure 4.3. of this book, shows that there were roughly three times as many fires involving 10 or more fatalities as there are fires involving twenty or more. On this basis, had Nash taken ten as his lower limit, he would have had to give particulars of more than a thousand fires and explosions. It does not follow therefore that twenty is necessarily the criterion of what constitutes a disaster; Nash was more or less compelled to use it for considerations of space.

It is also true that the number of incidents in the chemical and process industries of the kind discussed in this book which involved 20 or more fatalities is very few indeed.

18.5.2. CHOOSING A LEVEL.

It seems beyond dispute that an accident involving 20 fatalities would readily be accepted by practically everyone as constituting a major accident. The difficulty then lies in deciding how much lower to go. It might reasonably be argued that an accident involving 2 fatalities would not be so accepted and on this basis the level would seem to lie somewhere between 3 and 19. The present author is of the view that a figure half-way between, the figure of 10, would be an appropriate figure to constitute the point at which an accident becomes a major accident and this is the figure which will be assumed in the discussion below. But anyone wishing to assume a different level may still apply the methodologies which the author sets out and establish figures which are correspondingly higher or lower than those put forward by the author.

From the discussion it will be seen that a concealed assumption of a level of 10 fatalities appears to lie behind the considerations which have led to the fixing of some of the most important of the inventory levels in Annex III of the European Directive on major accident hazards.

It is one thing, however, to look at the historical record and to classify all accidents involving 10 or more as major accidents, and quite another to be able to predict whether or not a given hazard, if realised, would kill 10 or more people and thus be classified as a major accident hazard. This question is given detailed analysis below.

18.6. THE MORTALITY INDEX APPROACH.

18.6.1. INTRODUCTION.

The author introduced, in Marshall (1977), the concept of the "mortality index" as a means of determining the potential for fatality in the chemical and process industries. This index was the mean number of fatalities per tonne resulting from the realisation of the potential of given hazards. This was not necessarily a linear relationship; for example, it expressed the mortality index for high explosives as a power function of the quantity involved.

The author did not know when he wrote this paper that the concept had been developed in the Second World War as a means of assessing the efficacy of allied bombardment of Germany; this is discussed in Irving (1980). Irving notes that German bombs in the early months of the war had killed 40,000 persons in the UK per 50,000 tonnes of bombs. However this mass of bombs represents much less actual high explosive as the weight of the casing has to be deducted. Christopherson (1946) suggests that they would have had a charge weight of ca 60%, i.e. 60% of the mass was high explosive, 40% of the mass was casing, and thus the mortality index would have been 1.33. (British bombs were only about half the charge weight of German bombs at the beginning of the war. This was because of a mistaken assessment of the contribution of the bomb fragments to mortality; by the end of the war British bombs had a charge weight of 80%). Blackett (1941) estimated a mortality index of 0.2 per tonne for British bombs on Germany which would have been ca 0.6 for the mortality index measured against tonnes of explosive. Irving argued that Blackett's figures were out by a factor of 50 but this was true only for the qualitatively different fire storm raids, especially those on Hamburg and Dresden. Figure 18.1. shows the present author's interpretation of the bombing figures for the 22 most heavily bombed British towns, 3 German towns without fire storm raids, and 4 German towns with fire storm raids. The number of fatalities is plotted against the mass of bombs, not mass of high explosive. Blackett's figures for conventional bombing were too low but by no means as low as Irving claims.

Though mortality indices were used, but not under that name, in the Second World War, the author believes that he was the first to suggest that the concept could be used for quantifying major chemical hazards.

18.6.2. THRESHOLD LEVELS AND MORTALITY INDICES.

If it be accepted that threshold levels are based, if only implicitly, on the concept that they represent a certain level of fatality, it would follow that they are a mode of expression of mortality index. Thus if the implicit level of fatalities is 10 and the threshold level is Q tonnes, then the mortality index is given by:-

$$M_I = 10 / Q$$

where M_I = mortality index, fatalities per tonne

Though mortality indices in such papers as Marshall (1977) and Marshall (1978) were taken from the historical record, there is no essential connection between the concept of the mortality index and the concept of the historical approach. If alternative methods of deriving mortality indices which do not rely upon the historical record are available, they could be used also. The principles upon which the historical approach and alternative approach are based are discussed below.

18.6.3. THE HISTORICAL APPROACH.

In a sense, the fact that today major chemical hazards are being regulated derives from the historical approach. Had there not been a number of incidents of such a magnitude as to arouse the concern both of the state authorities and the public then it seems unlikely that legislation would have resulted. The magnitude of the incidents which gave rise to this concern is a measure, admittedly crude, which may be used as an index of what constitutes a major realisation. On the other hand such legislation will seek to regulate in areas in which there is, as yet, no historical experience of disasters and in order to prevent them from happening.

In some particular areas such as explosives there are extensive data, particularly if the experience of war is taken into account. In others, data are relatively few. In all areas there is a great deal of statistical variance, in some incidents hundreds of tonnes of agent have been involved without fatalities ensuing, in others there may have been multiple fatalities from quantities of less than a tonne. Clearly the level of population density in the vicinity is a factor of major importance here. Also, particularly for explosives under war time conditions, the degree of shelter is very significant.

18.6.4. THEORETICAL APPROACHES.

There can be no purely theoretical approaches to the prediction of fatality. An empirical relationship has to be derived which relates the level of human injury to some environmental change such as in temperature, pressure or chemical composition of the environment. It may then be possible to use this as a basis for the prediction of effects at various distances. In the area of toxic gases

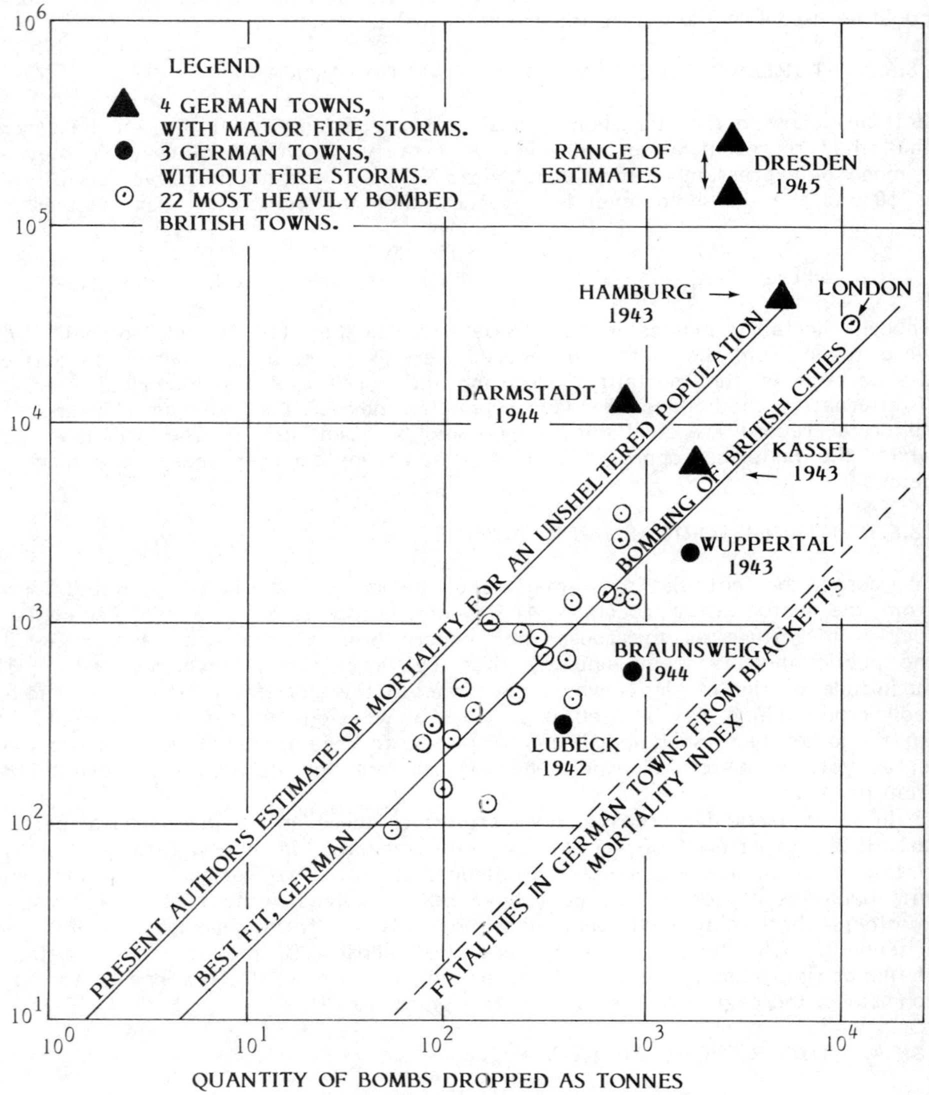

Fig. 18.1. FATALITIES AS A FUNCTION OF MASS OF BOMBS DROPPED, UK AND GERMANY 1940-45.

the difficulties inherent in this process have already been discussed in Chapter 14.

Such approaches might be expected to refine the historical approach by providing models which take account of such factors as population density, dispersion characteristics of vapour or gas clouds etc. It has to be said that though in the cases of toxic releases there has been of late a very marked improvement in techniques for predicting the behaviour of toxic clouds, there has not been, as yet, so marked an improvement in the prediction of the response of target populations. Some of the calculations have made predictions which are so far removed, by several orders of magnitude in certain instances, from the experience of actual releases, as to bring the technique into disrepute. However the author believes that what is needed is to subject the models to a more rigorous analysis which will expose the flaws in the methodology and correct them. In this way theoretical models will be produced which will be concordant with the historical experience and which will be more useful in predicting what is likely to happen in an actual situation.

18.7. THE MORTALITY INDEX OF EXPLOSIVES.

18.7.1. BLAST OVERPRESSURE AND FATALITY.

Injury to humans from explosions has been discussed in Chapter 10 and it is evident from the number of different paths there are to injury or death in explosions that considerable difficulties stand in the way of constructing a theoretical model based upon the effects of blast on human beings. Zuckermann, quoted in Irving (1980), confirmed this. Zuckermann had conducted experiments on live goats in the Second World War, and had concluded that overpressures of the order of 27 to 35 bars were needed to kill humans by pressure alone. (This compared with previous estimates of ca 1/3 of a bar). This is a different conclusion than that which would be drawn from Table 10.3., but this table relates to nuclear explosions where the duration is much longer than with higher explosives. People might therefore die at lower levels of overpressure from nuclear explosions. This may, in part, explain the high levels of mortality index for nuclear explosions. Extrapolation of Figure 10.2. would suggest that overpressures of 27 to 35 bars would correspond to a scaled distance of ca 8 metre $tonnes^{-0.333}$ or a distance of 8 metres from a 2 tonne bomb containing 1 tonne of explosive. Zuckermann observed that only a small proportion of those that had died in air raids on Britain had been within this distance.

18.7.2. CALCULATING THE MORTALITY INDEX FOR EXPLOSIVES.

18.7.2.1. Compilation of data.

Reference has already been made in Chapter 10 to a number of compilations of accidental explosions of conventional explosives. It will be convenient to list these in more detail as follows:-

Assheton (1930) gives 144 incidents up to 1930, with detailed case histories and lists 117 of these in ascending order of size of explosion.

Robinson (1944) gives 137 incidents up to 1943, in rank order of size of explosion. Some details on deaths, injuries and damage are given. It is largely a reproduction of Assheton's table with particulars of deaths etc taken from Assheton's text. No incidents are drawn from the 1930's; 9 incidents, identified by code number, come from the Second World War.

Healy (1965) gives 24 incidents from the Second World War with no apparent overlap with Robinson.

Biasutti (1985) has more than 800 accidents up to 1984 listed as case histories.

These are stated by the author to have occurred in the explosives industry but this seems not to be strictly true as a few of those listed do not fall into this category. (This is the second edition of Biasutti 1981).

18.7.2.2. Analysis of the data.

In Marshall (1977), 162 data points were obtained by combining the Assheton/Robinson/Healy data. These were divided into 12 classes by mass of explosive involved, each class lying between limits the higher of which was equal to the lower multiplied by $10^{0.5}$, i.e. by 3.17. The mortality index was calculated for each class and plotted on log/log paper against the mean mass for each class. The best fit line for this log/log relationship was calculated and it was noted that the number of incidents in each class appeared to have a quasi-Gaussian distribution around the class 1 - 3.17 tonnes. This was attributed to under-reporting of small incidents.

The author has now obtained a further 203 points from Biasutti (1985) covering the period 1935 to 1984. So far as he has been able to discover these are entirely independent of any of the points in the Assheton/Robinson/Healy set. These have been treated similarly to the way the data was treated in Marshall (1977).

The two sets of data are plotted in Figure 18.2. and they have been combined in Figure 18.3. Figure 18.4. shows the number of data reported in each class for the combined data set and demonstrates the quasi–Gaussian relationship which arises through under-reporting of small incidents. Table 18.1. shows the results for the linear regression analysis for the relationship, log mortality index as a function of log mass.

TABLE 18.1. REGRESSION ANALYSES OF MORTALITY DATA.

	No. of classes.	a_0	a_1
(1) Assheton/Robinson/Healy	9	4	-0.5
(2) Biasutti	12	4.7	-0.65
(2) plus (3)	12	6	-0.56
(4) Weighted data	7	3.5	-0.45

In order to form a more realistic appraisal of data in the range $\geqslant$1 tonne, which is the range with which major chemical hazards is concerned, a linear regression of the log/log relationship for this range has been conducted but with a weighting for the data points according to the number of incidents making up each class. In this way the tendency to overstate the mortality index for small incidents has been avoided, otherwise there would be bias in reporting incidents involving fatalities and, at the same time, ignoring non-fatal incidents.

The regression analysis shows consistency in the prediction of the number of fatalities associated with 1 tonne of high explosive. This lies in the range 3.5 to 6.0 fatalities. There is a much greater variance in the slope of the line which corresponds to the value of the exponent in the relationship $M_I = a_0 Q a_1$. This shows a variance between -0.45 and -0.65. It is however readily influenced by a few data for small, i.e. of the order of 1 kg, explosions. It is also probably true that some of the major explosions were probably a series of explosions rather than one single large explosion.

Theory would suggest a slope of -0.333. This follows from the concept that for a uniform population density, and assuming that fatality ensues at, or above, a given level of overpressure through whatever mechanism is responsible for fatality, the number killed will be those lying within the circumference

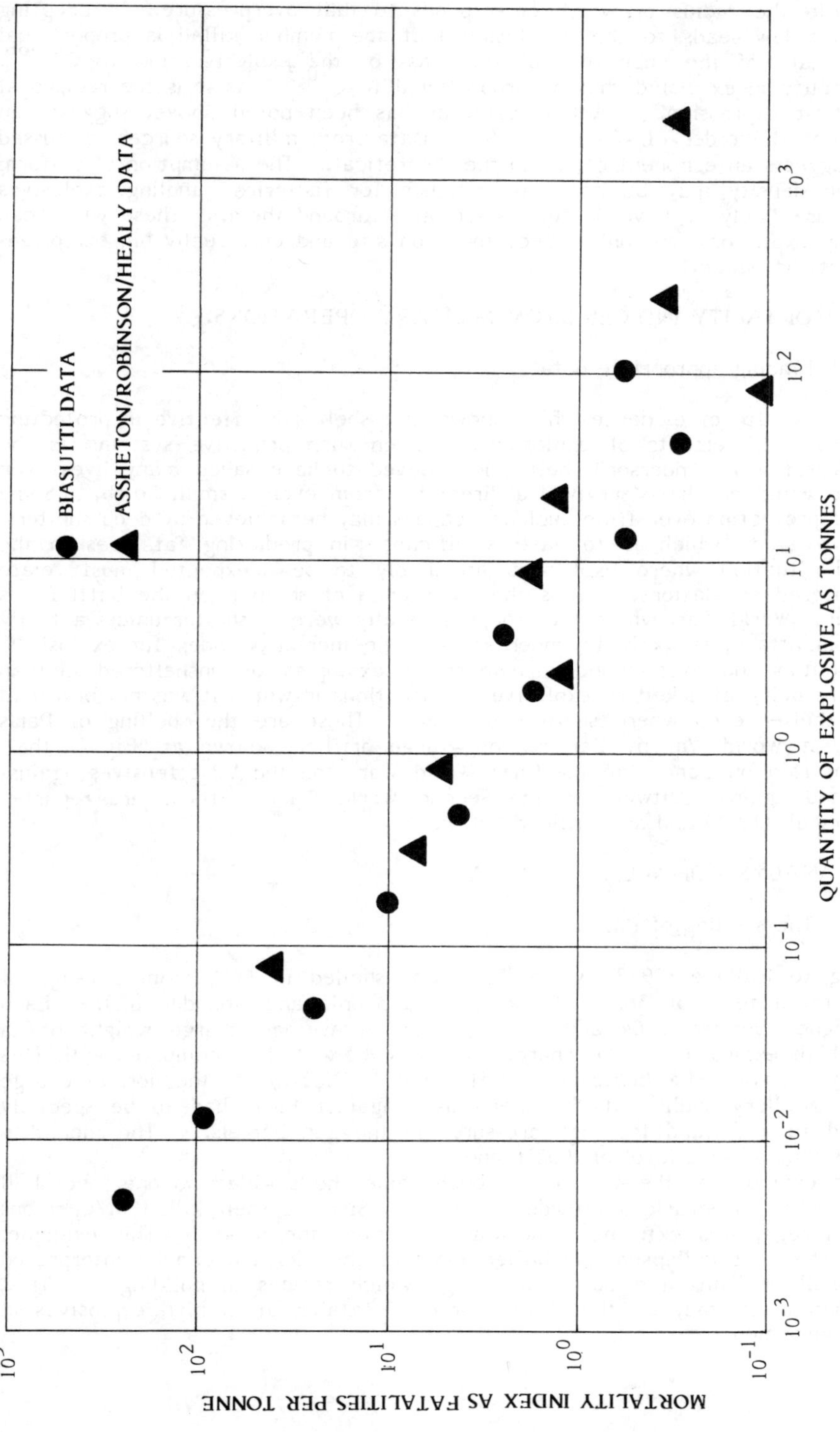

Fig. 18.2. MORTALITY INDEX v MASS OF EXPLOSIVE, TWO DATA SETS.

of a circle the radius of which corresponds to that overpressure. Accepting Hopkinson's law leads to the conclusion that the number killed is proportional to the square of the cube root of the mass of the explosive, i.e. to $Q^{0.666}$. The mortality index would thus be proportional to $Q^{-0.333}$ as it is the reciprocal of the first expression. Actual data, as has been noted above, suggests an exponent of the order of -0.45 to -0.65. Data from military sources, discussed below, suggests an exponent closer to the theoretical. The assumption of uniform population density may be open to criticism for factories handling explosives as these are likely to have a "cordon sanitaire" around them. These will mean that many explosions will only affect those on site and only really big explosions will affect the public.

18.7.3. MORTALITY INDICES FROM MILITARY OPERATIONS.

18.7.3.1. Finding appropriate data.

There is a wealth of evidence which shows that shelter is effective in protecting people from the effects of explosions. Even such primitive systems as the Second World War "Anderson" shelter is believed to have saved many lives even though it would not have survived a direct hit from even a small bomb. Some degree of protection even from nuclear weapons may be achieved by deep shelters. Hence any study which is to have significance in predicting fatalities in the industrial situation where explosions are likely to be unexpected must relate to unsheltered populations. Thus the experience of shelling on the battlefields of the First World War, where the troops generally were in shelter unless actually engaged in attack, is likely to underestimate the mortality index for explosives.

The author however found a number of examples of unsheltered civilian populations being attacked by explosives in situations in which it was not practical to take shelter even where shelters existed. These are the shelling of Paris in the First World War by German long-range artillery known as "Big Bertha"; a daylight raid on London in the First World War, and the V2 offensives against London and against Antwerp in the Second World War. These are reported on in Marshall (1978) and are outlined below.

18.7.4. ANALYSIS OF MILITARY DATA.

18.7.4.1. The shelling of Paris.

According to Hallade (1973), when Paris was shelled in 1918 from a range of ca 130 km, a total of 367 shells killed 256 people and wounded 620. Each shell weighed between 104 and 140 kg with an average charge weight of ca 8 kg of high explosive. The charge was thus 6.5%. This compares with 10% quoted by Encyclopedia Britannica (1951), (Vol 1, P.824d), as the normal charge ratio for artillery shells but the shells used against Paris had to be specially reinforced to withstand the high pressures in the gun barrels. The mortality index was thus 87 at a level of .008 tonnes.

As an example of the variance, though some shells killed no one, one shell killed 75 when it struck a crowded church. Since a shell killed 0.7 persons on the average, this extreme value was 100 times the mean. This example, in which the roof collapsed and buried most of the victims, can be interpreted as the shell realising the potential energy which resides in buildings. This, rather than blast, may be the chief source of fatality from high explosives in a built environment.

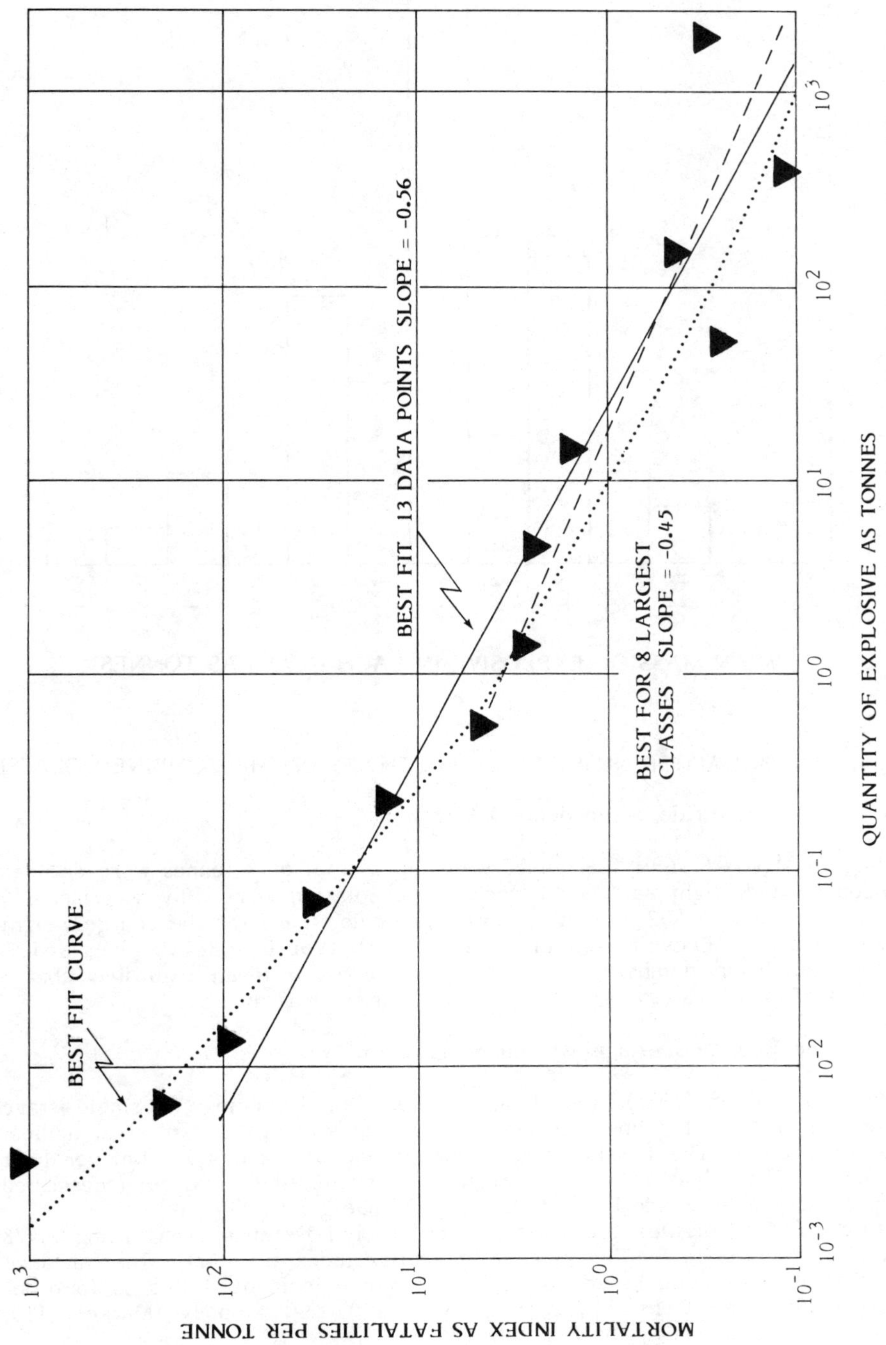

Fig. 18.3. MORTALITY INDEX v MASS OF EXPLOSIVE, COMBINED SETS.

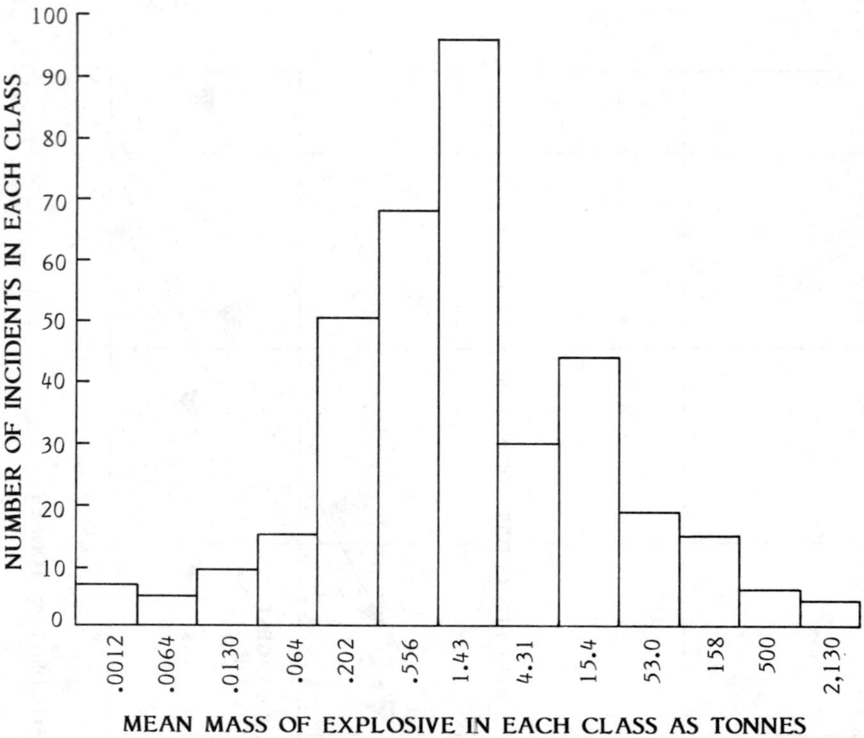

Fig. 18.4. HISTOGRAM OF NUMBERS OF INCIDENTS IN THE COMBINED CLASSES.

18.7.4.2. A daytime raid on London, 13/6/17.

According to Morris (1969), 126 bombs weighing a total of 4 tonnes were dropped on London in a daylight raid which took the population largely by surprise. 162 people were killed and 432 injured. Morris gives no figure for the charge weight of these bombs but Encyclopedia Britannica (1951), (Vol 1, P.882a), gives a 50% ratio for demolition bombs. Accepting this gives a mean explosive charge per bomb of .015 tonnes (15 kg). The mortality index was thus 81.

18.7.4.3. The V2 offensive against London 1944/45.

According to Collier (1976), this took the form of launching V2 single-stage rockets, each with a 1 tonne war head containing 750 kg of explosive, against the London area. The missiles struck the ground at some 4,000 km per hour (Mach 3.3). There was thus no warning. These missiles have been described in many works which include Irving (1980) and Dornberger (1954).

A total of 517 missiles fell on the London Civil Defence area killing 2,478 people, [Collier (1957)]. This gives a mortality index of 6.4. The variance was considerable. Some killed no one, other single incidents killed as follows:- New Cross (Shopping Centre) 168, Stepney (Flats) 134, Stepney (Market) 115, Islington (row of houses) 68, and Deptford (Flats) 52.

18.7.4.4. The V2 offensive against Antwerp.

According to Mathieu (1978) and Bouhon (1958), the city of Antwerp was attacked by German missiles in the period October 1944 to March 1945. A total of 1,923 missiles (including 211 V1s "flying bombs" which were sufficiently similar in their military effects as to be included with the much larger number of V2s) killed 3,592 people in the area of Greater Antwerp, including the inner city. In the inner city itself, 460 missiles killed 2,400 people. The respective mortality indices were 2.48 and 6.95.

18.7.5. INTERPRETATION OF THE MILITARY DATA.

18.7.5.1. Differences from data from accidents.

The principal differences from data from accidents are that (1) in the examples given above each was characterised by uniformity of charge size, (2) each was characterised by the possession of kinetic energy by the explosive charge and its casing. The former characteristic enables the mortality index/quantity relationship to be represented by a single point with a fair degree of accuracy.
In the case of both the Paris shells and the V2 rocket, the kinetic energy of impact was of the same order as the blast energy of the explosive it contained:-

$$K.E. = 1/2\ M \times V^2$$

For the Paris shells the impact velocity was 666 metres per second, [Hallade (1973)].

$$\begin{aligned} K.E. &= 1/2\ 120 \times 666^2 \ \text{J.} \\ &= 60 \times 0.44 \times 10^6 \ \text{J.} \\ &= 26.4 \times 10^6 \ \text{J.} \\ \text{Blast energy} &= 8 \times 4.23 \times 10^6 \ \text{J.} \\ &= 34 \times 10^6 \ \text{J.} \end{aligned}$$

Thus for the Paris shells the kinetic energy was 0.77 times the blast energy.
A similar calculation for the V2 based on an impact mass of 4 tonnes and a striking velocity of 1100 m/sec [Collier (1976)], shows the kinetic energy to have been 0.75 times the blast energy. However the kinetic energy would have been concentrated over a very small area equal roughly to the cross-sectional area of the missile. Collier (1976) comments (P.125) that the V1 with a comparable war head and a much slower impact did more damage than the V2 which buried itself in the ground before exploding. It seems therefore not necessary to correct for kinetic energy.
On the other hand industrial explosions are highly likely to occur at ground level and it may be that they will do more damage than an equivalent quantity of explosive which has buried itself in the ground.

18.7.5.2. The influence of population density.

Reference has been made above to a theoretical relationship between M_I and Q the mass of explosive. If the exponent is theoretically -0.333, what determines a_0 ? This, it has already been suggested in the paragraph in question, is a function of the population density. This is attested by the differing values for mortality index for inner Antwerp and Greater Antwerp. The author has therefore normalised the mortality indices for the military examples by relating the indices to a standard population density of 4.000 per square km after taking account of estimates of the actual population density at the time of the attack.

These data are set out in Marshall (1978) and are reproduced in Table 18.2. The figure of 4,000 per sq kilometre has been chosen as being reasonably representative of urban population densities current in the UK today.

TABLE 18.2. MEAN MORTALITY INDICES CORRECTED FOR POPULATION DENSITY.

Place	Weapon	Mean Mortality Index (M_I)	Estimated Population density (Thousands of persons per sq km)	Corrected Index (M_I per thousand persons per sq km)
London	Air Raid 1917	81	26.600	3.045
Paris	Shelling 1918	85	28.000	3.035
London (LCC area)	V2s 1944-5	6.39	8.125	0.78
Antwerp (Inner City)	V1s & V2s 1944-5	6.95	7.500	0.92

The un-normalised data fall close to the best fit line for the 12 data points on Figure 18.3. but the normalised points fall below that line. The best fit line for the normalised points is given by the equation:-

$$M_I = 3.1\ Q^{0.30}$$

The author has modified this slightly to take account of an anticipated somewhat higher level of fatalities from ground level explosions rather than from buried explosions, and to use the exponent derived from theory as follows:-

$$M_I = 4.0\ Q^{-0.333}$$

Alternatively, this may be expressed as:-

$$M_I = P_D \times Q^{-0.333}$$

where M_I = mean mortality index, fatalities per tonne
P_D = population density, thousands per sq km.
Q = mass of explosive.

18.7.6. THE NOMINAL LETHAL RADIUS.

An extension of the above arguments, given in Marshall (1978), is the calculation of a nominal lethal radius which may be defined as the radius of a circle drawn around the centre of an explosion such that the number surviving within that circle is equal to the number of fatalities outside of it. At the 1 tonne level and a population density of 1000 per sq km this gives 1000 sq metres per person. The radius of a circle of this area is ca 18 metres. This radius is invariant with population density and if the exponent of Q is 0.333 it is a scaled distance.

From a knowledge of the ratio between death and injury it is possible to extend this concept to cover injury also. Encyclopedia Britannica (1951), (Vol 1, P.459d), gives figures for deaths, serious injuries and slight injuries for both V1s and V2s. These have combined as follows:-

(a)	Deaths	8,938		
(b)	Serious injuries	25,504	Ratio (b) : (a)	= 2.75
(c)	Slight injuries	39,931	Ratio (c) : (a)	= 4.46
(a) +	(b)	33,442	Ratio [(a) + (b)] : (a)	= 3.74
(a) +	(b) + (c)	73,373	Ratio [(a) + (b) + (c)] : (a)	= 8.21

The appropriate radii for (a) + (b) and (a) + (b) + (c) are given by multiplying the nominal lethal radius by the square roots of 3.74 and 8.21.

The dead plus seriously injured nominal radius is thus 35 metres and the dead plus all injured nominal radius is 52 metres for 1 tonne.

They could well be rounded off to scaled distances of 20, 35, and 50 metres/tonnes$^{-0.333}$.

18.7.7. ADJUSTMENT FOR DIFFERENT EXPLOSIVES.

The discussion above has assumed a uniform blast energy equal to that of TNT. However explosives differ in specific blast energy and allowance should be made for this, where appropriate, in calculating potential death and injury.

Blast energies for a given explosive may vary according to the techniques used to determine it. There are a number of these techniques of which perhaps the most popular are the ballistic mortar test which measures the reactive pendulum swing of a suspended mortar, and the Trautzl block test which measures the enlargement of a standard cavity in a lead block when the test substance is detonated. Table 18.3. gives comparative blast energies for a number of explosive substances, the values of which are mainly derived from Kirk-Othmer (1980).

TABLE 18.3. COMPARATIVE BLAST ENERGIES.

Substance	Factor
Ammonium nitrate	0.56
Sodium chlorate	0.15
Nitromethane	1.34
Picric acid	1.06
Nitroglycerine	1.69
RDX	1.31
Lead azide	0.40
Mercury fulminate	0.50

Organic peroxides have recently been the subject of regulation by the EEC. These are discussed and some values of comparative blast energies are tabulated in Appendix IV.

Stull (1977) gives techniques, based on their thermochemical properties, for estimating the blast energies of unstable substances.

18.7.8. SUMMARY OF MORTALITY INDICES FOR EXPLOSIVES.

The analysis of two independent sets of data for accidental explosions and of four sets of data for attacks on unsheltered civilian populations shows a considerable measure of agreement for the mortality index for charges of 1 tonne of between 3 and 6. The former level applies to a normalised population density of 4,000 per square kilometre.

There is a great deal more variance in value of the exponent of Q. It is always negative but varies from -0.65 to -0.30, with a theoretical value of -0.333. The normalised values derived from bombardment of unsheltered civilians come nearest to the theoretical but these data cannot be more accurate than the estimates of population density on which the normalisation was based. The author has therefore decided to use the theoretical value for the exponent which, for values in the range 1 to 50 tonnes, is likely to give more conservative values for threshold inventories than would arise from taking most of the other values derived above for this exponent.

Individual values of mortality index show considerable variance for the same charge weight. These high values probably arise from two factors, high local population densities and the release of potential energy by collapsing structures. Using the figures given for V2 attacks, the probability that a given incident would kill 168 people or more, which was 37 times the mean value, was .0038, and the chance of killing 115 or more, which was 25 times the mean value, was .0057.

18.8. MORTALITY INDICES FOR VCEs AND FIREBALLS.

18.8.1. INTRODUCTION.

For tonnage releases of flammable vapours, it is difficult to predict whether they will lead, if ignited, to a fire ball or to a VCE. It seems reasonable therefore to examine the data for both and to use the more pessimistic of the two sets of data.

18.8.2. POSSIBLE LINES OF INVESTIGATION.

The first line of investigation would be to treat vapour cloud incidents in the way that accidental explosions have been treated above and that is to divide them into classes and to investigate whether a mortality index can be determined for each class on the basis of the historical record. The use of classes prevents a single large incident from obscuring the trend which would otherwise be disclosed by a number of small ones.

A second line of attack would be more theoretical and that would be to determine a nominal lethal radius for such clouds. By way of example, such a radius could be the actual radius of the cloud at the time of ignition coupled with the assumption that those caught within the cloud will die and that all those outside of it will escape. The adoption of such a radius would probably be understating mortality.

18.8.3. RESULTS FROM DIVISION INTO CLASSES.

Figure 18.5. shows the effect of dividing vapour cloud incidents into classes with $\sqrt[2]{10}$ multipliers and plotting log M_I against Q for each class. The best fit line shows that M_I is circa 30 for a one tonne release and that the slope of the trend line is almost exactly 1.0. This means that M_I and Q are inversely proportional. The conclusion to be drawn from this is that the number of fatalities is invariant with respect to the quantity of vapour in the cloud. This

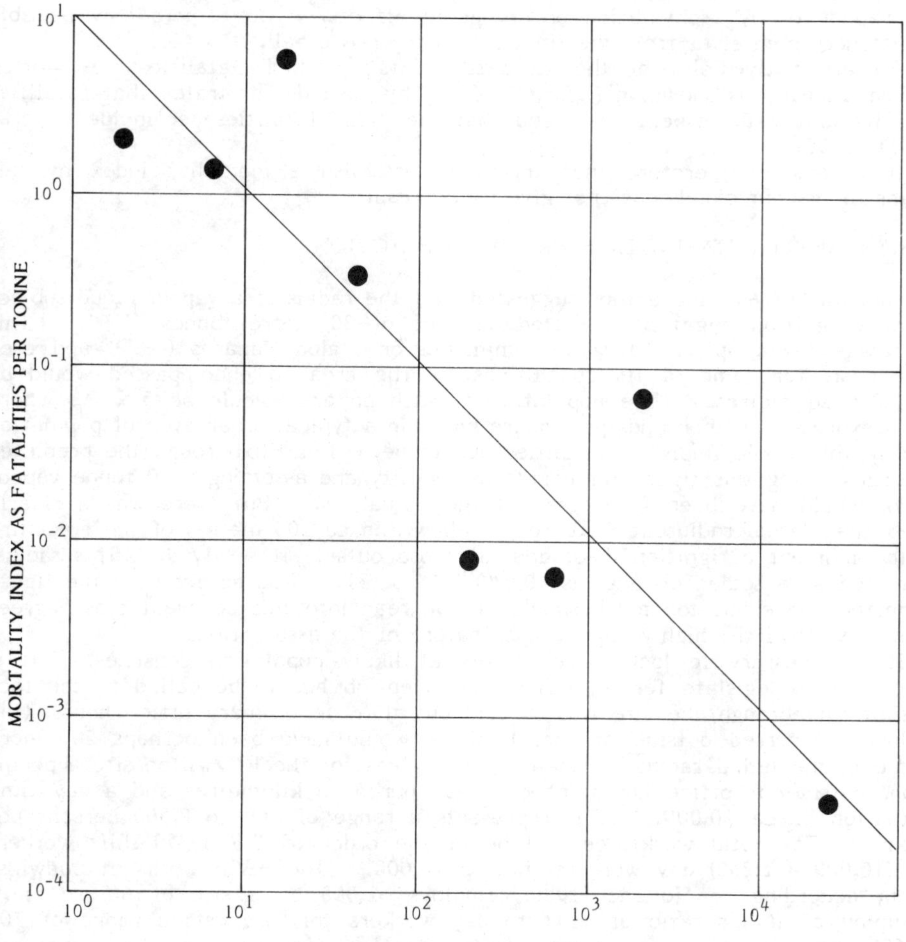

Fig. 18.5. MORTALITY INDEX VERSUS MASS SPILLED, VCEs AND FIRE BALLS.

is a very surprising result, and one which raises considerable difficulties for anyone trying to determine a threshold limit for the control of flammable liquefied gases.

There are many factors which make reliable statistical analysis difficult. The first is one which has been the subject of a number of comments above, that small incidents tend to be under-reported unless they involve fatalities. This leads to high values for the mortality index for small incidents. For very large clouds the time taken to form them may be appreciable thus giving time for people to run away. At Flixborough, where less than 1 minute elapsed between the collapse of the by-pass pipe and the explosion, 6 escaped by fleeing from the scene. A scenario which may be plausible with big leaks is that those immediately on the scene are killed by the pressure rupture which releases the vapour or by asphyxiation or cryogenic effects. But others may be able to distance themselves from the final explosion or fire ball.

An alternative plot of the 22 data points in which fatalities are plotted against quantity is shown in Figure 18.6. This also demonstrates that fatalities are invariant with respect to Q and that the mean fatalities per incident works out at ca 30.

It is evident therefore that trying to establish a mortality index by this means for vapour clouds does not give useful results.

18.8.4. MORTALITY INDEX BY RADIUS OF ACTION.

In Marshall (1976a), the author suggested that the radius of a vapour cloud subject to rapid ignition might be a scaled distance of 30 metre tonnes$^{0.333}$. This view was taken up in Advisory Committee on Major Hazards (1979) and used as a basis for some of its conclusions. The area so encompassed would be ca 3,000 sq metres. The population of such an area would be 3 x P_D where P_D is expressed in thousands per square km. In a typical urban area of population density this would imply 12 fatalities per tonne. For Flixborough the predicted fatalities, using this figure for population density and assuming a 40 tonne vapour cloud, would have been 41 instead of the actual 28. But there was a clearly established lethal radius at Flixborough, all within ca 105 metres of the epicentre at the moment of ignition died, and all those outside it survived. This radius, expressed as a scaled distance is $105/40^{0.333}$ or 31. This agrees with the figure suggested above but too much should not be read into this apparent close agreement in view of the highly approximate nature of the assumptions.

It is necessary to look more closely at likely population densities. It is impossible to legislate for all cases and attention has to be called to the fact that at Flixborough the site was so laid out that, in ordinary office hours, (the explosion occurred outside of these), there would have been perhaps 200 more people in the lethal radius. Challis (1977) gives for the ICI Wilton site a population, outside of office hours, of ca 1,250 per 4 sq kilometres and a day-time population of ca 10,000. This represents a range of 300 to 2,500 persons per sq km. The total workforce will be of the order of 4 x 1,250 shift workers and (10,000 - 1,250) day workers, i.e. ca 14,000. The BASF works at Ludwigshafen, according to Holland (1980), employs 52,000 on a site of 6.5 sq km. Assuming a similar ratio of shift to day workers this suggests a range of 700 to 6000 per sq km. The author has visited both factories and has concluded that Ludwigshafen seems to be much more densely populated. It might be remarked that ICI Wilton is a totally post Second World War development, whereas the Ludwigshafen factory has occupied the same site since the late nineteenth century. They may represent the two extremes of population density in chemical works. Taking the mean of the two densities for out of office hours working would give 500 per sq km and assuming that the process area population is doubled during office hours by maintenance workers etc, this would still suggest that

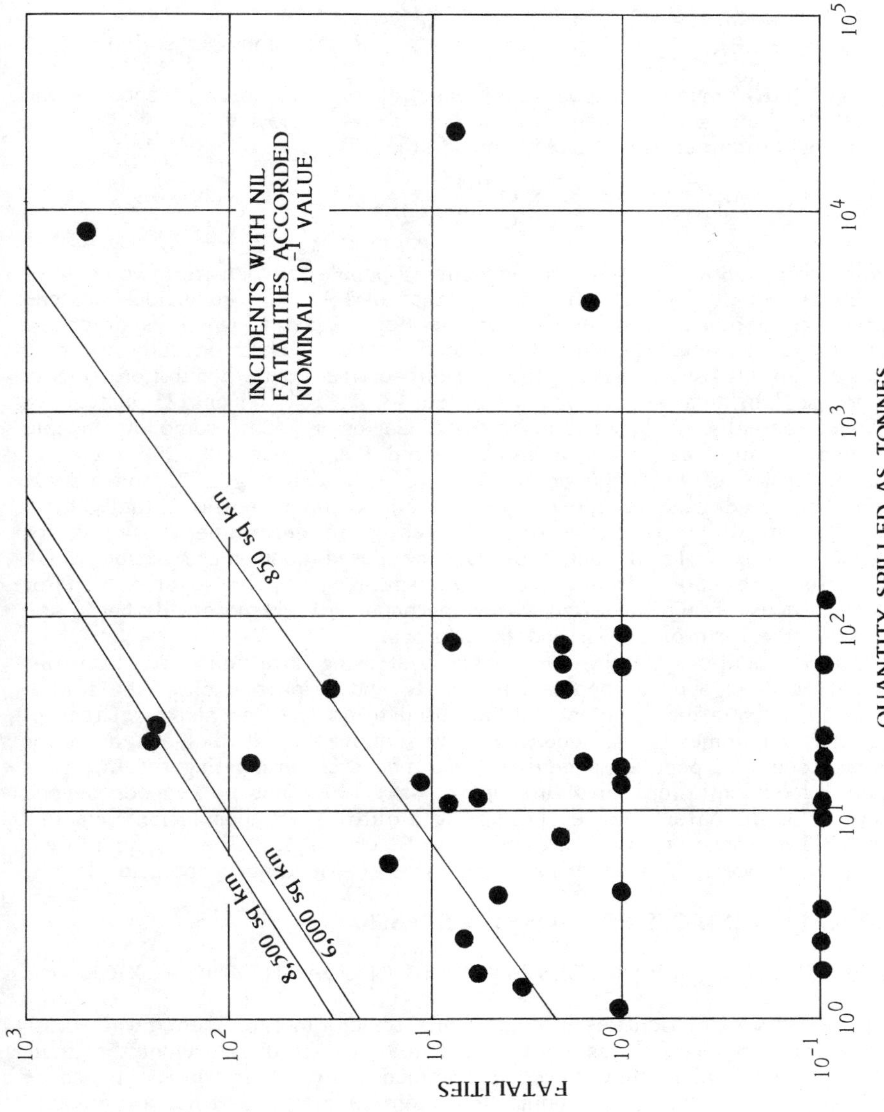

Fig. 18.6. FATALITIES v MASS OF SPILLAGE FOR FIRE BALLS AND VCEs AND FOR VARIOUS POPULATION DENSITIES.

the population density in the most vulnerable areas would be less than 1000 per sq km. The Flixborough site had a population of 70 on a Saturday afternoon. Its area was about 1/5 of a sq km giving a general density of 350 which is comparable with the ICI figure above. However a substantial portion of the site was taken up by derelict plant and there was a high concentration of people in the area where the explosion occurred because so many, 19 in all and more than a quarter of the site population, were in one building, the control building. And it is likely to be the case on most sites that the local population density around process units will considerably exceed the overall population density.

All in all, a figure of ca 850 per sq km of process area may be reasonable. On this basis, and for quantities of vapour of less than 100 tonnes, the following formulae might be used:-

$$M_I = 3\,P_D \times Q^{-0.333}$$

where P_D = population density in thousands per sq km.

This suggests that, tonne of vapour for tonne of high explosive, vapour cloud explosions or fire balls are three times as lethal as high explosive.

Thus for population density = ca 850 per sq km:-

$$M_I = 2.5 \times Q^{-0.333}$$
$$F = 2.5 \times Q^{0.666}$$

The basic equation above, and taking account of population densities would have predicted 29 deaths at Flixborough, under out of office hours conditions, instead of the 28 which actually occurred. For the Beek disaster, which is described in Beek (1976), it would have predicted 8 instead of the 14 which actually occurred. For Ludwigshafen (1948) and using the present office hours population density of 6,000 per sq km, this would have predicted $18 \times 30^{0.666}$ or 173 instead of the 209 which actually occurred. In the San Carlos incident, some 500 people were camping in an area which measured some 300 metres by 200 metres. This gives a density of ca 8,500 per sq km. This, assuming a 22 tonne leak, would lead to a prediction of 200 deaths instead of the 215 who actually died.

But a problem which is not easily resolved is to determine the basis for evaluating Q. For VCEs it should perhaps be based upon the fraction which flashed off, for fire balls perhaps the whole spillage may be involved. For the most hazardous of the liquefied gases, perhaps half or more will be in the vapour phase in the form of vapour and small drops.

These latter calculations show the weakness of using data points for fatalities which do not take account of population density and help to explain the failure to find a useful correlation from the method used to analyse the historical record.

The author recommends the equation given above which includes a factor which takes account of population density, as a means of forecasting the fatalities from vapour cloud explosions involving up to, say, 100 tonnes of vapour beyond which there is little data. Lines predicting fatalities for population densities of 850 to 8,500 persons per sq km are shown in Figure 18.6. The former represents the typical process area conditions, the latter dense urban population levels.

18.9. MORTALITY INDICES FOR TOXIC RELEASES.

18.9.1. THE SPECIAL DIFFICULTIES INVOLVED IN QUANTIFYING TOXICS.

That there are special difficulties in quantifying toxics emerges from the discussion in Chapter 14 where it was pointed out that toxics differ widely in their properties, the only thing they have in common being their ability to cause death or injury by attacking the human (or animal) body. It is not possible within the scope of this book to create a comprehensive scheme which would enable the hazards arising from all toxic substances to be quantified. Instead the scope of the discussion has been restricted to those which exercise their principal harm through causing death rather than by creating environmental damage through sub-lethal doses. Dioxin is the supreme exemplar of the latter.

Therefore only a few illustrative examples, chlorine, ammonia, mustard gas, and methyl isocyanate will be discussed.

18.9.2. THE MORTALITY INDEX FOR CHLORINE.

18.9.2.1. The historical record.

The historical record for chlorine falls into two distinct categories, war experience and industrial experience. These will be treated separately and some comparisons will be made.

18.9.2.2. War experience and the mortality index for chlorine.

The April 22nd Ypres gas attack has been described above in Chapter 15. The chlorine was discharged from a line of cylinders and escape for the vast majority was impossible. The quantity used, 168 tonnes, is well established but the fatalities are not so clearly established. The incident in fact is not well documented. The present author discussed, in Marshall (1977), the number claimed by the Germans as being 5,000 dead. However Haber (1986), which is based upon a detailed analysis of gas warfare, advances the opinion that the number was about 1,000. The number does fit reasonably well with the numbers, about 2 divisions, known to be in the 7 kilometre section of front at the time. 2 divisions could have been about 15,000 men of whom perhaps a half might have been in the front line or in close support. Such figures agree reasonably with data provided in Carrington (1970).

Accepting that the dead numbered 1,000 as Haber (1986) suggests, would give a mortality index of 6, but if the figure of 5,000 put forward by Prentiss (1937) and quoted in Marshall (1977) be accepted, it would be 30. These mortality indices apply to unprotected troops under extremely adverse conditions. The LD_{50} for chlorine is ca 3.5 mg/kg of body weight or about 1/5 gram per person. Thus the chlorine actually consumed in killing soldiers was about 200 grams to 1 kg out of 168,000 kg discharged. Put another way, the gas attack involved 8×10^{10} lethal doses.

Evidence from later attacks show much lower mortality indices. Figures on quantities and fatalities are provided in Prentiss (1937). These have to be taken with caution as it was difficult to tell exactly how many were killed by gas as opposed to shells, bullets etc; conditions of war precluded the possibility of post mortem examinations. For other cloud gas attacks, Prentiss attributes 2,312 deaths to 1041 tonnes of chlorine gas. This gives a mortality index of 2.23.

All in all, the figures suggest that with some degree of protection for the troops, but who were also at the time subject to the other harassments of war, the mortality index for chlorine was of the order of 2.25.

18.9.2.3. Industrial experience and the mortality index for chlorine.

It is a matter of some importance to be able to derive a mortality index for chlorine because of its widespread use and its toxicity. Such an index could perhaps then be capable of yielding mortality indices for other bulk chemicals based on their toxicity relative to chlorine. Such a step has to be taken with caution for the reasons set out in Chapter 14 but it could be at least indicative.

As will be shown below the death toll at Bhopal, after allowing for certain factors which intensified it, was to some degree predictable from the product of the mortality index for chlorine and the relative toxicity of methyl isocyanate and the quantity released.

One factor which requires early discussion is the question of scale effects. If for explosives M_I is a function of $M^{-0.333}$, what is the exponent of M for toxic clouds ? The author has been unable to find this discussed in the literature. McQuaid (1982c) has provided him with a set of calculated "lethal dosage envelopes" for different sizes of release, meteorological stability and windspeed. This calculation was based on the use of the DENZ computer program which is discussed in Chapter 7 above.

If it be assumed that all within these envelopes were to be killed and that the population density is the process area density of 850 per sq km which has been used earlier in this chapter, the death toll for 1 tonne of chlorine, i.e. M_I, would range from 51 to 289. These are figures which greatly exceed the mortality index for Ypres or any other recorded chlorine incident. The higher figure much exceeds that of Bhopal where the toxic agent was at least an order of magnitude more toxic than chlorine. The same may well be true of the population density.

The author has however looked at the scale effects which arise from the use of the model. The case with the lowest number of associated fatalities suggests that the exponent of M should be -0.222 and the worst case suggests that the exponent should be -0.06. Though the author has grave doubts as to the validity of the DENZ program as a predictor of fatalities and hesitates therefore to accept the concomitant scale factor. This, in either case discussed above, is a low-valued exponent and perhaps the safest course would be to assume that M_I is invariant with respect to mass.

Marshall (1977) concluded that the mortality index for chlorine in the industrial situation was ca 0.5. Advisory Committee on Major Hazards (1979) reduced this to 0.3. However, CECAR (1981), which makes use of the Marshall mortality index approach, suggests that for industrial releases it should be based upon data from the rupture of fixed store tanks rather than from transport vehicles. Table 18.4. lists incidents worldwide taken from Marshall (1977) and Advisory Committee on Major Hazards (1979) with the addition of the incidents at Mississauga, Canada, and Montana, Mexico. The index for all causes is 0.18 but for store tanks only it is 0.50. It may be sound therefore to set the mortality index for chlorine in fixed installations at 0.50.

18.9.3. THE MORTALITY INDEX FOR MUSTARD GAS.

18.9.3.1. The properties of mustard gas.

Based upon the treatment of mustard gas on Pages 342 and 343 above, the mortality index for mustard gas is given in two ways by Prentiss (1937). In the one, 27 kg produced one casualty but fatalities were only 1/50th of casualties. This would give 1,350 kg per fatality or M_I = 0.74. Prentiss also quotes 1,130 fatalities from 11,000 tonnes of gas or M_I = 0.1. On the other hand, the sinking of the John Harvey also referred to on P.343, gave an M_I of 0.83. A pessimistic view of M_I for mustard gas would be ca 0.80 making it, if suitably dispersed, somewhat more lethal than chlorine.

18.9.4. THE MORTALITY INDEX FOR AMMONIA.

Using the data in Table 15.2., for all cases, the M_I for ammonia works out at 41/2008 or 0.02. For the two store tank incidents and the ship storage connection it works out at 20/378 or .052. It is, in either case, much lower than chlorine

TABLE 18.4. WORLDWIDE CHLORINE VAPOUR CLOUD INCIDENTS OVER 60 YEARS.

Location	Released from	Tonnes Released	Fatalities
Montana, Mexico	Railcars	300	17
Baton Rouge, La.	Storage Tank	90	Nil
Mississauga, Toronto	Railcar	60	Nil
Rauma, Finland	Storage Tank	30	19
Cornwall, Ontario	Railcar	28	0
Griffith, Indiana	Railcar	27	0
La Barre, Louisiana	Railcar	27	1
St. Auban	Store Tank	24	23
Syracuse, NY.	Store Tank	24	1
Zarnesti, Rumania	Store Tank	24	60
Wynadotte, Mich.	Store Tank	17	1
Chicago, Ill.	Railcar	16	0
Niagara Falls, NY.	Railcar	15	1
Walsum, West Germany	Store Tank	15	7
Brandtsville, Pa.	Railcar	8	0
Mjodolen, Norway	Railcar	7	3
Freeport, Texas	Pipeline	4	0
Lake Charles, La.	Pipework	3	0
Johnsonburg, Pa.	Railcar	2	0

M_I, all cases = 133/720 = 0.180.
M_I, store tanks = 111/224 = 0.500.

and accords with ammonia's much lower toxicity (about 1/7 as calculated on P.333).

18.9.5. THE MORTALITY INDEX FOR METHYL ISOCYANATE.

Only one incident is available for estimating the mortality index for this substance and that is the Bhopal disaster of 3/12/84. According to New York Times (1985), there were some 45 tonnes of methyl isocyanate (MIC) in the store tank and afterwards 15 tonnes of polymer remained behind. The release could therefore be assessed at 30 tonnes. The death toll is stated to have exceeded 2,000 with 200,000 injured. If the death toll be taken as 2,500, then the mortality index would be ca 80. This compares with 6 to 30 for unprotected soldiers in the chlorine gas attack at Ypres.

In common with many of the substances studied in this book, data on the toxicity of MIC are rather few. As pointed out in Chapter 15, the figures in NIOSH suggest that the substance is most lethal when inhaled. NIOSH gives only one figure of significance and that is an LC_{50} for rats of 5 ppm for 4 hours. There is nothing in NIOSH to compare it with but Withers (1985) refers to a paper, Bell & Elmes (1965), which gives for chlorine a figure of ca 120 ppm for 3 hours using rats. This might be extrapolated to about 100 ppm for 4 hours. Allowing for the difference in molecular weights this would make MIC 100/5 x 71/57 or 25 times more toxic than chlorine.

If the Ypres model be set aside as having too many factors which set it apart from industrial circumstances and the mortality index of 0.50 which is recommended for chlorine be used, this would indicate a mortality index for MIC of 12.5. There is thus a ratio of 1:6.5 between the calculated and the actual mortality index. But, if population density be taken into account, the discrepancy largely disappears. The great majority of store tank incidents involving chlorine have occurred in chemical works with population densities in the neighbourhood which must have been far less than in the shanty towns in the immediate vicinity of the Bhopal plant. Press photos make it obvious that the homes of the unfortunate victims could not have afforded shelter from the cloud in the way that European or American houses would have done.

Bhopal may therefore be taken as an example of a worst case where the mortality index is about 6 times the calculated mean value for chlorine, largely because of a population density about 6 times that of European or North American housing.

18.10. THRESHOLD VALUES FROM MORTALITY INDICES.

Based on the isotraumic approach and on mortality indices, it is possible to calculate threshold values. These will be directly related to the assumed isotraumic fatality level. For purposes of the calculations below, this will be, in line with the assumption made in this book as to what constitutes a major accident, 10 fatalities.

For dense explosives, assuming a population density of 850 per sq kilometre and high explosive equivalent to TNT.

$$M_I = 0.85\ Q^{-0.333}$$

Since $M_I = F/Q$

where F = number of fatalities

$$Q_T = \left(\frac{10}{0.85}\right)^{1.5}$$

where Q_T = threshold quantity at isotraumic value of 10.

$$Q_T = 40$$

For VCEs, and assuming a population density of 0.85.

$$M_I = 2 \times P_D\ Q^{-0.333}$$

at $P_D = 0.85$

$$Q_T = \left(\frac{10}{1.7}\right)^{1.5} = 14.$$

This is tonnes of vapour and would correspond to a catastrophic spillage of, say, 40 tonnes.

For chlorine, the isotraumic threshold would be 20 tonnes based on $M_I = 0.50$ as determined above.

18.11. THE EQUAL PROBABILITY APPROACH.

18.11.1. INTRODUCTION.

This approach, which has been devised by the present author, seeks to determine the threshold quantity of any hazardous substance for which sufficient data exist, and for any assumed isotraumic level.

It consists, in essence, of using the historical evidence of incidents with the material and comparing the F v $\geqslant$N curves for N_Q = quantity and N_F = fatality. From a plot of F $\geqslant$ N then the frequency of occurrences of $\geqslant N_I$, where N_I is the isotraumic number, may be determined. The value of $\geqslant N_Q$

where $\geqslant N_Q$ is that quantity involved which has the same level of frequency as the isotraumic level be determined.

$$\text{Then } \geqslant N_I \equiv \geqslant N_Q$$

18.11.2. THE APPLICATION TO EXPLOSIVES.

The approach is illustrated in Figure 18.7., where the data are derived from the same sources as were used to determine the mortality index of chlorine. Here F is the frequency of explosions involving $\geqslant N_Q$ tonnes, or is the frequency of explosions killing $\geqslant$N persons. If the isotraumic level be taken as 10, the frequency corresponding to this is 0.43 per annum. The value $\geqslant N_Q$ corresponding to this frequency is 50 tonnes. This is close to the figure of 40 derived from considerations of mortality index, but too much significance should not be attached as the methods must be admitted to be inexact.

18.11.3. THE APPLICATION TO FLAMMABLES.

The approach may be applied to flammables also. Figure 18.8. shows F $\geqslant$ N curves for flammables derived from data in Wiekema (1984). This plot suggests that the equal probability level lies at 30 tonnes of vapour. On this basis, and using a factor of 1/2 for the ratio between the quantity of vapour and the total quantity spilled, the isotraumic threshold would be 60 tonnes.

18.11.4. THE APPLICATION TO CHLORINE.

When the approach is applied to chlorine using the data from Table 18.3. in Figures 18.9. and 18.10., the isotraumic levels are 30 tonnes for all spillages and 26 tonnes from store tanks. The agreement between these figures and those from the mortality index approach is reasonably good.

18.11.5. THE APPLICATION TO AMMONIA.

The same approach to ammonia, based on all classes of spillage and with data taken from Table 15.2., as shown in Figure 18.11., suggests that the isotraumic inventory for ammonia should be ca 500 tonnes.

18.12. ALTERNATIVE METHODS OF QUANTIFYING HAZARDS.

18.12.1. THE SIMMONDS et al APPROACH.

This approach, in Simmonds et al (1974), has formed the basis for a number of theoretical estimations of fatalities from toxic releases, especially chlorine releases. It incorporates a listing of major incidents provided by the Chlorine Institute as a private communication. It assesses the likely lethal area which would be covered by the catastrophic failure of a 90-tonne rail car on the assumption of an immediate flash off of 16 tonnes. For various windspeeds and conditions of atmospheric stability this is calculated to yield areas of lethal dosage ranging from 1 to 10 square kilometres. The population density assumed in the reference is not stated explicitly though a table of population density is given for the state of Ohio. Here 10 million people live on 40,000 sq miles giving an average density of 100 per sq kilometre. In terms of risk to the population this may or may not be representative as the population may be more concentrated round the railway tracks. The reference provides histograms of accident consequences for bands such as 50 - 100 fatalities which it rates as 10^{-2} per annum, and for 5,000 - 20,000 it rates at 10^{-4} per annum. Overall, and taking account of risk, it assesses the potential mortality as 13 per annum.

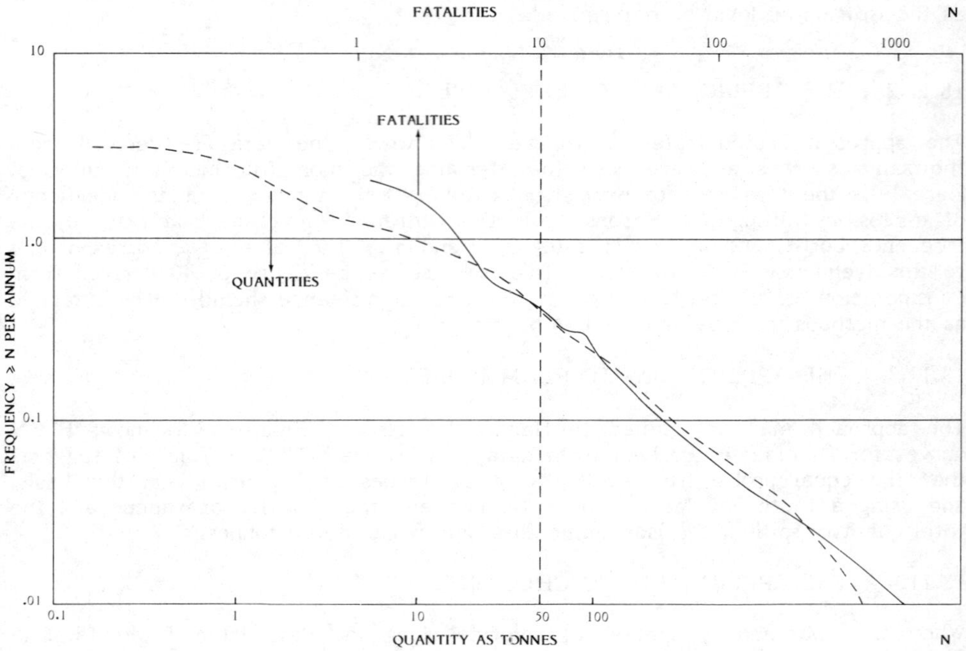

Fig. 18.7. EQUAL PROBABILITY APPROACH, EXPLOSIVES.

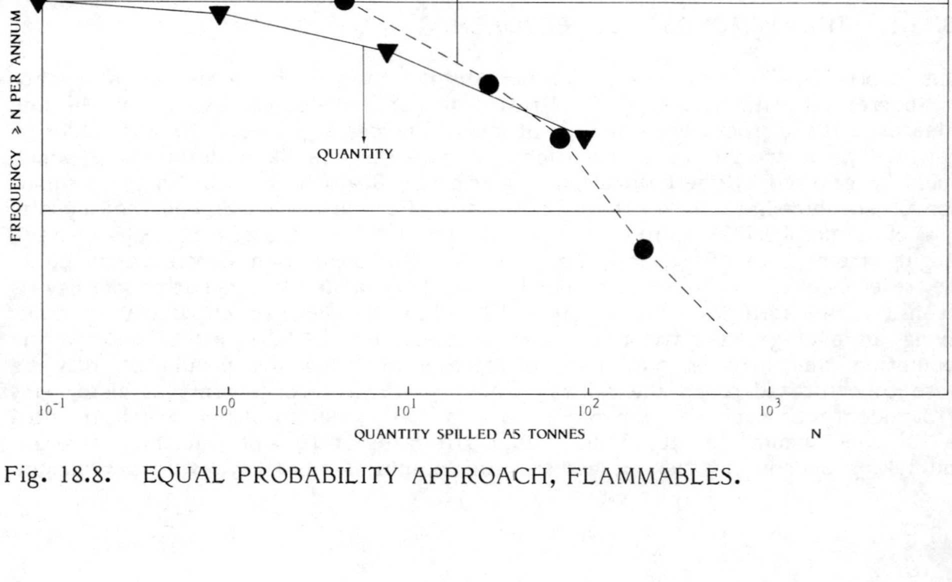

Fig. 18.8. EQUAL PROBABILITY APPROACH, FLAMMABLES.

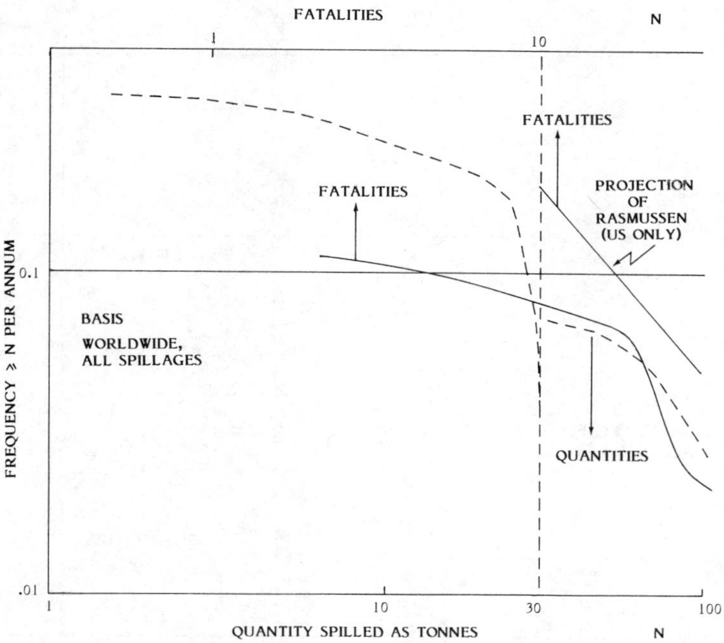

Fig. 18.9. EQUAL PROBABILITY APPROACH, CHLORINE, ALL SPILLAGES.

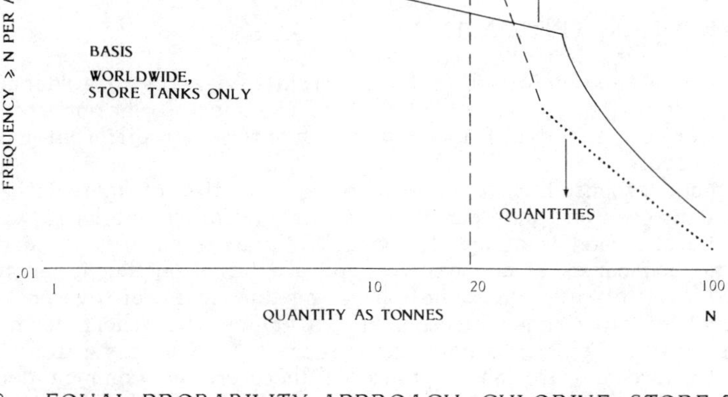

Fig. 18.10. EQUAL PROBABILITY APPROACH, CHLORINE, STORE TANK SPILLAGES.

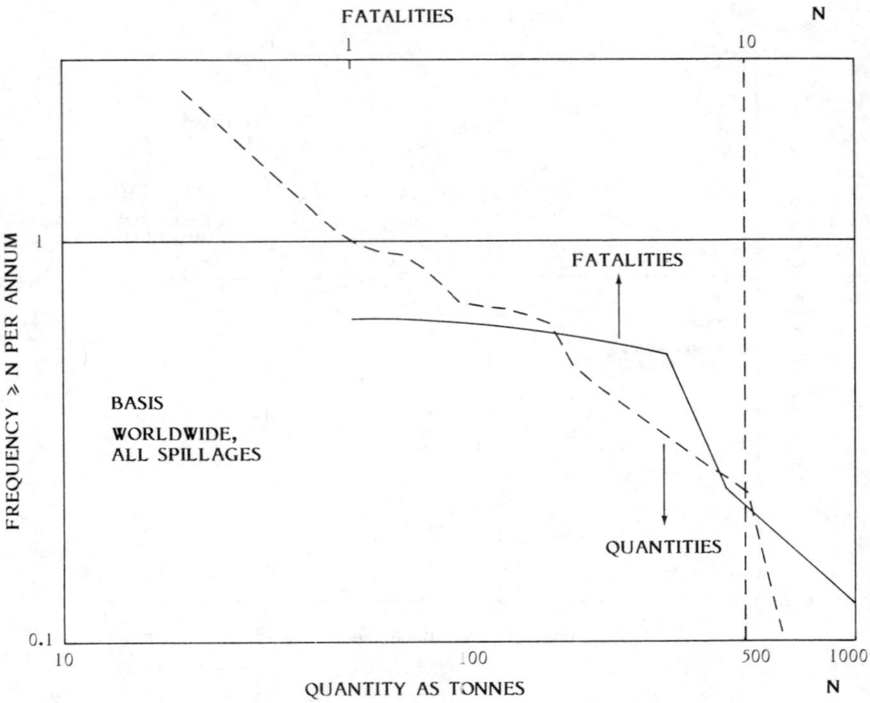

Fig. 18.11. EQUAL PROBABILITY APPROACH, AMMONIA.

In fact, as the reference admits, in 34 years there has only been 1 fatality from the rail transport of chlorine in the US. The ratio between predicted and historically determined fatalities is thus ca 450 : 1. The reference attributes this to evacuation measures. One figure used in the calculation which seems open to dispute is the assumption that the LD_{50} for chlorine is 1,000 ppm minutes for concentrations ⩾30 - 60 ppm. The discussion above in section 18.9.5. suggests that it is more like 24,000 ppm minutes but this is insufficient on its own to explain the discrepancy.

18.12.2. THE WILSON APPROACH.

This approach is set out in Wilson (1980) and relates hazards to a "damage zone". The reference claims that it is not possible to use the information on past accidents to compare in any detail the relative quantities of different gases which pose a similar hazard.

The technique proposed is to determine the quantity of material which will give rise to damage over an area of 1 sq km. Unfortunately it would seem that though the approach is designed to enable disparate hazards to be compared, in practice the boundaries of the hazard zone are based on the disparate criteria of a certain level of thermal radiation, a certain level of overpressure, and a certain level of toxic gas corresponding to twice the "short-term exposure limit" of the gas. For persistent toxics such as dioxin, the reference is not able to put forward a criterion. There is therefore no common denominator in the Wilson approach.

18.12.3. THE APPROACH OF POBLETE et al.

A recent paper, Poblete et al (1984), aims, according to its summary, to provide a model, based on a uniform population density, the inverse square law for the decay of the intensity of the physical effect and the lognormal distribution, or probit equation, for the relation between the causative, or injury, factor and the probability of injury.

The technique incorporates the concept of a radius r_{50} which is the radius at which the probability of sustaining a given level of injury is 0.5. This is rather different from the present author's nominal lethal (or injury) radius which has been defined above as the radius of a circle within which the number of those escaping that level of injury is equal to those who sustain it outside that radius. In some circumstances these radii may have approximately the same magnitude.

The inverse square approach seems to have physical validity for the decay of overpressure from explosions (the slope of log overpressure v log scaled distance is close to -2). But, as is demonstrated in Richmond et al (1968), the lethal shock pressure for mammals exposed to blast appears to vary with body weight. It might seem that the inverse square approach for thermal radiation is beyond argument as it is a basic law of physics. However, though this is a well established law for point sources in a vacuum, radiation from fire balls, for those at distances likely to lead to serious injury, arises from an object which deviates widely from a point source and for which the distance to the envelope may be more significant than distance from its centre. The influence of atmospheric attenuation has been discussed above, where the views expressed in Glasstone and Dolan (1980) and Moorhouse and Pritchard (1982) that such attenuation is significant, has to be noted.

The reference acknowledges that application to toxic releases raises difficulties, one being the highly directional nature of the release except under conditions of extreme calm. Unlike the models for explosions and radiation, the inverse square law has not as yet been proved to be applicable to toxic releases. One problem is that the location of most serious danger and hence of fatalities lies in the region of dense cloud formation and this does not seem to be susceptible to inverse square treatment. And other serious difficulties of applying simple laws to predict the mortality from toxic releases is discussed elsewhere in the book.

18.12.4. THE USE OF PROBIT EQUATIONS.

The use of probit equations by Richmond et al (1968), and other investigators, the general principles of which are discussed in Finney (1971), is an important development. However, some investigators of the effect on human populations have provided highly unrealistic values for the constants K_1 and K_2 in the probit equation:-

$$Y = K_1 + K_2 \ln X$$

More recent work as, for example, by Lees (1980), has improved the accuracy of these constants.

Chapter 19

LEGISLATIVE CONTROL

19.1. LEGISLATIVE CONTROL.

19.1.1. THE RELATIONSHIP.

Historically managerial control precedes legislative control. The question as to whether managerial control on its own is adequate is a matter of degree; clearly a business in which explosions are a daily occurrence will soon be bankrupt. But the commercial consequences of inadequate control by management are far from being an adequate criterion in most cases and central government may be compelled to intervene by legislation. Such intervention has occurred in the EEC through its Directive on Major Accident Hazards, [European Community Directive (1982)].

19.1.2. TREATMENT OF THE SUBJECT.

It has seemed to the author more convenient to treat the two subjects in a way which is at variance with their historical development. Chapter 19, therefore, will be devoted to the legislative control of major chemical hazards in the EEC and in the UK, with Chapter 20 in large part being devoted to a discussion of the managerial response. This is not intended to imply that the measures currently taken by management do not, in many cases, long predate legislative requirements.

19.2. CONTROL BY PLANNING.

19.2.1. EARLY WARNINGS.

Marshall (1983a) outlined the stages by which major chemical hazards had come under public control in the UK since 1972. Perhaps the first warning that there was need for control was that given in Her Majesty's Chief Inspector of Factories (1968):-

> "The scale of modern manufacture has resulted in the storage and use of very large quantities, often measured in thousands of tons, of potentially dangerous materials such as acrylonitrile, liquefied petroleum gas and liquid oxygen . . . The Inspectorate are constantly called upon to give advice on such matters, advice which must be realistic, having regard to the extreme remoteness of the risk, the scale of disaster which could ensue, and the cost of remedial measures."

This theme was taken up in Robens (1972), who recommended two steps to be taken in this area. The first was the need to set up a specialised Inspectorate to keep under special surveillance those hazards which had the potential for injuring the public. The second was that there should be a standing advisory committee to give advice on explosive and flammable substances. The recommendations did not refer to toxic substances.

The first of these recommendations was promptly acted upon but the proposed advisory committee was not set up until after the Flixborough disaster when, as the Advisory Committee on Major Hazards (ACMH), it was given wider powers than Robens recommended by having toxics included in its terms of reference. It commenced its work in January 1975.

19.2.2. CIRCULAR 1/72.

A few months before Robens (1972) was published, and in anticipation of its findings, action was taken to alert the local planning authorities in Great Britain on the need for consultation with the Factory Inspectorate before granting certain planning permissions. This took the form of Circular 1/72 which was issued by the Department of the Environment, [DOE (1972)]. It was concerned with the question of certain classes of "development", (the term development has a precisely defined legal meaning in the UK for which standard texts such as Heap (1982) should be consulted). The first was the question of new development on a site where this development might be judged to present a hazard to the public. The second was concerned with development, such as housing or building a school, in the vicinity of existing hazardous installations where it might be judged that intensification of population density was not desirable. There was also provision, where the local planning authorities were concerned about existing hazards, to consult with the occupiers of sites with a view to obtaining a voluntary agreement which would reduce the level of risk to the local population. Failing this there was the possibility of discontinuing an existing permission.

The Circular defined a major hazard as:-

"A situation where, if any incident occurred, there might be substantial loss of life or serious injury outside of the workplace".

There was an implicit assumption that "major hazard" meant major **chemical** hazard.

The Circular, and the comments of the Chief Inspector of Factories on which the circular was based, placed considerable emphasis upon the potential for dust explosions. More recent regulation, however, has discounted the major hazard potential of dust explosions, [Advisory Committee on Major Hazards (1984)].

The Circular provided a list which is reproduced as Table 19.1. The list was stated not to be exhaustive and that local planning authorities were at liberty to consult the Inspectorate over other substances which they might consider to be dangerous.

As with other aspects of planning law, there were appeals procedures with the possibility of the Minister of State "calling in" the application to make it the subject of a Public Inquiry. Such an Inquiry would be conducted by an Inspector appointed by the Minister, possibly assisted by a technical assessor.

19.2.3. THE ADVISORY COMMITTEE AND CONTROL BY PLANNING.

The ACMH, as an integral part of its activities, surveyed the whole question of planning in relation to major hazards. It took the general view that hazard formed only part, even if a very important part, of the overall problem of planning. Because there were many other factors it felt that decisions in this area should remain part of the responsibilities of the local planning authorities or, where major issues were concerned, with the decision of the Minister of State after

TABLE 19.1. MAJOR HAZARDS IN UK. CIRCULAR 1/72.

Industry	Materials Involving Risk	Total Storage Quantity requiring detailed investigation.
Petrochemical* and plastic polymer manufacture.	All	§
Other chemical works.	Acrylonitrile Ammonia Bromine Chlorine Ethylene Oxide Hydrogen Cyanide Phosgene Sulphur Dioxide	50 Tons 250 Tons 100 Tons 25 Tons 20 Tons 50 Tons 5 Tons 50 Tons
Fertiliser manufacture.	Ammonia	250 Tons
Aluminium and magnesium powder production.	All	§
Aluminium refining.	Chlorine	25 Tons
Paper pulp manufacture.	Chlorine Sulphur Dioxide	25 Tons 50 Tons
Air liquefaction plants and steel works.	Liquid Oxygen	135 Tons
Flour and sugar silos.	Flour Refined white sugar	200 Tons 200 Tons
All industries.	Liquefied Petroleum Gas	100 Tons

§ Economic size of plant would involve such quantities of materials that the risk would invariably be present.

* Petrochemical manufacture is defined as the manufacture of chemicals from an oil refinery product or from natural gas.

consideration of an Inspector's Report arising out of a Public Inquiry. It supported the view that there were two distinct situations, one of them the situation of the creation or intensification of hazard which should require planning permission, and the other the creation of new housing etc in the neighbourhood of a hazardous installation. It recommended that the informal arrangements put into effect by Circular 1/72 be formalised and that the views of the HSE be made known to the Minister at every stage. As a step in the formalisation, the ACMH recommended that Regulations be introduced to require occupiers of installations to notify the HSE if they had quantities of certain specified substances or classes of substances on their premises (Circular 1/72 had made no such requirement and it would have been beyond the powers of the Departments of State who issued the Circular to do so). Acting upon the recommendations of the ACMH, Regulations were introduced in 1983, [NIHHS (1983)]. These require notification of certain inventories of substances or classes of substances. The regulations also require re-notification if, at a later date, the quantity of substance has been increased by a factor of three.

It should be noted that these Regulations are purely domestic to Great Britain and were not made under any Directive of the European Community. The inventory levels, however, do have a relation, in many cases, to those prescribed by the Directive.

Where the matter of a Planning Application is referred to a Public Inquiry, it is the practice of the HSE to put forward a case based upon a detailed risk assessment. Advisory Committee on Major Hazards (1984), Appendix 11, gives the HSE's Major Hazards Assessment Unit's draft guidelines to planning authorities.

19.3. CONTROL BY REGULATION.

19.3.1. HISTORICAL REVIEW.

Reference has been made above under several headings to the historical evolution of major hazards control in Great Britain up to the setting up of the Major Hazards Advisory Committee. It is necessary, therefore, to provide only the briefest summary of this history at this point. In broad terms the major hazards problem came into existence after the Second World War and chiefly in the 1960s where it centred around technologies involving the use and storage of tonnage quantities of liquefied gases, especially flammables. The Factory Inspectorate became concerned about the matter recognising that the existing Factories Act was inadequate to regulate it. The Robens Committee, influenced by the Inspectorate, pressed for special measures to bring these hazards under control, [Robens (1972)]. Steps were taken to set up a specialised unit within the Inspectorate but the recommendation to set up an Advisory Committee was not acted upon until after the Flixborough disaster when speedy action was taken to set one up.

19.3.2. THE ADVISORY COMMITTEE ON MAJOR HAZARDS.

This Advisory Committee, the ACMH, which has been referred to previously, sat for nine years and produced three Reports, [Advisory Committee on Major Hazards (1976), (1979), and (1984)]. These recommended a comprehensive series of measures to control major hazards. Those relating to planning have already been discussed above.

The ACMH recommended that special measures of a legislative character were required, over and above the general provisions of the Health and Safety at Work Act 1974, [HASAWA (1974)], to take account of the need to safeguard the public. These measures were to be actuated by threshold levels of inventory of certain specified substances or classes of substances distinguished by chemical instability, flammability or toxicity.

The problem of how to identify hazardous substances and how to determine the level of inventory at which they become notifiable has been discussed in the previous chapter and will be discussed further later in this chapter.

It was anticipated that there would be two threshold levels for each hazardous substance or class of substances. The lower of the two levels, to be known as the Notification Level, would have, as its principal purpose, the alerting of the HSE to the existence of a substantial hazard to assist them in formulating advice to the Local Planning Authorities on developments in, and in the vicinity of, hazardous sites, as has been discussed above. These became the basis of the NIHSS Regulations discussed above.

The higher threshold level was to be set at ten times the Notification Level. This level was to be known as the Survey Level and it would actuate a series of obligations to be laid upon the occupier by the HSE. The chief of these was to be the requirement to conduct a safety survey, hence the title, Survey Level. A Consultative Document was drawn up, [Health & Safety Commission (1978)]. In it, in the form of hazard survey guidance notes, there was outlined the purposes of surveys and the procedures for carrying them out. The notes contained very useful information in the form of checklists.

This latter proposed legislation was not enacted because, in the aftermath of the Seveso affair, the question of protection against major hazards was referred to the Commission of the European Community.

19.3.3. THE EUROPEAN DIRECTIVE.

There was a delay in securing the adoption of a proposed Council Directive on "The major accidents hazards of certain industrial activities" due to the opposition of one of the Member States to one of its clauses. A Working Document was published on 3/12/81 and the Directive, 82/501/EEC, [European Community Directive (1982)], was promulgated on June 24th 1982. Thereafter, in Great Britain, Regulations were laid before Parliament on 18/12/84 to implement the Directive. These are the Control of Industrial Major Accident Hazards Regulations 1984, (The CIMAH regulations). As with the Directive, the Regulations identify major hazards with major chemical hazards by including a schedule of the industrial activities with which they are concerned. These, broadly speaking, are the chemical industry, both organic and inorganic, and petroleum refining.

It is evident that the Directive drew heavily on the British studies incorporated in the First and Second Reports of the ACMH. There are however some significant points of difference between the CIMAH Regulations and the 1977 Draft of British Regulations, which have arisen from the Directive. One is the requirement that the HSE shall notify the European Commission of major accidents. A second is that the CIMAH Regulations do not apply to pipelines even where these contain notifiable levels of specified hazardous substances. A third is that the CIMAH Regulations are a great deal more specific in the requirement that a manufacturer shall prepare an on-site emergency plan. A fourth is that they lay upon the local authority the duty to prepare an emergency plan and that the manufacturer is required to supply the relevant information to the local authority.

A fifth, and very significant change, is that the Regulations require the manufacturer to inform those living in the vicinity of the nature of any notifiable hazard, and of the safety measures and of the correct behaviour which should be adopted by the public should the hazard be realised. The Regulations provide for a manufacturer to enter into an agreement with the local authority that the local authority will disseminate the information which the manufacturer will provide; this will be at the expense of the manufacturer. It is envisaged that this is what will be done in most cases. Where the manufacturer is unable to enter into such an agreement, then the manufacturer is required to inform the public directly. In either case the area in which the people have to be notified shall be decided by the HSE. An example of such a statement by a local authority, issued some years ago, and before there was a legal requirement to do so, was the statement circulated by Castle Point District Council, Essex to householders of Canvey Island where, in addition to the hazards of industry, there were hazards of flooding, [Essex County Council (1978)].

19.3.4. THE SAFETY CASE.

Article 4 of the Directive requires that "all manufacturers are required to prove to the competent authority at any time that . . . they have identified existing major-accident hazards, adopted the appropriate safety measures, and provided the persons working on the site with information, training and equipment in order to ensure their safety. "Competent authority" refers to that organ of state within a Member State which has the responsibility for enforcing the national legislation which implements the Council Directive. In Britain this is the HSE but other organs of state corresponding to, say, the British Department of the Environment, the Ministry of Health, or the Home Office, may be the competent authority in other Member States.

This requirement is embodied in Regulations 4,7,8 and 9 of the CIMAH Regulations. The document which is required from a manufacturer in order to comply with these Regulations is known as a "Safety Case". The Regulations require that such a case must be updated whenever there has been any modification in the process which materially affects the previous report or at an interval of not greater than three years after submitting a report detailing any new knowledge which materially affects that report. It may be noted that quantitative assessment of the risks associated with the hazard is not specifically required as part of a safety case. Schedule 6 of the Regulations outlines the information required to be included in a safety case.

19.3.5. A TAXONOMY OF LEGISLATIVE CONTROL.

Table 19.2. is a taxonomy of control which identifies the level at which control is being exerted, the legislation which is appropriate to control at any particular level, and what is being controlled. It is taken from Marshall (1984a) in which the structure of control was seen as a hierarchy with the broad strategic measures being identified with the supra-national control exercised by the EEC and at the end of the chain the tactical measures being identified with on-the-spot plant management. The table epitomises the discussion above.

19.4. TECHNOLOGICAL PROBLEMS OF LEGISLATION.

19.4.1. INTRODUCTION.

Some of these problems have been touched upon earlier in the book but it seems to the author useful at this point to bring them together and to consider them along with related problems which have not been discussed previously. In Figure 4.5., attention was drawn to the main categories of occupational hazard and it was proposed that the upper left-hand quadrant of that diagram should constitute the area of major chemical hazards.

The Directive referred to below is the previously referred to EEC Directive on the major accident hazards of certain industrial activities, [European Community Directive (1982)]. The Directive, as has been previously noted, has been enacted in the UK as the CIMAH Regulations. These are set out with explanatory matter in Health & Safety Executive (1985a).

19.4.2. SCOPE.

The first problem is that of scope, i.e. which hazards are to be considered and which are to be excluded. In a previous reference to this problem it was noted that the UK Advisory Committee on Major Hazards had concluded that it would not be appropriate for it to attempt to advise on all major hazards which arise

TABLE 19.2. A TAXONOMY OF LEGISLATIVE CONTROL.

Level	Legislation	What Is Controlled
	STRATEGIC	
Supra-national, The EEC.	Council Directive of 24th June 1982.	General overview of legislation of Member States, Interstate relationships in area of major hazards. Data collection and analysis.
National level, Government of UK.		
Through Health & Safety Executive.	Health & Safety at Work Act, CIMAH Regulations.	Surveillance of management of major hazard sites.
Through Department of the Environment, advised by H.S.E.	Town and Country Planning Act, NIHHS Regulations.	Control of siting of hazardous installations and of development in vicinity through mechanism of Public Inquiries.
Local Planning Authorities, advised by H.S.E.	Town and Country Planning Act, NIHHS Regulations.	Control of siting of hazardous installations and of development in vicinity, by local decision in less contentious cases.
Local Emergency Committees, advised by H.S.E.	CIMAH Regulations.	Off-site emergency planning. Informing public of hazards.
Company level.	HASAWA. CIMAH Regulations. NIHHS Regulations.	Siting, layout, plant design. Safety policies. Hazard and risk analysis, hazard and operability studies. Notifications, on-site emergency plans. Informing public of hazards.
Plant management level.	As above.	Inspections, audits, labour discipline, permits to work etc.
	TACTICAL	

from human activities. Its terms of reference had already excluded nuclear hazards and the Committee itself excluded such activities as coal mining, dam failures, steam boiler explosions and accidents in the explosives industry. In any case the expertise of the members did not lie in these fields.

The EEC took a similar approach in its Directive where it dealt with the problem by requiring that "major accident hazards" were those hazards associated with operations in certain industrial establishments and which involved dangerous substances. In addition for a few specified substances their storage, even though in installations not classified as above, would also be covered by the provisions of the Directive.

Annex I of the Directive specified the industrial activities by stating that they included installations for the production and processing of organic or inorganic chemicals and where it particularised some 16 chemical process operations. These included such processes as condensation, hydrolysis and sulphonation. The Annex followed this by a broad requirement that installations processing chemical substances and using unit operations such as distillation, mixing etc be included. It also included among other processes petroleum refining, the production and processing of energy gases and the pyrolysis of coal. The list was open to criticism for lacking a fully consistent approach but the present author is of the view that it was better to produce an imperfect piece of legislation than none at all. The Directive in any case had provision for it to be revised in light of experience.

Furthermore the Directive made a number of specific exclusions by exempting nuclear and defence establishments, explosives and munitions works, mining, and waste disposal. It did not specifically exclude pipelines in the public domain but did not refer to them either.

In essence the major accident hazards of the Directive are the major chemical hazards of this book except that the book has concerned itself, to a degree, with the major hazards aspects of transport. This is a subject on which there is no comprehensive legislation in the EEC.

19.4.3. THE CONCEPT OF DANGEROUS SUBSTANCES.

Though it may seem to the public that no difficulty attaches to defining "dangerous substances" this is not true in practice. This shown by Annex III of the Directive in which the quantities quoted as threshold levels range from 1 kg to 5,000 tonnes. On this basis it may seem that one "dangerous substance" can be five million times as dangerous as another "dangerous substance".

The properties that determine the level of hazard associated with a dangerous substance seem to include both intrinsic and extrinsic properties. Chemical composition, state of matter, temperature and pressure are clearly intrinsic properties in that they are totally independent of the quantity of the material present. The quantity of the material present may be regarded as an extrinsic property. However those properties which most clearly are associated with hazard, such as flammability, explosibility and toxicity do not seem to be fully intrinsic as they are dependent upon the quantity of the material present. For all of these properties it may be demonstrated that there has to be a threshold quantity of the substance present before the properties are manifested.

The Directive in Annex IV takes account of this by defining indicative criteria for "very toxic substances" in terms of dosage as LD_{50} and LC_{50} values. However for flammables and explosives in Annex IV there is an assumption that there is a sufficient quantity present to exhibit these properties. This observation has a particular relevance to explosive substances some of which have to be present in quantities greater than 100 tonnes before detonation is possible.

19.4.4. THRESHOLD VALUES.

19.4.4.1. The question of a scale.

Examination of the quantities which are quoted in the Directive shows that, with few exceptions, the thresholds fall on what may be termed a "coinage scale" of 1, 2, 5, 10, 20, 50, 100, 200.. This is a scale in which the numbers arise from a hidden multiplier of $10^{0.333}$, which is equal to 2.154, but which is then rounded off to the nearest whole number. It seems to the author that it is essential to use a logarithmic scale of this type. But he believes that any finer division such as $10^{0.25}$, which is equal to 1.778, is not justified by the reliability of the data and can only serve to give a spurious impression of the accuracy of the calculations. It seems to him, for similar reasons, difficult to justify the six departures from the "coinage scale" in Annex II and the one in Annex III of the Directive.

19.4.4.2. A critique of threshold values in the EEC Directive.

Chapter 18 discussed quantification and in it threshold values, based upon an isotraumic value of ten fatalities, were calculated for TNT, flammable gases, chlorine and ammonia together with an indicative value for mustard gas, which is known as bis (2 chloroethyl sulphide) in the Directive. These values are tabulated together with their threshold values in Annex III in Table 19.3. It will be seen from this table that there is good agreement for TNT and ammonia, fair agreement for chlorine and rather poor agreement for flammable gases. However for mustard gas there is absolutely no agreement at all.

TABLE 19.3. THRESHOLD LEVELS IN ANNEX III OF DIRECTIVE COMPARED WITH PREDICTION.

Substance.	By Mortality Index.	By Equal Probability.	Annex III.
TNT	40	50	50
Chlorine (Store tanks)	20	26	50
Ammonia	200	500	500
Mustard Gas	125	-	1 kg

These are all in tonnes except for the Annex III threshold for mustard gas.

With the exception of mustard gas, the predicted values lie within 1 scale point of the Annex III threshold levels.

It is the present author's view that for most of the 30 substances which are given a threshold of 1 kg in Annex III, this extremely low threshold level is quite unjustified. Some of them, such as 2 naphthylamine, are known carcinogens but their hazards arise from long-term occupational exposure and they are not suitably controlled by legislation which is concerned with acute events. He would express a similar view of some of the other highly toxic substances which are given threshold limits of 10 or 100 kg. It is not solely that their toxic effects are overestimated but, as was pointed out in Marshall (1982) and in Marshall (1983b), they do not possess inherent dispersive energy like liquefied gaseous toxics or highly volatile toxics.

There are also anomalies in the threshold values given to explosive substances.

Even if there are difficulties in assigning an absolute value to the threshold for TNT it seems hard to see why the various values given for different explosives should not at least relate to those for TNT in inverse ratio to their relative blast energies some of which are set out in Table 18.3. But this is not so. Though mercury fulminate has a lower blast energy than TNT it is assigned a limit of 10 tonnes. Lead azide with a similar blast energy to mercury fulminate is given a limit of 50 tonnes. The high inventory given to ammonium nitrate, which does not accord with its relative blast energy, arises from a judgement that this is the minimum inventory at which an explosion is conceivable.

The Directive does provide a mechanism for its revision and it seems likely that some of these anomalies may be removed in the near future.

19.4.4.3. Problems for the future.

Annex III lists approximately 180 substances and classes of substances with their threshold levels. However some of these have lost their former significance whereas new substances of a toxic nature are continually appearing. There is probably no area where there is a more rapid development of toxic materials as there is in the production of pesticides which, if they were not toxic to at least one form of life, would not be produced. It is not only in the finished products that the hazards lie, the Bhopal disaster was not caused by the product of the plant which was sevin but by an intermediate, methyl isocyanate. (This is listed in Annex III with a current level of 1 tonne which suggests that it constitutes a hazard which is only one thousandth of that attributed to 30 substances in the Annex !). The pesticides industry, in which it is possible to produce variants by substitution of one chemical grouping by another in endless variety, seems to the author to be almost impossible to control by prescriptive legislation. In his view it would be better to make the production and formulation of pesticides constitute a "major accident hazards industry" with the possibility of a manufacturer being granted exemption where proof can be afforded that the particular process being conducted is non-hazardous.

A similar consideration might be applied to any process which has polychlorinated aromatic hydrocarbons as raw materials, intermediates or finished products. Here too it is very difficult, if not impossible, to prescribe all the possible chloracnegens which may arise.

Chapter 20

THE ROLE OF MANAGEMENT IN THE CONTROL OF MAJOR CHEMICAL HAZARDS

20.1. LEGISLATION THROUGH COOPERATION BETWEEN STATE AND INDUSTRY.

In the eyes of some sections of the mass media and some members of the public, the only effective means of controlling major chemical hazards lies in the legislative process; without such legislation, in their view, companies will do nothing.

But this picture is one which distorts the reality of the relationship between legislative and managerial control. Such a total lack of concern by management would not make economic sense as the realisation of major chemical hazards is, in the long run, a costly business. However it must be agreed that some companies have failed always to realise the long-term consequences of their actions. This may reflect ignorance or it may represent risk taking on the basis that it may never happen to the firm in question or if it does it will be sometime in the distant future. Sometimes it has taken the form of "taking a free ride", i.e. leaving it to the competition to spend their resources on research into safety and then to cash in on their efforts. But, even with the most responsible of managements, state intervention may still be necessary to get the balance right and to ensure uniformity of standards. Uniformity of standards is beneficial to the more responsible firms in that it provides protection against unfair competition from the "free riders".

On the other hand the view that the organs of state are, of their essence, far-sighted, and capable of controlling major chemical hazards on their own, may be open to challenge. As has been pointed out in Chapter 17, the UK Parliament has played only a minor role in the shaping of major hazards legislation. Few of its members have had the interest, and fewer still the expertise, to instigate Parliamentary action in this area.

So far as the Civil Service was concerned it was clear in 1974, at the time of the UK Flixborough disaster, that the Factory Inspectorate, though much concerned over the development of major chemical hazards, lacked the expertise and resources to cope with them. In this they were not alone as corresponding agencies in other European countries were in the same position. It is only in the last decade, as the result of cooperation in the UK between the state, the chemical and process industries, the Universities, and the professional institutions, and through international cooperation in the EEC, that a comprehensive system of legislation has been evolved.

The view that the control of chemical hazards should solely be the responsibility of the state is one which, in any case, runs completely contrary to the UK principle of "self-regulation". (This is not the same as "self-enforcement"). It is an impractical solution as the state could never be in the business of writing safety manuals for chemical plant. What many well-meaning people sometimes fail to grasp is that such prescriptive legislation just does not correspond to the realities and complexities of present-day industrial processes.

A further aspect of considerable significance is the extent to which the best companies in the UK and in other countries have anticipated legislation. There is scarcely a feature of the CIMAH Regulations which had not already been practised in some measure by UK chemical and process industries. By way of example, the drawing up of emergency plans for application both on-site and off-site has been the standard practice of many of the major companies who have operated in cooperation with police, fire and ambulance services. As an example of this, in 1975, long before legislation was envisaged, Millbank Films, at that time the film-making unit of ICI Ltd, but now an independent company, made a film "Incident rendered safe" which was concerned with cooperation between industry and the emergency services, [Millbank (1975)]. This has since been updated as "Incident under control", [Millbank (1985)].

20.2. STRATEGIC CONTROL OF MAJOR CHEMICAL HAZARDS.

20.2.1. THE BASIS OF THE STRATEGIC APPROACH.

The author drew attention in Marshall (1975) to the distinction to be drawn between tactical and strategic measures of control of major chemical hazards. In broad terms, he drew an analogy between strategy in the military sense and strategy in control, and argued that in the past a proportion of chemical manufacturers had sought to cover up strategic weaknesses in design by stringent methods of tactical control which depended on imposing a high measure of discipline on the work-force and on placing an excessive reliance on personal protection which he dubbed "the goggles, gloves and gas-masks" approach. This was coupled with the relegation of the control of safety to relatively junior staff who lacked the necessary basic knowledge of the process and the necessary authority to take safety into the Board Room.

The strategic approach, to be successful, must start as early as possible in the sequence of events through which the process for manufacturing a given chemical is conceived and translated into hardware. The approach has to seek to avoid the "fait accompli" as far as possible. Such a situation has been common in the past where plants have been built and their safety has been "added on" at the commissioning stage or even later.

The Flixborough disaster, which has been described in Chapter 13, was an example of this approach. There was a serious inherent weakness in the process. This incorporated an inventory of flammable liquefied gas in excess of 100 tonnes in the system and also had large diameter pipework which, in the event of its failure, could permit the release of tens of tonnes of vapour in less than a minute. The principal responsibility for safety lay in the hands of a junior executive who lacked both knowledge and authority. When the management undertook a serious and fatal modification to the equipment they did not even consult him.

20.2.2. HAZARD AND INVENTORY LIMITATION.

The development of a strategic approach to plant design, especially through inventory limitation, owes much to Kletz who has produced a series of papers around this theme and which have now been consolidated in Kletz (1984). The principal theme of Kletz is designing for inherent safety by substituting less dangerous materials or, if this be impracticable, by minimising their inventory.

Deliberately contrived inventory limitation during the processing of certain substances has been practised for some time. It is an obvious strategy where substances are highly expensive and where the inventory may account for the greater part of the capital cost. This has found special applicability in the manufacture of certain drugs, especially when newly developed. A further example lies in the final stages of processing precious metals.

Inventory limitation has also found applicability in nuclear processing where the prevention of criticality is paramount. In the explosives industry where batch nitration was formerly the rule, this has been superseded by continuous processing with very much smaller inventories. Such processes are associated with the names Biazzi and Schmid-Meissner. Figure 20.1. shows a diagram of a batch nitrating pot taken from Read (1942); according to Kletz such pots, which were in use until about 1950, had an inventory of 1 tonne. The temperature at which the batch was nitrated was supervised by operators who were provided with a one-legged stool on which they could sit but not sleep. Biasutti (1981) shows a photograph of the aftermath of an explosion in a modern continuous nitration pot where the inventory is stated to have been 1 kg. Kletz illustrates the N.A.B. process for manufacturing nitroglycerine in which the residence time of the reactant is reduced to 2 minutes in place of the 2 hours required in the batch process. Here the glycerol is mixed with the acid in an injector.

Kletz gives numerous other examples of new designs which reduce inventories. These are generally continuous versions of what had formerly been batch processes. In a number of them a feature is the achievement of intensified mixing by injectors or in the bodies of pumps.

Section 20.3. which follows gives examples of the application of strategic principles to the realisation of a chemical process.

20.3. STAGES IN THE REALISATION OF A PROCESS.

20.3.1. THE SCENARIO.

For the purposes of this discussion, 9 stages are envisaged as below. Of these, (b), (c), (d), (e), and (f) will be subjected to detailed discussion.

(a) The identification of a commercially desirable product.
(b) The devising of a process for making it.
(c) The creation of a flow sheet.
(d) The specification and design of the equipment and the control system.
(e) The choice of a location.
(f) The devising of a site layout.
(g) The design of interconnecting piping and layout of instrumentation connections.
(h) The procurement, erection and installation of the plant and equipment.
(i) The commissioning and routine operation of the plant.

Though the above is set out as a chronological sequence, in practice there will almost certainly be looping back. For example, at the flow sheet stage, questions may arise as to whether the process needs to be modified and this may result in further research leading to the desired modification. Such looping back may have nothing to do with safety; for example, a separation process may present difficulties and may be replaced by another.

20.3.2. THE ROLE OF CONTRACTORS.

20.3.2.1. General.

The greatest part of chemical and process plant throughout the world is built by specialist contractors as even the largest companies do not usually build their own plant. Contractors may perform a number of functions including process development, process design, detailed plant design, procurement,construction

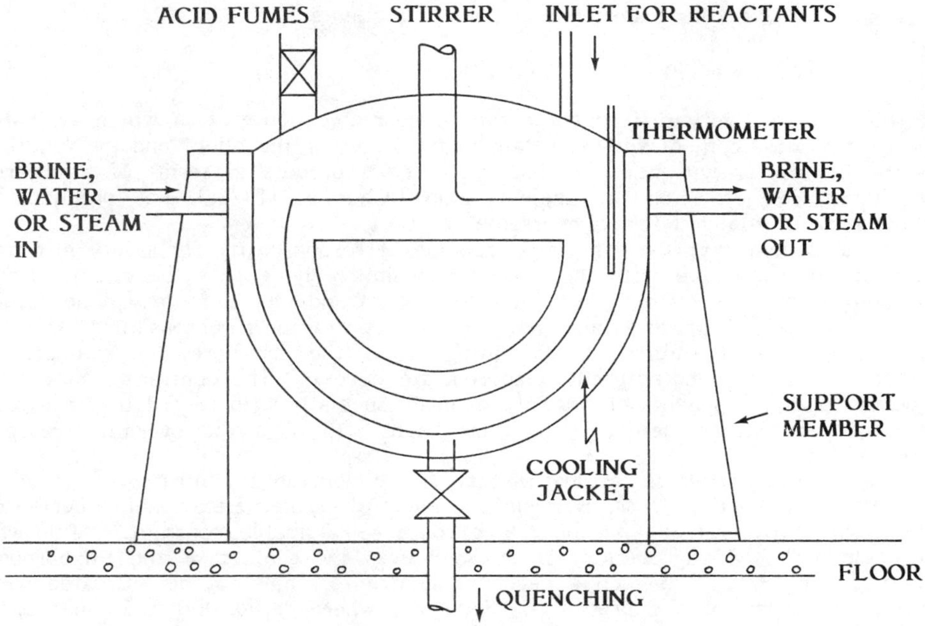

Fig. 20.1. BATCHWISE PRODUCTION OF EXPLOSIVES.

and commissioning. Typically the principal contractor does not manufacture equipment which is the subject of sub-contracts and, in principle, other functions described above may also be sub-contracted. The role played by a contractor in any particular contract may be thought of as lying on a spectrum between a "lump sum" contract and a "reimbursable" contract.

20.3.2.2. Lump sum contracts.

In the former case the contractor will perform all the functions described above and will tender for the job for a fixed sum. In the extreme case of a lump sum contract, the contractor will possess all the "know-how" and the client will be totally dependent on the contractor for it. The conditions of contract which set out the relationship between client and contractor are the subject of I.Chem.E. (1981) which, though it is concerned with UK law, nevertheless establishes principles which may be of validity elsewhere.

Typical examples of processes which form the subject of lump sum contracts are the manufacture of sulphuric acid, nitric acid, caustic soda, chlorine, and some organic compounds. Such plants are often provided as adjuncts to chemical complexes, i.e. as "captive" plants.

So far as safety is concerned, the contractor has to shoulder most of the burden and therefore the strategic considerations described below fall largely on the contractor. However the securing of a suitable location for the plant will lie upon the client. The contractor will have the responsibility of providing the client with the appropriate safety manuals to provide for safe operation after the contractor has left the site.

The great advantage of a contractor in such a situation lies in the experience which has been gathered over many years of constructing such plant and of access to the operating experience of numerous clients.

20.3.2.3. Reimbursable contracts.

Matters may be very different for the reimbursable contract in which typically the process development will have lain in the hands of the client and the function of the contractor will be to realise the client's process in terms of hardware. The Institution of Chemical Engineers, in I.Chem.E. (1976a), has produced a model form of contract for reimbursable contracts.

It is evident that there must be division of responsibility for safety in reimbursable contracts, but with the chief responsibility lying with the client. Here the contractor may know little about the process and the client may deliberately restrict access to process know-how. The contractor's responsibility will be restricted to responsibility for the integrity of the hardware, for example to ensure that it is manufactured to appropriate codes. The contractor may well produce some components of a safety manual especially with regard to the avoidance of personal accidents. But the client will have the overall, strategic responsibility.

Two factors operate against safety in reimbursable contracts. One is that it is the new process, without previous full-scale operational experience, which will most typically be the subject of a reimbursable contract. This was the case with the caprolactam plant which was the subject of the Flixborough explosion of 1974. In such a case the contractor may not be entrusted with a full knowledge of the process. Secondly, where responsibility is shared, it is possible for an area to be overlooked by both of the parties.

20.3.3. DEVISING THE PROCESS.

There are usually a number of chemical processes which will produce a given product. Some of these are ruled out immediately on grounds of economics but others will require detailed investigation before the most advantageous route can be established. It is during these early stages that the first strategic decisions have to be made. These will arise from considerations of the hazardous nature of raw materials, intermediates and waste products. When the hazards have been established consideration needs then to be given as to whether any of them constitute unacceptable hazards and, if so, can these be eliminated by substitution of safer materials. Kletz (1984) discusses the problem of making toluene di-iso-cyanate by processes which do not require the use of highly toxic phosgene, but concludes that the alternative route proposed has not been very successful.

The next alternative Kletz discusses in this case is inventory limitation and he reports that processes have been developed which reduce the phosgene inventory by using it directly from the phosgene manufacturing unit.

An alternative to inventory limitation is to handle hazardous substances at such a temperature as to minimise the hazard. An example of this is the development of Moss Morran in Scotland where liquefied gases are handled under refrigeration so that, in the event of leakage, flashing will not occur and the rate of evaporation will be governed by the rate of heat input from the surroundings.

It is very important to note that these decisions involve looping back between the stages of flow sheet creation and of equipment specification and the process selection stage. They represent a complete break from the past in which the process left the laboratory as a fait accompli and where the task of the engineers was to realise it, not to criticise it.

20.3.4. CREATING A FLOW SHEET.

The development of a flow sheet for hazardous plant also requires a break from tradition. The flow sheet needs to show the measure of hazard at each stage of the process whether from flammables, explosives, toxics, or the other hazards discussed in Chapter 16, together with their inventories. In addition such features as bunding need to be an integral part of the flow sheet and not features to be added on at the engineering design stage.

One of the principal strategic decisions to be taken is whether the process is to be conducted batchwise or continuously. In some processes there is usually a break point at which increasing scale of production dictates the substitution of continuous for batch processing. Since batch processes usually imply higher inventories this break point is likely to occur, in hazardous processes, at a lower level of production than would be the case if economics alone were to be the ruling factor.

The function of Hazard and Operability studies has been discussed in Chapter 17. It is at this point that they would be applied.

20.3.5. SPECIFYING THE EQUIPMENT.

Kletz also discusses a number of aspects of equipment selection which bear upon the possibility of inventory reduction of which the design of distillation columns may be quoted as an example. Kletz advocates narrow bases to columns to avoid unnecessary liquid hold up in the reboiler and favours low hold-up column packing. Kletz's book gives further examples. He also concludes that a single plant is, overall, less of a hazard than two half-sized plants.

A further point raised by the present author is that hazardous plant should be accorded more generous factors of safety than would be accorded to non-hazardous plant.

20.3.6. CHOICE OF LOCATION.

Under UK legislation manufacturers, as was discussed in Chapter 19, do not have total liberty to select locations for hazardous plant. Location is subject to planning permissions and the planning authorities will seek and be given advice from the HSE. It will be a telling point in seeking planning permission, to be able to demonstrate that all possible steps to achieve inventory limitation have been taken.

20.3.7. THE DEVISING OF A SITE LAYOUT.

Layout is a stage which follows on from decisions of location (siting) because the layout of a site is determined by its shape and other physical features.

There is some discussion of siting considerations in Rudd and Watson (1968) but the horizon of this work, which is devoted to an overall strategic approach to process engineering, is largely limited to the "escalating disaster" of which it gives a case history. A more detailed approach is that of Mecklenburgh (1973) which discusses the layout of sites partly from the point of view of safety distances. This reference pointed out the need for access for emergency services and gave some safety distances. However, by the time a second edition of this book was contemplated, the climate of opinion had changed greatly. Mecklenburgh (1982) reflected this change and set out a comprehensive philosophy of the application of hazard assessment to plant layout. This noted that formerly plant layout was the responsibility of non-graduates as it was governed by the application of codes rather than an assessment of the needs of the plant. Mecklenburgh proposed the need to take account of the information which currently exists on VCEs and toxic releases as a constraint upon plant layout and to procure adequate safety distances based on the known effects of these hazards being realised.

One of the prime strategic considerations lies in where to site the most hazardous areas. If, for example, local population densities are not uniform, it may be advantageous to locate the most hazardous areas on that part of the site which is most remote from the local population. Alternatively, the direction of the prevailing wind may be a determining factor.

The location of control buildings, fire stations, medical and emergency centres relative to the most hazardous areas will be discussed in Chapter 21.

As pipework may account for about a third of the capital costs of a chemical plant, there are strong incentives to reduce its overall length. This is facilitated by computer-aided design and the net effect is to produce a very compact layout. A further influence favouring such a compact layout is that it is far less necessary than it used to be to provide facilities for extracting for repair plant items such as columns, at ground level. These may now be plucked out by tower cranes. (These are, in themselves, a hazard to neighbouring plant should they collapse). As a consequence of dense layout, fire spread may be aided. Where flammable vapours may be released, dense packing of plant items may favour escalation to a vapour cloud explosion, (see Chapter 12).

One aspect of the planning of layout is concerned with fire fighting such as the layout of water mains, stationing of foam appliances, location of sprinkler systems etc. One of the principal aims of fire fighting is the prevention of the spread of fire but it will not be further considered; it is a specialist subject with a vast literature of its own.

20.4. STRATEGIC APPROACHES TO MITIGATION.

20.4.1. PREVENTION AND MITIGATION.

Clearly the primary strategy for controlling major chemical hazards is to prevent their realisation and the discussion in the book has been largely centred up till now around prevention.

Some have argued that prevention is all important and that, therefore, all available resources should be concentrated upon prevention and none upon mitigation. This is an extreme view and one which ignores certain realities. Of these the chief is that mitigation is also a form of prevention. If, by mitigation, a spillage may be contained, then this may prevent the consequences from being realised. It also ignores the cardinal principle that, if hazard cannot be reduced to zero, neither can risk.

In order to distinguish between the two concepts, prevention may be regarded as being those measures which are taken to avoid a loss of containment, or the release of blast energy, in the case of dense explosives. Mitigation, on the other hand, constitutes the series of measures which are taken to minimise the consequences of such loss of containment or release of blast energy.

Both prevention and mitigation are activities to which strategic approaches may be applied.

20.4.2. STRATEGIES OF MITIGATION.

20.4.2.1. Introduction.

There are a number of areas in which a strategic approach may be applied to mitigation. They may be divided into two main areas, physical measures which include secondary containment, cloud dispersal facilities, provision of detection equipment, and the hardening of facilities which may be subject to blast. The latter are discussed in Chapter 21. The other measures, which are partly physical and partly organisational, are the preparation of on-site and off-site emergency plans.

20.4.2.2. Secondary containment.

One measure of secondary containment which has long been adopted, is to install "bunds" around tanks containing liquids. (A bund is a confining wall; it is often combined with a pit). This may be required even in cases where the liquid is, in the ordinary sense, innocuous as, for example, water, as the damage which can be done by a tank failure can include drowning.

The design of bunds is perhaps not as simple as it may seem. Decisions have to be made as to whether a bund should be able to accommodate the entire contents of all the tanks within its confines or only some of them. It also has to be of a geometry suitable for preventing liquid escaping at any level from a tank from having a trajectory which will carry it over the top of the bund. Sudden surges of liquids can overflow bund walls even when the bund is theoretically capable of accommodating all the liquid. Arrangements have to be made to discharge rain water without allowing spilled contents to escape.

20.4.2.3. Cloud dispersal.

The installation of steam jets, air jets, and water sprays have been proposed as a means of preventing the horizontal spread of hazardous vapours. The use of steam jets is discussed by Seifert et al (1984). Steam pipes can be run horizontally around a bund wall with the jets pointing vertically upwards. The reference claims that such barriers are effective in dispersing vapour clouds and that static electricity build-up, which could give rise to ignition of flammable gases, can be prevented by suitable earthing. Rulkens et al (1984) concludes that though air curtains may be successful in theory, unlike steam curtains they may not be practicable on a large scale. However the availability of steam varies from site to site. Deaves (1984) reports experimental work on water sprays and concludes that forward-facing and vertical-facing sprays are more effective than backward-facing sprays.

20.4.2.4. On-site emergency planning.

All work-places require some measure of emergency planning even if this is of an elementary nature such as designating a person to telephone for an ambulance and for someone to be in charge of first aid. The scope of emergency planning depends upon the size of the installation and upon its hazards. As has been noted above, voluntary action by the industry has often anticipated legislation and the subject under discussion is an example of this. The UK Chemical Industries produced a booklet on the subject in 1976, [Chemical Industries Association (1976)].

The kind of installation to be discussed below is one which would qualify for notification under legislation designed to control major chemical hazards. As such, the drawing up of an on-site emergency plan is obligatory in the EEC. Such an installation could come under three categories, (a) having flammables only, (b) having toxics only, and (c) having both toxics and flammables.

All these cases would require the provision of a suitably equipped emergency centre. The siting of such a centre will be discussed in Chapter 21 below. The centre could be a complex of buildings including a fire station for cases (a) and (b), a medical centre (which usually has a day-to-day function of medical supervision and of attending to personal accidents), and a communications room. The author has had an opportunity to inspect such emergency centres at a number of sites controlled by different companies and has noted a strong generic similarity between them which has obviously sprung from a common philosophy. The emergency medical provisions may differ between the different categories but what is common is that the centre is designed to cater for multiple accidents of a type which might be anticipated according to the activities conducted on the site. For example, one chlorine handling installation the author inspected had the capability of administering oxygen to 20 persons at once. It is obviously of importance that sufficient stocks of known antidotes be maintained.

Where the works maintains stocks of flammables at major hazard level most companies maintain a works fire brigade for first aid fire fighting. Such a brigade, in the case of most fires, has the fire under control before the public brigade arrives. However once a fire requires the public brigade to bring it under control, the public brigade takes charge. Works fire brigades differ in the number of full-time firemen they employ. They normally have some professional firemen, sometimes recruited from firemen who have retired from the public service. As part of their duties, they inspect and maintain the works fire fighting provisions, conduct fire practices and give training to the process workers. The rest of the brigade are part-time, being otherwise employed on the site, perhaps as fitters, electricians, etc.

The organisation for dealing with emergencies differs from one site to another and the precise details of who does what will vary. What is common is the recognition that such matters have to be decided in advance and not just improvised when the emergency happens.

20.4.2.5. Off-site emergency planning.

The arrangements discussed above for on-site planning are only a part of the overall exercise of emergency planning of which off-site planning forms the other component. But there is a crucial difference between on-site and off-site planning. This is that on-site planning is totally under the control of the site operator, whereas off-site planning is not, as the site operator has no authority over what happens outside of the site boundary. Off-site emergency planning makes imperative the closest liaison between site management and the public emergency services. Diggle (1976) has described such arrangements; some 50 resources which have to be called upon are listed, both internal and external.

Marshall (1979a) gave an account of the operation of the emergency services in response to the Flixborough disaster. It drew attention to the overall authority exercised by the police in the UK who alone have the authority to close roads or to organise evacuation of population at risk. How far, in law, they may compel evacuation is open to doubt but clearly most people will respond to police advice. There may be need for emergency feeding and shelter and this is provided by the local authority. Each area has an Emergency Planning Officer whose primary functions are concerned with civil defence but clearly the planning which goes in to this area can be activated for purely civil emergencies. There is a very detailed account in Humberside Police (1974) of the operation of the police in the Flixborough emergency, which includes information not available to the general public on such matters as the recovery of bodies, procedures for identification, emergency mortuary arrangements and the arrangements for post mortem examinations.

Emergency plans of the kind discussed above have existed for a decade or more in the vicinity of large chemical and refinery complexes in the UK and are now compulsory for CIMAH sites. A feature has been regular exercises which in the case of a number of major fire headquarters can take the form of "table top" exercises. These are conducted in "real time"; that is the exercise lasts for, perhaps, a whole day and a redeployment of a fire appliance is not effective until a lapse of time corresponding to the time it would have actually taken to move it from one location to another. The author has had the opportunity to discuss these exercises with a number of chief fire officers and they consider them to be very cost-effective. However they all agree that occasional field exercises, though they are far more expensive to mount, are indispensable as a check on the physical resources available and upon their operability.

The obligations which have newly arisen through the recent CIMAH Regulations, whereby occupiers have to inform the public of the hazards of their sites and of the steps which the public need to take in emergencies, are a component part of the overall scheme of off-site emergency planning. It is of interest that a number of companies told the author that they were very worried as to the response that sending out these letters would provoke. They were very surprised that, in the event, there was little public reaction.

Chapter 21

THE SITING AND PROTECTION OF BUILDINGS IN HAZARDOUS AREAS

21.1. INTRODUCTION.

Effective control of chemical and process plant is a prime necessity for the prevention of fires, explosions and toxic releases. Such effective control demands the correct design and siting of control systems and of the buildings which house such systems. Control buildings (alternatively known as "control rooms") are peculiarly liable to damage, and the personnel inside to injury, because such buildings usually have to be sited near the centre of hazardous areas for considerations which will be described below.

There have been a number of vapour cloud explosions which have caused severe damage to such control buildings. One example is the Flixborough explosion, which has been described above as a case history of a vapour cloud explosion and which resulted in the total collapse of the building. A second example is the severe damage which occurred to the control building at Beek, The Netherlands, on November 7th, 1975. This explosion has not been described in this book as a case history but the circumstances of the damage to its control building will be described below.

The section will be concerned with an appraisal of the present-day functions of control buildings. This is in order to assess the possibilities which exist for a different approach to the siting and functions of control buildings in hazardous plant as compared with the approach for control buildings in relatively non-hazardous plant.

The section will be concerned secondly with design of explosion-resistant buildings. Though control buildings may be required to resist penetration by toxic releases, this subject will not be discussed as the author has not been able to give it the necessary detailed study.

Though many kinds of buildings have suffered various degrees of damage in areas in the vicinity of vapour cloud explosions, only the protection of control buildings will be treated as these are the most vulnerable to damage and the most vital to protect.

21.2. BRIEF HISTORY OF THE EVOLUTION OF CONTROL BUILDINGS.

There does not appear to be any comprehensive work which analyses the history of the development of the techniques of controlling chemical and process plant. Bennett (1979) is an account of the history of control engineering but it only covers developments up to 1930 and, even so, it excludes the control of chemical and process plants. The bibliography of this work mainly relates to the 19th century and earlier. The author has spoken to Bennett who told him that no history of the control of chemical and process plant had been written up till

1980.

Studies by the author suggest that in the half century under review, the following trends may be discerned:-

(a) There has been an enormous growth in the stream capacities of process plant.
(b) The growth in scale made it no longer feasible to house process units in buildings.
(c) This has tended to engender separate control buildings even though initially their functions were limited to the display of process variables.
(d) The growth in scale made pipe lines so big that the control of flow by valves had to be mechanised.
(e) This led to the use of servo-mechanisms for the regulation of valves.
(f) These servo-mechanisms then made possible the centralised control of flow.
(g) The development of computer control has strongly consolidated centralised control.
(h) Though, generally speaking, there has been a tendency for electronic signals to displace pneumatic signals (in 1982 half of UK chemical plant was electronically controlled) pneumatic systems are still popular as they are cheap and they are intrinsically safe in atmospheres liable to contain flammable gases. However the growth of computerised control might be expected to favour the substitution of pneumatic by electronic signals, at least in larger plants. However, because of the recent recession there has been a dearth of new installations, certainly in the UK, and this makes it difficult to judge present trends.

21.3. THE PRESENT-DAY FUNCTIONS OF CONTROL BUILDINGS.

There have been a number of papers which have dealt with the topics of the anticipated level of overpressure and the design of control buildings to withstand these levels of overpressure. However, little has been published which has sought to analyse the functions of control buildings. The analysis conducted below has two aims, the first is to determine how essential it is to site such buildings in close proximity to hazardous plant. The second aim is to establish which functions are absolutely essential to discharge in a control building. By eliminating all functions not deemed to be absolutely essential, it is possible to reduce the number of persons inhabiting them to a minimum and hence reduce the number of potential casualties. One effect of such elimination is to minimise the size of control buildings; if they are expensive to construct because of special requirements, money can be saved by making them as small as possible.

Marshall (1975b) drew attention to the need for plant to be monitored by human operators through the media of sight, sound and smell which require the stationing of operators near the plant. It is true however today that closed circuit television has done much to reduce the need for visual observation and gas detectors have reduced the need to monitor smells. Maintenance also requires close communication with the control function. The author's discussion with the management of a number of major plants has reinforced the view that remote siting of control buildings, though desirable from the point of view of disaster, may be detrimental to routine operation. As Marshall (1975b) points out, the contradiction is that, from the point of view of routine operation, the ideal control building would be like an airfield control tower and it would be sited in the heart and centre of the plant which it controls. But, designed from the point of view of disaster survival, a control building would be of massive construction, without windows and it would be sited at a place remote from the plant. It is obvious that there has to be some measure of compromise between the two concepts.

The question of the place of the control building in the overall management of chemical plant was taken up in Marshall (1976a), which distinguished four aspects of the control function:-

(a) Forward control.
(b) Analytical control.
(c) Management control.
(d) Emergency control.

"Forward control" was defined in this paper as the centre at which input signals from the process generate output signals to control the process and at which data is logged. The generation of output signals may be by human intervention or by some form of automatic control of which computer control is the most sophisticated. It was noted that "signals" may include visual, aural or olfactory signals and that in order to pick up these latter signals, there needed to be periodic patrolling by the operators. "Process plant is not yet an 'automated battlefield', patrols are still needed".

By "analytical control" was meant control measures which arise from the chemical analysis of process streams. Much of today's analysis is by techniques such as gas/liquid chromatography. A great deal of it is on-line and local to the process unit generating the process stream. Accordingly, unless a special case can be made out, the permanent stationing of laboratory assistants in the control building should not be permitted.

The reference claimed that there was no justification for the management control function to be discharged from the control building or indeed in proximity to the plant. In general, only those personnel whose presence there is absolutely indispensable, should be housed in the control building.

Finally, the paper analysed the location of disaster control. It is obvious that the control building has a vital rcle to play in the case of mishap; that role is to minimise the effects of the mishap as far as possible by measures of control. However, in the author's view, it is wrong to locate the disaster centre in the vulnerably sited control building. It needs to be as remote as is feasible from the likely sites of explosion or fire and, other things being equal, a location near the main point of entry to the site for emergency vehicles seems indicated.

The points noted above were later incorporated in Advisory Committee on Major Hazards (1979) and are now conventional wisdom in the chemical and process industries. As will be seen below, the design philosphy of the Flixborough plant violated virtually all of them. This is true also of many control buildings built in, or before, the early seventies.

21.4. THE LOCATION AND DESIGN OF EXPLOSION-RESISTANT BUILDINGS.

21.4.1. THE HAZARDS TO BE GUARDED AGAINST.

The hazards considered in this chapter are those which arise from explosions. The author considers that the measures taken to protect these buildings from explosion are likely to be adequate, in the main, to protect them also from fire balls. The hazard of toxic release will not be discussed. Protection against physical explosions and dense explosions will be considered, but the main emphasis will be on protection against VCEs.

21.4.2. PROTECTION AGAINST DENSE EXPLOSIVES.

21.4.2.1. The analysis of Christopherson.

A basic work in this field is Christopherson (1946). This treats, in detail, the effects of high explosive on buildings. However it is directed principally to the military aspects of the problem and much of this is not relevant to the present study, for example it considers air-raid shelters and armour-piercing shells.

One part of the work cited which is of particular interest is a section which deals with dimensional theory and damage to buildings. The conclusion is drawn that, by and large, Hopkinson's Law (discussed in section 10.5.4.) may be used to predict damage to full-scale buildings on the basis of damage sustained by model buildings. Though Christopherson draws attention to a number of scale effects, i.e. departures from Hopkinson's Law, these seem to cancel each other out. One of the points made is that gravity cannot be scaled; this complicates prediction of damage to buildings because full-scale buildings, though the initial damage pattern may be the same as in the model, may show more tendency to ultimate collapse because of the "debris load" on the upper floors. The discussion confines itself to the situation in which a damage to a model is compared with damage to full scale; it does not consider the scale dependence which arises when a characteristic dimension of the target does not bear a constant ratio to the radius which corresponds with a given level of overpressure. This will be discussed below. A further point made by Christopherson is that Hopkinson's Law takes no account of the time dependency of the stress/strain relationship in materials such as reinforced concrete.

21.4.2.2. Safety distances.

Healy (1965) is a Report which reviewed, in the light of the experience of the Second World War, the safety distances for conventional explosives stored in bulk. The subject lies outside the field surveyed by this thesis but the Report contains much material of general interest.

One of the methods adopted by Healy was to attempt to establish the radius at which a given level of damage was sustained by dwelling houses arising from a given mass of explosive. The source of the blast was assumed to be a standard TNT-filled bomb fused to produce maximum blast effect. Results are expressed in terms of the mass of explosive, not the mass of the bomb.

The damage categories were as follows:-

(A) Almost complete demolition.
(B) Damage so severe as to necessitate demolition.
(Cb) Not feasible to render it habitable during the war.
(Ca) Feasible to render it habitable during the war.
(D) Requiring urgent repairs but still inhabitable.

Healy claimed that "These damage classes have precise definitions". It is difficult to believe this and the plates illustrating the Report show a wide range of what constituted B damage. One illustration shows a building with 75% demolition of external walls, the other shows a building with all the walls standing but heavily cracked.

The radii attributed to these damage categories occurred were then expressed as a multiplier of the radius (RB) at which B class damage would occur. These multipliers are given as follows:-

0.675	x	RB	=	A class damage.
1.74	x	RB	=	Cb class damage.
5.0	x	RB	=	90% window damage.
10	x	RB	=	50% window damage.
20	x	RB	=	5% window damage.

The degree of precision with which A and Cb radii are expressed is, in the present author's view, spurious.

A formula is given for calculating the mean RB. The present author has modified this formula, to bring it into line with the units generally used in the book, as follows:-

$$RB = 56\,Q^{0.333} / [1 + (100/Q)^{0.5}]^{0.166}$$

RB = Radius of B class damage in metres.
Q = Mass of explosive in tonnes.

The Report draws attention to the fact that for explosive charges greater than ca 4 tonnes, this approximates to:-

$$RB = 56\,Q^{0.333}$$

i.e. in this range RB is a Hopkinson scaled distance. [Hopkinson's Law is not mentioned in Healy (1965)].

Though Healy does not say so, for small values of Q the equation can be shown to reduce to:-

$$RB = 38\,Q^{0.666}$$

The exponent for Q thus lies, according to the scale of the explosion, between 1/3 and 2/3. This may account for the use of an exponent of 1/2 for Safety Distances in the British Magazine Regulations of 1934 and in the French and American Regulations which were current when Healy (1965) was written.

In the American Regulations, however, the value of the "constant" K in the equation:-

$$RS = K\,Q^{0.5}$$

changes with the value of Q. The form of Healy's equation in which the exponent of Q changes with the scale of Q is mathematically more sound. One effect of this, as compared with the original basis for British Safety Distances, is to reduce these distances for large values of Q. The Committee which produced the Report had concluded, on the basis of the experience of the Second World War, that practice had shown also that the Safety Distances of 1934 were excessive for large values of Q.

It is of interest to explain why RB is apparently scale-dependent, i.e. it is not a scaled distance, at low values of Q. The explanation, in the present author's view, lies in the fact that, at such values of Q, the dimensions of houses become significant when compared with RB. For 1 kg of explosive, RB = 5.6 metres. Even if such a charge be placed against one of the walls of a semi-detached house, i.e. at zero distance from the house, much of the house still lies outside the RB. Thus for charges less than a few tonnes, RB is less than would be deduced for a scaled distance based on RB for tonnage quantities. For charges of less than, say, 100 kg, RB has little meaning for a house as a whole. Figure 21.1. shows values of RB v mass of explosive.

There are considerable grounds for scepticism about the accuracy of these radii. Healy speaks of "mean radii" without providing data on variance. Variance would be expected on account of a number of factors:-

(a) Uncertainty about the calibre of the bombs.
(b) The possibility of overlap between adjacent bomb blasts. Small bombs were dropped as "sticks" from racks and could land about 100 metres apart. If the plane dropping them flew at ca 450 km/hr, they would burst at 1-second intervals.
(c) Differences between houses, size, whether detached, semi-detached, two-storey or bungalow, age, standard of construction etc. Houses are not very satisfactory as scientific instruments.
(d) Subjectivity of assessment of damage category. The people responsible were hardly likely to be viewing this as a scientific exercise; their job was

to deal with urgent human needs. For example assessment of "habitability" may have been influenced by the availability of alternative accommodation.

21.4.2.3. Damage as a measure of overpressure.

A further development, not attempted by Healy, is to attribute to the various radii, values of overpressure as deduced from the data obtained by tests on military proving grounds. To do this is to assume that the overpressures produced by impact-fused bombs are identical with those produced on the proving grounds. The graph given on Page 221 (Fig. 10.2), gives such data based on bare, hemispherical charges of TNT, which may not be comparable with bombs with an equal charge. Taking these figures at their face value gives the following:-

$$\text{A class damage} \equiv 7 \times 10^4 \text{ pascals (0.7 bar)}$$

Figure 21.2. shows some of these nodal values on a plot of overpressure v scaled distance.

This combination of Healy's concept of damage radii, with the overpressure v scaled distance relationship, has been used by investigators to estimate the TNT equivalence of VCEs. It must be apparent from the discussion above that such a technique is imprecise. It may, even so, be the best method presently available.

TABLE 21.1. CLASS OF DAMAGE v OVERPRESSURE.

A Class damage	7	x	10^4	pascals	0.7	bar
B Class damage	3.33	x	"	"	0.33	"
Cb Class damage	2.5	x	"	"	0.25	"
90% Window damage	0.4	x	"	"	0.04	"
50% Window damage	0.02	x	"	"	0.02	"
5% Window damage	0.005	x	"	"	0.005	"

21.4.2.4. Protection against missiles.

Large-scale blast damage to control buildings, arising from the detonation of dense explosives in chemical and process plant, as distinct from that arising in explosives manufacture, is unlikely. Such blast damage may arise from the detonation of organic peroxides but these possess a relatively low TNT equivalent, perhaps about 20%. Further possibilities arise from runaway reactions or from physical explosions such as arise from the pressure rupture of pressure vessels.

Where protection against blast has to be provided, the considerations discussed below under the protection of buildings against VCEs will be applicable to such cases.

Dense explosions are much more likely than VCEs to generate missiles because of their much higher initial pressures; this subject is discussed in High (1980). High claims that in the case of fracture into multiple fragments about 30% of the energy in a pressure vessel appears as kinetic energy of the particles. In some cases where, for example, a thermocouple pocket is projected, energy transfer may reach a figure as high as 60%. This energy represents the maximum which may be imparted to a target on impact, in practice aerodynamic drag will exercise an effect which will increase with distance.

According to Christopherson, tests on reinforced concrete using incendiary bombs with varying mass and varying velocity, but a constant diameter of ca 50 mm, yielded the following correlation which the present author has transformed into metric units:-

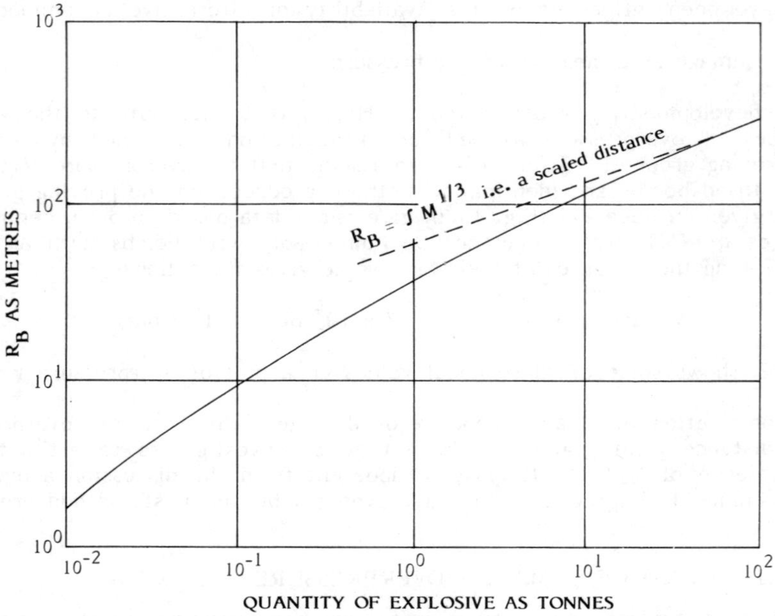

Fig. 21.1. VALUES OF RB VERSUS MASS OF EXPLOSIVE.

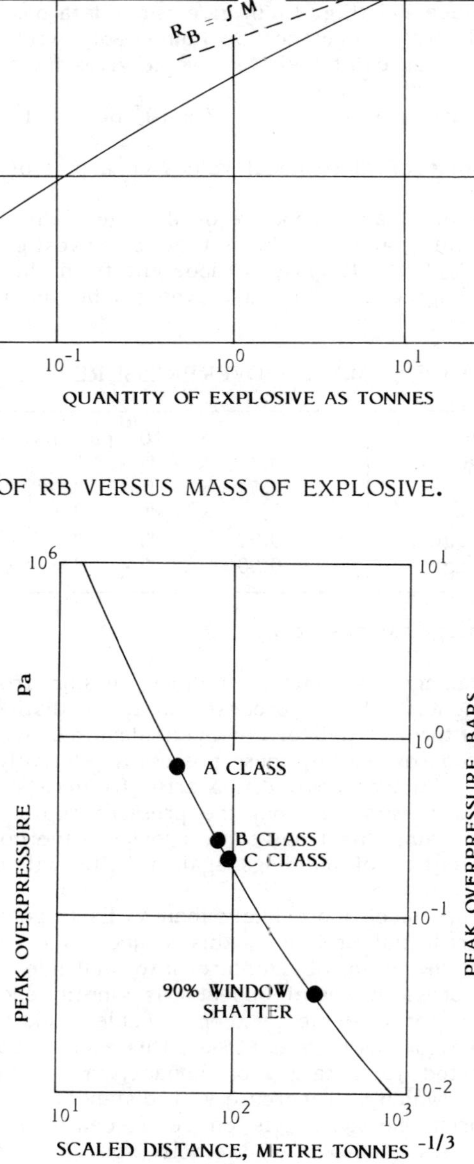

Fig. 21.2. NODAL VALUES OF OVERPRESSURE VERSUS SCALED DISTANCE.

$$D_p = 3.46\,(1/2\;M\;V^2)^{0.4}$$

where D_p = Depth of penetration in millimetres
M = Mass of missile in kg
V = Velocity of missile in m/sec.

Where the mass = 1 kg and the striking velocity = 60 m/sec, the depth of penetration = ca 70 mm.

It should be noted that this represents the situation of maximum damage. The equation is not being put forward as a design equation but one that indicates that significant penetration can result from the impact of small missiles.

Christopherson calls attention to the phenomenon of "scabbing" in concrete and claims that more than half of the penetration arises from this cause. When a missile strikes a slab of concrete a shock wave travels ahead of it through to the back of the slab. This may be capable of producing a crater from which broken concrete is projected violently rearwards. If the missile possesses sufficient kinetic energy the front crater and the back crater coalesce and the missile at that point encounters no further resistance to its passage, though it will have been slowed down. A military solution is to fit a "scabbing plate". This is a steel plate covering the rear of the wall which, though it cannot prevent scabbing, may prevent the projection of the broken concrete. Whether it is a practical proposition to fit scabbing plates in control rooms seems doubtful on grounds of cost.

Missiles can weigh many tonnes and may be hurled hundreds of metres. Brasie and Simpson (1968) quotes a 1-tonne turbine projected ca 1,500 metres and 60-tonne pressure vessel fragment which was hurled ca 400 metres. Some of the case histories in this book bear evidence that substantial portions of road vehicles have been projected, as the result of purely physical explosions, for distances of the order of 400 metres.

21.5. PROTECTION AGAINST BLAST.

21.5.1. THE LITERATURE.

There is a far more extensive literature on the protection of control buildings against blast than exists for protection against missiles. The literature divides itself into four interlinked themes:-

(a) the assessment of the hazards of blast,
(b) protection by location,
(c) general papers on the design of buildings to resist blast,
(d) papers specially concerned with the protection of control buildings from blast.

An early paper, now mainly of historical interest, is Brasie and Simpson (1968). This was concerned to warn of the possibilities of blast from dense explosives and from confined explosions which have damage effects on their surroundings. The paper comments that "essentially gas cloud blasts" may occur without causing craters. Data is given on the relationship between overpressure and distance and between overpressure and damage effects.

Later papers to develop this theme include Marshall (1975b), (1976b), and Advisory Committee on Major Hazards (1979) which incorporated and enlarged upon these references which laid particular stress upon the problems posed by VCEs.

Gugan (1979) contains material on damage to buildings, including photographs.

There have been numerous general papers on the effects of blast on buildings and on design methods which minimise damage. These have mainly originated from research conducted in the United States into the effects of nuclear weapons.

One non-military paper is Allan and Athens (1968). This paper analyses the dynamics of the interaction of blast waves with buildings distinguishing between the initial "diffraction" loading and the subsequent "drag" loading. Diffraction loadings are the forces which result from the direct and reflected pressures which arise during the initial phase. The reflected pressure may be two to eight times the direct of "side on" pressure. The pressure wave then bends or diffracts round and over the structure. After diffraction is complete the structure is momentarily subject to a force field in which the pressures are substantially equal otherwise known as the "stagnation" pressure. The time required to reach stagnation pressure is given as 3H/V where H is the height or half the width of the building, which ever is the less, and V is the blast velocity. Following this comes the "drag" phase in which the strong transient winds operate. The duration of this phase is the residue of the duration of the positive phase of the explosion. The task of the designer is to evaluate the magnitudes of these pressure regimes and to calculate the response of the building. The loading duration for VCEs may be sufficiently high that it is comparable with the time required for the dynamic response of the building. This will be unlike the situation obtaining for the blast from dense explosives where the duration is likely to be much less than the response time of the building, a situation described as "impulse" loading. The reference goes on to discuss the impulsive response of buildings to blast from dense explosions.

An early, if not the earliest, paper devoted to the blast protection of control buildings is Bradford and Culbertson (1967). This paper contains an early reference to a VCE, that at Raunheim, West Germany on 16th January 1966. [Case 49 in Gugan (1979a)]. In this instance, a brick-walled control building situated some 30 to 50 metres from the plant was subjected to an ethylene VCE. The explosion blew in the windows and walls and rendered the control building inoperative. Bradford and Culbertson emphasise the economic element, stressing the rarity of such incidents. They argue that, where loss of the control building can appreciably add to the time that a refinery is out of action, new control buildings for hazardous plant should be constructed to withstand 0.2 bar overpressure and that they should be located at least 30 metres from process equipment. Large window areas and multistorey construction would not be permissible for such cases. As to existing control buildings the reference recommends that, since such upgrading would be very expensive, it should only be undertaken on a basis of the examination of each individual case and where it can be shown that serious problems would ensue if the building were to be seriously damaged. The reference claims that a new reinforced concrete building, built to the higher standards, would cost 15% more than a conventional building.

Langeveld (1976) set out the then current position of the Shell organisation which had been called into existence after the Pernis, Netherlands, explosion of 1968. Shell had reached the conclusion that it was difficult to define the characteristics of a standard explosion and that it is better to concentrate on a strong building "than to lose oneself in sophisticated dynamic calculations for which the assumptions are highly disputable". He contrasted the Shell philosophy with that of the US organisation, Humble, as set out in Burns (1967), but concluded that whatever the differences in philosophy the end results would prove to be very similar. The paper claimed that the detailed design took account of American Society of Civil Engineers (1961) and was based upon an overpressure on the walls of 1 bar, and on the roof of 0.25 to 0.5 bar. The building would be a monolithic concrete box type structure but would be provided with special windows whose area would not exceed 7% of the area of the wall in which they were placed.

A detailed analysis of the blast-proof building design within the context of the problems of control buildings, is given in chapter 4 of Oyez (1981). Here there is an assumption of an overpressure of 70 kN/m^2 and a duration of 20 millisecs. It is argued that design should be based upon impulsive loading and upon the momentum so imparted to the building.

21.5.2. CODES OF PRACTICE.

21.5.2.1. Some existing codes.

There are a number of codes of practice to consider here. Two of them have been compiled by industry, the British Code, Chemical Industries Association (1979), and the American Code, Manufacturing Chemists Association (1978). One has been compiled by the Netherlands Government, Ministrie van Sociale Zaken (1977). In making the comparison, reference should be made also to the recommendations of the UK Advisory Committee on Major Hazards (1979). The summary which follows is a slightly modified version of a chapter which the present author wrote for Oyez (1981). This summary, which does not discuss the purely civil engineering aspects, compares six features of location and design. For brevity the year of publication is omitted from the references quoted below in this paragraph.

21.5.2.2. The meaning of "withstand".

Control buildings must be designed to "withstand" blast. The various meanings attached to "withstand" are:-

ACMH. The control building may finish up very heavily damaged, with walls cracked and leaning and roof sagging but, so long as those inside are alive and well, and enough of the equipment remains serviceable to let other plant be shut down without allowing another disaster to follow, the building has, in our sense, "withstood" the catastrophe.

CIA. The control building will survive one incident arising from an "at or near ground level" explosion . . . the aim is that occupants must be protected and equipment must be sufficiently preserved to be in working condition after the explosion (although it may be necessary to rebuild the structure).

MCA. To afford adequate protection for the people in the control house. Deformation limits are set for structural components.

MvSZ. After it has been subjected to the maximum stress as the result of an explosion the whole construction may be damaged beyond repair. The only condition is that the structure shall remain standing.

21.5.2.3. General principles.

ACMH. Control buildings should be of single-storey construction. There should be no heavy equipment on roof nor structures that could fall on it. The personnel housed in building should be restricted to essential categories.

CIA. The design is mainly for single-storey buildings. Criteria could be applied to multistorey buildings but would probably make them uneconomic. The number of personnel housed should be minimum required for operation.

MCA. Single-storey construction is preferred on grounds of cost and design simplicity though it is accepted that two-storey construction may sometimes be necessary owing to lack of space. Housing is to be restricted to essential personnel.

MvSZ. The building is to be detached. There is no recommendation as to number of storeys. The housing to be restricted to essential personnel.

21.5.2.4. Location and blast pressures.

ACMH. The building is to be sited outside of any likely cloud if possible. If not, it should be able to withstand a peak overpressure (side on) of 1 bar for 30 ms. Buildings outside likely clouds may be designed to withstand lower overpressures based on scaled distance as given in a graph.

CIA. For high hazard (CIA's Category 1) situations, where buildings are likely to be engulfed in a cloud or partially confined by other buildings, CIA considers that overpressure may be one bar for 30 ms. For situations outside a vapour cloud it is assumed to be 0.7 bar for 20 ms. If the cloud could engulf the roof to a depth of 15 m, the roof should be designed to resist same pressure as walls. Building to be as far as practicable from sources of 15 tonnes of vapour or more and not less than 30 m from them.

MCA. For high hazard (MCA's Category C) situations, they should be sited not less than 30 m distance. It is assumed that the minimum distance for the building to remain unaffected is 60 m from the source of the explosion. Likely blast parameters are 0.7 bar for 20 ms. For less severe deflagrations 0.2 bar is tolerable.

MvSZ. It is assumed to be on outskirts of installation. External walls are to be designed to withstand 0.3 bar reflected overpressure and the roof to withstand 0.2 bar unless the building is likely to be engulfed in which case the assumed overpressure is 0.3 bar.

21.5.2.5. Materials of construction.

ACMH. To be in heavily reinforced concrete or steel frame.

CIA. To be in reinforced concrete. Unreinforced concrete, masonry or brick are excluded.

MCA. To be built as a monolithic reinforced concrete structure or in reinforced concrete with structural steel frames or other ductile walls. No brittle materials are to be used anywhere.

MvSZ. To be built in a ductile material, e.g. reinforced concrete or steel or material of equivalent ductility.

21.5.2.6. Doors, windows and escape routes.

ACMH. Windows are desirable on balance but could be eliminated given the provision of closed-circuit television. Windows if provided are to be restricted in number and in size and very firmly mounted.

CIA. Individual maximum window frames are to be 0.25 m^2 with catchbars. Doors to be as few as possible and have steel frames. Total area of the openings is not to exceed 0.0067 m^2/m^3 of building volume and of this not more than half are to be in one face of the building. Ready pedestrian access must be provided.

MCA. Windows are not to be allowed unless they and their sashes possess energy absorption capacity, prior to collapse, equal to twice absorption capacity required to resist design blast loading. Window areas are not to exceed 0.46 m^2 or 5% of wall area whichever is the less. At least two exit doors on opposite sides with panic bolts. These doors are to open outwards, be flush with walls and be supported by door frames. Total areas of openings not to exceed 0.0067 m^2/m^3.

MvSZ. Some windows are obligatory. Individual windows are not to exceed 0.25 m^2. Total window area is to be less than 7% of wall area. Performance of external windows not to be inferior to that of laminated glass of two sheets each 3 mm with 1.9 mm sheet of plastic in between. Catch bars are to be provided on external windows with no french windows or skylights. Not less than two doors are to be provided on opposite walls. Escape routes must be provided.

21.5.2.7. Special features.

ACMH. The report advised on principles of emergency control, contending that the emergency control centre must not be in a vulnerable control building. There is a chapter devoted to the regulation of personnel in relation to hazard and damage, especially glass damage to other buildings, including dwelling houses.

CIA. Provides an appendix on the design of control buildings subject to toxic hazards.

MCA. Provides an outline design for a one-storey blast-resistant building of box-type shear-wall construction and for a steel frame building with reinforced concrete cladding.

21.6. CASE HISTORIES.

21.6.1. INTRODUCTION.

The case histories to be drawn on are those of Flixborough, UK, 1/6/74, and Beek, The Netherlands, 7/11/75. In choosing these case histories, the author does not wish to imply that there have not been other VCEs which have damaged control buildings, of which there have been a number of cases. The case histories are included because they display features typical of severe damage to control buildings where these have been inadequately protected, or located too close to the site of an explosion, or some combination of these factors.

21.6.2. FLIXBOROUGH, UK, 1/6/74.

21.6.2.1. Introduction.

The VCE which destroyed the control building at Flixborough has been described in Chapter 13. Though there has been a considerable literature on the VCE, little of it has been concerned with a detailed analysis of the damage to certain key buildings on the site. The sub-section will discuss damage not only to the main control building but also to the main office block and associated buildings, and the oleum plant control room.

21.6.2.2. The main office block.

This building is shown on the plan, Figure 21.3. It will be seen that the building was located barely 50 metres from the point of escape of the vapour. Tucker and Webb (1975) has this to say:- "The main office block was reportedly a three-storey building. It appears to have been of concrete frame and brick construction but had been totally demolished. The other buildings (weighbridge, gate house and security office, VM) were presumably similarly demolished - their remains were not noticeable". The main office block is shown to have been of three-storey construction in Figure 21.4. which is a pre-explosion aerial view of the site taken from the south east.

When the present author inspected the remains of this building on 6/6/74, it was just a pile of shattered reinforced concrete beams and slabs and rubble, largely heaped over its original foundations. His recollection of its mode of construction from earlier visits he made to the site agrees with that of Tucker and Webb. In common with other such buildings, which include reception areas for visitors, there were large areas of glazing on the ground floor. The author has not been able to find a captioned photograph of the wreckage of this building but has been able to identify two of the photos in his possession as featuring it. These are shown in Figures 21.5. and 21.6. The second figure is clearly a close-up view of the extreme right-hand section of the first figure, the location of which can be readily established. The photo accords with the known three-storey construction of the building.

The class of damage to the main office building and its associated buildings can only be described as class A but this damage occurred to a building which was, at least so far as its main building elements were concerned, of much stouter construction than a dwelling house. A class damage is defined as "more than 75% of the external brickwork demolished" and is associated with overpressures of the order of 0.7 bar. Though no detailed civil engineering calculations have been published, to the author's knowledge, on the overpressure necessary to have brought about the degree of damage suffered by this building, it seems reasonable to suppose that the building had been subjected to an overpressure in excess of 0.7 bars. The estimate, published in Marshall (1976a), of a maximum overpressure of 1 bar arising from VCEs, may therefore not be unreasonable. It may be said with confidence that the building did not "withstand" the blast.

The single-storey temporary project buildings to the west of the main office block which are shown in the works plan and in the aerial photograph, totally disappeared in the explosion.

The population of the main office block during office hours was of the order of 100 persons. The state of the wreckage was such that, had the accident occurred in office hours, it is difficult to believe that any of these would have survived. It is true that, with the exception of Beek, the other disasters occurred outside of office hours; however for continuously operating plant there is only one chance in four that a disaster would occur in office hours. At Beek there was only one small office building on the site. At Ludwigshafen 1948, which has not been treated as a case history on account of lack of information on damage to control rooms, the heavy death toll must have been greatly augmented by the disaster occurring at mid-afternoon. The list of those killed there includes 13 women which is possibly evidence that office workers were killed.

21.6.2.3. The oleum plant control building.

Tucker and Webb (1975), who describe it as "Sulphate plant control house", have this to say:- "R.C. frame with brick infill. Near walls blown in, far walls blown out. Frame appeared undamaged". This building was 150 metres from the point of escape and was of two-storey construction. The present author

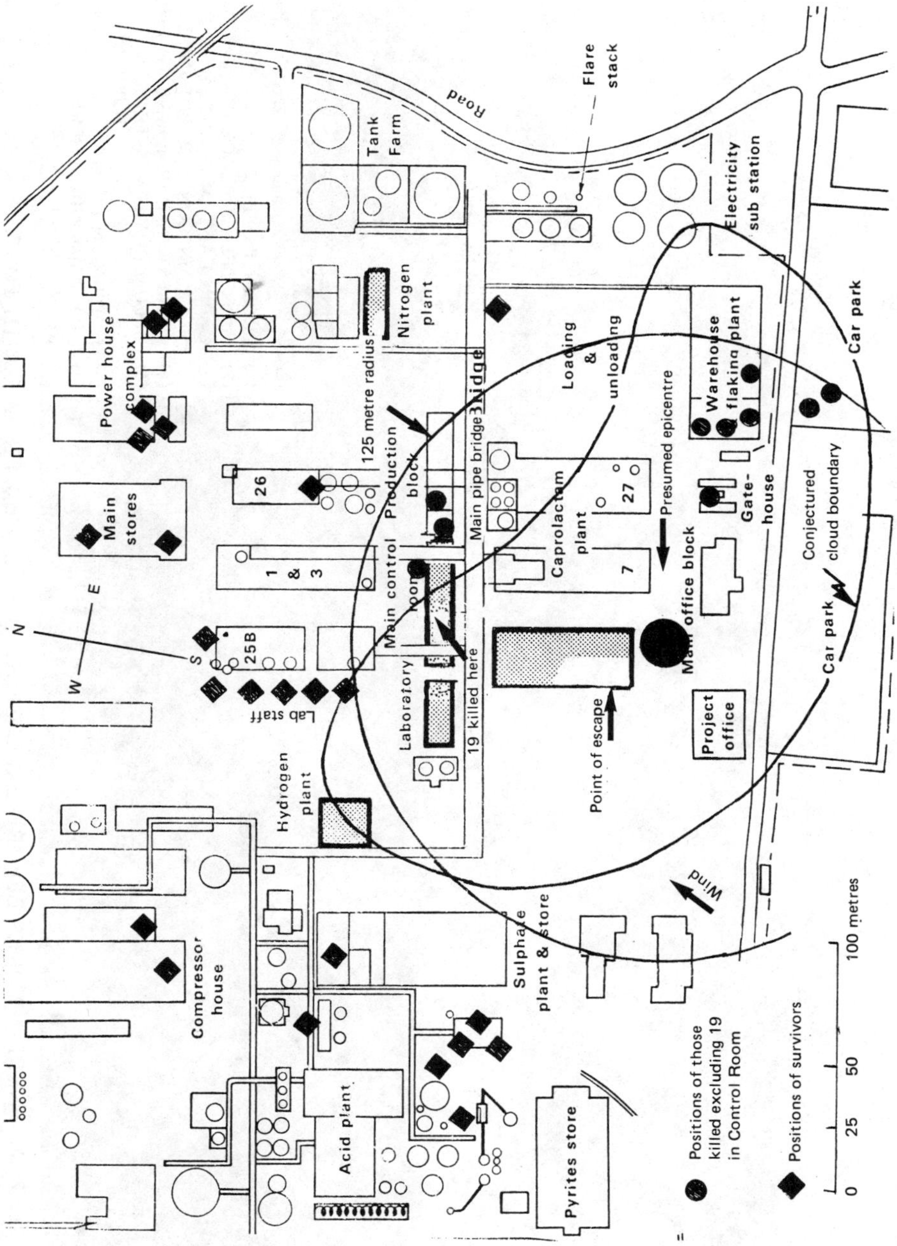

Fig. 21.3. SIMPLIFIED PLAN OF NYPRO WORKS.

Fig. 21.4. NYPRO WORKS BEFORE EXPLOSION, AERIAL VIEW FROM SOUTH EAST.

Fig. 21.5. VIEW FROM WEST OF CYCLOHEXANE OXIDATION UNIT WITH WRECKED MAIN OFFICE BLOCK AT EXTREME RIGHT.

Fig. 21.6. CLOSE UP VIEW FROM WEST OF WRECKAGE OF MAIN OFFICE BLOCK.

was able to make a brief inspection of it after the explosion and also possesses photographs which are reproduced as Figures 21.7. and 21.8. He would prefer to say that some of the panels on the south and east faces of the first floor, which were each at 45° to the presumed line of blast, were blown in as was a small portion of west wall where it joined the south wall. The roof had not fallen in and the frame was largely undamaged.

There appear to have been at least two men in the building at the time of the initiating event. One ran out on hearing the initial bang and was knocked over by the blast in the open, [Manderstam (1974), P.83]. According to Humberside Police (1974), Para. 27, a witness, unidentified in the reference, states:- "We both went back to the site and made for oleum control. We started shouting 'Harry' and we heard him answer from inside the control building which was badly damaged. We climbed in through a broken window. We found him near the door under a pile of rubble. We started to dig him out and were joined by the works firemen". The author has identified the rescued man with a man who is listed in Humberside Police (1974) as having been treated in hospital but not detained.

If the distance of the control room be taken as 150 metres and the TNT equivalence as 32 tonnes ground burst, then the scaled distance of the control building would have been 47 metres/tonnes$^{0.333}$. This would correspond to an overpressure of ca 0.5 bar and damage to a dwelling house between class A and class B. Allowing for the stronger construction of the control building, this seems to be in reasonable agreement. Generally speaking the mode of construction of this building, which was built without anticipation of explosion, appears to have been similar to that of the totally collapsed main office block.

Could it be claimed that the building "withstood" the blast ? It seems a borderline case but the issue is complicated by the possibility that it was shielded by a sulphate store. Had the walls of the building been panelled with reinforced concrete, perhaps it might have withstood the blast quite well. In fact the building was not pulled down but was subsequently repaired and rehabilitated. During the late seventies it functioned again as the oleum plant control building.

For some reason neither this building nor the main office block feature in the analysis of Sadee et al (1977).

21.6.2.4. The caprolactam plant control building.

Tucker and Webb (1975) simply describe it as "control room" and says:- "Reportedly of monolithic concrete construction. Completely destroyed by explosion, fire and the collapse of the pipe bridge". Sadee et al (1977), Table 1, says:- "Control room - Demolished, Distance from ground zero 105 metres". Beyond this, apart from generalities, little analysis has been published.

Tucker and Webb was in error in describing it as being of monolithic concrete construction. It was in fact of similar construction to the other buildings described above, as is clearly demonstrated in Figure 21.9. which shows the wreckage of the building with heaps of brick rubble among the concrete slabs and pillars. The author has several photographs of this wreckage in his possession which also make it clear that the infilling must have been of brick.

The building was, in its centre section, ca 10 metres high but was only of the order of 4 metres high at each end as shown in the sketch, Figure 21.10. Figure 21.11. shows the building in section looking from the west. This shows it to be of two and a half storeys as there was a mezzanine floor which carried electricity cables. The building, had it not been in a hazardous area, would have been an example of successful economical design. It combined under one roof the control room and the switch room for distributing three-phase power and the lighting circuits for the caprolactam complex. Transformers

Fig. 21.7. DAMAGE TO OLEUM PLANT CONTROL BUILDING.

Fig. 21.8. DAMAGE TO OLEUM PLANT CONTROL BUILDING.

F g. 21.9. VIEW FROM NORTH OF WRECKAGE OF MAIN CONTROL BUILDING.

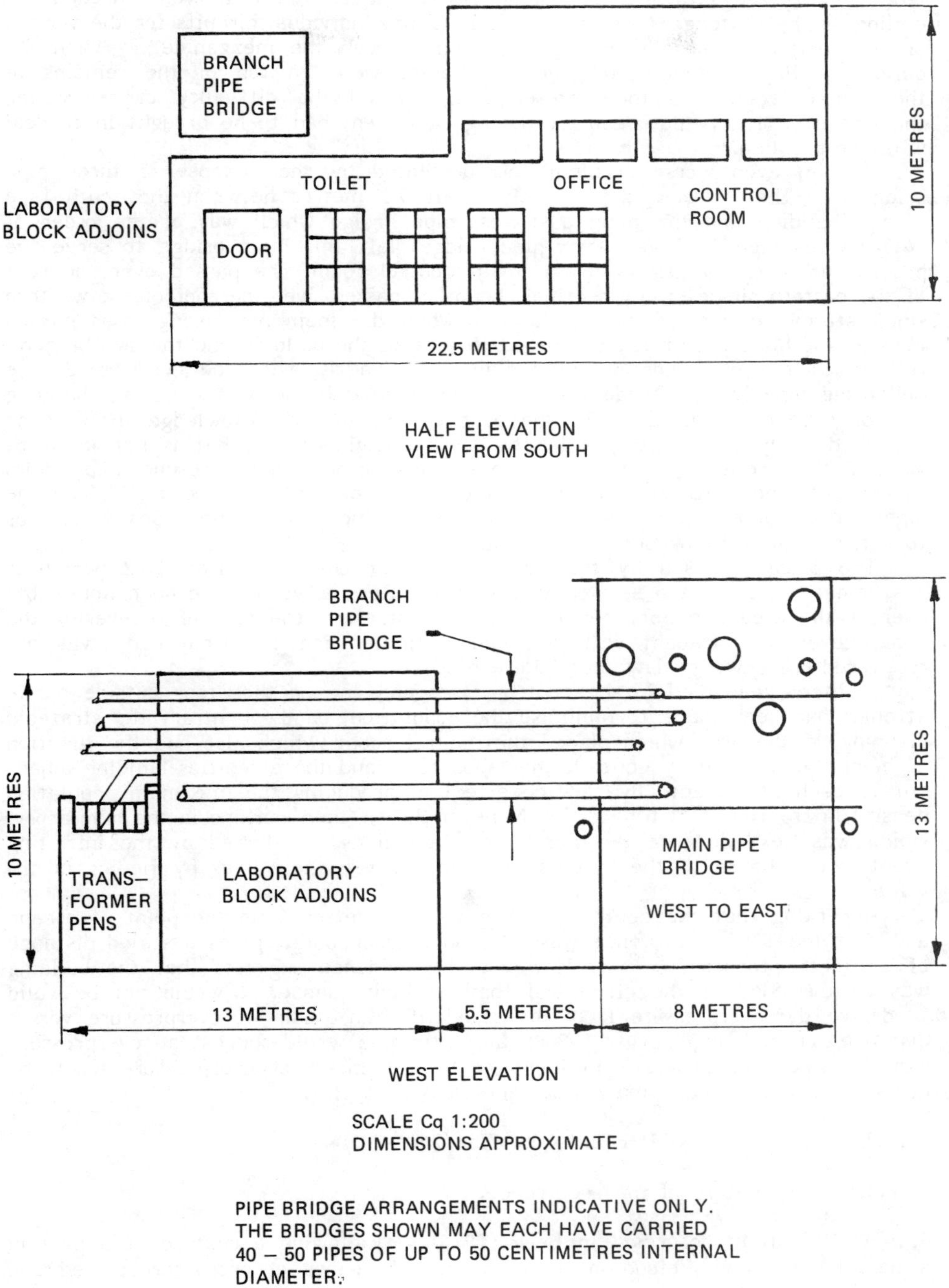

Fig. 21.10. ELEVATIONS OF CONTROL BUILDING AND PIPE BRIDGE.

were housed in pens on the north side of the building with their low tension output being fed into the mezzanine floor. From there it went through the ceiling to the switchgear which then split it into individual circuits for the various process units. These cables also passed through the mezzanine. When the building collapsed the switchgear and cables were on top of the remains of the control room and their presence aggravated the difficulty of recovering the bodies. Heavy duty cable cropping equipment had to be brought in to deal with the problem.

But an even worse problem was presented by the collapse of three pipe bridges. There was a gap of only about 5.5 metres between the south face of the building and the main east/west pipe bridge which was of the order of 16.5 metres high. Two branch pipe bridges left this main bridge to serve the process units to the north of the main control room, one passed over the roof of the eastern single-storey section, the other passed over the roof of the western single-storey section of the building. When the main pipe bridge was pushed over by the blast it fell against the south face of the building and the two branches fell on to the two ends of the building. Exactly what the contribution the collapsing pipe bridges made to the catastrophic collapse of the control building can only be conjectured. To the best of the author's knowledge there is no published analysis of this question by a civil engineer. What is not in doubt is that the presence of these collapsed pipe bridges made rescue impossible. Apart from the mechanical problems the pipes were full of process fluids, some highly flammable, others corrosive or toxic, which made the pipes hazardous to cut, especially by oxyacetylene flames.

The problem posed by the wreckage is set out in Figure 21.12. which is a schematic sketch of a section through it. Rescue would have been impossible even if the area had not been engulfed in flames. The task of recovering the bodies even was fraught with extreme difficulty and the final body was not recovered until twenty days after the explosion.

Eighteen men died in the control room and one on the steps outside. No attempt had been made to minimise the population, on the contrary the strategic planning of the area which placed the control room (which also had the function of being the emergency centre), the laboratory, and the amenities building adjacent to each other, actually had the effect of producing the maximum population density in an area of high hazard. None of the six men working in the laboratory, which was next door to the west, lost their lives. Alerted by the initiating event, the collapse of the by-pass pipe, they saved their lives by fleeing northwards.

Accepting that the control building was 90 metres from the point of escape and that the TNT equivalence was 32 tonnes groundburst, gives a scaled distance of 21 metre tonnes$^{0.333}$. However the evidence suggests that the building was on the edge of the cloud and that, in consequence, it would not be valid to derive the overpressure from the standard relationship of overpressure versus distance for high explosives. Such a relationship would suggest an overpressure well in excess of 1 bar. Perhaps the best estimate of overpressure would be circa 1 bar which would imply a damage class of A plus.

21.6.3. BEEK, THE NETHERLANDS, 7/11/75.

21.6.3.1. The nature of the occurrence.

This VCE is fully described in Beek (1976). Following a rupture of a pipe in a naphtha cracker a cloud made up of ca 5.5 tonnes of propylene ignited and the resulting VCE killed 14 men, all on the site, 6 of them in the control room. 104 persons were injured on-site and 3 off-site. The TNT equivalence was estimated by the organisation TNO to have been ca 2.2 tonnes at ground level.

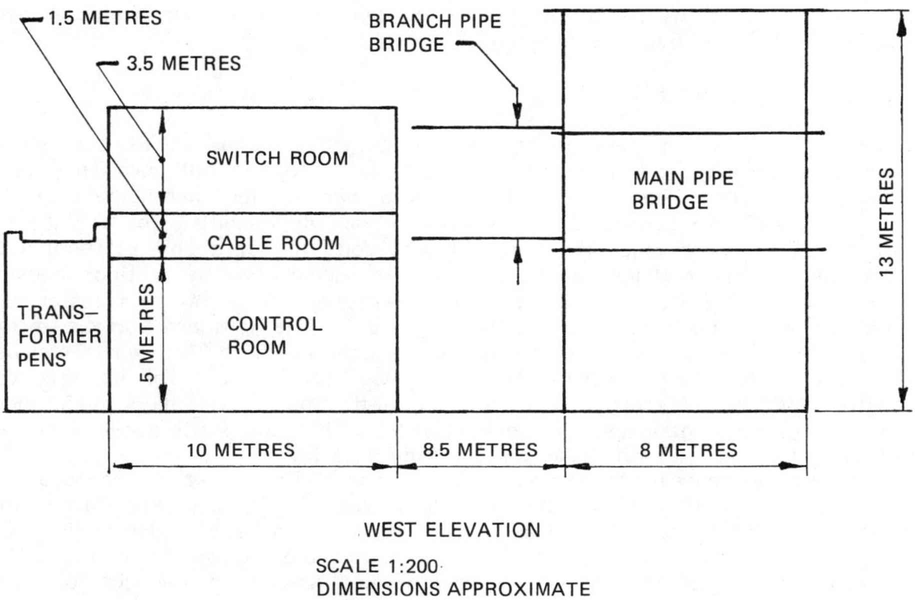

Fig. 21.11. SCHEMATIC SECTION OF CENTRE OF CONTROL BUILDING.

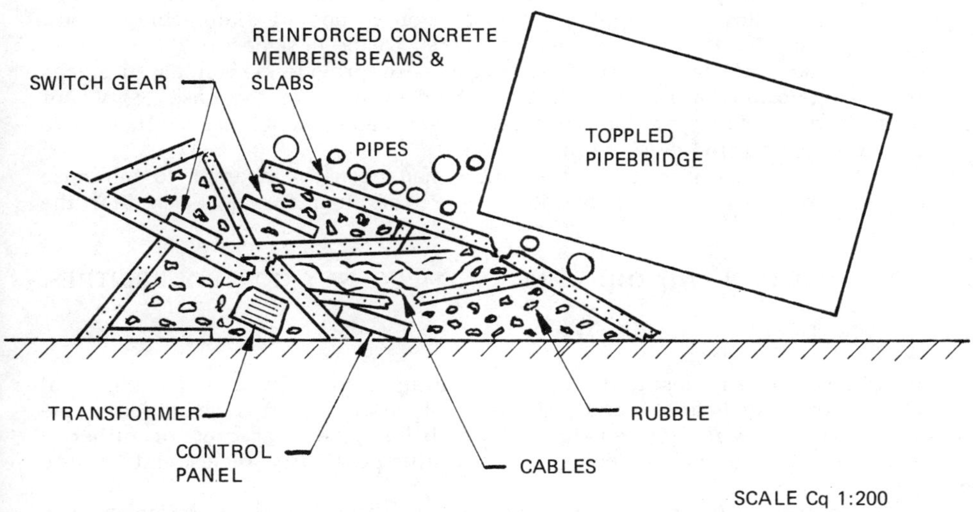

Fig. 21.12. SCHEMATIC SECTION OF WRECKAGE.

TNO concluded also that overpressures of up to 1 bar had been generated. The explosion and subsequent conflagration caused serious damage. The control building was severely damaged both by the explosion and by the fire which ensued. It was not demolished by the explosion.

21.6.3.2. The control building.

The control building at Beek resembled the control building at Flixborough in its design philosophy and mode of construction. This is not surprising as it was owned by DSM (Dutch State Mines) who were major shareholders in the Nypro plant and the source of the know-how on which that plant was based.

Figure 21.13. shows that the northern part consisted of a control room with a mezzanine cable distribution floor above it surmounted by a floor housing low-tension switchgear, exactly as at Flixborough. There were transformers located along the northern face of the building. The southern portion of the building, which was mostly of the same height as the northern part, was somewhat differently arranged and housed a process control computer. The building was 34 metres long by 17 metres deep by 11.5 metres high. A major difference, evident from the photographs in Beek (1976), is that the infill panels were of concrete, not brick as at Flixborough. See Figure 21.14.

There was no general collapse such as occurred at Flixborough. The quantity of vapour involved was much less and there was no pipe bridge collapse onto the building even though there was one on its western face. However there was a severe internal fire involving the electric cables in the building. This fire destroyed the computer and its records and spread to the control room itself which was partly burned.

The building could hardly have been more vulnerable on account of its location. Its western wall was only 20 metres from the point of escape and it had a pipe bridge immediately adjacent to its western face.

21.6.4. GENERAL CONCLUSIONS FROM CASE HISTORIES.

The most general conclusion is that all the buildings examined suffered damage through being too close to a potential source of escape of flammable vapour. It does not appear that this problem even occurred to the designers.

This being so, the failure of the designers to provide a building designed to withstand explosion naturally follows. As does also, in the case of two of the buildings, the use of the buildings for power cable switching. These two buildings were also far too close to pipe bridges.

Finally the case histories show that there had been no attempt to minimise the vulnerable population of the buildings at Flixborough and Beek, in fact the contrary is true.

21.7. THE FUNCTION AND THE SITING OF EMERGENCY CONTROL CENTRES.

21.7.1. FUNCTION.

There is a need in any installation which contains within it a major chemical hazard, to have a dedicated emergency control centre. This would have the function of dealing with any emergency which has the character of either a major incident or one which seems to have the possibility of escalating into one.

Such a centre would have the capability of directing the emergency and maintaining communication with the emergency services. It would direct the on-site remedial measures and, at the same time, provide a communication centre for the off-site measures under the control of police, fire brigade,ambulance

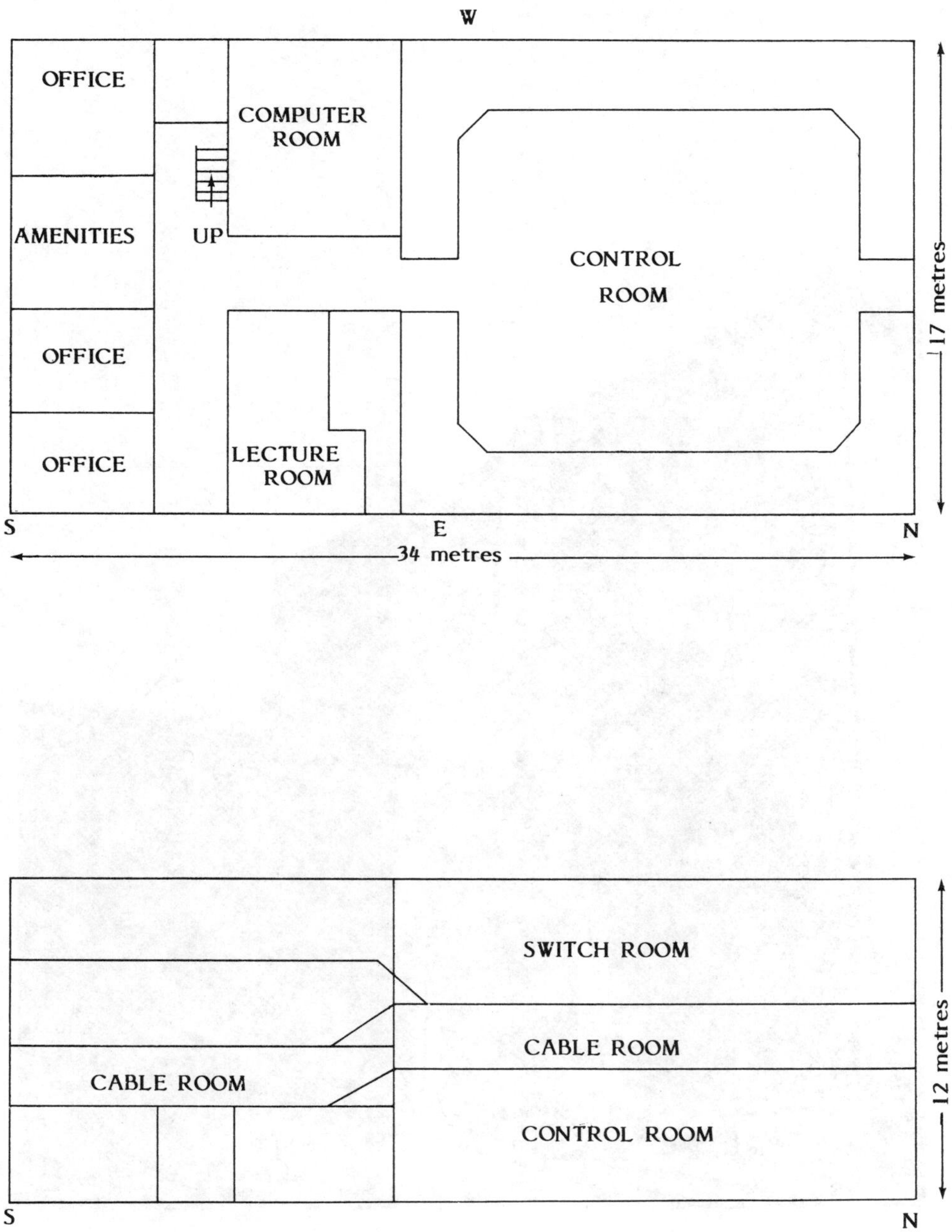

Fig. 21.13. SKETCH OF BEEK CONTROL BUILDING.

Fig. 21.14. DAMAGE TO NORTHERN FACE OF BEEK CONTROL BUILDING.

services and the local authorities. It may also function as the centre for public relations and the channel of communication with the media. Where the installation forms part of an industrial complex, it would handle communications with neighbouring installations; this would be of especial importance if there was reason to apprehend a "domino" disaster.

21.7.2. EQUIPMENT OF THE CENTRE.

Such an emergency centre is described in Appendix 5 of Chemical Industries Association (1976). The reference sets out a number of prime requirements of which the most important are concerned with ease of communication both internally and with the world outside. Such a centre should be capable of housing the appropriate representatives of management, the Health and Safety Executive, the Police and the Fire Service.

In the decade since Chemical Industries Association (1976), there have been major developments in information technology which considerably up-grade the facilities of an emergency centre. Today there are available computerised systems which provide not only instant information retrieval but also such information as the prediction of cloud dispersion based upon signals from gas monitors and from a meteorological tower. Such a system is described in Lashover (1985) and is illustrated in Figure 21.15.

21.7.3. STRATEGIC CONSIDERATIONS.

It does not seem appropriate to repeat here all the topics covered by Chemical Industries Association (1976); instead there will be discussion of those issues which are most relevant to the strategic approach to safety.

The first of these is the question of location of the emergency centre which the reference recommends should be as far as possible from the areas of greatest hazard and near to the main entrance. However the location of the main entrance is a strategic question which arises when site layout is considered. The greatest numbers of people who enter and leave will use this entrance but only a small proportion will have essential duties in the most hazardous areas. It is thus very important that the hazardous areas are so located that the fewest numbers have to pass them or work close to them.

This principle was grossly violated on the Flixborough site where virtually every employee either worked within 50 metres of the most hazardous area, or had to pass through it on the way to their duties. As has been pointed out in Chapter 20, had the Flixborough disaster occurred during office hours, it seems probable that at least 100 more people would have been killed. It was also true that at Flixborough the employees were required, in the event of an emergency, to report to the control building which, in the event, was totally destroyed by the explosion. But there is no evidence that this building contained a room provided with the facilities recommended in Chemical Industries Association (1976).

A further strategic question is the relationship between the location of the emergency centre and the location of related facilities. These include the works fire station, the medical centre, the security office and the telephone exchange. All of these appear to have the same requirements of remoteness from hazardous areas and proximity to the main entrance.

The need to locate the works fire station at a point remote from hazardous areas seems obvious, yet this principle was violated at the Ayrshire nitro-cellulose fire (Page 207) and at Mexico City (Page 202). The author was shown some years ago a fire station at a works in Belgium where, when there was an explosion on a neighbouring plant, and it was decided to turn out the fire appliances on a precautionary basis, it was found impossible to open the doors as they had

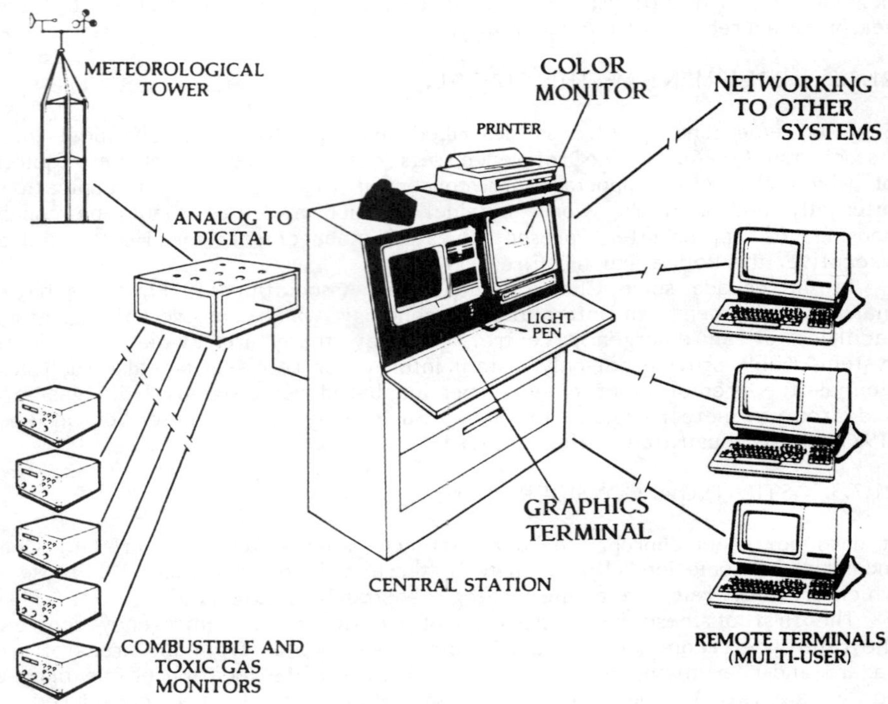

Fig. 21.15. EMERGENCY CONTROL CENTRE EQUIPMENT.

been buckled by the blast wave. The station now has a different design of door and the ability for the appliances to leave by the rear of the building also.

There are arguments for having the emergency centre adjacent to the medical centre. They have the same constraints as to location but there also seem positive advantages in having the medical personnel in close proximity to the emergency centre.

The subject of security as opposed to safety has not previously been discussed. The function of security is to protect company property, both physical and intellectual. This entails preventing theft, arson, and terrorism. Depending on company policy, some safety functions may be discharged by security staff. These may include enforcement of "no smoking" regulations, and general site safety, where this has little or no technological content, of which road safety is an example. But the most important function of security, and which has most bearing on the implementation of emergency plans, is that it is the job of security to control access to the site and to be the point of initial reception of visitors. It may be expected, in the event of serious emergency, that the overall control of the access to the site will lie in the hands of the police but the security staff would be more able than the police to identify those who would be authorised to enter.

Chapter 22

SAFETY PROFESSIONALS AND THEIR TRAINING

22.1. THE STRUCTURE OF SAFETY MANAGEMENT.

The aim of this chapter is to discuss that structure of management which can lead to the most effective control of major chemical hazards. There are some problems in trying to present this, as policies of management differ from one company to another in the same country, and different countries have different management cultures.

What will be attempted will be firstly, to examine the relationship between the control of major chemical hazards and the control of other occupational hazards which may affect employees and the public, secondly to examine the functions which have to be discharged by the responsible persons to ensure effective control, and thirdly to classify these functions into a hierarchy.

22.2. MAJOR CHEMICAL HAZARDS IN RELATION TO OTHER OCCUPATIONAL HAZARDS.

22.2.1. RESPONSIBILITIES TO THE PUBLIC.

The relationship between major chemical hazards and certain other hazards has been discussed previously in Section 6 of Chapter 4, where it was illustrated in Fig. 4.5. Management has other problems of hazards arising from the installations of the chemical and process industries and these are indicated in Figure 22.1. which is an extension of Figure 4.5. and which illustrates certain hazards to persons not employed by the company. Hazards to such persons include:-

(a) Bodily injury or damage to the property of members of the public arising from the realisation of major chemical hazards. Such hazards have already received detailed discussion in the book.

(b) Bodily injury or damage to the property of members of the public arising from the realisation of personal accident hazards. This could include, for example, the collapse of scaffolding on the outside of a factory wall.

(c) Bodily injury or damage to property of members of the public from chronic discharges. There have been, for example, deaths resulting from persistent discharges of air-borne asbestos in the UK and from the taking home of asbestos contaminated overalls. These, for example, occurred at Acre Mill, Yorkshire, UK, [Acre Mill (1976)]. There have been numerous cases of the deposition of harmful or annoying substances on the neighbours by chemical works. These cases fall generally under the heading of environmental pollution and include contamination of land, water and air. There are also problems arising from noise, bright or flashing lights or possibly interference with radio or television reception. The question of ionising radiation is excluded from this book. The

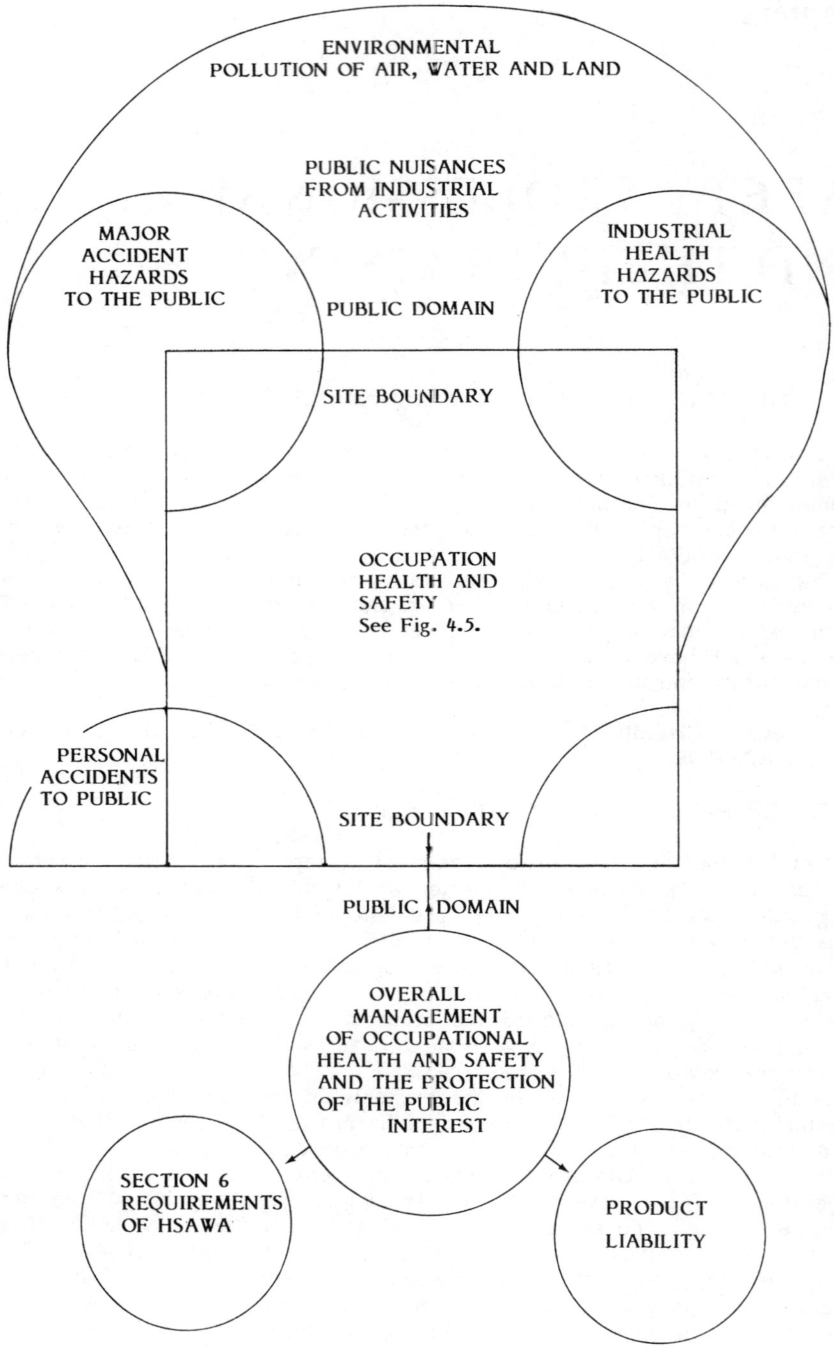

Fig. 22.1. RELATIONSHIP WITH PUBLIC AND EMPLOYEES.

question of allergies may also arise whereby distress may occur to a few people at very low levels of contamination of the atmosphere.

(d) An area not previously discussed is that of product liability. In the UK, when chemical substances are sold "for use at work", i.e. to other manufacturers, the manufacturer is responsible for ensuring that "the substance is safe and without risks to health when properly used", [Section 6, Health and Safety at Work Act (1974)]. When sold to the general public, there are various consumer protection acts which control the quality, packaging and labelling of such goods.

22.2.2. RESPONSIBILITIES TO EMPLOYEES.

That management has the responsibility to protect its employees from the realisation of major chemical hazards has been a central theme of this book.

However the great majority of employees who are killed or injured in chemical and process plant suffer these injuries through personal accidents. Of these personal accidents, including laboratory accidents on site, only about one-third to one-quarter arise specifically from the central technology of such installations. The majority are caused by falls, collisions, electric shock, by machinery, and by machine or hand tools.

Of the occupational diseases caused in chemical works, some are not caused by the central technology and may arise through noise, poor lighting, or stress.

22.2.3. THE FUNCTIONS TO BE PERFORMED.

The functions to be performed require people to handle the broad categories below, (hazards from ionising radiation are excluded).

The Public.

1. MAJOR CHEMICAL HAZARDS.
(a) Information in preparation for realisation of major hazards.
(b) Information in the event of realisation including information provided to the news media.

2. PERSONAL ACCIDENTS.
Prevention of personal accidents to the public.

3. ENVIRONMENTAL POLLUTION.
(a) Prevention of pollution of land, water and air.
(b) Prevention of nuisance from noise, smell, light, non-ionising radiation.
(c) Public relations in connection with above.

4. CONSUMER PROTECTION.
(a) Duties to other employers.
(b) Duties to the public.
(c) Public relations in connection with above.

Employees.

5. MAJOR CHEMICAL HAZARDS.
(a) Physical measures of protection.
(b) Training and information.

6. PERSONAL ACCIDENTS, NON-CHEMICAL.
(a) Physical measures of protection.
(b) Training and information.

7. PERSONAL ACCIDENTS, CHEMICAL.
(a) Physical measures of protection.
(b) Training and information.

8. OCCUPATIONAL DISEASES.
(a) Physical measures of protection.
(b) Monitoring of working conditions.
(c) Medical surveillance of personnel.
(d) Training and information.

22.2.4. THE GROUPING OF THE FUNCTIONS.

22.2.4.1. Influence of size of company.

For a very large company there could be people with responsibilities for each of the separate sections above or even for some of the sub-sections. In smaller companies one individual may handle several of the functions.

22.2.4.2. Broad groupings.

These groupings may represent suitable functions to be performed by members of the board of directors or by individual members of the higher management.
(1) Public safety and environmental protection.
(2) Consumer protection.
(3) Works safety and occupational health.
(4) Training.

22.3. THE HIERARCHY.

Within each function there will be a hierarchy of management with the strategic functions in the hands of the most senior people and with the tactical functions being entrusted to juniors. Such a hierarchy for works safety for a large company is illustrated in the table below:-

TABLE 22.1. A HIERARCHY OF SAFETY MANAGEMENT.

LEVEL	TYPICAL FUNCTIONS.
Board [Safety Director]	Formulation of company policy. Resource allocation. Overall supervision of safety functions.
Senior Management [Safety Adviser]	Advice on the formulation of company policy and on compliance with safety legislation. Process appraisal. Preparation of safety cases. Liaison with Health and Safety Inspectors.
Middle Management [Safety Officer]	Monitoring of compliance with company policy; safety audits. Issuing of permits to work. Training of personnel.
Technical	Monitoring of atmosphere for flammable vapours. Checking of safety instrumentation.
Technical	Responsibility for fire prevention and first aid fire fighting.
Foreman	Stores. Issuing of safety equipment and protective clothing.

Similar hierarchies may be drawn for the functions of public safety and of consumer protection. In all but the largest companies one person may discharge more than one function. Such combinations may be horizontal, i.e. works safety may be combined with public safety, or they may be vertical so that the functions of safety adviser and safety officer may be combined.

Table 22.1. does not cover occupational health nor does it cover the security function. These functions may also be combined with the safety function in some situations. Safety advisers will generally have oversight of occupational health as well as occupational safety. Some companies combine the safety function with the security function. However, in the author's view, this is not a very satisfactory combination. The two functions are by no means compatible and the training and experience required for each are very different. Security officers are often retired policemen.

In a large company it may be considered desirable to divide the safety function into a chemical safety function and a non-chemical safety function and to appoint two people for these jobs. The author's view is that at safety adviser level the two functions should be discharged by the same person.

22.4. THE QUALIFICATION AND EXPERIENCE OF SAFETY PROFESSIONALS.

22.4.1. THE OPERATION OF CHEMICAL AND PROCESS PLANT.

To a large and increasing extent the operation of chemical and process plant today lies in the hands of chemical engineers. However other professions such as chemists are represented and in the petroleum industry, petroleum refining specialists may be the plant operators. It seems to the author that it is from the ranks of such qualified people with experience of plant operation that the safety professionals in the chemical and process industries should mainly be drawn. It should be made clear that safety professionals, in the terminology of management science, perform a "staff" and not a "line" function.

22.4.2. THE SAFETY PROFESSIONALS.

The need for safety professionals in the chemical works has long been recognised. The UK chemical employers association recommended, in 1950, that all chemical manufacturers should appoint one or more safety officers and that for factories employing more than 500 persons, there should be at least one full-time safety officer, [ABCM (1950)].

As has been discussed above, there is a considerable area of safety in the industry which is non-chemical and which includes workshop safety and the safety of such workers as fitters, electricians, and instrument technicians when they are working out on the site.

Persons qualified as in the paragraph above have no special expertise in non-chemical safety which is a specialist subject. They should not conclude that this area is in any way elementary; if called upon to include it in their duties they need to undergo specialised training.

The author is strongly of the view however that whereas, if due diligence be applied, a person qualified in chemical engineering, chemistry, or in petroleum engineering, may without difficulty become competent in non-chemical safety, it is very much harder for a person trained in non-chemical safety but without appropriate basic qualifications and plant experience, to be trained for the responsibilities of chemical safety, especially on major hazard plant.

In short, it is highly desirable, if not to say essential, for a safety professional in chemical safety to be possessed of suitable technological qualification and experience. Added to this, such a person must receive appropriate specialist training. In the discussion below, the term "safety professional" will be shorthand for a specialist in major chemical hazards. This does not in any way imply that people in other areas of safety are not entitled to be regarded as professionals.

22.4.3. THE TRAINING OF SAFETY PROFESSIONALS.

22.4.3.1. Avenues for specialist training.

The paragraph above refers to the need for specialist training. This may be achieved through a combination of a number of inputs.
(1) A safety component in a first degree course at University level. In the UK such courses may be given at either Universities or Polytechnics.
(2) A master's degree course in which safety is a significant component.
(3) Short courses at University level.
(4) Attendance at appropriate conferences.
(5) On-the-job training by senior colleagues.
(6) Private study.

22.4.3.2. Bachelor's degree courses.

Until comparatively recently there was no general requirement for UK University courses, leading to a Bachelor's degree in chemical engineering, to include any material on chemical plant safety. Some courses did include such a component, and this reflected an interest in the topic by members of the faculty.

The new model syllabus put forward by the Institution of Chemical Engineers includes a component on Safety and Loss Prevention. The appropriate section reads:-

> "13. Process Equipment and Plant Design.
> 13.1.3. Safety and Loss Prevention. Legislation. Management for safety. Systematic identification and quantification of hazards, including hazard and operability studies. Pressure relief and venting. Emission and dispersion. Fire, flammability characteristics. Explosion. Toxicity and toxic release. Safety in plant operation, maintenance and modification. Personal safety". [I.Chem.E. (1983)].
> The above forms part of the core syllabus.

There is no compulsion on the part of a UK University department to implement such recommendations but, in practice, the pressure to conform is strong as without it graduates of such a non-accredited course would find difficulty in obtaining the status of Chartered Engineers.

There are, broadly speaking, two routes to implementation in Bachelor's degree courses. These routes are those of formal teaching and of project work. Project work may take the form of an essay or a dissertation or case study, or as a component of a design project.

Since there is a growing emphasis on "learning by doing", some combination of formal teaching with project work seems to be the best solution. Where there is a design project some safety input seems imperative. Students should not have to wait till they get into industry to find out that safety is a design constraint. However it needs to be borne in mind that only a small minority of chemical engineers are destined to become safety professionals and the safety component in the course therefore represents what a newly graduated general practitioner in chemical engineering should be expected to know in the field of safety. And that level of knowledge is only a beginning.

The American Institute of Chemical Engineers, the A.I.Ch.E., has set up its "Center for Chemical Plant Safety" on March 25th, 1985. It is proposing among other things to establish Fellowships or a Chair to develop case studies for college level courses in Chemical Engineering, [A.I.Ch.E. (1986)]. The aims and objects of the Center will be discussed further below.

22.4.3.3. Master's degree courses.

Though desirable in principle, there does not appear to be an adequate level of demand for a Master's degree course in the UK in the field of process safety. Indeed it has not proved possible to maintain courses which have a much wider remit and which cover all aspects of occupational safety and health. There has, however, been discussion recently of the possibility of organising an EEC-based course in safety and reliability in which chemical and process plant safety would be a significant component.

22.4.3.4. Short courses.

Short courses on various aspects of safety and loss prevention have been very successful. These generally last for four days and in the UK they are typically a joint activity between the I.Chem.E. and an Institution of University level. The current programme for 1986 comprises 14 such courses, [I.Chem.E. (1985b)]. The tutors are usually a mix of academics, industrial safety professionals and consultants in the field. One of these short courses is organised in conjunction with DECHEMA (Deutsche Gesellschaft für chemisches Apparatewesen Frankfurt).

However, there are also a number of freelance agencies involved in the field. These tend to employ the same people as have been referred to above and for this reason also achieve high standards.

The short courses under discussion are primarily for safety professionals but do also cater for the needs of plant operators who wish to up-grade their knowledge of safety. A recent development is that I.Chem.E. are now offering "in-company" courses which include Safety and Loss Prevention as one of the subjects offered.

Short courses differ from conferences, which are the next to be discussed, in that they are more clearly didactic and they are more highly structured. There is not the same necessity for the material presented to be an original contribution to knowledge.

22.4.3.5. Conferences.

Conferences are distinguished by their emphasis upon new knowledge. Though the papers are usually grouped under general subject headings they are not always closely related to each other in their themes. Conferences tend to attract those who research safety and the more senior members of the safety profession.

The I.Chem.E. organised its first major conference in the field in 1971, [I.Chem.E. (1971)], and since that time has organised a great number of such conferences. In addition to regional and national conferences, the I.Chem.E. has participated in 4 international conferences organised by the European Federation.

The A.I.Ch.E. has organised annual conferences in the field of Loss Prevention.

22.4.3.6. On-the-job training.

Such training, in which a safety professional learns the job from senior people presupposes the existence of a hierarchy of safety personnel and a ladder of promotion. Such is the state of affairs in the larger companies. In some cases staff may be seconded from one company to another to gain such experience.

In some cases secondment may take place between the academic world and industry, or it may be practised by professional institutions or research associations.

Training of this sort, where it can be given, is obviously of great and unique value. The use of training modules in which a tutor leads a discussion based upon material in the module and illustrates the themes by slides, videos or films, is rapidly growing in importance. A number of these are produced by I.Chem.E. Such modules may be suitable for training at all levels from the shop floor to the level of safety professionals.

Training in the form of exercises is now made much easier by the availability of computer programs and videos which can simulate incidents.

There is a clear need for training modules for the training of process workers. Where these are in the form of videos they can be shown in the control room when things are quiet. However, heed should be taken of the cautions expressed elsewhere about having too many people in vulnerable areas.

22.4.3.7. Private study.

All safety professionals must engage in private study though the extent of the role which it plays in training varies a great deal. In some instances, it may have to substitute for other forms of training.

This form of study may include the use of books, journals, videos or films.

22.4.4. THE QUESTION OF STATE ACCREDITATION.

As is well known in most, if not all countries, there are a number of professions which can only legally be practised by individuals who are accredited by the state. Generally this takes the form of the recognition of an approved degree course and some form of public registration of practitioners. In the UK for example, the General Medical Council supervises medical education and enters on a register, open for public inspection, the names of those who become qualified. Misconduct can lead to a person's name being struck from the register which amounts, for all practical purposes, to the person being unable to practise.

In the US, similar functions are discharged by boards of licensure in each state.

No such restrictive regulation of the profession of engineering exists in the UK where anyone can call themselves an engineer. However, the recently established Engineering Council regulates those who are entitled to call themselves "Chartered Engineers".

There are a number of branches of engineering in the UK where there is state certification. Marine engineers are certified and graded and managers of coal mines have to be certificated. The author understands too that those who manage waterworks have to be certificated.

But there are no such formal requirements in the chemical and process industries. The CIMAH Regulations do not make a specific requirement that there shall be a safety professional on the site, but schedule 6 of the Regulations require, as part of a safety case, that information be provided on the management system for controlling the industrial activity including the staffing arrangements and the name of the person responsible for safety on the site.

There is a presumption that the HSE would have the right to enquire into the competency of those so designated on grounds of qualification, experience or training.

Whether formal state accreditation is necessary or desirable is a very debatable issue. It would have to be considered in the context of safety professionalism in general and the UK would be reluctant to take steps which would not accord with the policies of its EEC partners and the EEC itself. There do not seem to be any indications that the EEC favours further extension of prescriptive professional accreditation.

On the other hand, should such formal state accreditation of safety professionals in the chemical and process industries be contemplated in the UK, the most appropriate professional institution to prescribe the qualifications, training and experience required would appear to be the Institution of Chemical Engineers.

22.4.5. THE INSTITUTION OF CHEMICAL ENGINEERS REGISTER OF SAFETY PROFESSIONALS.

In 1985, the I.Chem.E. set up an indicative register of Safety Professionals serving the Chemical and Process Industries. To quote the Foreword to the Regulations governing admission to the Register:-

> "The Register is an indicative list, open to public inspection, of Chartered Chemical Engineers, who are deemed, by the judgement of their peers, to be suitably qualified by their training and experience to be recognised as specialists in Process Safety and Loss Prevention. These specialists may be employees of the Chemical and Process Industries, or be independent consultants, or be engaged in teaching and research in the field, or be employed by those agencies of Government which are charged with responsibility for enforcing safety legislation in the Chemical and Process Industries. Regarding future developments, it is appreciated that probably there are large numbers of professional people active in Safety and Loss Prevention who are not members of the Institution of Chemical Engineers. The Institution will be keeping this matter under review and would consider ways of possibly accommodating such people on the Register if there was sufficient call. Furthermore, when other professional bodies establish similar Registers for their own members, a cooperative approach is envisaged to ensure the establishment of a comprehensive register system". [I.Chem.E. (1985c)].

The Register is not prescriptive and the Institution does not intend to imply that anyone not so registered is thereby adjudged not to be competent either to assume the responsibilities or to fulfil the duties of a safety professional.

Chapter 23

THE ROLE OF PROFESSIONAL INSTITUTIONS IN THE CONTROL OF MAJOR CHEMICAL HAZARDS

23.1. INSTITUTIONS AS LEARNED SOCIETIES.

There are a considerable number of learned societies in the UK and these have their counterparts abroad. These are societies or institutions whose members are "united in some common interest, organised to pursue some definite line of research and issue periodical publications concerned with the progress of study in such spheres of human knowledge". [Encyclopedia Britannica (1951)]. In the UK the preeminent learned society is the Royal Society of London which was founded in 1660. This Society has published, in Royal Society (1981) and Royal Society (1983), two major contributions to the debate on "Risk".

In the UK the principal institutions concerned with the sciences and with engineering and technology such as the Royal Society of Chemistry and the Institution of Chemical Engineers, are learned societies though they perform other functions in addition. Important learned societies outside of the UK which are concerned with the subject area of this book are the American Institute of Chemical Engineers and, in Germany, the Deutsche Gesellschaft für chemisches Apparatewesen (DECHEMA). There is a European Federation of Chemical Engineering (EFCE) which is an association of 54 European technical and scientific societies concerned with chemical and process engineering, [see ACHEMA (1985)].

All of the chemical engineering societies are concerned with safety and loss prevention and many have been active in promoting research and the exchange and publishing of information in this area. The I.Chem.E. has organised nine major conferences on hazards in the process industries since 1971 and has, in addition, cooperated in the organisation of four international conferences.

The EFCE organised its 5th International Conference at Cannes in 1986 under the title of "Loss Prevention and Safety Promotion in the Process Industries".

The A.I.Ch.E. through its journal "Chemical Engineering Progress", has published two Symposium Series. The first of these CEP Technical Manuals is "Ammonia Plant Safety (and related facilities)", which was formerly entitled "Safety in Air and Ammonia Plants". Vol. 25 in this series was published in 1985. The second series of Technical Manuals is the more broadly based series "Loss Prevention" of which volume 14 was published in 1980, (now discontinued).

However, the organisation of conferences is only one aspect of the dissemination of knowledge by these learned societies. Reference has already been made to the work of the I.Chem.E. in fostering the continuing education of its members and other interested persons in process safety and loss prevention. A further publication by I.Chem.E. is its Loss Prevention Bulletin which was founded in 1974, and is available on a subscription basis. It brings together case histories of incidents made available by its subscribers on a purely voluntary basis, and which are usually published anonymously, analyses of past incidents, together with the theoretical treatment of problem areas, and also includes annually a list of serious incidents in the chemical and process industries which have occurred throughout the world.

23.2. THE ETHICAL FUNCTION.

The Institutions under discussion play an important ethical role. This they do by formulating and promulgating their codes setting out the professional responsibilities to the public which they require of their members.

The I.Chem.E. in its Charter declares that its objects are to:-

> "promote, foster and develop the general advancement of the science of chemical engineering in all its branches . . . to the benefit of the community at large".

In its By-Laws and Rules of Professional Conduct, it declares that its members have the duty to:-

> "safeguard the public interest in matters of safety, health and otherwise" and that "For situations not specifically encompassed by the rules . . . in any conflict between a member's personal interests and the fair and proper interests of the community, the member's duty to the community should prevail". "A member shall take all reasonable care in his work to minimise the risk of death or ill health to any person and of damage to property. In his work, a member shall respect all laws and statutory regulations applicable to the design, operation and maintenance of chemical and processing plant. In addition, a member shall have due regard for the need to protect working and living environments . . . "

The Institution's rules provide for disciplinary action to be taken against members who violate its professional code and provide for warnings, reprimands, suspension or expulsion of members found guilty of violation of the above and for other breaches of professional conduct.

The A.I.Ch.E. requires that "Engineers shall hold paramount the safety, health and welfare of the public in their professional duties".

Flores (1983) reviews the problem of the potential conflict between professional ethics and the commercial interests of employers.

> "In the real world where engineers practice, the ethical responsibilities of engineers must be predicted on a realistic assessment of the power that engineers have to control the outcomes of their professional practice".
> and
> "Surprisingly, nearly three fourths of respondents [employed by a company with a high reputation for safety,VM] had never taken a college course in which safety was a major topic . . . their education had failed to properly expose them to industry and government safety standards . . . "

It seems clear from this that Institutions have an important job to do in educating their members and making up for what the Universities are not able to do. But it may be said that it is difficult for first degree level students sometimes to grasp the significance of what they are taught as the hazards of industry and their ethical overtones may seem very remote from the lecture hall.

23.3. THE INSTITUTIONS AND THEIR ROLE IN INFLUENCING POLICY.

Members of professional institutions play an important role in formulating public policy by which major chemical hazards may be regulated both through "self-regulation" and through legislative control.

Though the I.Chem.E. was not represented as such on the Major Hazards Advisory Committee, four of its senior members served on the Committee. The Institution is represented through its members on a number of official and semi-official bodies concerned with the establishment of safety standards.

The I.Chem.E. has a number of Committees, Panels, and Working Parties concerned with the establishment of standards and with formulating the Institution's views on matters of safety. It is of interest here that the Institution welcomes the participation of specialists who are not its members in these deliberations.

The Center for Process Safety, recently established by the A.I.Ch.E., declares that it has the aims, as paraphrased by the present author, of:-

(1) Conducting research and providing information.
(2) The professional development of engineers involved in process plant safety.
(3) The collation of existing safety procedures.
(4) Information exchange.
(5) Dialogue with trade associations and other interested organisations.

Though not explicitly stated in its aims, there seems little doubt that the Center will also influence public policy.

Chapter 24

RESEARCH AND CONSULTANCY

24.1. THE ROLE OF RESEARCH.

The control of major chemical hazards contains within it, as a key element, the acquisition and transmission of information on mechanisms of release, dispersion of energy or materials, effects on persons and property, and the various means of prevention and mitigation.

Research in the field of major chemical hazards, though it has almost all been conducted in the last twenty years and most intensively in the last decade, has already generated an enormous literature which it is quite impossible to review in the space available. The discussion which follows can be, at best, only indicative and can only outline work in a few of the countries which have contributed to it.

It is of some interest to note that the mass media, even including the popular scientific press, is largely unconscious of the effort which has been expended and when things go wrong, it gives greatest prominence to the views of amateur commentators and allows them to denigrate the industry and the enforcement authorities by accusations of apathy.

It seems convenient to the author to divide research into three aspects, recognising however that these overlap and interact. They are:-

(a) Practical research.
(b) Theoretical research.
(c) Research through scholarship.

Practical research includes laboratory research as, for example, assessing toxicity by animal experiments, or gas dispersion in wind tunnels, or determining burning velocities and flame speeds. It also includes field trials which are conducted in the open air and which may be on a very large scale as in the Thorney Island trials.

Theoretical research includes the development of models based on first principles, or the results of trials, or some combination of the two. The development of computer programs today is a major element in theoretical research.

Scholarship consists, for example, of the study and analysis of past events or the synthesis of broad principles from the results of practical and theoretical research. Scholarship has especial significance as the control of major chemical hazards is multidisciplinary in its nature. Though the findings of research into physical and chemical principles plays a vital part, nevertheless other sciences such as management sciences, also have an indispensable role. So too has the study of legislation.

The UK Advisory Committee on Major Hazards in its reports laid great stress upon the importance of research. Much of the research which has been conducted in the UK on dispersion arose from its recommendations.

24.2. RESEARCH INSTITUTES.

The UK Health and Safety at Work Act lays upon the Health and Safety Commission the duty:-

> "to make such arrangements as it considers appropriate for the carrying out of research, the publication of the results of research and the provision and training and information in connection with those purposes and to encourage research and the provision of training and information in that connection by others". [Section 11 (2)(c)].

The HSE, which is the executive arm of the Commission, has research laboratories of which the most important, from the point of view of major chemical hazards, are the Safety Engineering Laboratory, Sheffield, which supervised the Thorney Island trials, and the Explosion and Flame Laboratory, Buxton. Research into dust explosions is conducted at the Fire Research Station, Borehamwood, Herts., which is part of the Building Research Laboratories of the Department of the Environment.

Extensive research programmes are conducted by bodies such as British Gas and the major petroleum companies which have included large-scale field trials. Work on modelling dispersion through wind tunnels has been conducted at Warren Spring Laboratory.

Theoretical modelling and risk assessments are carried out by the Safety and Reliability Directorate (SRD) of the United Kingdom Atomic Energy Authority (UKAEA), which was responsible for the two Canvey Island Reports and which also undertakes contract work in the field.

In the Netherlands the organisation TNO, which is the Netherlands organisation for applied scientific research, maintains major laboratories at Delft and Apeldoorn.

In the US, a major role is played by the Coast Guard who are much concerned about dangerous cargoes but whose interests also have significance for land-based major hazards.

24.3. OTHER AGENCIES OF RESEARCH.

The research input of UK Universities into the control of major chemical hazards reflects all three aspects of research. The University institutions concerned have included the Universities of London (Imperial College), Manchester (UMIST), Aston, Loughborough, and Bradford and the Polytechnics of the South Bank (London) and Teesside.

The I.Chem.E. has encouraged scholastic research. As an example, Gugan's book "Unconfined Vapour Cloud Explosions" [Gugan (1979)] arose from a Fellowship awarded by the Institution. In addition to publishing conference proceedings, the I.Chem.E. has publications on specific aspects of safety and loss prevention. The current publications catalogue lists thirteen such monographs in addition to seven hazard workshop training modules. All of these have been compiled by drawing on the collective experience of its membes and of other specialists not in membership of the Institution, [I.Chem.E. (1986)].

24.4. THE ROLE OF CONSULTANTS.

Consultants are independent professional partnerships whose principal function is to tender advice to clients who are their source of income. They do not engage in large-scale practical research though they may maintain testing laboratories. Consultants today are heavily engaged in the assessment of hazard and risk both of existing and projected installations. Their business normally cover a much wider field than major chemical hazards and may include, for example, process design or the assessment of environmental impact. Generally the partners and associates have between them a wide range of qualifications and experience.

The UK Association of Consulting Scientists, [ASC (1984)], lists three Members who act as consultants in the field of major chemical hazards. These are Cremer and Warner, Dr. J.H. Burgoyne and Partners, and Technica Ltd. All three are active internationally but are based in the UK. There are other consultancies in the UK who are not members of this Association.

The UK firms of consultants listed above are young compared with the US Arthur D. Little Inc. This is a wider-ranging consultancy organisation which is currently celebrating its centenary. This organisation, which was founded by A.D. Little who discovered the concept of "unit operations", employs 1,300 consultants at its Head Office; 150 of them are chemical engineers. It gives advice on major chemical hazards among many other topics, [Caruana (1986)]. Another US consultancy of note is Battelle Laboratories.

The research contribution of consultants lies mainly in the field of theoretical modelling and of scholarship. The primary output of consultancies is in advice to clients, but there has been a measure of general publication through conference papers and articles in journals. In addition, some reports produced by consultants have been published by their clients. This was the case, for example, with the Rijmond Study, [CECAR (1981)].

In the present author's view, it would be very undesirable for a client operating a CIMAH site to be totally reliant upon the services of consultants to prepare a safety case. However, valuable services can be provided for such clients by consultants because of the wide range of experience and expertise which such consultants can bring to bear.

Chapter 25

ADDENDUM

P.21. Davenport (1984) now provides an overview of vapour cloud explosions which is more comprehensive than that previously presented in Gugan (1979).

P.26. Witcoff and Reubens (1980) contains a useful discussion of the economies of scale.

P.60. Fletcher (1982) provides evidence that for water and Refrigerant 11, flashing does not occur in pipes where L/D is less than 0.33. A liquefied gas, flowing as a liquid through a hole in a tank wall, thus flashes only when it reaches the open air.

P.67. The Chapter excludes from consideration the specialised topic of the failure to operate relief systems such as safety valves and bursting discs. It is of some interest to note that, at the time of writing, there is no general legal requirement in the UK to fit such devices except in specified services such as steam boilers and air receivers.

P.71. Smith (1985) reviews the possibility of cold, whole vessel failure, to establish whether "leak before break" is more likely than catastrophic failure and concludes that suitably stress-relieved vessels should leak before major failure.

P.75. Propane may be carried safety in "thin-walled" structural steel tanks provided that they meet the specifications laid down in BS 1510 (1958) as amended 20th March, 1962, as well as to BS 5500 (1982).

P.79. It should be noted that a vessel completely filled with water may weigh considerably more than the vessel does in normal service. Allowance must be made for this in the design of the supporting structure.

P.81. Hunt and Ramskill (1985) and Moodie et al (1985) give descriptions of how LPG storages behave when engulfed in fire.

P.89. Appendix II of Alderson (1985) concludes that for an on-shore sited sphere, likely seismic effects could considerably outweigh wind-loading effects.

P.99. It is of interest that there are today commercially available computer systems with hardware costing under £1,000 which model, using a moderately sophisticated program, the dispersion of a spillage of liquefied gas and in which the substance and the parameters of the source term may be selected from a menu. A much more sophisticated facility, which costs a great deal more, is described on Page 487.

P.100. Havens (1985) updates the analysis of the dispersion of heavy gas clouds and takes the view, which is one which accords with the view of the HSE, that three-dimensional hydrodynamic models make excessive demands on computer time.

P.116. According to NFPA (1985), in 1984 there were 343 fires in the US with losses exceeding $1 million. If allowance be made for inflation the numbers of large fire losses in the US have been fairly constant in the decade 1975-84.

P.128. The Health and Safety Executive has produced a revised guidance note on LPG. This is CS4 ISBN 0 11 883539 4 (1986).

P.138. Otway (1985) quotes a statistic that, in the US in 1836, 14 boiler explosions between them cost 496 lives. These figures are attributed to a paper, Burke (1966), which discusses boiler explosions and the public interest which these engendered.

P.139. The figure deduced for the loading of flammables at the Mexico City installation, which is given on Page 205 of this book, was 450 kg/m^2. This considerably exceeds the figure put forward by Carrier et al as the minimum to sustain a fire storm, who postulate a similar time of burning. No fire storm was recorded at Mexico City but the area of the conflagration was only a tiny fraction of the 10 km^2 postulated by these authors as being necessary for the instigation and maintenance of a fire storm.

P.139. (Further reference). Later information, based on M & M (1986), suggests that the possibility of flame impingement on the storage tanks had been envisaged. The intention had been to cool the tanks by means of monitors which are mobile jets which can be set up in any location and left then unattended. Unfortunately the monitors were kept at a point on the site which was engulfed by the initial fireball.

P.142. UEG (1983) provides a diagram, reproduced as Fig. A.1, which summarises the limits of comfort of the human body. These limits are expressed in terms of skin temperature and core (central body) temperature.

P.210. Bartnecht (1981) is an important monograph on explosions.

P.254. Though Wiekema (1984) claims that motor cars have never been recorded as being overturned by a VCE, the present author recalls seeing a sulphuric acid tanker at the gatehouse which had been overturned as a result of the Flixborough explosion. The driver was one of the dead.

P.258. Hawksley (1986) reviews and analyses a number of incidents involving hydrogen and hydrogen-rich gases and concludes that VCEs involving tonnes of gas are unlikely to occur. However there have been explosions involving kilogram quantities of the gas in the open air.

P.327. The Health and Safety Executive has now issued a guidance note on bulk storage installations for chlorine. This is ISBN 0 11 883872 5 (1986).

P.331. Fire Prevention (1986) records an incident in Cleveland, England, on 23/9/85, in which a faulty non-return valve allowed hydrogen and chlorine to mix in an electrolysis plant. This led to spontaneous combustion and explosion. There was no record of injury but the damage was in excess of $500,000.

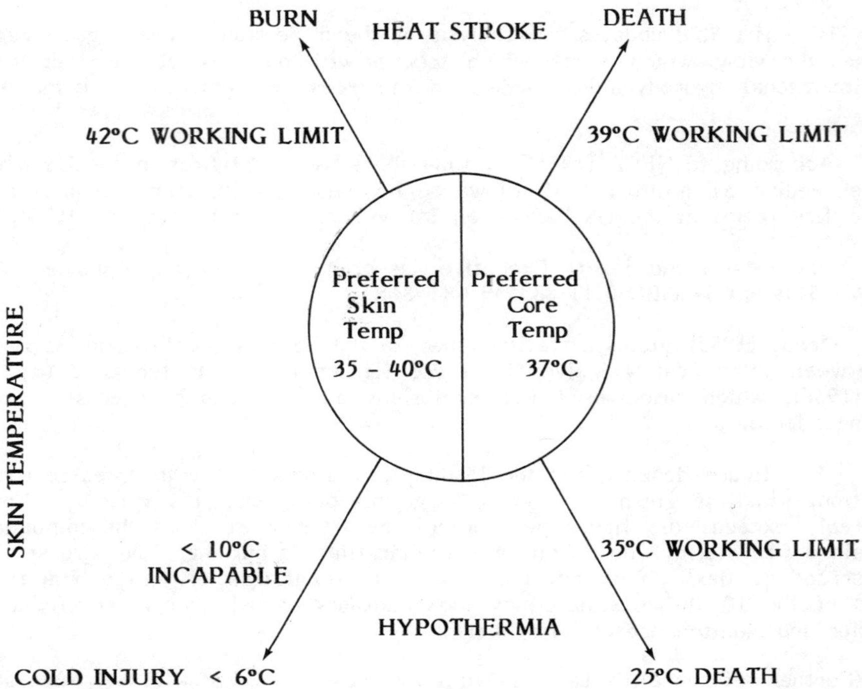

Fig. A.1. LIMITS OF COMFORT OF THE HUMAN BODY.
(By courtesy of UEG, 9 Storeys Gate, London SW1P 3AU).

P.346. The recent decision by the US to stockpile nerve gases envisages that this would take the form of manufacturing binary weapons. Chemistry in Britain (1986) shows a sketch of such a weapon in which the reactants would be phosphonyl difluoride and isopropanol, the latter containing a "booster", presumably this means a catalyst.

P.347. A major new publication is the book "Dioxin, Agent Orange, the facts" [Gough (1986)]. This is a highly detailed and largely sceptical view of the problem. It notes that production of both 2,4,5-T and 2,4,5-trichlorphenol have now ceased in the US.

P.368. The Chemical Engineer (1986) in an article concerned with monitoring dioxin in the environment, claims that the dioxin content of soil in the UK varies widely according to locality. It claims that dioxin has been detected in a soil sample dating from 1877.

P.384. The tank contained about 12,000 tonnes of molasses.

P.386. The view that flesh may undergo self-sustained combustion in an oxygen-rich atmosphere appears to be confirmed in a case quoted in Harvey (1986). This reference claims that after an occurrence in a ship under repair, and in which it appeared that two men had been working in an oxygen-rich atmosphere, "nothing remained of the two workmen who had been in the tank at the time". The Health and Safety Executive now recommends adding a stench agent to oxygen used in shipyards and possibly other places, CS7 ISBN 0 11 883578 5 (1986).

P.390. A further derailment involving a tank car containing yellow (white) phosphorus is reported as having occurred on Tuesday, July 8th, 1986, at Miamisburg, Ohio, US. Fifteen cars of a forty-four car train were derailed. The car containing the phosphorus exploded. 20,000 people were evacuated as a precaution and 60 had to be given medical treatment though none were detained. The incident is reported in New York Times (1986). Though reported on the BBC teletex service (CEEFAX) it received little or no attention in the UK press.

P.391. An event of a similar character to the Morley incident, but one which has had far more serious consequences, occurred on the night of 30th October/1st November 1986. This occurred at the works of Messrs. Sandos, Basle, Switzerland, where a fire in a warehouse containing ten sorts of pesticide, with a total inventory of ca 1200 tonnes, resulted in the pollution of the Rhine from Basle to the sea. The greatest part of the chemicals were organophosphorus compounds toxic both to people and to fish. A very toxic contribution also arose from 12 tonnes of an organic mercury fungicide containing 1.9 tonnes of mercury. This latter is a compound which decomposes slowly to yield mercury oxides.

The quantity of contaminated fire-fighting water initially released into the Rhine has been estimated at 10,000 cu metres. (This would cover an average football pitch to a depth of 1.5 metres). Extensive precautions had to be taken by the countries bordering the river, such as ceasing to abstract drinking water. Many fish were killed.

P.392. The UK Health and Safety Commission [HSC (1986)] approved the issue of guidelines on the release of genetically manipulated organisms for agricultural and environmental purposes. Its advisory body is now called the Advisory Committee on Genetic Manipulation (ACGM).

P.394. Though the new Public Prosecutions Office has taken over criminal prosecutions which were formerly instigated by the police, in England and Wales the Factory Inspectorate will continue to instigate prosecutions for breaches of the Health & Safety at Work Act. In Scotland the Procurator Fiscal will continue to be the prosecuting authority for breaches of this Act.

P.407. A recent report, CECIF (1986), written from the point of view of an association of European chemical manufacturers, is severely critical of quantitative assessment of risks. It states:-

> "The quantification of 'off-site' risks from chemical industry installations to persons living nearby is at present regarded as increasingly attractive in certain circles and by the regulatory and planning authorities of several European countries. Whilst not denying its possible role in the decision-making process, CECIF has serious misgivings about too extensive and rigid a use of this form of risk assessment.
> These misgivings mainly stem from the unreliability of the numerical figure finally obtained - one to several orders of magnitude - and the complexity of value judgements necessary for comparing this figure with numerical criteria of acceptability".

Joschek (1983) is also extremely critical of quantitative risk assessment and argues that the chief cause of accidents is human error whereas fault tree and similar techniques are concerned with hardware failure.

P.416. A mortality index approach is now being used to compare fatalities from various routes to electrical generation where deaths are being related to a gigawatt year, [Hill (1983)].

P.443. In 1985 an installation, now owned by Chemoxy International, on Teesside, England, which was sited in a heavily populated area and which was a major user of ethylene oxide, was bought out by the local authority and the operations involving ethylene oxide ceased. The lccal authority was compensated by the state.

P.457. The second edition of Mecklenburg's book has now been published by Godwin in 1985. It treats the safety aspects of layout in detail with worked examples.

P.458. Kletz (1985) shows a photograph of a crane which fell onto chemical plant as the result of over-extending its jib.

P.495. A modular programme for the training of all categories of inspectors employed by the HSE, leading to a Diploma of the University, has been established in the Health and Safety Unit of Aston University, Birmingham, UK. It covers the whole subject area of occupational health and safety but one of its core subjects is "Risk Assessment and Management". An interesting feature is that it will be largely taught "off campus" by means of video tapes, computer-based learning and recommended books.

P.498. The A.I.Ch.E. journal, "Loss Prevention", ceased publication at volume No. 14 which reported the Loss Prevention Conference of 1980. It was succeeded in 1982 by an A.I.Ch.E. quarterly publication "Plant/Operations Progress" which covers the same ground as "Loss Prevention".

Chapter 26

SUMMARY AND CONCLUSION

26.1. THE NATURE OF THE PROBLEM.

Though the problem had its origins in the long-established explosives industry, it was considerably aggravated by the development of large-scale chemical processes in the first half of this century. These were continuous processes and their process streams, by their nature, had no discernible upper limit of size. This growth in the size of process streams was spurred on by the considerable economic advantages which reside in scale. As a result, inventories of hazardous substances grew, bringing with them the hazards of major fires, explosions, toxic releases and perhaps other hazards to the public and to the environment.

The growth of such processes accelerated after the Second World War as supplies of cheap petroleum-based products displaced coal-based products. It became an increasingly common practice to handle liquefied petroleum gases and other liquefied gases as well as cryogenic fluids such as methane, oxygen, nitrogen and ethylene.

The growth took place continuously and, to an extent, imperceptibly. It was however noted, with concern, by the UK Factory Inspectorate. But lack of concrete evidence of potential harm inhibited special measures of control which were not to come into being in the EEC until twenty years later and then after some very serious incidents had occurred.

Gathering reliable information on incidents worldwide presents considerable problems because only the most serious incidents are given international coverage and each country treats its own incidents far more seriously than those of other nations. The book draws together many of the compilations which have been produced by other authors but it concludes that the reporting is incomplete especially as far as the smaller incidents are concerned.

26.2. THE NATURE OF HAZARDS AND RISKS.

Rational discussion of the problem has not been made easier by lack of an agreed nomenclature and a recent publication on this theme by the Institution of Chemical Engineers is welcome. The book uses its terms in a manner which is consistent with the usages recommended by the I.Chem.E. An illustrative example of the meanings of the principal terms is provided. The author strongly argues for such an agreed terminology.

The relationship between hazard and risk is examined and it is concluded, based on examples, that over a sufficiently wide range it cannot be inverse, i.e. the frequency must fall off more rapidly than the magnitude of the consequences.

26.3. THE HANDLING AND STORAGE OF LIQUIDS AND GASES.

The problem of analysing the behaviour, on spillage, of liquids and gases is referred to in the phase diagram for the substance. Based upon the temperature of storage, six categories of liquids are recognised of which the most dangerous are the liquefied gases which, if containment be lost, will flash off an appreciable fraction of the liquid as cold, dense vapour. The phenomena of flashing are analysed and the significance of the geometry of the ruptured system is emphasised.

26.4. CAUSES OF LOSS OF CONTAINMENT.

Serious rupture of pressure vessels, perhaps amounting to catastrophic failure, is recognised as a rare event but it is not so rare that it can be ignored. The probability of failure versus vessel life may be plotted as a "bathtub" curve with a high initial probability of failure chiefly due to defects in design, or in fabrication. The probability falls to a minimum level after the first year but later begins to rise again through causes such as corrosion, erosion, fatigue and creep.

The management of pressure vessels is a multi-disciplinary activity in which mechanical and chemical engineers and metallurgists all play a role.

The importance of operating within specified temperature limits is stressed. Some materials are seriously embrittled by low temperatures and all are weakened by high temperatures. Thus it is imperative that pressure vessels be protected against fire. Attention is called to the need to avoid hydraulic rupture which almost certainly caused one serious disaster and is suspected of causing two others. The associated pipelines are much more vulnerable than pressure vessels to mechanical shocks, as are pumps and valves.

The need to treat modifications with the same care as the original design and fabrication cannot be too heavily stressed.

26.5. VAPOUR CLOUDS.

The specialised branch of fluid mechanics which studies and analyses the formation and dispersion of vapour clouds is hardly fifteen years old. Yet, in the period since the work of van Ulden in the early seventies, immense strides have been made. A combination of systematic field trials, with theoretical models and with wind tunnel experiments has brought us near to a comprehensive theory in which the characteristics of the source term can be matched to the meteorological conditions, including both wind velocity and direction and the stability characteristics of Pasquill, to give a prediction of the variations of three dimensional concentration contours (isopleths) with time.

Highlights in this progress have been the trials at Maplin Sands, Porton Down and Thorney Island, the development of theoretical models such as DENZ and the wind tunnel experiments of Warren Spring. As a result of this work it may now be said with some confidence that the wide variations between the prediction of the various models have been evened out. Even so, the present models are only applicable to flat terrain without obstruction.

It may also be said that, though the ability to predict isopleths for toxic releases is an important first step, much remains to be done before the models can reliably predict likely levels of human harm. Some models have been proposed which predict levels of fatality which far exceed those observed to happen in actuality.

26.6. CATACLYSMIC FIRES.

Major chemical fires represent only a portion, though a significant proportion, of major fire losses; this is shown by an analysis of major fires both in the United Kingdom and worldwide. Fire fatalities per million of population vary by a factor of 10 from 3 in Switzerland to about 30 in North America, with the UK having 18. Losses as a percentage of Gross National Product vary by a factor of 5 between Switzerland at 0.08% to Norway at 0.40%. The UK stands about half way at 0.18%.

LPG is seen as the most dangerous source of fire based upon a notional partition fraction between the quantity of the flammable substance in the vapour phase and that in the liquid phase.

Pool fires in which the liquid burns with a relatively stationary flame are analysed for factors such as burn-up rate, flame appearance and geometry and for radiative flux. An account of a number of trials is given.

Fire balls are considered and are seen to be associated with massive vapour clouds. They are seen to have an enormous burn-up rate of the order of tonnes per second and to be highly dangerous sources of radiative flux. The question as to how closely experience with rocket fuels can be related to the behaviour of hydrocarbon vapours mixed with air is discussed and a cautionary note is struck.

The term BLEVE is criticised for being imprecise and for meaning too many things.

The question as to whether "fire storms", such as occurred in the bombing of some cities in the Second World War, could happen in major chemical complexes is considered and it is concluded that such an event would be most unlikely to occur especially in a well laid out site with appropriate separation distances.

A section is devoted to the problems of anaerobic fires which do not require external supplies of oxygen.

The subject of thermal radiation and its effect on people is analysed in detail. This analysis is conducted with especial reference to the likely effects of fire balls and the calculation of safety distances. The injury pattern from a number of case histories is examined and it is concluded that the analysis of actual incidents suggests that some theoretical approaches have overstated the risk of fatality. The use of data from nuclear explosions is criticised and it is suggested that rotation of the body and running away may account for some of the discrepancies.

26.7. CASE HISTORIES OF CATACLYSMIC FIRES.

The case histories of Cleveland, Ohio, (1944), Feyzin, France, (1966), the fire which succeeded the explosion at Flixborough (1974), the San Carlos disaster in Spain (1978), and the Mexico City LPG disaster (1984) are examined in detail, together with four short case histories of anaerobic fires.

The Cleveland disaster was an LNG fire and the author finds no evidence that it took the form of a vapour cloud explosion. Explosions which occurred in the sewers, in the author's view, were purely physical explosions arising from pressurisation of the sewers due to the rapid vapourisation of LNG spilled into them. The cause of the disaster seems to have lain in the use of an unsuitable material of construction, a nickel steel which displayed embrittlement at LNG temperatures. The siting of the works in a densely populated area was chiefly responsible for the high death toll of 128.

The Feyzin disaster was caused by a design of a sampling system in such a way that when the sample valve could not be closed, a fire started under a spherical storage holding about 450 tonnes of propane. Though the relief valve operated, the top of the vessel was seriously weakened by flame impingement and underwent a petal fracture causing a severe death toll among the fire fighters.

Though the event at Flixborough will always be remembered as a vapour cloud explosion, the explosion was succeeded by a most serious conflagration which made rescue work impossible. Some 50 appliances were employed in total but, even so, the fire burned itself out consuming 1,600 tonnes of flammable liquids over two days in a somewhat congested site.

The San Carlos disaster, which killed more than 100 people immediately and a total of 215 people, arose from the rupture of an overloaded propylene road tanker. Controversy has surrounded the exact circumstances of the rupture but the author expresses the view that it was almost certainly a hydraulic rupture. He discounts the view that the vehicle brought down a length of wall and entered the camp and he also discounts the view that, prior to rupture, the tanker was engulfed in flames. The disposition of the various parts of the vehicle affords a picture of their orientation at the time of rupture and leads to the view that the rear portion ejected a slug of some 12 tonnes of vapour/liquid mixture which knocked down the wall and sprayed the scantily clad campers with liquid propylene at sub-arctic temperatures before igniting.

The Mexico City fire, which cost more than five hundred lives, arose in a well constructed installation which represented a highly congested site in a highly populated area.

It is noted that the lessons of Feyzin, that storage tanks containing liquefied gases need to be protected against flame impingement, appear to have been ignored and that the fire fighting facilities were sited too close to the centre of the site.

26.8. EXPLOSIONS.

Explosions are defined and some key reference sources are identified. They are classified into two main divisions that lie between dense explosives such as TNT or nitroglycerine, and rarefied explosives such as dust explosions and vapour/air explosions. Rarefied explosions in their turn are classified as confined explosions and open air explosions.

The energetics and the chemistry of dense explosions are analysed as are the properties of blast waves. The relationship between mass and overpressure is set out and the significance of Hopkinson's Law is recognised.

26.9. CASE HISTORIES OF DENSE EXPLOSIVES.

Because the author has decided to exclude a detailed study of the effects of the accidental explosion of conventional explosives from the book, only two case histories are provided and both are concerned with ammonium nitrate.

The principal conclusion drawn is that ammonium nitrate, unmixed with conventional explosives such as TNT, is capable of detonating if a sufficient quantity, of the order of thousands of tonnes, is present.

26.10. RAREFIED EXPLOSIONS.

Dust explosions have long been recognised as being capable of giving rise to severe effects under conditions of confinement. Evidence is presented that relatively small dust explosions are possible in the open air. The conditions which can give rise to dust explosions are set out together with preventative measures.

Confined gas/vapour explosions have also a long history but recent research sheds new light on them. There are some 100 explosions causing appreciable damage each year in the UK in domestic premises and, on the average, they kill ten people p.a. However such explosions are comparatively small scale.

But a few have occurred in large buildings with correspondingly more severe effects. The effects of such explosions are, even so, much smaller than those which can arise from some of the open air explosions discussed below.

Open air vapour cloud explosions were, until twenty years ago, regarded as impossible but a number of serious events enforced their recognition. Though formerly described as "unconfined", analysis has cast doubt on whether blast effects would be observed in the complete absence of turbulence-creating obstacles and of the semi-confinement produced by the presence of plant and equipment. However the Port Hudson explosion, described as a case history, seems to cast doubt on this, though it seems to have been initiated by a small confined explosion. Perhaps a certain critical mass, (of the order of tonnes of hydrocarbon), is needed to make possible an open air explosion in the presence of obstructions or partial confinement and a much higher critical mass, (of the order of hundreds of tonnes), may make possible an open air explosion even when these features are absent.

The quantification of vapour cloud explosions remains a vexed question as vapour cloud explosions differ physically from the detonation of dense explosives. In particular the latter generate overpressures up to ca 100,000 bars and, at ground level, expend about half their energy in the production of craters. Vapour cloud explosions produce levels of overpressure of the order of 1 bar and do not produce craters. But, in the far field, the effects are much less easily distinguishable. The duration of the pressure pulse from VCEs is far longer than for high explosives and hence a VCE will have a lower value of overpressure but a higher value for duration than an equal impulse for a high explosive.

The author criticises the approach of selecting worst cases of damage as criteria for the level of overpressure at any radial distance.

26.11. CASE HISTORIES OF RAREFIED EXPLOSIONS.

Case histories of three major dust explosions are provided. Two of these explosions occurred in grain elevators and one in a flour mill. The author is critical of management practices in that major strategic errors were made which led to greater loss of life. One case history of a possible open air explosion is provided and it is noted that an open air explosion may have occurred in one of the major case histories.

Four case histories are provided for confined vapour/gas explosions in store tanks which can be very large especially if the cargo spaces of super-tankers be included in this category. Explosions in store tanks, unless of the floating roof type, are far more severe in empty than in full tanks.

Two case histories of serious vapour/gas explosions in buildings are recorded. The case history of Abbeystead presented some unusual features.

Five case histories of VCEs are provided. The first two, which occurred in the same site in the 1940s, were not recognised as VCEs till later. The second case history is examined in detail and the author points out that the generally accepted theory, that it was caused by a hydraulic rupture, was not proven by those who investigated the explosion.

Port Hudson was a very large VCE which did not produce human injury. It provides little supporting evidence for the theory that obstructions and partial confinement are necessary conditions for the initiation of such explosions.

Flixborough has been studied in much greater detail than any other VCE and it provides lessons in many areas. It was caused by a series of errors in a plant which had a large inventory of cyclohexane boiling at about 8 bars. Safety management on the site was purely tactical. The overpressure pattern as deduced from the damage caused was far from symetrical and this may be related to the disposition of the process units. The layout of the site concentrated too many people in highly vulnerable areas.

Decatur, reported by some investigators as having a TNT equivalence greater than Flixborough is shown by the present author to have been on a smaller scale than Flixborough.

The history of hydrogen-filled airships is examined for evidence of VCEs but little evidence is found for them. The great majority of those which caught fire burned without generating appreciable overpressure.

26.12. TOXIC RELEASES.

Unlike fires and explosions, where there are considerable generic similarities within each class, the action of toxic substances is diverse. There are many modes of attack on the human body and agents also differ widely in the intensity of their toxic properties. For gases, lethal dosage is not simply a product of concentration and time as is demonstrated by the existence of threshold values of concentration below which indefinite exposure does not produce harm.

The classical case of poisoning, and the one most familiar to the public, arises through the ingestion of a solid or liquid poison. Most of the data in the literature of toxicology relate to this circumstance. However, in the context of major chemical hazards, such data has little applicability. Here the hazard arises through the inhalation of toxic vapours or gases and, in most though not all cases, these arise through the loss of containment of liquefied gases.

Data on toxicity rarely arise through direct observation of toxic effects on human subjects and rely, instead, upon animal experiments. The translation of the results of animal experiments is difficult. Even closely related species of animals may show widely different values of LD_{50} and LC_{50}.

There is surprisingly little data available on the toxicity of common substances like chlorine and ammonia and what there is suffers from lack of consistency as to exposure times. For acute events, LC_{50} related to 30 minutes' exposure would seem to be the most appropriate data to have available.

When an attempt is made to combine prediction of the dispersal of vapours with toxic data, the difficulties are multiplied because of the possibilities of escape, of improvised protection, and the mitigating effects of medical treatment on human subjects.

26.13. CASE HISTORIES OF TOXIC RELEASES.

Because of the diverse nature of toxic hazards, the book presents modes of attack and toxic data for each substance along with case histories. Some of these case histories are derived from military experience.

For chlorine, which is an intensely irritant gas produced on a large scale, the quantity of data applicable to acute releases is surprisingly small even though there has been large-scale use of the gas in war. This is the data which is reviewed in I.Chem.E. (1985a). Accidental releases of chlorine (18 in all) are tabulated and 7 are given as case histories. Most fatalities occurred within 250 metres and one occurred more than 400 metres from the point of release.

Ammonia is an irritating gas produced on a large scale. Data on its toxicity are few but there is no doubt that it is much less toxic than chlorine. Ammonia vapour clouds, contrary to what might be expected on account of the density of the pure gas, when initially formed are heavier than air. It has had no military use. There are 11 incidents tabulated with 3 case histories.

Phosgene is highly toxic, much more so than chlorine. It has similar effects to chlorine but there is a significant delay before they are manifested. It has been extensively used in war where it was more lethal than chlorine. There is only one major industrial release of phosgene on record and this is made the subject of a case history.

Carbon monoxide has caused many deaths in personal accidents. It is a gas of considerable industrial significance. There is a brief case history of a multiple accident.

Mustard gas was highly significant in the First World War where it caused many casualties though relatively few died. It causes severe blisters. Though listed in Annex III of the EEC Directive, it has no industrial significance.

Organophosphorus compounds are of considerable industrial significance as pesticides. Some of them are highly toxic of which parathion is an example. Closely related chemically are the nerve gases which were manufactured but not used in the Second World War. Some are being stockpiled today as "binary weapons", i.e. only the precursors are actually manufactured.

Dioxin appears to have engendered more controversy than any other substance this century. It is extremely toxic to animals, though the toxicity varies by orders of magnitude from one species to another. Evidence of human fatality arising from it is ambiguous. In humans the most clearly defined symptom is chloracne. The literature on it is voluminous with 14 major works being cited in the text and 1 in the Addendum. There have been a number of significant releases of which 6 are tabulated and of which the Seveso incident is described in detail. The incident was the subject of an official Italian Report which the present author criticises for failing to discover the causation of the release and for excessive attention to the chemical and toxicological aspects at the expense of the engineering aspects. It seems clear from the present author's analysis that the initial injuries were caused by caustic substances and that the chloracne appeared later. Not all those with caustic burns developed chloracne and not all those who developed chloracne were initially treated for caustic burns. Prompt evacuation would have considerably reduced the numbers of those injured.

Methyl isocyanate (MIC) caused the worst chemical disaster in history at a factory for making the insecticide, Sevin, in Bhopal, India. The extremely toxic properties of this highly volatile liquid were available before the disaster occurred and showed it to be more dangerous than phosgene, chlorine or hydrogen cyanide. In spite of this the plant was designed to store up to 120 tonnes of the material in three store tanks. It appeared that water found its way into one of these tanks and that an exothermal reaction developed. Though the tank had a refrigeration system, this had been shut off six months before for undisclosed reasons, so there was nothing to slow down the runaway reaction. There was an escape of about 30 tonnes which killed around 3,000 people and injured 200,000 more.

26.14. OTHER MAJOR HAZARDS.

Though most serious accidents have fallen into the category of fires, explosions or toxic releases, attention is called to injuries from hot liquids, the effects of low temperatures, asphyxia, excessive oxygen concentration, and corrosive substances. It is recognised that there may be environmental disasters which do not injure people. There is a brief discussion of the possible ill-effects arising from the use of genetic manipulation by the chemical industry.

26.15. METHODS OF CONTROL.

The general principles of hazard control are discussed and it is demonstrated that the control of major chemical hazards has many features in common with the control of many other, diverse, hazards with an interaction between the state and those who originate the hazards. The many problems of public perception of risk are reviewed and the author concludes that the chemical industry has a low perceived necessity even though the products of the industry are ubiquitous and a vital component of Western lifestyles. Though all risk assessment is subjective, factors such as fear of the unknown and the concept that the risk is "imposed" rather than being voluntarily assumed, play a major role in shaping public attitudes.

The balancing of risk against benefit objectively is difficult though everyone, every day, makes subjective judgements of this character. The balance requires a common denominator which is lacking. Using the financial value of human life has many critics though the courts have to do it every day.

An alternative is to use the historical experience of subjective public attitudes where there is clear evidence that the public accepts certain risks such as death on the roads because these risks seem to be outweighed by palpable benefits. The present author, though he has criticisms of the concept, puts forward a modified approach based upon the risk of dying from all causes at various times of life. Various levels of risk can be indicated on this "bath tub" curve and he suggests that a level of 10^{-5}, which is one-tenth of the risk of a child of 10 dying from all causes, be the maximum permitted. The author puts forward the view that the historical record suggests that the mean level of risk to the public of the UK from chemical accidents is less than 10^{-7}.

The chapter reviews techniques of risk assessment but notes that methods such as fault tree analysis do not, of themselves, distinguish between minor events and major events.

The techniques of hazard and operability studies are briefly reviewed.

26.16. QUANTIFICATION OF HAZARDS.

It is not easy to determine the point at which a hazard becomes a major hazard but it is not possible to legislate against major hazards without developing criteria for this purpose. There is general agreement that one characteristic of major hazards, among others, is that they may cause deaths among the general public as well as among employees. Though there are a number of measures of hazard, the most obvious seems to be the potential for causing fatalities. Though there may be other criteria which apply in certain cases, the potential for fatality remains the central criterion. It is of course relatively unambiguous whereas the criterion of injury has an infinite spectrum of severity.

If this criterion be accepted, then any hazard with the potential under average conditions of causing a given number of fatalities will be equal to any other hazard with the same potential under mean conditions. Such hazards, to use a term coined by the author, are "isotraumic". The question of the isotraumic level remains to be determined and the author advances the view that this should be 10 fatalities.

Two methods of assessing the isotraumic level of 10 are put forward by the author. The first is the "mean mortality index" approach in which the ratio fatalities : size of release is determined. The ratio may have to be normalised to take account of variations in population density. Though in the book the historical approach is the one used, mortality indices could be computed from theoretical models. Indices are derived for explosives and for chlorine and ammonia which, when compared with EEC Directive threshold levels, appear to approximate to an isotraumic threshold of around 10.

An alternative approach, also developed by the author, plots $F \geqslant N$ curves for frequency versus size of incident and for frequency versus number of fatalities. The isotraumic level is that level which has the same frequency as the frequency of killing 10 or more people. Thresholds for explosives and for chlorine and ammonia so obtained are reasonably close to those derived by the mortality index approach. In addition, the equal probability approach yields a value for flammable vapour clouds.

Alternative approaches by other investigators are also examined.

26.17. LEGISLATIVE CONTROL.

Managerial control of major chemical hazards came first but was shown, in some cases, to be inadequate as evidenced by serious incidents. However much work was done by non-statutory bodies before legislation emerged.

The UK Factory Inspectorate gave warnings in 1968 and these led to some measure of planning control in 1972. The Robens Committee recommended action on major chemical hazards but the Advisory Committee was only set up in the aftermath of the Flixborough disaster of 1974. Its recommendations were largely taken over by the EEC and incorporated into the European Directive of 1982 which was also brought into being as a result of public concern following the Seveso disaster of 1976. The UK has, however, purely domestic Regulations dealing with the planning aspects of major chemical hazards.

The European Directive is based upon the recognition of installations or isolated storages containing dangerous substances. These are given indicative criteria and their control is by specifying threshold inventories. These inventories, with few exceptions, lie on a 1,2,5,10,20,50, . . . scale. The inventories of certain key substances are analysed in the book and are shown, in some cases, to be related to an apprehended mean fatality level of 10 persons. However some toxic substances have very low inventories of as low as 1 kg, inventories for which the author can find no justification. Though the general principles of the Directive are sound, there is a need to revise the details.

26.18. COOPERATION BETWEEN STATE AND INDUSTRY.

The view that only legislation can control major chemical hazards is false. Many companies took some measures before legislation required them to. But legislation curbs the activities of free riders and prevents unfair competition by those who take risks. Legislation itself has been shaped by cooperation between the state, industry, the universities and the professional institutions.

The present author is strongly of the view that the hazards can only be managed successfully if there is an overall strategic approach, one which is the opposite of the "add on" approach. The process has to be analysed for its safety aspects from the conceptual stage. Fait accomplis must be avoided and the design must be criticised by those who have to realise it.

A vital and integral part of this strategic approach is the avoidance of dangerous substances where this be possible; and where this is unavoidable, the plant should be designed for minimum inventory of such substances. The site needs to be laid out to separate people as far as possible from the process. Process plant contractors, particularly those offering "lump sum" tenders, have a major role to play in this.

As well as applying a strategic approach to prevention, it is necessary to apply it to the mitigation of the consequences. This requires, among other steps, the drawing up of emergency plans both for on-site and for off-site. Such planning is now compulsory in the EEC though many UK companies have conducted such planning for a decade or more. A properly conceived and sited emergency centre is a necessity. Under the CIMAH Regulations in the UK which implement the European Directive, it is required by law that installations so designated by the Regulations must inform those living in the neighbourhood of the hazards and the action to take should they be realised.

26.19. SITING AND PROTECTION OF BUILDINGS IN HAZARDOUS AREAS.

Control buildings have to be sited near to hazardous plant and there have been a number of cases of such buildings which have either been totally destroyed or have been badly damaged as the result of VCEs.

As has been mentioned above, it is vital to restrict the number of people housed in control buildings to those who absolutely have to be there. However, it does not seem practical to site the buildings themselves very far from the plant they seek to control. To do so would be to sacrifice the day-to-day good running of the plant for a fairly remote eventuality.

The general view today is that control buildings in areas where there is an explosion hazard should be single-storey and of ductile construction, i.e. without brick infill, and that windows should be specially designed to resist blast and their area should be kept to a minimum. The building should be sited so that columns and pipe bridges cannot fall upon it. The building should be able to withstand, without collapse, side-on overpressures of 0.3 to 0.7 bar. Two case histories with illustrations are provided. The case history of Flixborough considers the fate of the main office block and the oleum plant control room, as well as the main control room which totally collapsed killing 18 men inside it.

Emergency centres need to be as remote as possible from hazardous areas and this may dictate that they are situated at the main entrance which should itself be remote from hazardous areas so as to reduce exposure of personnel. The equipment required is discussed and attention is called to the availability of computers which can simulate the behaviour of any vapour cloud which may be released on site.

26.20. SAFETY PROFESSIONALS AND THEIR TRAINING.

There is a need for a carefully thought out hierarchial structure for safety management which bears a proper relation to other management functions which deal with such matters as occupational health, environmental protection, and security.

Within the safety function, there are two principal areas, non-chemical safety such as the avoidance of falls, electric shock and lacerations from hand-tools, and there is chemical safety which includes both the avoidance of personal accidents and the control of major chemical hazards. It is a strongly held view of the author that appropriate academic qualification, together with experience and training, are absolutely essential for those responsible for chemical safety. It may be said, however, that non-chemical safety is a specialised area and one in which chemical engineers and chemists have no special expertise.

It is highly desirable that the safety professional, meaning here the chemical hazard safety professional, should have had experience of plant operation before moving into the safety area.

Training in the UK may include acquiring some basic knowledge of safety at undergraduate level. Until recently this was not a component of all courses but it is now included in the I.Chem.E's new model syllabus, so shortly all chemical engineering graduates will know at least the rudiments of the subject. The A.I.Ch.E., through its new "Center for chemical plant safety", is also seeking to upgrade college education in safety.

For the specialist or for senior production management, there are numerous short courses and regular conferences on safety and loss prevention. There are also available workshop training modules.

There has been some discussion of state accreditation for safety professionals in the UK, but this does not seem likely to come to pass. The I.Chem.E. has recently established an indicative register of those of its members who are safety professionals.

26.21. THE ROLE OF PROFESSIONAL INSTITUTIONS.

The I.Chem.E. and the A.I.Chem.E., together with corresponding institutions in other countries, are learned societies. As such they are concerned with the generation and dissemination of all aspects of knowledge which concerns the profession. Safety and loss prevention are part of this corpus of knowledge. The institutions in the field thus disseminate journals and bulletins and organise courses, seminars, and conferences for these purposes.

The institutions have also an ethical function. It is a condition of membership that the members subscribe to a code of ethics and the institutions possess disciplinary powers.

Furthermore, the institutions, either by virtue of their general activities or by conscious pronouncement, are able to influence public policy. This they may do as a corporate body or through the participation of their members in policy-making bodies. Close liaison is maintained between the institutions and the appropriate trade associations.

26.22. RESEARCH AND CONSULTANCY.

Research, be it practical, theoretical, or scholastic, has played, and will continue to play, a vital part in increasing the understanding of major chemical hazards.

Some of the practical research into the hazards has to be conducted on a very large scale and is only possible to conduct with the aid of international funding.

All three branches of research overlap and interact. The control of major hazards is an interdisciplinary subject involving as it does physical science, biological science, and the social and managerial sciences. In the UK the largest effort comes from the Health and Safety Executive but there are major inputs from the Fire Research Station, from Warren Spring, from British Gas, and the major chemical and petroleum companies. The SRD specialises in theoretical modelling and in risk assessment. There are major centres of research in the Netherlands and in the US.

Consultants play an important role in promoting safety as they can bring to their clients a very wide range of experience. There are major firms of consultants in the UK and the US and in other countries.

26.23. POSTCRIPT.

In its Second Report, published in 1979, the UK Advisory Committee on Major Hazards commented that the incidence of serious vapour cloud incidents appeared to be increasing on an exponential basis. Nearly a decade later the picture seems less frightening but how far this is due to higher safety standards and how far it is due to the recession is difficult to say. Probably both have had their effect. Even so the two worst incidents in the history of the chemical and process industries occurred less than two years ago.

It is true that most of the plant being operated today was built before current attitudes to safety had evolved. May the next generation of chemical and process plants be conceived and designed with safety considerations playing a major and integral role from their very inception.

Appendix I

GLOSSARY OF TERMS

Where terms in this glossary appear in I.Chem.E. (1985), in virtually all such cases the definition given in that reference is used. Such words are identified by an asterisk.

ACUTE* Immediate, short-term. Relating to exposure conditions which develop rapidly and may cause harm in a short time. Relating to effects, effects which appear promptly after exposure.

ALVEOLI Air pockets in the lung in which oxygen and carbon dioxide are exchanged.

AT RISK Exposed to a significant level of individual or societal risk in the event of the realisation of a specified hazard or hazards.

ATTENUATION The loss of power suffered by radiation as it passes through matter.

AUSTENITIC Steel containing iron carbide in solution.

BIFURCATED Having two prongs.

BLEVE* Used to describe the sudden rupture of a vessel system containing liquefied flammable gas under pressure due to flame impingement. The pressure burst and the flashing of the liquid to vapour creates a blast wave and potential missile damage and immediate ignition of the expanding fuel-air mixture leads to intense combustion creating a fire ball.

BRISANCE The shattering power of an explosive.

BRITTLENESS Lacking resistance to the propagation of cracks.

BUND, BUNDING A system of containment for catching liquid spillage. It may take the form of a low wall surrounding the area or a shallow pit, or some combination of the two.

CATASTROPHIC FIRES Fires which are distinguished from fires ordinarily encountered by their intensity, their rapidity of spread or their size, or some combination of these factors.

CAVITATION A form of erosion, particularly to be found in pumps, connected with the collapsing of vapour bubbles.

CHARLES LAW One of the Gas Laws; it states that, for an ideal gas, at constant pressure, all gases expand by 1/273 of their volume at $0^{o}C$ for each $1^{o}C$ rise in temperature.

CHEMICAL HAZARD* A hazard involving chemicals or processes which may realise its potential through agencies such as fire, explosion, toxic or corrosive effects.

CHRONIC* Persistent, prolonged and repeated. Relating to exposure; frequent or repeated or continuous exposure to substances. Relating to effects; when physiological effects appear slowly and persist for a long time or with frequent recurrences.

COMPLEX A group of neighbouring plants on the same site.

CONFIDENCE LIMITS A term used in statistics to denote the end values of certain intervals within which there is a specified limit of expectation of finding a statistic.

CONFINED EXPLOSION* An explosion of fuel-oxidant mixture inside a closed system (e.g. vessel or building).

COMMISSIONING The initial programme of starting up a chemical plant.

COMPETENT AUTHORITY The Authority set up, or appointed by, a Member State to fulfil the requirements of Article 7 of the EEC Directive on Major Accident Hazards.

CONDITION MONITORING Surveillance of the condition of plant and equipment by visual observation and, more particularly, by instrumentation, on a periodic or continuous basis, in order to obtain early warnings of the development of defects or malfunctions.

CREEP Elongation in linear dimensions produced by prolonged loading.

CRITICAL PRESSURE That maximum level of pressure which, below the critical temperature of a substance, will maintain it in the liquid phase.

CRITICAL TEMPERATURE The temperature above which a substance cannot exist in the liquid state regardless of the pressure.

CRYOGENIC FLUID A gas liquefied by refrigeration at temperatures below those encountered in nature on the earth. Examples are liquefied natural gas and liquid oxygen.

DECISION TREE See Fault Tree.

DEFLAGRATION* The chemical reaction of a substance in which the reaction front advances into the unreacted substance at less than sonic velocity. (Sonic velocity is the velocity of sound in the material, VM).

DETERMINISTIC MODELLING A system of modelling in which any event is rigorously determined by a past event, or rigorously determines a future event.

DETONATION* An explosion caused by an extremely rapid chemical reaction in which the reaction front advances at a velocity at greater than sonic velocity.

DEVELOPMENT A legal term used in the United Kingdom and having the meaning "Carrying out of building, engineering, mining or other operations in, on, over, or under land, or the making of any material change in the use of any buildings or other land".

DIMENSIONAL ANALYSIS The use of the dimensions of physical quantity to check relationships betwen them. The fundamental dimensions used in this book are length, time, mass and temperature.

DIMENSIONLESS NUMBER The ratio between two groups which have the same dimensions expressed as time, length, mass or temperature.

DIAPHRAGM VALVE A type of valve in which closure is effected by pressing a flexible diaphragm on to a seat.

DIOXIN Dioxin is the common name for 2,3,7,8-tetrachlorodibenzo-p-dioxin. It was the principal cause of harm in the Seveso disaster in 1976. It is a solid substance which, in minute quantities in contact with the skin, gives rise to the disfiguring disease chloracne. It is highly toxic to animals and is extremely persistent.

DISASTER In terms of this book it is generally an event which causes ten or more fatalities.

DRY ADIABATIC LAPSE RATE The rate of fall in temperature of a mass of ascending dry air which is not exchanging heat with its surroundings. A typical value is $1^{o}C/100$ metres.

DOMINO EFFECT A situation in which a major accident on one site gives rise to consequential accidents on a neighbouring site.

DUCTILITY Having the capacity to flow; resistant to the propagation of cracks.

DUST EXPLOSION An explosion which results from the ignition of a mixture of finely divided combustible solids and air. The effect of such an explosion is comparable with that which results from the ignition of an equal volume of air mixed with flammable vapour.

ENTHALPY Heat content. Is equal to the sum of the internal energy of the system plus the product of the pressure - volume work done on the system.

EDDY DIFFUSIVITY Diffusion of molecules by eddies in turbulent flow. As opposed to molecular diffusion.

ENTRAINMENT The carry over of liquid droplets or foam during boiling.

ENTROPY A measure of the extent to which the energy of a system is unavailable. A mathematically defined thermodynamic function of state, the increase in which gives a measure of the energy of a system which has ceased to be available for work during a certain process.

EPICENTRE* The ground location beneath the inferred centre of a vapour cloud explosion. (Originally used to denote the point on the earth's surface directly over the point of origin of an earthquake).

EQUATION OF STATE An equation which connects the pressure, volume and temperature of a substance.

EQUIPMENT (OR EQUIPMENT ITEM) The basic items from which process units are assembled such as pumps, valves, pipe fittings etc.

EXPLOSIBLE Of a dust suspended in air, the ability to give rise to an explosion. This depends, among other things, upon the chemical nature of the substance, its particle size, the oxygen content of the air and the concentration of the dust.

EXPLOSION A rapid release of energy which causes a pressure discontinuity or shock wave, which then moves away from the source at a rate determined partly by the pressure differential and partly by the properties of the medium through which the shock wave is propagated. This pressure discontinuity and the subsequent shock wave are termed the blast wave.

FATAL ACCIDENT FREQUENCY RATE See Fatal Accident Rate.

FATAL ACCIDENT RATE* (Previously known as FAFR). The number of deaths that have occurred, or are predicted to occur, in a defined group, in a given environment, during 10^8 hours of exposure.

FATIGUE In metals the ability to fracture under repeated or fluctuating stresses at a stress much below their nominal ultimate tensile stress.

FAULT TREE ANALYSIS* A method for representing the logical combinations of various system states which lead to a particular outcome (top event).

FIRE BALL* A fire burning sufficiently rapidly for the burning mass to rise into the air as a cloud or ball.

FIRE STORM* An extremely large area fire resulting in a tremendous in-rush of air which may reach hurricane force.

FLAME SPEED The speed with which a flame front propagates relative to a fixed point.

FLASH FIRE* The combustion of a flammable vapour and air mixture in which flame passes through that mixture at less than sonic velocity, such that negligible damaging overpressure is generated.

FLASHING The phenomenon whereby, when the pressure on a liquefied gas is reduced, partial vaporisation occurs. The latent heat of vaporisation is supplied at the expense of the sensible heat of the liquid.

FLASH POINT The minimum temperature at which the vapour in equilibrium with a flammable liquid is equal to or exceeds the lower flammable limit.

F/N CURVE* A plot showing, for a specified hazard, the frequency of all events causing a stated degree of harm to N or more people against N. (see below).

F/N CURVE A plot showing, for a specified class of events, the frequency (F) of the occurrence of such events and all events of the same class but of greater magnitude ($\geqslant N$). [Present author's more general definition]. (see above).

FREON A substituted hydrocarbon in which some of the substituting atoms are those of fluorine. Generally nonflammable and of low toxicity. Used mainly as a refrigerant.

GAUSSIAN DISTRIBUTION Otherwise known as "normal" distribution. A type of statistical distribution followed, for example, by the same measurement made several times when the variation of a quantity about its mean value is entirely random.

HALOGEN-TYPE SUPPRESSOR A halogenated hydrocarbon which may be injected to suppress a dust explosion. It is often used to protect electrical apparatus.

HAZARD* A physical situation with a potential for human injury, damage to property, damage to the environment, or some combination of these.

HAZARD ANALYSIS* The identification of undesired events that lead to the materialisation of a hazard, the analysis of the mechanisms by which these undesired events could occur and usually the estimation of the extent, magnitude and likelihood of any harmful effects.

HAZARD ASSESSMENT See Hazard Analysis.

HIGHLY FLAMMABLE LIQUID A liquid of flash point (q.v.) of $0^{o}C$ or less and capable of supporting combustion. [Legally defined for the UK in the Highly Flammable Liquids and Liquefied Petroleum Gases Regulations (1972)].

HOPKINSON'S SCALING LAW That self-similar blast (shock) waves are produced at identical scaled distances (qv) when two explosive charges of similar geometry and the same explosive but of different mass are detonated in the same atmosphere.

IDEAL GAS Otherwise known as a perfect gas. A gas which would obey the gas laws exactly. In such a gas the volume occupied by the molecules would be negligible and so would the attractive forces between them.

IDLH Immediately Dangerous to Life or Health. Conditions such that an acute exposure will lead to acute or chronic effects.

INDUSTRIAL COMPLEX A group of neighbouring industrial sites.

INSOLATION Exposure to solar radiation.

INVENTORY The quantity of a specified material or materials which an installation contains when all equipment is filled to its designed capacity.

ISOPLETH* A surface joining points of equal concentration, i.e. a three dimensional contour in a gas cloud.

ISOTRAUMIC Expression coined by the present author to denote hazards with an equal capacity for inflicting injury.

KINEMATIC VISCOSITY The ratio of the dynamic viscosity of a fluid to its density.

LAMINA A thin sheet.

LAPSE RATE A meteorological term denoting the rate at which temperature falls with height.

LC_{50} That steady concentration of a toxic gas in air which, if breathed by a defined animal population for a stated time, will result in 50% fatalities within a stated time. Expressed as parts per million or mg per litre.

LD_{50} That dose, received otherwise than by inhalation, which will cause the death of 50% of an entire defined animal population within a defined period. Expressed as mg per kg of body weight.

LEAST SQUARES REGRESSION A method of obtaining a "best fit" line. For a linear regression it is that line from which the sum of the squares of the deviating points is a minimum.

LETHAL DOSAGE ENVELOPE A surface which encloses a volume engulfed by a toxic cloud and within which a high probability of fatality is assumed.

LOGIC TREE (OR DIAGRAM) See Fault Tree.

LOG MEAN TEMPERATURE DIFFERENCE A mean temperature difference where the difference is subject to considerable variation along a conductor. Obtained by dividing the difference between the extreme values by the natural log of their ratio.

LOSS PREVENTION* A systematic approach to preventing accidents or minimising their effects. The activities may be associated with financial loss or safety issues.

LOWER FLAMMABLE LIMIT* The lowest concentration of a flammable vapour or gas in air at which a self-sustaining flame may be initiated.

MAJOR ACCIDENT HAZARD The subject matter of the European Directive of 24th June 1982. Synonymous with Major Chemical Hazard.

MAJOR CHEMICAL HAZARD A chemical hazard which has the potential to give rise to a major accident.

MAJOR HAZARD A hazard which has the potential for major harm to employees, the public or the environment.

MASS TRANSFER In general, the process of diffusion of molecules from a region in which they are in high concentration to a region in which they are in a lower concentration.

MORTALITY INDEX A term coined by the present author to denote the number of fatalities per tonne of agent associated with the realisation of the potential of a hazard.

NEUTRAL BUOYANCY Having the same density as the surrounding air.

NON-DESTRUCTIVE TESTING Methods of testing metals or other materials of construction which neither break nor permanently distort them.

NORMALISING A form of heat treatment for steel in which it is heated to 950°C and cooled in air.

NOTCH BRITTLENESS An index of brittleness measured by impacting to destruction a notched material by, for example, a swinging pendulum.

NOTIFIABLE INSTALLATION An installation which is required to be notified to the HSE under the NIHHS Regulations (1982).

OBLATE HEMISPHEROID A flattened hemisphere. (The earth is an oblate spheroid).

OPTIMISTIC Of assumptions, those which lead to the conclusion that the consequence of the realisation of a hazard will be less than the true consequences. Opposite of pessimistic or conservative.

PARAMETER A quantity that, when varied, affects the value of another.

PEAK SHAVING In the US gas industry, the provision of stocks of gas usually, but not necessarily, in a liquefied form to accommodate peak demand.

PITOT TUBE A device for measuring flow in which a thin tube is inserted into a pipe with the mouth pointing into the flow stream. The differential between the pressure at the mouth and the pressure at the wall of the pipe is a measure of the rate of flow when taken in conjunction with such variables as the position of the tube, diameter of the pipe and the physical properties of the fluid.

PLANT An assemblage of process units which produces a definable product or products.

POLYMERISATION Most generally the union of one or more molecules of the same compound to form a larger molecule. Can include unions between different molecules to give a "copolymer".

POOL FIRE* The combustion of material evaporating from a layer of liquid at the base of a fire.

PRECOMMISSIONING The initial stage of commissioning before chemicals or other raw materials are introduced into the plant. Includes hydraulic or other forms of pressure testing.

PROBIT EQUATION An equation of the form $Y=K_1 + K_2 (\ln x)$ where Y is the probit (which may be expressed as fraction or a percentage), K_1 and K_2 are constants, x is a normalised injury factor.

PROCESS UNIT An assembly of equipment which performs a definable function in a process. Corresponds often with the concept of a 'unit operation'.

PROTOTYPE The first, or original, type on which something is modelled. The model will, in general, differ in its dimensions from the prototype.

RADIOGRAPHIC TESTING Taking X-ray photographs of metal, especially at welds.

REDOX A reaction in which reduction and oxidation take place simultaneously. Combustion reactions are a common example but in some redox reactions the halogens, for example, can take the place of oxygen.

RELIABILITY* The probability that an item will perform a required function under stated conditions for a stated period of time or for a stated demand.

RIGHT CYLINDER Having its axis at right angles to its base.

REDUNDANCY* The performance of the same function by a number of identical but independent means.

RISK* The likelihood of a specified undesired event occurring within a specified period or in specified circumstances. It may be either a frequency (the number of specified events occurring within unit time) or a probability (the probability of a specified event following a prior event) depending on the circumstances.

RISK ASSESSMENT* The quantitative evaluation of the likelihood of undesired events and the likelihood of harm or damage being caused together with the value judgements made concerning the significance of the results.

RISK, INDIVIDUAL* The frequency at which an individual may be expected to sustain a given level of harm from the realisation of specified hazards.

RISK, SOCIETAL* The relationship between frequency and the number of people suffering from a specified level of harm in a given population from the realisation of specified hazards. Usually quantified by an F/N curve.

SCALED DISTANCE The distance from the centre of an explosion source at which a particular effect occurs divided by the cube root of the mass of the explosive.

SCALING The problem, when changing such dimensions as those of length between a model and a prototype, of preserving the value of the dimensionless ratios. This is in general impossible and compromises have to be reached.

SCATTERING When radiation traverses a medium, scattering may take place, either through random reflection from suspended particles, or by diffraction when the particle diameters are small compared with the wavelength of the radiation.

SHOCK WAVES* A pressure pulse formed by an explosion in which a sharp discontinuity in pressure is created as the wave travels through a fluid medium at greater than sonic velocity.

SHOCK FRONT VELOCITY The velocity, measured from a fixed point, of the propagation of a shock (blast wave). Initially supersonic, it decays to a sonic wave.

SITE (OR FACTORY OR WORKS) An area of land with a plant or plants with their associated buildings built on it and enclosed by a common security fence.

SONIC VELOCITY The speed of sound in the medium.

SOURCE TERM* The quantitative description of a release required as input to a consequence model, i.e. quantity or rate of concentration, density, temperature etc.

SPECIFIC IMPULSE In rocketry, the thrust of a rocket motor (force) divided by the rate of fuel consumption (mass/sec). Misleadingly expressed as "seconds".

SPILL OVER Otherwise known as "flame drag". Horizontal displacement at the base of a pool fire in the direction of the wind.

STOCHASTIC MODELLING A system of modelling which incorporates elements of probability or chance.

STOICHIOMETRIC MIXTURE One in which the reactants are present in correct theoretical proportions for complete chemical reaction. In the case of a cloud of hydrocarbon mixed with air a stoichiometric mixture would contain just sufficient oxygen to burn the carbon in the compound to carbon dioxide and the hydrogen in the compound to water.

STORAGE UNIT A unit for storing particulate solids, liquids or gases or containers for such storage.

STRAIN ENERGY The energy locked up in stressed material. It is unevenly distributed and is enhanced by the presence of sharp corners.

STRESS RAISER A feature, such as a sharp corner, which creates high local stress concentrations or strain energy.

THICK-WALLED VESSEL A vessel, the ratio of the thickness of its walls to its diameter being sufficiently large that when internally pressurised, significant differences exist between the tensions at the inside and outside of the walls.

THIN-WALLED VESSEL A vessel, the ratio of the thickness of its walls to its diameter being sufficiently small that, when internally pressurised, the differences in tension between the inside and outside of the walls may be neglected.

TLV An American Conference of Governmental Industrial Hygienists (ACIGH) recommended upper limit or time weighted average concentration of a substance to which most workers can be exposed without adverse effect.

TNT EQUIVALENT* The amount of TNT (trinitro-toluene) which would produce the same damage effects as the explosion. For non-dense phase explosions, the equivalence has meaning only at a considerable distance where the nature of the blast wave arising is comparable with that of TNT.

TORUS Shaped like a doughnut, or a tyre inner tube.

TOUGHNESS The opposite of brittleness.

TRANSDUCER A device for transforming a physical property into a signal. For example, a thermometer is a transducer which transforms a rise in temperature into a visible change in the length of a column of mercury.

TRANSITION TEMPERATURE That temperature (or range of temperature) at which the material changes from ductile to brittle behaviour.

ULLAGE Empty space above the liquid contained in a vessel.

ULTRASONIC TESTING Testing using sound waves of higher than audio frequency. Of particular value for examining boundary conditions such as cracks.

UNCONFINED VAPOUR CLOUD EXPLOSION An imprecise term. "Vapour cloud explosion" is now preferred.

UPPER BOUND Extreme value for some specific statistic.

UPPER FLAMMABLE LIMIT* That concentration in air of a flammable material above which combustion will not propagate.

VAPOUR Is a gas which can be liquefied by the application of pressure alone without reduction in temperature.

VAPOUR CLOUD EXPLOSION* The preferred term for an explosion in the open air of a cloud made up of a mixture of a flammable vapour or gas in air.

VOLATILITY A measure of the ability of a liquid to evaporate at a given temperature. May be expressed as the concentration of the substance in the vapour phase in equilibrium with the liquid.

YIELD POINT STRESS The minimum stress to which a material must be subjected to produce a permanent change in its dimensions after the removal of the stress.

YOUNG'S MODULUS The ratio of stress to longitudinal strain.

Appendix II

ABBREVIATIONS

ACMH	Advisory Committee on Major Hazards.
AFF	Adiabatic Flashing Fraction.
AGA	American Gas Association.
AIChE	American Institute of Chemical Engineers.
ASME	American Society of Mechanical Engineers.
ASTM	American Society for Testing Materials.
AWRE	Atomic Weapons Research Establishment. (UK)
BASF	Badische Anilin und Soda Fabrik.
BLEVE	Boiling Liquid Expanding Vapour Explosion.
BS	British Standard.
CIA	Chemical Industries Association. (UK)
CIMAH	Control of Industrial Major Accident Hazards. (UK Regulations)
DALR	Dry Adiabatic Lapse Rate.
FAR	Fatal Accident Rate.
FAFR	Fatal Accident Frequency Rate.
FDR	Federal Republic of Germany. (West Germany)
HASAWA	Health and Safety at Work Act. (UK)
HSC	Health and Safety Commission. (UK)
HSE	Health and Safety Executive. (UK)
IChemE	Institution of Chemical Engineers. (UK)

ICMESA	Industrie Chemiche Meda S.A. (Italy)
IDLH	Immediately Dangerous to Life or Health.
LFL	Lower Flammable Limit.
LG	Liquefied Gas.
LNG	Liquefied Natural Gas.
LOX	Liquid Oxygen.
LPB	Loss Prevention Bulletin. (IChemE publication)
LPG	Liquefied Petroleum Gas.
MCA	Manufacturing Chemists Association. (US)
MCH	Major Chemical Hazard.
MIC	Methyl Isocyanate.
MvSZ	Ministry of Labour and Social Security. (Netherlands)
NDT	Non-destructive Testing.
NIHHS	Notification of Installations Handling Hazardous Substances. (UK Regulations)
NTSB	National Transportation Safety Board. (US)
PCDD	Polychlorinated dibenzodioxins.
PPDS	Physical Properties Data Service. (IChemE)
RARDE	Royal Armament Research and Development Association. (UK)
SOE	Shorter Oxford English Dictionary.
SRD	Safety and Reliability Directorate. (UK)
TAFF	Theoretical Adiabatic Flashing Fraction.
TCDD	Tetrachloro-dibenzo-dioxin. ("dioxin")
TCP	Trichloro-phenol.
TLV	Threshold Limit Value.
UFL	Upper Flammable Limit.
UKAEA	United Kingdom Atomic Energy Authority.
UVCE	Unconfined Vapour Cloud Explosion.

VCE	Vapour Cloud Explosion.
VCM	Vinyl Chloride Monomer.
VM	Initials of present author in interpolations.

Appendix III

ACCESS TO DATA BASES ON CHEMICAL ACCIDENTS

III.1. AIM OF APPENDIX.

The aim of the Appendix is to provide a ready guide to the existing data bases in the field of chemical accidents, and to discuss their content, format, and means of access.

III.2. NATURE OF DATA BASES.

The expression "data base" is used as a general term in this appendix to mean any systematically organised, retrievable, collection of information. Unfortunately, however, this is a field where there seems to be some lack of agreement on terminology. Some prefer to use the term "data bank" to refer to data in printed form and to reserve the term "data base" for describing a computerised data bank. But one modern dictionary, Chambers Everyday Dictionary, gives both the same meaning, that of data stored in a computer. A key report quoted below uses data base as the general term. To avoid ambiguity the specific term "computerised data base" will be used below where this is appropriate.

Data bases of relevance to this discussion may be conceived as lying on a spectrum. At one end are data bases made up of files each of which relates to recorded data which may be a book or paper, or possibly a film or video, of which the contents are abstracted and coded, and which relates to the general field of the data base. These may be conveniently described as "literature data bases". At the other end of the spectrum are data bases where the files relate, for example, to specific incidents or to the properties of specific substances. Such data bases may be described as "incident data bases" or "substance data bases". It is this latter type of data base which, when stored in a computer and suitably coded, is most readily subject to analysis.

The general question of access to data bases of relevance to chemical engineers is discussed in a recent publication, Jankowski and Selover (1986).

Information provided on any data base given below is believed by the author to be correct in September 1986.

III.3. THE STATUS OF DATA BASES IN EUROPE AND THE US.

A recent report, TNO (1985a), has surveyed the availability of industrial accident data bases in five West European countries and in the United States. It should be noted that this survey was concerned with industrial accidents in general and not with major chemical accidents in particular. It was concerned with all forms of data base, however maintained.

The general conclusion that emerged from it was that the establishment of such bases was, at that time, only in a formative stage, this being especially true of publicly accessible computerised data bases. For instance, of the situation in the US, the Report states "Despite the number of government bodies concerned, no one does actually operate a wide scope national data base for industrial accidents. Valuable data collection schemes are operated by private institutions, i.e. NFPA (fires), by the Industrial Risks Insurers, and by the American Petroleum Institute, on special subjects". However attention is called below to the current public availability of the NIOSH data base. The Report gives a survey of the literature on data bases which is much more comprehensive than it is feasible to give here.

III.4. PRINTED WORD DATA BASES.

Though a number of compilations of chemical disasters have been referred to in the main text, these all, inevitably, have had a "once and for all" character. The purpose of this section is to draw attention to those printed publications which have the function of providing a continuous up-dating of the information which is available in the field.

The Institution of Chemical Engineers, through its "Loss Prevention Bulletin", publishes annually a list of serious chemical accidents in the chemical and process industries and in transport, including marine transport. The entries include particulars of human injury and financial loss where these are known, [LPB (Series)].

The Chemical Industries Association of the UK produces a quarterly bulletin of accidents, including laboratory accidents, in the chemical and process industries, [Chemical Industries Association (Series)].

The Royal Society of Chemistry publishes monthly, "Chemical Hazards in Industry". This provides, among other items, case histories of chemical accidents in the form of abstracts from the literature. Each issue contains about 200 items, [RSC (Series)]. It is also available as a computerised data base as noted below.

Marsh McLennan publish annually an up-date on the hundred most serious fires and explosions which have occurred in the previous thirty years in the hydrocarbon-chemical industries and including marine and other forms of transport handling chemicals and hydrocarbons. The nine editions to date span almost forty years. Losses in US dollars are given, corrected for inflation to the year of publication. Human injury is not recorded, [M & M (Series)].

The UK journal, Fire Prevention, gives a monthly case book of serious fires in the UK and an annual analysis. Major fires and fire statistics from abroad are also reported. The fires recorded are of all types but they include chemical fires, [FPA (Series)].

The National Fire Protection Association (US) assembles and maintains the Fire Incident Data Organisation (FIDO) which is concerned mainly with large fires of all types. Major fires are analysed and the results published in "Fire Journal" which also provides a bimonthly case book of major fires. Incident documentation is supplied to the public for a fee, [Fire Journal (Series)].

III.5. COMPUTERISED DATA BASES.

III.5.1. GENERAL.

Data bases may be classified as "private", i.e. compiled and operated by an organisation solely for its own use, and which are not further discussed below, "public off-line" where information is publicly available as a print out on request by letter, phone or telex, "public on-line" where the information may be directly accessed via a network using the appropriate equipment, and "stand alone" where the data base is made available to the public in the form of a disc which can be accessed in-house using suitable equipment.

A point to be borne in mind is that data bases on health and safety, especially those on the narrow field of major chemical hazards, are likely to be of interest to only a limited clientele. Hence public access is only likely to arise as a spin-off from the internal use which is made of the data base by a major organisation.

III.5.2. ACCESS TO COMPUTERISED DATA BASES.

Access in the UK to public on-line data bases may be obtained through two types of system, both of which use the telephone network. There is the Prestel system which requires only a modified television set coupled to a telephone line by an adaptor which includes a keyboard. For the other systems access is by means of a minicomputer equipped with a suitable screen and/or printer and which is connected to the telephone system through a modem or an acoustic coupler.

The details of how to gain access to the on-line systems listed may be obtained by contacting the referenced organisation.

III.5.3. COMPUTERISED LITERATURE DATA BASES.

HSE Prestel programme. This data base supplies information through the Prestel system which displays a logically connected sequence of index pages. It covers general information on the organisation and structure of the Commission and Executive, on the law, on news releases and on the publications of the HSC and HSE. It also provides access to HSELINE which is described below. The system can be used to order HSE publications and leaflets, [HSE (Series)].

NIOSH. This is a comprehensive data base on occupational health and safety compiled in the US which contains 110,000 abstracted references with a yearly rate of up-date of 12,000 references per year. The records date back to the 19th century. Obviously only a small proportion of the literature stored in the data base is directly relevant to the field of major chemical hazards. NIOSH may be accessed through Pergamon InfoLine, [Pergamon (1986)].

HSELINE. This is a comprehensive data base on occupational health and safety compiled in the UK which contained 60,000 references in January 1985 and which goes back to 1977. New entries are being added at the rate of around 12,000 per annum. As with the NIOSH data base, only a small proportion of the publications are directly relevant to the field of major chemical hazards. As well as being accessible through Prestel as noted above, HSELINE may also be accessed through the European Space Agency Information Retrieval Service (IRS-Dialtech) or through Pergamon InfoLine, [Pergamon (1986)].

Chemical Hazards in Industry. This data base consists of the back file of the printed version published by the Royal Society of Chemistry which is referred to above. It is available from Pergamon InfoLine [Pergamon (1986)] and from Data Star [Data Star (1986)]. It is expected that it will be accessible in Germany from late in 1986, [RSC (Series)].

OSH-ROM. This is the trade mark of a "stand alone" system which offers three complete data bases on a compact disc. The data bases are HSELINE and NIOSH which are referred to above, and CISDOC which is the data base of the United Nations occupational safety and health organisation operated by the International Labour Organisation. In total there are 240,000 citations with abstracts or key words taken from over 500 journals. Updated discs are provided at regular intervals. Access to the data requires the customer to have an IBM compatible machine, a suitable disc drive, and, if hard copy is required, a suitable printer, [Health & Safety Executive (1986)].

III.5.4. COMPUTERISED INCIDENT DATA BASES.

MHIDAS (Major Hazard Incident Data Service). This is a UK data base, launched in September 1986, which has been prepared jointly by the HSE and SRD. The incidents covered are incidents involving hazardous materials which either resulted in, or had the potential to produce, off-site impact. Transport incidents are included. The data, broadly speaking, covers the last 25 years but includes some earlier well documented incidents. The coded information is stored under 24 separate fields. It is currently available only as a public off-line data base, [SRD (1986)].

RM Consultants. This organisation maintains a computerised data base which is available off-line to the public. It includes off-shore incidents. The organisation also acts as the UK agent for the TNO FACTS data base, [RMC (1986)].

TNO. The computerised data base FACTS maintained by TNO appears to have characteristics similar to the computerised data bases referred to above. It is available off-line in the UK through RM Consultants Ltd (see above).

III.5.5. COMPUTERISED SUBSTANCE DATA BASE.

PPDS. This is a UK physical properties data base operated by the National Engineering Laboratory in association with the Institution of Chemical Engineers. The primary function of this data base is to provide physical data to the designers of plant in the chemical and process industries and of plant for raising and utilising steam. The data provided is that which is required for general technological purposes. However the accuracy of physical data is also an imperative for safe design. The service is of value also for the forensic analysis of incidents as it enables the design parameters of plants involved in accidents to be checked.

The data base provides data on more than 850 compounds. These are mainly organic but some elements and inorganic compounds are included. Some 38 properties of each are listed and data is also provided for mixtures. The base is available to the public both off-line and on-line world wide, through a variety of agencies. A recent development has been to make available a "stand alone" electronic data module to be used in-house by the customer. Details of the facilities offered may be obtained through PPDS (1986a).

Appendix IV

NOTES ON THE TNT EQUIVALENCE OF ORGANIC PROPERTIES

IV.1. COMMERCIAL SIGNIFICANCE.

Most organic peroxides, of which there are more than 65 commercially available, with over 100 formulations, are used as initiators of polymerisation reactions. The exception is peroxyacetic acid (peracetic acid) which is used in the manufacture of pharmaceutical intermediates, epoxy compounds, as a bactericide, as a fungicide and as a sterilising agent. Peroxyacetic acid is manufactured on a larger scale, worldwide, than other organic peroxides.

Organic peroxides, generally speaking, are produced on only a modest scale compared with that of many of the hazardous chemicals discussed in the main text. The annual rate of production of the more important organic peroxides is shown in Table IV.1.

TABLE IV.1. PRODUCTION RATES OF ORGANIC PEROXIDES IN TONNES p.a.

Substance	US (1978)	UK (1985) (approximate figures)
Dibenzoyl peroxide	3,600	900
Dialkyl peroxides	2,242	700
Methyl ethyl ketone peroxide	3,250	800
Peroxyesters	2,378	250
Total production of organic peroxides excluding peroxyacetic acid.	13,263	3,000

Sources:- Kirk-Othmer (1982) for US figures; private communication to present author for UK figures. The annual production of peroxyacetic acid in the UK is of the order of thousands of tonnes per annum.

IV.2. HAZARDOUS PROPERTIES.

Many organic peroxides are unstable and some are capable of explosion; their hazardous properties are described in Castrantas et al (1965).

The hazardous properties of peroxides as a class arise from the presence of the unstable grouping:-

$$X - O - O - Y$$

where X or Y may be a metal, or hydrogen or an organic group. Where X and Y are both metals, the hazards of instability are relatively small. Where X and Y are hydrogen the substance is the well known compound hydrogen peroxide. Hydrogen peroxide is available commercially in admixture with water in all concentrations up to and exceeding 90%. Under the name of "High test hydrogen peroxide" it has been extensively used in rocket technology as a monopropellant or in a mixture with a reducing agent. It has been used, after catalytic decomposition, as a working fluid to drive the pumps in liquid fuelled rockets. Hydrogen peroxide is extensively used as the starting point for manufacturing organic peroxides.

When hydrogen peroxide undergoes decomposition the heat released at 20°C and 1 bar is approximately 24×10^6 J per kg mole or 7×10^5 J per kg. However where organic material is available for reaction with the liberated oxygen the heat released is far greater. In the presence of a stoichiometric quantity of formic acid the heat release is 351×10^6 J per kg mole of hydrogen peroxide, [Stull (1977)]. It will be seen from this that high concentrations of hydrogen peroxide are potentially capable of sustaining anaerobic fires or of bringing about explosions in confined spaces.

Where X or Y or both are organic groupings the liberated oxygen may take part in a redox reaction with the organic group producing a much higher level of heat release than occurs in the decomposition of hydrogen peroxide. The level of heat release per mol is limited by the availability of oxygen and remains relatively constant with increase in, for example, the length of an aliphatic chain attached to the peroxide grouping. However compounds with more than one peroxide grouping will have higher rates of heat release per mole than those with a single peroxide group.

However the value of the heat release is not the only factor in determining whether a peroxide is explosive as this depends also on the rate at which the heat is released.

IV.3. THE DETERMINATION OF TNT EQUIVALENCE.

The TNT equivalence of organic peroxides has been determined by a number of investigators and for Table IV.2. below values have been taken from two of these reference sources, Clancey et al (1970) and Yoshida et al (1985). There are a number of standard test methods for testing conventional explosives, Clancey et al used two of these, the Ballistic Mortar and the Trautzl lead block and Yoshida et al used a ballistic mortar only. Table IV.2. below has been derived by the present author from the ballistic mortar test results of the two investigators and expresses the blast energy released as a fraction of the blast energy released by an equal mass of TNT and, in some cases, as blast energy per kg mole.

The ballistic mortar test in which the substance is detonated by a powerful detonator under conditions of confinement is a very severe test and it may be reasonably concluded that the figures tabulated below are pessimistic in relation to actual industrial conditions. Nevertheless such materials are likely to give rise, under certain conditions, to severe anaerobic fires or to confined explosions.

The dangerous properties of organic peroxides are greatly reduced by dilution with water or with organic solvents. Many of the formulations in current use, which are set out in Interox (1985), represent a compromise between considerations of safety and considerations of convenience, as dilution makes the peroxides less convenient to use.

Annex III of the European Directive on Major Accident Hazards lists a number of organic peroxides, some of which are in the table below, and applies a threshold inventory of 50 tonnes to all of them. This is the same level of inventory that the Annex applies to TNT and to some other conventional explosives.

The discussion above, taken in conjunction with Table IV.2., seems to indicate that the threshold level for organic peroxides in Annex III is excessively stringent.

TABLE IV.2. BLAST ENERGY EQUIVALENCES OF ORGANIC PEROXIDES.

Substance	Ref	Blast energy kJ/kg mole x 10^6	Blast energy as a fraction of that of TNT
Dibenzoyl peroxide (pure)	1	113	0.09
t-Butyl-peroxyacetate (70%)	1	150	0.17
t-Butyl-peroxypivalate	1	112	0.14
t-Butyl-peroxy maleate (pure)	1	108	0.14
Methyl-ethyl-ketone peroxide (60%)	1	359	0.26
Cyclohexanone peroxide (60%)	1	137	0.13
Peroxyacetic acid (40%)	1		0.05
tert butyl peroxybenzoate	2		0.40
Dibenzoyl-peroxy benzoate	2		0.25
di-tert-butyl peroxide	2		0.38

Notes:- Ref. 1 is to Clancey et al, Ref. 2 is to Yoshida et al. Inclusion of a substance in the table should not be taken to indicate that it is supplied commercially in the formulation in which it was tested.

IV.4. CASE HISTORY.

Stull (1977) gives a case history, undated but prior to 1966, of an incident which occurred in the US in which 17.5 tonnes of containers of benzoyl peroxide, methyl-ethyl-ketone peroxide, lauroyl peroxide, and tertiary butyl hydroperoxide caught fire on a lorry. What started as a small fire escalated to a detonation and four firemen were killed. Two buildings were destroyed and others suffered structural damage.

IV.5. ACKNOWLEDGEMENT.

The author wishes to thank Messrs Laporte Industries Ltd., Widnes, UK, and Interox Chemicals Ltd., Warrington, for assistance in compiling this appendix. The opinions expressed in it are, however, entirely his own.

REFERENCES

Numbers at the end of references refer to the pages in the text in which they appear.

ABCM (1950). Association of British Chemical Manufacturers. "Safety rules for use in chemical works". ABCM, London, 1950. 493.

Aberfan (1967). "Report of the tribunal appointed to inquire into the disaster at Aberfan on 21st October, 1966. HMSO, 1967. 384.

ACHEMA (1985). "ACHEMA Jahrbuch, Vol 1". DECHEMA, Frankfurt, 1985. 489.

Acre Mill (1976). "Report by the Parliamentary Commissioner for Administration into a complaint by Mr. J.P. Buick". Report to Mr. Max Madden, M.P. at the House of Commons, 23rd March, 1976. 489.

Advisory Committee on Major Hazards (1976). "First Report". HMSO, 1976. 2, 445.

Advisory Committee on Major Hazards (1979). "Second Report". HMSO, 1979. 2, 4, 5, 21, 62, 70, 117, 127, 129, 245, 252, 256, 280, 284, 313, 328, 333, 430, 434, 445, 464, 469, 471.

Advisory Committee on Major Hazards (1984). "Third Report". HMSO, 1984. 140, 142, 207, 236, 391, 443, 445.

A.I.Ch.E. (1986). American Institute of Chemical Engineers. "Center for chemical process safety". A.I.Ch.E., New York, 1986. 495.

Alderson, M.A.H.G. (1985). "Seismic design criteria and their applicability to major hazard plant within the United Kingdom". UKAEA Safety & Reliability Directorate, 1985. 89, 504.

Aldis, D.F. and Lai, F.S. (1979). "Review of literature related to engineering aspects of grain dust explosions". US Dept. Agriculture Misc. Publication No.1375, 1979. 235.

Ale, B.J.M. and Bruning, F. (1980). "Unconfined vapour cloud explosions". The Chemical Engineer, Jan. 1980. 256.

Allan, D.S. and Athens, P. (1968). "Influence of explosions on design". CEP Loss Prevention 1968, Vol.2. 470.

American Gas Association (1973). "Consequences of LNG spillages on land". LNG Safety Program Phase II Project IS-3-1, 1973. 124, 125.

American Gas Association (1974). "LNG safety program - interim report on Phase II work". Report No. IS-3-1, 1974. 126.

American Society of Civil Engineers (1961). Manual of Engineering Practice 42. "Design of structures to resist nuclear weapons effects". USA, 1961. 470.

Annual Register (1759). A UK periodical, published annually, from 1758 onwards. Now the Annual Register of World Events. Longmans. 4.

ANON (1966). Account of Feyzin Disaster. Fire International 12, 1966. 175, 178, 180.

ANON (1975). "Disaster". Phoebus Publishing Co, BPC Publishing Ltd, Poland Street, London, 1975. 22.

ASC (1984). Association of Consulting Scientists. "Members and services". Buntingford, Herts, 1984. 503.

Assheton, R. (1930). "History of explosions" (on which the American table of distances was based). Institute of Makers of Explosives, Wilmington, Delaware, USA, 1930. 211, 419.

Association of Petrochemicals Industries Japan (1974). High Pressure Polyethylene Committee. "Safety in high pressure polyethylene plants". AICE 74th Annual meeting, Tulsa, Oklahoma, March 1974. 254.

Atkinson et al (1980). Atkinson, R., Bull, D.C., Shuff, P.J. "Initiation of spherical detonation in hydrogen/air". Combustion and Flame, The Combustion Institute, NY, 1980. 258.

Backmann, E. (1974). Letter to Dr. N. Harnby, Bradford University. Rhone Poulenc, 1974. 175.

Baker, W.E. (1973). "Explosions in air". University of Texas Press, Austin and London, 1973. 209, 210, 218, 220, 251.

Baker et al (1983). Baker, W.E., Cox, P.A., Westine, P.S., Kulesz, J.J., Strehlow, R.A. "Explosion hazards and evaluation". Elsevier, Amsterdam, 1983. 217, 218, 220, 222, 230, 231, 234, 236.

Ball, W.L. (1970). Review of atmospheric ammonia storage research study. CEP Technical Manual (Safety in air and ammonia plants), Vol.12. A.I.Ch.E, New York, 1970. 333.

Barrell, A.C. (1981). Pheasant Wood inquiry, Sept. 1981, proof of evidence. HSE, 1981. 99.

Barrell, A.C. and McQuaid, J. (1985). "The HSE programme of research and model development on heavy gas dispersion". The assessment of control of major hazards, I.Chem.E, Rugby, Symposium series No. 93, 1985. 105, 108.

Bartnecht, W. (1979). "Prevention and protection (from dust explosions)". OYEZ Conference, The hazards of industrial explosion from dusts. OYEZ, IBC Ltd, London, 1979. 213, 230, 233, 234.

Bartnecht, W. (1981). "Explosions; course, prevention, protection". Springer-Verlag, 1981. 505.

Barton, J. and Nolan, P. (1984). "Runaway reactions in batch reactors". I.Chem.E. Symposium series No. 85. I.Chem.E, Rugby, 1984. 83.

BASF (1921). Badischen Anilin und Soda Fabrik. Werkzeitung der BASF (in German), 9th year, No. 10. Ludwigshafen, October 1921. 22

BASF (1948). Badischen Anilin und Soda Fabrik. "Explosions katastrophe in Ludwigshafen am Rhein" (in German). Published by Badischen Anilin und Soda Fabrik, 1948. 22, 271.

BASF (1958). Badischen Anilin und Soda Fabrik. "Firmengeschichte". BASF, Ludwigshafen, December 1958. 224.

Beattie, J.R. (1963). "An assessment of environmental hazards from fission product releases". AHSB (S), R64. HMSO, 1963. 96.

Bednar, H.H. (1981). "Pressure vessel design handbook". Van Nostrand Reinhold, New York, 1981. 69.

Beek (1976). English summary of official Dutch report of inquiry into the explosion at DSM Beek (Limburg), 7th Nov. 1975. Published by the Directorate General for Labour, Ministry of Social Affairs, P.O. Box 69, Voorburg, Holland. 432, 482, 484.

Bell et al (1965). Bell, G.H., Davidson, J.N., Scarborough, H. "Text book of physiology and biochemistry". E.S. Livingstone, Edinburgh & London, 1965. 339.

Bell, D. and Elmes, P. (1965). "The effects of chlorine gas on the lungs of rats". J.Path.Bact.84, 1965. 435.

Bennett, S. (1979). "A history of control engineering - 1800 to 1930". Peter Peregrinus Ltd, Stevenage, UK & New York in association with the Institution of Electrical Engineers, London,1979. 461.

Berthelot, M. (1892). "Explosions and their power". Translated from the French by Hake, C.N. and Macnab, W. John Murray, London, 1892. 245.

Beveridge, G.S.G. (1981). "Note on spillage of refrigerated ammonia on land". Loss Prevention Bulletin No. 45, I.Chem.E, Rugby, 1981. 333.

Biasutti, G.S. (1981). "History of accidents in the explosives industry". Published by the author, Case postale 312, 1800 Vevey, Switzerland, 1981. 209, 420, 454.

Biasutti, G.S. (1985). "History of accidents in the explosives industry" 2nd edition. Published by the author, Case postale 312, 1800 Vevey, Switzerland, 1985. 4, 21, 211, 224, 419, 420.

Birchon, D. (1975). "Non-destructive testing". Oxford University Press, 1975. 78.

Blackett, P.M.S. (1941). "A note on certain aspects of the methodology of operational research". (Quoted in Irving 1980, P.34). 1941. 416.

Blackmore et al (1982). Blackmore, D.R., Herman, M.N., Woodward, J.L. "Heavy gas dispersion models". Journal of hazardous materials, Vol.6, Elsevier, Amsterdam, 1982. 99, 100.

Blackmore et al (1982a). Blackmore, D.R., Eyre, J.A., Summers, G.G. "Dispersion and combustion behaviour of large spillages of LNG and LPG onto the sea". Trans.I.Mar.E. (TM), Vol.94, Paper 29, 1982. 122, 123, 124, 125, 130.

Bodurtha, F.T. (1980). "Industrial explosion prevention and protection". McGraw-Hill, New York, 1980. 210, 230.

Booth, N. (1973). "Industrial gases". Pergamon Press, 1973. 385.

Borkin, J. (1979). "The crime and punishment of I.G. Farben". Andre Deutsch, London, 1979. 345.

Bouhon, J. (1958). "Antwerp at the liberation". Antwerp, Belgique exposition, 1958. Editions du Marais, Brussels, 1958. 425.

Bradford, W.J. and Culbertson, T.L. (1967). "Design of control houses to withstand explosive forces". A.I.Ch.E. meeting, Houston, Texas, Feb. 21st, 1967. 470.

Bradley, D. (1978). "Unconfined vapour cloud explosions". Fire Prevention Science and Technology No. 21, 1978. 212, 256.

Brasie, W.C. (1976). "The hazard potential of chemicals". CEP Loss Prevention, Vol.10, P.135, 1976. 129, 131.

Brasie, W.C. and Simpson, D.W. (1968). "Guidelines for estimating damage from chemical explosions". CEP Loss Prevention 1968, Vol.2, P.91. 225, 244, 469.

Bremen (1979). "Fire and explosion, Roland Mill, Bremen, 6/2/79". Bremen City and Police Office, Bremen, 1979. 264.

Bretherick, L. (1979). "Handbook of reactive chemical hazards" 3rd edition. Butterworths, 1979. 388.

Bretherick, L. (1981). "Hazards in the chemical laboratory". The Chemical Society, London, 1981. 387, 388.

Bridges, J.W. (1984). "The problems with toxic chemicals". Symposium on European Major Hazards, Sept. 1984. OYEZ, London, 1984. 318, 319, 321, 323, 353.

British Cryogenics Council (1970). "Cryogenics safety manual". I.Chem.E, Rugby, 1970. 258, 382, 383, 385.

British Gas (1983). Letter to author from M.F. Jusuf (statistician), 11th August, 1983. 341.

Britter, R.E. and Griffiths, R.F. (Editors), (1982). Journal of Hazardous Materials, Vol.6. Elsevier, Amsterdam, 1982. 99.

Brossard et al (1984). Brossard, J. and seven co-workers. "Experimental analysis of unconfined explosions of air-hydrocarbon mixtures". I.Chem.E. Symposium Series No. 80, 1984. 246.

Brown, R.F. (1978). "Injury by burning". In "The pathology of violent injury". (Ed. J.K. Mason). Arnold, 1978. 386, 388, 390.

Brown, V.K. and Bridges, J.W. (1983). "The problems of highly toxic substances". Symposium on implementing the Seveso directive, October 1983. OYEZ, London, 1983. 353.

Browning, J.B. (1985). "After Bhopal". In "The chemical industry after Bhopal". IBC Technical Services, 56 Holborn Viaduct, London, 1985. 370, 371, 379.

Bryan, A. (1975). "The evolution of health and safety in mines". A Mine and Quarry publication, London, 1975. 227, 402.

Bryant, P.M. (1964). "Methods of estimation of the dispersion of windborne material". UKAEA Report AHSB (RP) R42, Code No. 91-3-15-93. HMSO, 1964. 96.

BS 1510 (1958). British Standard. "Steels for use in the chemical petroleum and allied industries, low temperature supplementary requirements". British Standards Institution, 1958. 504.

BS 2955 (1958). British Standard. "Glossary of terms relating to powders". British Standards Institution, 1958. 231.

BS 4422 (1969). British Standard. "Glossary of terms associated with fire". British Standards Institution, 1969. 110.

BS 131 (1972). British Standard. "Methods for notched bar tests, part 2. The Charpy V-notch impact test on metals". British Standards Institution, 1972. 75.

BS 4547 (1972). British Standard. "Classification of fires". British Standards Institution, 1972. 110.

BS 5908 (1980). British Standard. "Fire precautions in chemical plant". British Standards Institution, 1980. 128.

BS 5500 (1982). British Standard specification for unfired fusion welded pressure vessels. British Standards Institution, 1982. 69, 70, 77, 78, 79.

Bulkley, W.L. and Jacobs, R.B. (1966). "Hazard of atmospheric releases of large volumes of hydrogen". Facilities sub-committee, API (US), 4/10/66. 258.

Burgess et al (1970). Burgess, D.S., Murphy, J.N., Zabetakis, M.G. "Hazards of LNG spillage". US Bureau of Mines report S.4105, Pittsburgh, 1970. 101k.

Burgess et al (1972). Burgess, D.S., Biordi, J., Murphy J. "Hazards of spillage of LNG onto water". US Bureau of Mines report, PMSRC Report No. 4177, Pittsburgh, 1972. 101.

Burgess, D.S. and Zabetakis, M.G. (1972). "Detonation of a flammable cloud following a propane pipeline break". US Bureau of Mines, Report of Investigations R.I.7752, 1972. 261, 279, 281.

Burke, J.G. (1966). "Bursting boilers and the federal power". Technology and Culture 7, 1-13, 1966. 505.

Burns, B.W. (1967). "Design control centres to resist explosions". Hydrocarbon Processing 46, No. 11, 257, 1967. 470.

Buschmann, C.H. (1975). "Experiments on the dispersion of heavy gases and the abatement of chlorine clouds". Proc. Symposium Transport of Hazardous Cargoes by Sea, Jacksonville, Fl, USA, 1975. 102.

Bush, S.H. (1975). "Pressure vessel reliability". Journal of Pressure Vessel Technology, Transactions of AMSE, 54, Feb 1975. 71, 73.

Canning, J. (1976). "Great disasters". Octopus Books, London, 1976. 22, 182.

Canvey (1978). Health & Safety Executive. "Canvey, an investigation of potential hazards in the Canvey Island/Thurrock area". HMSO, 1978. 407.

Canvey (1981). Health & Safety Executive. "Canvey: A second report". HMSO, 1981. 117, 119, 137, 210, 407.

Cardillo et al (1984). Cardillo, P., Girella, A., Ferraiolo, G. "The Seveso case and the safety problem in the production of 2,4,5 trichlorophenol". Journal of Hazardous Materials, Elsevier, Amsterdam, Aug 1984. 349, 350, 362.

Cardillo, P. and Anthony, E.J. (1979). "Dust explosions and fires, guide to the literature". Stazione Sperimentale per Combustibili, San Donato, Milaneso, 1979. 235.

Carrasco, D.J.A. (1978). Report and technical study on the causes of an accident involving a tanker truck at Los Alfaques. Translation by HSE, 1982. 184, 186, 188, 195, 196.

Carrington, C. (1970). "Soldier from the wars returning". Arrow Books, 1970. 433.

Carrier et al (1985). Carrier, G.F., Fendell, F.E., Feldman, P.S. "Firestorms". Journal of Heat Transfer, Vol 107/19. ASME New York, Feb 1985. 139, 140.

Carter et al (1975). Carter, C.D., Kimbrough, R.D. and others. "Tetrachlorodibenzodioxin:An accidental poisoning episode in horse arenas". Science Vol. 188, May 1975. 347, 349.

Caruana, C.M. (1986). "Arthur D. Little Inc. Innovator, celebrates centennial". Chemical Engineering Progress, June 1986. 503.

Cashdollar, K.L. and Hertzberg, M. (1983). "Infrared temperatures of coal dust explosions". Combustion and Flame Vol. 51. The Combustion Institute (US), 1983. 232.

Castrantas et al (1965). Castrantas, H.M., Banerjee, D.K., Noller, D.C. "Fire and explosion hazards of peroxy compounds". ASTM, STP 394, 1965. 537.

Cattabeni et al (1978). Cattabeni, F., Cavallora, A., Galli, G. [Editors]. "Dioxin, toxicological and chemical aspects". SP Medical and Scientific Books, Halsted Press, New York & London, 1978. 347, 349, 351, 352.

CECAR (1981). Central Environmental Control Agency Rijnmond. "Risk analysis of six potentially hazardous industrial objects in the Rijnmond area". Reidel Publishing Co, Dordrecht, Holland, 1981. 434, 503.

CEFIC (1986). Conseil Europeen des Federations de l'Industrie Chimique. "CEFIC views on the quantitative assessment of risks from installations in the chemical industry". CEFIC, Avenue Louise 250, Bruxelles, 1986. 507.

Central Statistical Office (1980). "Monthly digest of statistics". HMSO, Sept. 1980. 44, 257.

Challis, E.J. (1977). Private communication, 26/10/77. 430.

Chambers (1973). Chambers Encyclopedia. London, 1973. 27, 131.

Chambers (1979). Chambers English Dictionary. Chambers, 1979. 209.

Chan et al (1981). Chan, S.T., Gresho, P.M., Ermak, D.L. "A three dimensional, conservation equation model for simulating LNG vapour dispersion in the atmosphere". Lawrence Livermore National Laboratory Report UCID 19210, Sept.1981. 102.

Chandler, T.J. (1965). "The climate of London". Hutchinson, London, 1965. 96.

Chemical Defence Establishment (1976). Private communication by A. Spence, Information Officer. Porton Down, 27/9/76. 328.

Chemical Engineer (1986). "A clean bill of health for ReChem". The Chemical Engineer, June 1986. 507.

Chemical Engineering (1971). News Report on Floral incident. Chemical Engineering, McGraw-Hill, July 1971. 334.

Chemical & Engineering News (1985). Special issue "Bhopal report". 11th Feb, 1985. 373.

Chemical Industries Association (Series). "Chemical Safety" Summary published quarterly, available on subscription. Chemical Industries Association, Albert Embankment, London. 534.

Chemical Industries Association (1975). "Code of practice. Anhydrous ammonia in bulk by rail in the UK". Chemical Industries Association, London, 1975. 77, 278.

Chemical Industries Association (1975a). "Codes of practice for chemicals with major hazards - chlorine". Chemical Industries Association, London, 1975. 83.

Chemical Industries Association (1976). "Handling major emergencies". Chemical Industries Association, London, 1976. 459, 487.

Chemical Industries Association (1977). "Hazard and operability studies". Chemical Industries Association, London, 1977. 410.

Chemical Industries Association (1978). "Guide to safe practice in the use and handling of hydrogen fluoride". Chemical Industries Association, London, 1978. 389.

Chemical Industries Association (1979). "Process plant hazard and control building design". Chemical Industries Association, London, 1979. 471.

Chemistry in Britain (1986). "Chemical weapons made comeback" P.507. Chemistry in Britain, June 1986. 507.

Chicken, J.C. (1975). "Hazard control policy in Britain". Pergamon Press, Oxford, 1975. 394, 395.

Christopherson, D.G. (1946). "Structural defence 1945". Ministry of Home Security, Research & Experiments Department. Jan 1946. 416, 464.

Chuse, R. (1977). "Pressure vessels - the ASME code simplified". McGraw-Hill, New York, 1977. 69.

Clancey, V.J. (1963). "The explosive and fire hazard characteristics of ammonium nitrate fertilisers". Part 1, Introduction and critical literature survey. Royal Armament Research & Development Establishment (UK), RARDE Memorandum (X) 48/63. Sept 1963. 4, 141, 207, 211, 224.

Clancey et al (1970). Clancey, V.J., Owen, A.J., Taylor, A.J. "Organic peroxides, Part 1 assessment of the hazards of organic peroxides as offered for transport". Royal Armament Research & Development Establishment, Memo 21/70, 1970. 538.

Clark, D.A.R. (1953). "Materials and structures". Blackie & Son, London, 1953. 75.

Clarke, R. (1968). "We all fall down". Penguin Press, 1968. 342.

Colenbrander, G.W. (1980). "A mathematical model for the transient behaviour of dense vapour clouds". Proc. 3rd International Symposium, Loss Prevention & Safety Promotion, Basle 1980. 103.

Collier, B. (1957). "The defence of the United Kingdom". HMSO, 1957. 424.

Collier, B. (1976). "The battle of the V weapons" [Appendix]. Elmfield Press, Morley, Yorkshire, 1976. 286, 424, 425.

Commercial Union (1976). Commercial Union Risk Management Ltd, Technical Report No. 2. "Fire and explosion risk control in the petrochemicals industry". London, 1976. 175, 245.

Cook, M.A. (1966). "The science of high explosives". Reinhold Publishing Corporation, New York, 1966. 210, 214, 218, 252.

Cottrell, A.H. and Swann, P.R. (1976). "A metallurgical examination of the failures in the 8" diameter stainless steel pipe". In I.Chem.E. "The technical lessons of Flixborough", Rugby, 1976. 293.

Coulson, J.M. and Richardson, J.F. (1956). "Chemical engineering" Vol.2. Pergamon Press, London, 1956. 62.

Coulston, F. and Pocchiari, F. (1983). [Editors]. "Accidental exposure to dioxins: human health aspects". Academic Press, N.Y., 1983. 349, 350, 365, 366, 368.

Council for Science & Safety (1977). "The acceptability of risks". Published by Barry Rose, London, 1977. 39.

Cox, J. (1976). "On the warpath". Oxford University Press, 1976. 346.

Craven, A.D. (1976). "Fire and explosion hazards associated with the ignition of small scale unconfined spillages". I.Chem.E. Symposium Series No. 47, Rugby, 1976. 122.

Crawley, F.K. (1982). "The effects of the ignition of a major fuel spillage". I.Chem.E. Symposium Series No. 71, Rugby, 1982. 122, 129, 146, 150, 154, 156.

Cremer and Warner (1974). Second report to Flixborough disaster court of inquiry, 4th Oct 1974. 63, 65, 295.

Cremer and Warner (1974a). "Proof of evidence of Professor Sir Frederick Warner". Document available to the Flixborough court of inquiry, Sept 1974. 295, 300, 301.

Crow, K. (1978). "Chloracne: the chemical disease". New Scientist, London, 13/4/78. 347, 349, 353, 355.

CW (1971). See SIPRI (1971). 323.

Cybulski (1975). "Dust explosions and their suppression". US Dept of Commerce, NTIS Springfield, Virginia, Translation TT 73-54001, 1975. 228.

Dagani, N. (1985). "Data on MIC's toxicity are scant". Chemical and Engineering News, 11th Feb 1985. 370, 372, 373.

Daniels, F. (1982). "The bombing of Japan". History Today, P.14, Vol.32, Jan 1982. 139.

Data Star (1986). "Chemical hazards in industry". Data Bank. Data Star, Plaza Suite, 114 Jermyn Street, London SW1Y 6HJ, 1986. 535.

Davenport, J.A. (1977). "A study of vapour cloud incidents". Paper 24a, A.I.Chem.E, Houston, Texas, March 1977. 175, 245, 304, 306.

Davenport, J. (1984). "A study of vapour cloud incidents - an update". I.Chem.E. Symposium Series No. 80, Rugby, 1984. 245, 246, 257, 258, 271, 279, 281, 304, 306, 308, 504.

Davis, G. (1901). "Handbook of chemical engineering". Manchester, 1901. 7.

Davis, L.N. (1979). "Frozen Fire". Friends of the Earth, San Francisco, 1979. 168, 170, 184, 186, 266.

Dawson, T.H. (1983). "Offshore structural engineering". Prentice-Hall, 1983. 89.

Deaves, D.M. (1984). "Experimental and computational assessment of full scale water spray barriers". I.Chem.E. Symposium Series No. 180, Rugby, 1984. 459.

Deighton, L. and Schwartzman, A. (1978). "Airshipwreck". Holt, Reinhart & Winston, NY, 1978. 259, 310.

Diggle, W.M. (1976). "Major emergencies in a petrochemicals complex; planning action by emergency services". I.Chem.E. Symposium Series No. 47, Rugby, 1976. 460.

DOE (1972). Department of the Environment and the Welsh Office. "Development involving the use or storage in bulk of hazardous material". Circular 1/72, HMSO, 1972. 142, 443.

Dornberger, W. (1954). "V2". Hurst & Blackett, 1954. 424.

Du Pont (1969). "Freon compounds and safety". E.I. DuPont de Nemours & Co, Wilmington, Delaware, USA, 1969. 382, 383, 385.

Dupuy, R.E. and Dupuy, T.N. (1970). "The encyclopedia of military history". Macdonald & Co, London, 1970. 327.

East Anglia Area Factory Inspectorate (1981). "British Gas Corporation methane terminal Canvey Island". Report by East Anglia Area, HM Factory Inspectorate, Sept 1981. 57.

Efficient Use of Fuel (1964). "The efficient use of fuel" 2nd Edition. HMSO, 1964. 62.

Eilers, L. (1976). "Report on Baton Rouge chlorine incident". Interview given to the newspaper "Morning Advocate", Baton Rouge, Dec 1976. 328.

Eisenberg et al (1975). Eisenberg, N.A., Lynch, C.J., Breeding, R.J. "Vulnerability model : A simulation system for assessing damage resulting from marine spills". US Coast Guard Report No. CG-D-137-75, 1975. 146, 162, 163, 165.

Elliott et al (1946). Elliott, M.A., Subel, C.W., Brown, F.N., Artz, R.T., Berger, L.B. "Report on the investigation of the fire . . . at Cleveland, Ohio . . ." US Bureau of Mines, RI 3867, Feb 1946. 168, 170, 172.

Encyclopedia Britannica (1872). Ninth Edition. Published from 1872 onwards. 259.

Encyclopedia Britannica (1951). Encyclopedia Britannica Ltd, Chicago & London, 1951. 5, 231, 319, 422, 424, 427, 498.

Enger, T. and Hartman, D. (1972). "Rapid phase transformation during LNG spillage on water". 3rd LNG Conference, Washington Hilton, Washington DC, 24th-28th Sept 1972. [Reprinted University Microfilms]. 104.

England et al (1978). England, W.G., Teuscher, L.H., Hauser, L.E., Freeman, B. "Atmospheric dispersion of natural gas vapour clouds using SIGMET . . ." Washington State University, June 1978. 104.

ENR (1919). Brown, B.S. "Details of the failure of a 90ft molasses tank". Engineering News Record, Vol.82, No.20, May 15th 1919. 384.

ENR (1920). Anon. "Boston molasses tank trial". Engineering News Record, Vol.85, No.15, October 7th 1920. 384.

Ens, H. (1986). "The Spanish camp site disaster, a new scenario". Loss Prevention Bulletin No. 68, I.Chem.E, Rugby, April 1968. 199.

Ermak et al (1982). Ermak, D.L., Chan, S.T., Morgan, D.L., Morris, L.K. "A comparison of dense gas dispersal model simulations with burro series LNG spill test results". Journal of Hazardous Materials, Vol.6, Elsevier, Amsterdam, 1982. 102, 104.

Essex County Council (1978). Castle Point District Council. "Canvey Island, advice to householders". [The booklet is undated, estimated publishing date 1978]. Chelmsford, 1978. 446.

European Community Directive (1982). "On the major accident hazards of certain industrial activities". 82/501/EEC. Official Journal of the European Communities No. L203/1, 1982. 214, 326, 342, 343, 391, 442, 446, 447.

Everard, S. (1949). "The history of the Gas Light and Coke Company". Benn, London, 1949. 236.

Farmer, F.R. (1984). "Quantification, experience and judgement". European Major Hazards Seminar, Sept 1984. OYEZ, London, 1984. 407.

Fay, J.A. (1980). "Gravitational spread and dilution of heavy vapour clouds". Proceedings 2 IAHR Symposium, Stratified Flows, 1980. 99.

Fay, J. (1981). "Scale effects in liquefied gas vapour dispersion". DOE (USA) Contract DE-AC02-77, EVO4204, Sept 1981. 100.

Ferraiolo, G. (1979). Title not known. Chimica e Industria, 61, 1979. 357, 359.

Field, P. (1982). "Dust explosions". Elsevier Scientific Publishing Co, Amsterdam, 1982. 211, 228, 230, 232, 234, 235.

Finney, D.J. (1971). "Probit analysis". Cambridge University Press, 1971. 441.

FIRE (1973). "A 'ball of fire' covered storage compound". FIRE, Dec 1973. 207.

Fire Command (1975). "Coordinated attack limits post-blast damage". Fire Command Vol.42, No.7, USA, July 1975. 304, 306.

Fire Journal (Series). "Fire Journal". National Fire Protection Association, Quincy, Mass. 534.

Fire Prevention (1973). "One dead and 83 injured in extensive fire at Ayrshire chemical complex". Fire Prevention No.103, 1973. 207.

Fire Prevention (1982). "International fire losses". Fire Prevention, Oct 1982. 112.

Fire Prevention (1982a). "Dust explosion creates havoc on the A1 trunk road". Fire Prevention, London, December 1982. 266.

Fire Prevention (1986). FPA Casebook of fires. Chemical manufacturer, Wilton, Middlesbrough, Cleveland. Fire Prevention, May 1986. 505.

Firth, G. (1977). "The development of the Northbrook chemical works, Bradford". Industrial Archaeology Review, Vol.II, No.1. Oxford University Press, Autumn 1977. 12.

Fitt, J.S. (1981). "Pressure piling, a problem for the process engineer". The Chemical Engineer, May 1981. 256.

Fletcher, B. (1982). "Sudden discharge of a superheated fluid to atmosphere". I.Chem.E. Symposium Series No. 71, April 1982. 504.

Flixborough (1975). "The Flixborough disaster - report of the court of inquiry". HMSO, 1975. 23, 63, 76, 77, 85, 86, 140, 147, 180, 182, 249, 284, 286, 289, 291, 293, 294, 295, 300, 301, 363, 399.

Flores, A. (1983). "Safety in design - an ethical viewpoint". Chemical Engineering Progress, Nov 1983. 499.

Fordham, S. (1980). "High explosives and propellants". Pergamon, 1980. 387.

FORTUNE (1986). Fortune Magazine. "How much money is your life worth?". March 3rd, 1986. 403.

Fire Protection Association (Series). "Fire Prevention". Published monthly. Fire Protection Association, Aldermary House, Queen Street, London EC4N 1TJ. 114, 534.

Fries, A.A. and West, C.J. (1921). "Chemical warfare". 1921. 314.

Fryer, L.S. and Kaiser, G.D. (1979). DENZ - a computer programme for the calculation of the dispersion of dense toxic or explosive gases in the atmosphere. SRD. R152. UKAEA, Culcheth, Warrington, 1979. 93, 99, 335.

Fu, W.B. and Eyre, J.A. (1980). Letter to the Editor, The Chemical Engineer. Aug/Sept 1980. 256.

Galloway, R.L. (1898). "Annals of coal mining and the coal trade". The Colliery Guardian Company, London, 1898. 228.

Gans, M. (1976). "The A to Z of plant start up". Chemical Engineering, Vol.83, March 15th. McGraw-Hill, 1976. 79.

Gee, D.E. (1981). Exposition at lecture at Shell research centre, Thornton, Cheshire, 12/6/81. 103.

German Daily Weather Report (1948). US Zone, July-Dec 1948. [from Meteorological Office Archives, Bracknell, Berks]. 277.

Giesbrecht et al (1981). Giesbrecht, H., Hemmer, G., Hess, K., Leuchel, W., Stoeckel, A. "Analyse der potentiellen explosionswirkung von kurz zeitig in der atmosphere freigesetzen brenngasmengen". [In German]. "Analysis of the potential explosion effect of flammable gases released into the atmosphere over a short time". Part 2, Chem.Ing.Tech.53, No.1, Weinheim, 1981. 256, 271.

Glasstone, S. (1962). "The effects of nuclear weapons". 2nd Edition. United States Atomic Energy Commission, 1962. 162.

Glasstone, S. (1964). "Die wirkungen von kernwaffen". (The effects of nuclear weapons). S. Heymann, Verlag A.G., Koln, 1964. 271.

Glasstone, S. and Dolan, P.J. (1980). "The effects of nuclear weapons". Prepared by the US Department of Defense and the US Department of Energy. Castle House Publications, Tunbridge Wells, UK, 1980. 139, 161, 165, 220, 222, 252, 441.

Goran, M. (1967). "The story of Fritz Haber". University of Oklahoma Press, Norman, USA, 1967. 109, 317.

Gordon, J.E. (1968). "The new science of strong materials". Penguin Books, London, 1968. 70.

Gough, M. (1986). "Dioxin, agent orange. The facts". Plenum Press, 1986. 507.

Grayson, J. (1975). "The Flixborough disaster. The lessons for the British labour movement". Publisher unknown, 1975. 284.

Green, A.E. and Bourne, A.J. (1981). "Reliability technology". John Wiley & Sons, 1981. 84.

Gretener, M. (1968). "Attempts to calculate the fire risk of industrial and other objects". 3rd International Fire Protection Symposium, Eindhoven, 1968. 127.

Griffith, A.A. (1921). "The phenomenon of rupture and flow in solids". Philosophical Transactions, Royal Society Vol.221, P.163, 1921. 70.

Griffiths, R.F. (1981). "Problems in the use of risk criteria". Chapter in "Dealing with risk". Manchester University Press, 1981. 39.

Griffiths, R.F. and Fryer, L.S. (1979). "World wide data on the incidence of multiple fatality accidents". Safety & Reliability Directorate, UKAEA, Culcheth, Warrington, UK, April 1979. 19.

Gugan, K. (1979). "Unconfined vapour cloud explosions". I.Chem.E, Rugby, 1979. 21, 116, 175, 211, 245, 246, 248, 255, 256, 257, 258, 279, 281, 284, 295, 304, 469, 502, 504.

Gugan, K. (1979a). "Unconfined vapour cloud explosions". Letter to Advisory Committee on Major Hazards, 24/7/79. 304, 470.

Guinness Book of Records (1971). 18th Edition, 1971. 231.

Guinness Book of Records (1983). 30th Edition, 1983. 52.

Haber, L.F. (1986). "The poisonous cloud - chemical warfare in the First World War". Clarendon Press, Oxford, 1986. 314, 317, 327, 328, 337, 342, 433.

Hake, C.N. and MacNab, W. (1892). "Explosives and their power". Translated from the french of M. Berthelot. John Murray, London, 1892. 5.

Haldane, J.B.S. (1927). "Possible worlds". Chatto & Windus, London, 1927. 317.

Haldane, J.B.S. (1938). "A.R.P" (Air Raid Precautions). Gollancz, London, 1938. 314.

Hall, L.J. (1978). "Insurance company views". International Symposium on grain elevator explosions. Vol.1. National Materials Advisory Board, Washington DC, 1978. 230.

Hall, D.J. (1979). "Comparison of a wind tunnel model of a denser than air gas with field experiments". Nato CCMS 10th International Technical Meeting, Rome, 23-26th Oct 1979. 106.

Hall, D.J. (1979a). "Further experiments on a model of an escape of heavy gas". Report LR 312 (AP), Warren Spring Laboratory, Stevenage, Herts, UK, 1979. 106.

Hall, D.J. (1982). "Wind tunnel modelling of heavy gas spills". Lecture notes. van Karman Institute for fluid mechanics, Rhode St. Genese, Belgium, March 1982. 94, 106.

Hall et al (1974). Hall, D.J., Barrett, C.F., Ralph, M.O. "Experiments on a model of an escape of heavy gas". Warren Spring Laboratory, Report CR 882 (AP), 1974. 106.

Hall et al (1982a). Hall, D.J., Hollis, E.J., Ishaq, M. "A wind tunnel model of the Porton dense gas spill field trials". Report LR 394 (AP), Warren Spring Laboratory, Stevenage, Herts, UK, 1982. 105, 106, 108.

Hallade, J. (1973). "Big Bertha bombards Paris". Parnell's history of the World Wars - The big guns. BPC Publishing, London, 1973. 422, 425.

Hammerton (1916). (Editor). "The war illustrated". Vol.3. Hammerton, London, 1916. 109.

Hanslian, R. (1937). "Der chemische krieg". Verlag von E.S. Mittler und Sohn, 1937. 328.

Hardee et al (1977). Hardee, H.C., Lee, D.O., Benedick, W.B. Combustion Science Tech. 17, 1977. 134.

Hardee, H.C. and Lee, D.O. (1975). Accident analysis and prevention. No.7, 91-102, 1975. 136.

Harris, N.C. (1981). Proof of evidence to Pheasant Wood inquiry, Sept 1981. Department of the Environment document, 1981. 327, 328, 331, 332, 333, 337, 339.

Harris, R.J. (1983). "Gas explosions in buildings and heating plant". E & F.N. Spon, London, 1983. 195, 238, 240, 241, 242, 243, 244.

Harris, N.C. (1985). "A user's view of data obtained and those which may still be required". Journal of Hazardous Materials, Vol. 11, Elsevier, Amsterdam, 1985. 105.

Harvey, J.F. (1974). "Theory of design of modern pressure vessels". Van Nostrand and Rheinhold, New York, 1974. 69.

Harvey, K. (1986). "Fires in ships 1974-1984, Part 2". Fire Prevention, May 1986. 507.

HASAWA (1974). Health & Safety at Work Act 1974. Cap. 37. HMSO, 1974. 44, 395, 445, 491.

Hasegawa, K. and Sato, K. (1978). "Study on the fire ball following steam explosion of n-pentane". Proceedings, 2nd International Symposium on Loss Prevention, Heidelberg, 1977. Dechema, Frankfurt, 1978. 129, 131, 132, 134.

Havens, J.A. (1978). "An assessment of the predictability of LNG vapour dispersal from catastrophic spills onto water". Proc. 5th Int. Symposium on the transport of dangerous goods by sea or inland waterways, Hamburg 1978. [Published under the same title, Journal of Hazardous Materials 3, 1980. Elsevier, Amsterdam, 1980]. 100.

Havens, J.A. (1982). "A review of mathematical models for prediction of heavy gas atmospheric dispersion". The Assessment of Major Hazards, April 1982. I.Chem.E. Symposium Series No. 71. I.Chem.E. Rugby, 1982. 93, 99, 100.

Havens, J.A. (1982a). "A description and computational assessment of the SIGMET LNG vapour dispersion model". Journal of Hazardous Materials 6, Elsevier, Amsterdam, 1982. 104.

Havens, J.A. (1985). "The atmospheric dispersion of heavy gases : an update". The Assessment and Control of Major Hazards. I.Chem.E. Rugby Symposium Series No. 93, 1985. 504.

Hawksley, J.L. (1986). "Unconfined vapour cloud explosions involving hydrogen rich gases". I.Chem.E. Loss Prevention Bulletin No. 68, April 1986. 505.

Hay, A. (1982). "The chemical scythe". Plenum Press, 1982. 318, 324, 349, 350, 351, 352, 353, 355, 356, 390.

Hazardous Cargo Bulletin (1980). "The safety belt syndrome". Hazardous Cargo Bulletin, Sept 1980. 238.

Health & Safety Commission (1978). Consultative Document. "Hazardous Installations (notification and survey) regulations". HMSO, 1978. 446.

Health & Safety Commission (1986). "Genetic manipulation guidelines". HSC Newsletter, June 1986. 507.

HSE (Series). Health & Safety Executive. "Prestel programme". Page guide updated quarterly. HSE Prestel Unit, HSE, Baynards House, 1 Chepstow Place, London W2 4TF. 535.

Health & Safety Executive (1976). The explosion at Appleby-Frodingham steelworks, Scunthorpe, 4/11/75. HMSO, 1976. 23, 381, 382.

Health & Safety Executive (1976a). "The storage of liquefied petroleum gas at factories". HMSO, 1976. 128, 180.

Health & Safety Executive (1977). Health & Safety Statistics 1975. HMSO, 1977. 44, 341, 387.

Health & Safety Executive (1979). "Manufacturing and service industries 1977". HMSO, 1979. 341.

Health & Safety Executive (1980). A translation of the official report of the (Italian) Parliamentary commission of enquiry. HMSO, 1980. 23, 350, 360, 363.

Health & Safety Executive (1981). Manufacturing and service industries report 1979. HMSO, 1981. 387.

Health & Safety Executive (1984). Guidance Note EH43. "Carbon monoxide". HMSO, August 1984. 339,370,371.

Health & Safety Executive (1985). "Deadly maintenance". HMSO, 1985. 87.

Health & Safety Executive (1985a). "A guide to the control of major accident hazards regulations 1985". HMSO, 1985. 141,447.

Health & Safety Executive (1985b). "The Abbeystead explosion". HMSO,1985. 270.

Health & Safety Executive (1985c). "Fires and explosions due to the misuse of oxygen". HSE Leaflet No. 8. HSE (Nominal date), 1985. 386.

Health & Safety Executive (1986). "OSH-ROM, Information on a silver platter". HSE Library & Information Services, Broad Lane, Sheffield S3 7HQ, UK, 1986. 536.

Healy, F. (1965). "Notes on the basis of outside safety distances for explosives involving the risk of mass explosion". Explosives Storage & Transport Committee of the UK, March 1965. 146,211,251,419,465,466.

Heap, D. (1982). "An outline of planning law". Sweet & Maxwell, London 1982. 443.

Heath, D.F. (1961). "Organophosphorus Poisons". Pergamon Press, Oxford, 1961. 324,343,345.

Heinrich, H.W. (1950). "Industrial accident prevention". 3rd Edition. McGraw-Hill, New York, 1950. 414.

Her Majesty's Chief Inspector of Factories (1968). Annual Report for 1967, Cmnd 3745. HMSO, 1968. 442.

Her Majesty's Chief Inspector of Factories (1975). Annual Report for 1974. HMSO, 1975. 267.

Hersh, S.M. (1968). "Chemical and biological warfare". MacGibbon & Kee, London 1968. 342,343.

Hewison, C.H. (1983). "Locomotive boiler explosions". David & Charles, Newton Abbot, 1983. 138.

High, R.W. (1968). "The Saturn fireball". Annals New York Academy of Sciences, Vol.152, Part 1, Page 441-51, 1968. 129, 130, 131, 132, 134, 136, 137, 157, 159.

High, W.G. (1980). "The design and siting of buildings to resist chemical hazards". OYEZ Conference, Waldorf Hotel, London, June 1980. 467.

High, W.G. (1982). "Compressor house explosion". Verbal presentation. UK Explosion Liaison Group, British Gas, Solihull, 29/9/82. 84, 85, 268.

High Pressure Technology Association (1975). "High pressure safety code". c/o Dr. R.A. Duckett, Dept. of Physics, University of Leeds, 1975. 69, 189.

Hill, J. (1983). "Nuclear power". In "The energy debate". Benn Technical Books, 1983. 508.

History of the Great War (1923). Medical Services, Vol. 11. HMSO, 1923. 328.

Holdgate, M.W. (1981). "Risks in the built environment". In "The assessment and perception of risk". The Royal Society, London, 1981. 403.

Holland, C. (1980). "People important in giant German firm". Bradford Telegraph & Argus, 10/4/80. 430.

Holmstedt, B. (1980). Prolegomena to Seveso. Archives of Toxicology 44, 1980. 349, 350, 352, 355, 356.

Homberger et al (1979). Homberger, E., Reggiani, G., Sambeth, J., Wipf, H.K. "The Seveso accident : Its nature, extent and consequences". Annals of Occupational Hygiene, Vol.22, 1979. 347, 349, 350, 365.

Home Office (1938). Home Office (United Kingdom) ARP Handbook No. 1. "Personal protection against gas". HMSO, 1938. 314, 321, 337, 342.

Hopkinson, B. (1915). British Ordnance Board minutes No. 13565, 1915. 220.

Houwink, R. (1976). "Sizing up science". The Scientific Book Club, London, 1976. 217.

Humberside Police (1974). "Explosion, Nypro UK Ltd chemical factory, Flixborough". Published by Humberside Police, Kingston upon Hull, 1974. 23, 182, 281, 284, 302, 304, 306, 460, 478.

Hunt, D.L.M. and Ramskill, P.K. (1985). "The behaviour of tanks engulfed in fire - the development of a computer programme". I.Chem.E. Rugby Symposium Series No. 93, 1985. 504.

Hunter, D. (1977). "George Stephenson". Quartet Books, London, 1977. 402.

Hutchings, F.R. (1978). "Service failures in the chemical industry". British Engine Insurance Ltd, Manchester. Technical Report, Volume XIII, 1978. 74, 80.

Hymes, I. (1982). "The physiological and pathological effects of thermal radiation". SRD UKAEA (Draft, unpublished), 1978. 146, 184, 186.

Hymes, I. (1985). "Phosphorus railcar incident". I.Chem.E. Loss Prevention Bulletin No. 061, Feb 1985. 389.

I.Chem.E. (1971). Institution of Chemical Engineers. "Major loss prevention in the process industries". A symposium at Newcastle upon Tyne, UK. I.Chem.E. 1971. 495.

I.Chem.E. (1974). Institution of Chemical Engineers. "Commissioning of chemical and allied plants". Unpublished but photocopies available on application. I.Chem.E. Rugby, April 1974. 79.

I.Chem.E. (1976). Institution of Chemical Engineers. "The technical lessons of Flixborough". A Symposium Dec 1975. I.Chem.E. Rugby, 1976. 284.

I.Chem.E. (1976a). Institution of Chemical Engineers. "Model form of conditions of contract for process plants suitable for reimbursable contracts". I.Chem.E. Rugby, 1976. 456.

I.Chem.E. (1978). Institution of Chemical Engineers. "Guide notes on the safe use of stainless steel in chemical process plant". I.Chem.E. Rugby, 1978. 76, 77.

I.Chem.E. (1979). Institution of Chemical Engineers. Symposium on run-away reactions, unstable products and combustible powders. N.W. Branch, I.Chem.E, Chester, March 1979. 83.

I.Chem.E. (1981). Institution of Chemical Engineers. "Model form of conditions of contract for process plant suitable for lump-sum contracts in the UK". I.Chem.E. Rugby, 1981. 455.

I.Chem.E. (1983). Institution of Chemical Engineers. "A scheme for a degree course in chemical engineering". I.Chem.E. Rugby, 1983. 494.

I.Chem.E. (1984). Institution of Chemical Engineers. "Nomenclature for hazard and risk assessment in the process industries" (Draft). I.Chem.E, Rugby, 1984. 83, 138, 312, 410.

I.Chem.E. (1985). Institution of Chemical Engineers. "Nomenclature for hazard and risk assessment in the process industries". I.Chem.E, Rugby, 1985. 34, 520.

I.Chem.E. (1985a). Institution of Chemical Engineers. "The toxic effect of chlorine". Loss Prevention Bulletin 064. I.Chem.E, Rugby, 1985. 326, 332, 371, 514.

I.Chem.E. (1985b). Institution of Chemical Engineers. "Continuing education programme". I.Chem.E, Rugby, 1985. 495.

I.Chem.E. (1985c). Institution of Chemical Engineers. "Register of process safety and loss prevention specialists serving the chemical and process industries". I.Chem.E, Rugby, 1985. 497.

I.Chem.E. (1986). Institution of Chemical Engineers. "Publications catalogue 1986". I.Chem.E, Rugby, 1986. 502.

ICI (1979). "ICI ammonia synthesis technology". Imperial Chemical Industries Ltd, Billingham, Cleveland, UK, 1979. 14.

IISS (1981). International Institute for Strategic Studies. "The military balance 1981-82". IISS, London, 1981. 403.

Inglis, C.E. (1913). "Stresses in a plate due to the presence of cracks and sharp corners". Proc. Inst. Nav. Arch, 1913. 70.

International Labour Organisation (1983). "Encyclopedia of occupational health and safety". Geneva, 1983. 314, 323, 382.

Interox (1985). "Organic peroxides, safe handling and storage". Interox Chemicals Ltd, Warrington, UK, 1985. 538.

Irving, D. (1980). "The destruction of Dresden". Futura Publications, London, 1980. [First published 1966 by William Kimber & Co Ltd]. 139, 416, 419, 424.

Jankowski, D.A. and Selover, T.B. (1986). [Editors]. "Chemical engineering data sources". A.I.Ch.E. Symposium Series Vol.82, 1986. 533.

Johnstone, E. and Thring. (1957). "Pilot plants, models and scale up methods in chemical engineering". McGraw-Hill, New York, 1957. 106, 107.

Jones, C. and Sands, R.L. (1981). "Great balls of fire". Published privately c/o 26 Exhall Green, Exhall, Coventry, UK, 1981. 184.

Jones, T.B. and Spracklen, C.T. (1974). "Ionospheric effects of the Flixborough disaster". Nature Vol.250, P.719-20, Aug 30th, 1974. 249, 287, 294.

Joschek, H.I. (1983). "Quantitative risk assessment is of little practical value to plant managers or design engineers". Plant/Operations Progress, Jan 1983. 507.

Joyner, R.E. and Durel, E.G. (1962). "Accidental liquid chlorine spill in a rural community". Journal of Occupational Medicine Vol.4, No.3, March 1962. 331.

JRUSI (1928). Journal of Royal United Services Institute. Anon. "Explosion of phosgene at Hamburg". JRUSI No.73, 1928. 337.

Kates, R.W. (1977). [Editor]. "Managing technological hazard". University of Colorado, 1977. 402.

Kauffman, C.W. (1980). "Analysis of recent explosions". Grain elevator and Processing society annual conference, Oregon, Feb 1980. 263.

Kellermann, O. (1968). "Present views on recurring inspection of reactor pressure vessels in the F.D.R". I.A.E.A. No.81, 1968. 73.

Kennett, F. (1975). "The greatest disasters of the 20th century". Marshall Cavendish, London & New York, 1975. 22, 182.

Kharbanda, O.P. (1985). "Lessons from Bhopal". In "The chemical industry after Bhopal". IBC Technical Services, 56 Holborn Viaduct, London EC1A 2EX, 1985. 373.

King, P.J. (1977). [Chairman]. "Report of the inquiry into serious gas explosions". HMSO, 1977. 238.

Kingery, C.N. and Pannill, B.F. (1962). "Peak overpressure versus scaled distance for TNT surface bursts (hemispherical charges)". BRL Report 1518. Aberdeen Proving Ground, Maryland, USA, 1962. 220, 296.

Kirk-Othmer (1964). Kirk-Othmer Encyclopedia of Chemical Technology, 2nd Edition, Vol.5. John Wiley & Sons, 1964. 153.

Kirk-Othmer (1979). Kirk-Othmer Encyclopedia of Chemical Technology, 3rd Edition, Vol.5. John Wiley & Sons, 1979. 345, 346.

Kirk-Othmer (1980). Kirk-Othmer Encyclopedia of Chemical Technology. "Explosives and propellants". 3rd Edition, Vol.9. John Wiley & Sons, 1980. 211, 212, 218, 255, 427.

Kirk-Othmer (1981). Kirk-Othmer Encyclopedia of Chemical Technology, 3rd Edition, Vol.13. John Wiley & Sons, 1981. 343, 345, 370.

Kirk-Othmer (1982). Kirk-Othmer Encyclopedia of Chemical Technology, 3rd Edition, Vol.17. John Wiley & Sons, 1982. 537.

Kirk-Othmer (1985). Kirk-Othmer Concise Encyclopedia of Chemical Technology. John Wiley & Sons, 1985. 210, 391.

Kletz, T.A. (1977). "Some myths on hazardous materials". Journal of Hazardous Materials, Vol.2, No.1, Elsevier, Amsterdam, 1977. 62.

Kletz, T.A. (1979). Private communication, 1979. 245.

Kletz, T.A. (1984). "Hazop and hazan, notes on the identification and assessment of hazards". I.Chem.E, Rugby, 1984. 410, 453, 456.

Kletz, T.A. (1985). "An engineer's view of human error". I.Chem.E, Rugby, 1985. 86, 88, 508.

Kletz, T.A. and Turner, E. (1979). "Is the number of serious accidents increasing". C.I.A, London, 1979. 313.

Kluwer-Harrap (1974). "Handbook of risk management". London, 1974. 27.

Kneebone, A. and Prew, L.R. (1974). "Shipboard jettison tests of LNG onto the sea". Fourth Int. Conference on liquefied natural gas, Algiers, June 1974. 57, 94, 101.

Kniel et al (1980). Kniel, L., Winter, O., Stork, K. "Ethylene, keystone to the petrochemical industry". Marcel Dekker, New York, 1980. 16.

Knight, R.W. (1938). "The Hindenburg accident". Dept. of Commerce, Bureau of Air Commerce, Safety & Planning Division Report No.11, Washington, Aug. 1938. 310.

Kogarko et al (1966). Kogarko, S.M., Aduskin, V.V., Lyamin, A.G. "An investigation of spherical detonations of gas mixtures". Inst.Chem.Eng. 6, P.393, 1966. 255.

Koopman et al (1980). Koopman, R.P., Bowman, B.R., Ermak, D.L. "Data and calculations on $5m^3$ LNG spill tests". Report P, liquefied gaseous fuels, Safety & Environmental Control Assessment Programme Second Status Report. US DOE/EV 0085, 1980. 57, 102.

Koopman et al (1982). Koopman, R.P. and nine co-workers. "Analysis of burro series 40 m^3 LNG spill experiments". Journal of Hazardous Materials 6, Elsevier, Amsterdam, 1982. 102, 103.

Kunreuther, H.C. and Linnerooth, J. (1983). "Risk analysis and decision processes". Spinger-Verlag Berlin, 1983. 407.

Lamb, H.H. (1964). "The English climate". English Universities Press, London, 1964. 95.

Langeveld, J.M. (1976). "Structural design of control buildings". I.Chem.E. Symposium Series No.47, 1976. 470.

Langhaar (1951). "Dimensional analysis and the theory of models". Wiley, New York, 1951. 106, 107.

Lapin, A. and Foster, R.H. (1967). "Oxygen diffusion in the atmosphere from liquid oxygen pools". Adv.Cryo.Eng.13, 1967. 101.

Lashover, J.H. (1985). "Computerised emergency system provides rapid warning of hazardous chemical leaks". Chemical Processing, Sept 1985. 487.

Lawley, H.G. (1974). "Operability studies and hazard analysis". C.E.P., April 1974. 410.

Lee, T.R. (1981). "The public's perception of risk and the question of irrationality". In "The assessment and perception of risk". The Royal Society, London, 1981. 396, 413.

Lees, F.P. (1980). "Loss prevention in the process industries". Butterworths, London, 1980. 441.

Legge, T. (1934). "Industrial maladies". Oxford Medical Publications, O.U.P., London, 1934. 314, 327, 333, 337, 341.

Leonetti, F. and Brats, G. (1966). "Report on Feyzin incident to the Ministère des Affaires Sociales". Caisse Nationale de Securité Sociale 9ème Bureau, Paris 1966. 175, 178.

Lihou, D.A. and Maund, J.K. (1982). "Thermal radiation hazard from fireballs". I.Chem.E, Rugby, Symposium Series 71, 1982. 122, 129, 137.

Little, A.D. (1979). "Experiments involving pool and vapour fires from spills of LNG on water". Report proposed for US Coast Guard No. CGD 55-79, 1979. 124, 125.

Long, A.J. (1948). "Inquiry into the fire at the premises of Alfred Harris & Co Ltd on 4th Sept, 1947". Cmd 7440. HMSO, 1948. 207.

Lonsdale, H. (1975). "Ammonia tank failure - South Africa". CEP Technical Manual Vol.17. A.I.Chem.E, New York, 1975. 335.

LPB (Series). Loss Prevention Bulletin. Six issues per year available on subscription. Institution of Chemical Engineers, Rugby, UK. 534.

Loss Prevention Bulletin (1977). "Plant commissioning hazards". No.018. I.Chem.E, Rugby, 1977. 79.

Loss Prevention Bulletin (1979). "Pressure testing". No.026. I.Chem.E, Rugby, 1979. 78.

Lowrance, W.W. (1976). "Of acceptable risk". William Kaufmann Inc, Los Altos, Cal, USA, 1976. 39.

Luckritz, R.T. (1977). "An investigation of blast waves generated by constant velocity flames". University of Illinois, Ph.D. Thesis, 1977. 253.

Malpas et al (1984). Malpas, R., Anderson, T., Mitchell, J.M. "The elegant use of oxygen". Royal Society of Chemistry, London, 1984. 385.

Manderstam (1974). "Report into the causes of the Flixborough disaster on behalf of Nypro (UK) Ltd". L.H. Manderstam & Partners (UK) Ltd. Flixborough Inquiry Document, 1974. 478.

Manufacturing Chemists Association (1978). "Siting and construction of new control houses for chemical manufacturing plants". Safety Guide SG22, Washington DC, 1978. 471.

Marsh and McLannan (1985). "One hundred largest losses, a thirty year review". Marsh & McLennan, Chicago, May 1985. 268.

Marshall, V.C. (1974). Nypro (Flixborough) Explosion. A preliminary report to the trade unions, June 1974. 286.

Marshall, V.C. (1975). "The strategic approach to safety". The Chemical Engineer, April 1975. 26, 453.

Marshall, V.C. (1975a). [Editor]. "Comminution". Institution of Chemical Engineers, Rugby, 1975. 70.

Marshall, V.C. (1975b). "Process plant safety - a strategic approach". Chemical Engineering, Dec 22, 1975. 462, 469.

Marshall, V.C. (1976). "Seveso and Flixborough". The Chemical Engineer, Oct. 1976. 256.

Marshall, V.C. (1976a). "The siting and construction of control buildings". I.Chem.E. Rugby, Symposium Series No. 47, 1976. 256, 430, 464, 474.

Marshall, V.C. (1976b). "Safety of control rooms - a strategic view". Institut Belge de Petrole, 30th March 1976. 469.

Marshall, V.C. (1977). "How lethal are explosions and toxic escapes ?". Chemical Engineer, Aug 1977. 21, 313, 327, 328, 415, 416, 417, 420, 433, 434.

Marshall, V.C. (1977a). "Safety in the laboratory". Chemistry & Industry, 2nd July 1977. 45.

Marshall, V.C. (1977b). "Chemical conurbations, the domino danger". "Chemical engineering in a hostile world". Birmingham, 20th June 1977. 109, 129, 133, 134, 136, 137, 147, 149, 175, 245.

Marshall, V.C. (1978). "Physical implications of major hazards". Proc. Seminar on Major Chemical Hazards, April 1978. Published by UKAEA, Harwell, Oxon OX11 0RA, 1978. 417, 422, 426.

Marshall, V.C. (1979). "The hazards, their probable outcome and frequency, bulk storage and handling of flammable gases and liquids". OYEZ Conference Intelligence Report, London, 18/4/79. 8, 19, 256, 257, 313.

Marshall, V.C. (1979a). "Disaster at Flixborough". A. Wheaton & Co. Ltd, Exeter, 1979. 22, 182, 284, 460.

Marshall, V.C. (1980). "The Decatur anomaly". Chemical Engineer, Feb 1980. 249, 281, 302, 304.

Marshall, V.C. (1980a). "Seveso, an analysis of the official report". The Chemical Engineer, July 1980. 350, 357, 362.

Marshall, V.C. (1980b). "The public, the mass media and the process industries". Eurochem 80. EFCE event. No. 239, 1980. 400.

Marshall, V.C. (1980c). "Historical and theoretical approaches to the prediction of hazard and risk". EFCE, 3rd Int. Symposium on Loss Prevention and Safety Promotion in the Process Industries, Basle, Sept 1980. 409.

Marshall, V.C. (1981). "Hazard . . . risk . . . which?". Health & Safety at Work, London, March 1981. 30.

Marshall, V.C. (1982). "The equivalence of major hazards". Unpublished internal paper of Advisory Committee on Major Hazards, May 1982. 415, 451.

Marshall, V.C. (1983). Letter to Chemistry and Industry, 1983. 318.

Marshall, V.C. (1983a). "Flixborough, Seveso and after". Conference transcript, OYEZ Publishers, London, 1983. 415, 442.

Marshall, V.C. (1983b). "A critique of the EC Directive on major accident hazards". I.Chem.E, Rugby, Symposium Series No. 79, 1983. 451.

Marshall, V.C. (1984). "Living with risk". Seminar on European Major Hazards. OYEZ, London, Sept 1984. 400.

Marshall, V.C. (1984a). "Shaping a strategy of control". The Chemical Engineer, June 1984. 447.

Mason, J.K. (1978). [Editor]. "The pathology of violent injury". Edward Arnold, London, 1978. 142, 145.

Mathieu, G. (1978). Private communication drawing on official sources, 8/1/78. 425.

Matsui, H. and Lee, J.H. (1978). In 17th Symposium (International) on combustion, 1978. 258.

May, G. (1973). "Chloracne from the accidental production of tetrachlorodibenzo-dioxin". British Journal of Industrial Medicine Vol.30, 1973. 347, 349, 355, 356, 361, 365.

Mecklenburgh, J.C. (1973). "Plant layout". Leonard Hill in Association with I.Chem.E., Aylesbury, Bucks, 1973. 127, 457.

Mecklenburgh, J.C. (1982). "Hazard assessment of layout". I.Chem.E, Rugby, Symposium Series No. 71, 1982. 457.

Merriam-Webster (1958). "The Merriam-Webster pocket dictionary". Cardinal Books Inc, New York, 1958. 231.

Meyer, P. (1980). "Luftschiffe" [In German]. Verlag Wehr & Wissen, Koblenz, 1980. 259.

Microinfo (1982). Microinfo Ltd. "Health & Safety". Vol.5, No.11, Page 1, November 1982. Alton, Hampshire, England, 1982. 116.

Middlebrook, M. (1984). "The battle of Hamburg". Penguin Books, 1984. 139.

Millbank (1975). "Incident rendered safe" (Film). Millbank Films, 1 Adam Street, London, 1975. 453.

Millbank (1985). "Incident under control" (Film). Millbank Films, 1 Adam Street, London, 1985. 453.

Milnes, M.H. (1971). "Formation of 2,3,7,8-tetrachlorodibenzodioxin by thermal decomposition of sodium 2,4,5-trichlorophenate". Nature, Vol.232, August 1971. 347, 349, 351, 356, 361, 362.

Ministrie van Sociale Zaken (1977). Directoraat General van de Arbeid van het Ministrie van Sociale Zaken. "Explosiebestendige kontrole-gebouwen in de procesindustrie". The Netherlands, October 1977. 471.

Mitchell, F. (1980). Private communication, June 1980. 14.

Mizner, G.A. and Eyre, J.A. (1982). "Large scale LNG and LPG pool fires". I.Chem.E, Rugby, Symposium Series No. 71, 1982. 119, 122, 123, 124, 125, 126, 130, 154.

M. & M. (Series). "One hundred largest losses, a thirty year review of property damage losses". M. & M. Protection Consultants, 222 South Riverside Plaza, Chicago, USA, 1986. 508, 534.

Monji, N. and Businger, J.A. (1972). "Stability dependence of temperature, humidity and vertical wind velocity variances in atmospheric surface layer". J.Met.Soc. Japan, Ser.II, Vol.50, 1972. 98.

Moodie, K. (1982). "Assessment of failure sequence for volatile liquids stored without pressure relief, with particular reference to liquid propane". HSE, R. & L.S. Division, Red Hill, Sheffield, 1982. 184, 189, 196.

Moodie et al (1985). Moodie, K., Billinge, K., Cutler, D.P. "The fire engulfment of LPG storage tanks". I.Chem.E, Rugby, Symposium Series No.93, 1985. 504.

Moorhouse, J. (1982). "Scaling criteria for pool fires derived from large scale experiments". I.Chem.E, Rugby, Symposium Series No.71, 1982. 122, 123, 124, 125.

Moorhouse, J. and Pritchard, M.J. (1982). "Thermal radiation hazards from large pool fires and fireballs - a literature review". I.Chem.E, Rugby, Symposium Series No. 71. Published as an addendum to "The assessment of major hazards". I.Chem.E, Rugby, 1982. 122, 124, 126, 129, 137, 150, 155, 157, 161, 162, 441.

Moo-Young, M. (1985). "Comprehensive biotechnology". Pergamon, 1985. 392.

Morozzo, Count. (1785). "Account of a violent explosion in a flour warehouse in Turin". The Repertory of Arts & Manufacturing. 2, 416, (1785). 228.

Morris, J. (1969). "The German air raids on Great Britain 1914-18". H. Pordes, London, 1969. 259, 424.

Morton, F. (1970). "Report of the inquiry into the safety of natural gas as a fuel". HMSO, 1970. 237.

MSA (1975). "Experiments with freon 12 and chlorine". Netherlands Ministry of Social Affairs, Voorburg, April 1975. 94.

Muir Wood, R. (1985). "Problems of seismic hazard at British and mediterranean petrochemical installations". I.Chem.E, Rugby, Symposium Series No.93, 1985. 89.

Munday, G. (1975). "Unconfined vapour explosion". The technical lessons of Flixborough. Symposium organised by the Institution of Chemical Engineers, Nottingham University, 16/12/75. 298.

McAdams, W.H. (1954). "Heat transmission" 3rd Edition. McGraw-Hill, New York, 1954. 179.

McGraw-Hill Encyclopedia of Science and Technology (1982). 5th Edition. McGraw-Hill, New York, 1982. 141, 210.

McKee, A. (1972). "Black Saturday". Nel Books, London, 1972. 206.

McKee, A. (1982). "Dresden 1945". Souvenir Press, London, 1982. 139.

McMullen, G. (1976). "A review of the May 11th ammonia truck accident". City of Houston Health Dept, Houston, Texas, 1976. 335.

McQuaid, J. (1979). "The behaviour of gases released to the atmosphere; a progress report on field trials commissioned by the HSE 1976-78". Safety Engineering Laboratory, Sheffield, UK, 1979. 93, 94, 98, 99, 104.

McQuaid, J. (1981). Pheasant Wood inquiry. Proof of evidence. HSE, 15/9/81. 99.

McQuaid, J. (1982). "Future directions of dense gas dispersal research". Journal of Hazardous Materials 6, Elsevier, Amsterdam, 1982. 93, 94, 99.

McQuaid, J. (1982a). "The HSE large scale experiments". von Karman Institute for Fluid Mechanics, Belgium. Lecture series 1981-82. 93, 94, 99.

McQuaid, J. (1982b). "The HSE large scale experiments". Lecture course, von Karman Institute, Belgium, March 1982. Safety Engineering Laboratory, HSE Sheffield, 1982. 99.

McQuaid, J. (1982c). Private communication on lethal dosage envelopes for chlorine. HSE Safety Engineering Laboratory, Sheffield, 18/2/82. 434.

McQuaid, J. (1985). "Heavy gas dispersal trials at Thorney Island". University of Sheffield, April 1984. Journal of Hazardous Materials Vol.11, Elsevier, Amsterdam, June 1985. 105.

Nash, J.R. (1976). "Darkest hours". Nelson Hall, Chicago, 1976. 4, 5, 19, 41, 43, 168, 170, 224, 225, 236, 415.

National Fire Protection Association (1957). "Report of important dust explosions". National Fire Protection Association, Boston, Mass, 1957. 5, 21, 228, 235.

National Fire Protection Association (1985). "Large fire losses". Fire Journal, November 1985. 504.

National Transportation Safety Board (1969). "Southern railway company train derailment with fire and explosion". Laurel, Mississipi 25/1/69. Report dated 6/10/69. 189.

National Transportation Safety Board (1972). Pipeline accident report. Phillips Pipeline Company propane gas explosion, Franklin County, Missouri, Dec 9th 1970. NTSB Report No. PAR 72/1, Washington, DC, 1972. 261, 279, 280.

National Transportation Safety Board (1973). "Multiple vehicle collision followed by propylene cargo-tank explosion, New Jersey Turnpike Exit 8, 21/9/72". NTSB-HAR-73-4, 1973. 189.

National Transportation Safety Board (1975). "Hazardous materials accident in the railroad yard of the Norfolk and Western Railway at Decatur, Illinois, July 19th, 1974". NTSB Report RAR-75-4, Washington, DC, 1975. 304, 308.

National Transportation Safety Board (1976). "Surtigas S.A. tank, semi-trailer overturn . . . Eagle Pass, Texas, 29/4/75". NTSB Report HAR-76-4, Washington, DC, 1976. 189.

National Transportation Safety Board (1980). "Collision of Peruvian freighter . . . and US butane barge . . . Good Hope, Louisiana, Aug 30th 1979". NTSB Report HAR-80-7, Washington, DC, 1979. 147, 148.

Newland, D.E. (1975). "Report on an investigation of possible causes of failure of the 20 inch by-pass assembly at Flixborough". Report presented in evidence to the Flixborough inquiry, 1975. 291.

New York Times (1985). "The Bhopal disaster - how it happened". New York, 28/1/85. 372, 373, 377, 435.

New York Times (1986). Report Friday, 11th July, 1986. 507.

Nichols, R.W. (1969). [Chairman]. "Report of the committee of enquiry on pressure vessels". HMSO, 1969. 67, 69.

NIHHS (1983). Health & Safety Commission. "The notification of installations handling hazardous substances regulations 1983". HMSO, 1983. 445.

NIOSH (1978). National Institute for Occupational Safety and Health. 1978 Registry of toxic effects of chemical substances. US Dept of Health, Education and Welfare, Cincinnati, Ohio, 1979. 314, 318, 319, 323, 326, 333, 337, 339, 342, 343, 390.

NIOSH (1982). National Institute for Occupational Safety and Health. 1982 Registry of toxic effects of chemical substances. US Dept of Health, Education and Welfare, Cincinnati, Ohio, 1982. 371.

NIOSH/OSHA (1978). US Dept of Health and Human Services and US Dept of Labor. "Pocket guide to chemical hazards". US Govt. Printing Office, Washington, 1978. 314, 370, 371, 388.

NMSA (1975). "Experiments with freon 12 and chlorine". Netherlands Ministry of Social Affairs, Voorburg, April 1975. 97, 102.

OPCS (1979). Office of Population Censuses and Surveys. Series DH4, No.5, Table 4. HMSO, 1979. 341.

Otway, H. (1985). "Regulations and risk analysis". In "Regulating industrial risks". Butterworths, 1985. 505.

OYEZ (1981). "Design and siting of buildings to resist explosions and fire". OYEZ Publishing Ltd, London, 1981. 470, 471.

Ozisik, M.N. (1985). "Heat transfer, a basic approach". McGraw-Hill, New York, 1985. 56.

Palmer, K.N. (1973). "Dust explosions and fires". Chapman & Hall, London, 1973. 5, 211, 230, 232, 234, 235.

Pasquill, F. (1961). "The estimation of the dispersion of windborne material". Met. Mag. 90, P.33-39, 1961. 96.

Pasquill, F. and Smith, F.B. (1983). "Atmospheric diffusion". 3rd Edition. Ellis Horwood, UK, 1983. 95, 96.

Patty (1978). Patty's industrial hygiene and toxicology Vol.1, 2nd Edition. Wiley, 1978. 314.

Peabody Holmes (1978). "Safety with inert gas". Publication 61, Holmes Oil Fired Inert Gas Generators. Peabody Holmes Ltd, Turnbridge, Huddersfield HD1 6RB, UK, 1978. 242.

Pergamon (1986). "InfoLine data base". Pergamon InfoLine Ltd, 12 Vandy Street, London EC2A 2DE. Pergamon InfoLine Inc, 1340 Old Chain Bridge Road, McLean, VA 22101. Pergamon Press of Canada, Suite 104, 150 Consumers Road, Willowdale, Ontario, Canada M2J 1P9. 1986. 535.

Pernis (1968). "Rapport over oorzaak explosie bij Shell Pernis 20/1/69". (In Dutch). Ministerie van Sociale Zaken en Volksgezondheid, The Hague, 1968. 213.

Perry, R.H. and Chilton, C.H. (1973). "Handbook of chemical engineering". 5th Edition. McGraw-Hill, 1973. 16, 54, 58, 62, 63, 69, 107, 151, 294, 382.

Phillips, H. (1981). Comments on Gugan's book "Unconfined vapour cloud explosions". The Chemical Engineer, June 1981. 253, 256.

Physical Properties Data Service (1984). I.Chem.E, Rugby, 27/7/84. 276.

Physical Properties Data Service (1986). "Density of propene (propylene)". I.Chem.E, Feb 1986. 199.

Physical Properties Data Service (1986a). "Information Pack". I.Chem.E, 165 Railway Terrace, Rugby, CV21 3HQ, UK, 1986. 536.

Picknett, R.G. (1978). "Field experiments on the behaviour of dense clouds". Report No. Ptn/IL 1154/78/1. Chemical Defense Establishment, Porton. HSE, Sheffield, 1978. 94, 104.

Pietersen, C.M. (1985). "Analysis of the LPG disaster in Mexico City". Loss Prevention Bulletin No.64, I.Chem.E, Rugby, Aug 1985. 202.

Pikaar, M.J. (1984). "Unconfined vapour cloud dispersion and combustion". An overview of theory and experiments. I.Chem.E, Rugby, Symposium Series NO.80, 1984. 248, 254.

Poblete et al (1984). Poblete, B.R., Lees, F.P., Simpson, G.B. "The assessment of major hazards - estimation of injury and damage round a hazard source". Journal of Hazardous Materials 9, 1984. 441.

Powers, G.J. and Tompkins, F.C. (1973). "Synthesis strategy for fault trees in chemical processing systems". CEP Loss Prevention. A.I.Chem.E, Vol.8, 1973. 409.

Prentiss, A.M. (1937). "Chemicals in war". McGraw-Hill, New York, 1937. 314, 319, 321, 323, 337, 342, 433, 434.

Preston et al (1972). Preston, S.H., Keyfitz, N., Shoen, R. "Causes of death - life tables for national populations". Seminar Press, New York & London, 1972. 403, 405.

Price, D.J. and Brown, H.H. (1922). "Dust explosions". NFPA, Boston, 1922. 235.

Puttock et al (1982). Puttock, J.S., Blackmore, D.R., Colenbrander, G.W. (1982). "Field experiments on dense gas dispersion". Journal of Hazardous Materials 6, Elsevier, Amsterdam, 1982. 100, 101, 103, 104.

Raj, P.P.K. (1977). "Calculation of thermal hazards from LNG fires - a review of the state of the art". American Gas Association transmission conference, St. Louis, USA, 18/5/77. 119.

Raj et al (1979). Raj, P.P.K., Moussa, A.N., Aravamudan, K. "Experiments involving pool and vapour fires from spills of LNG on water". US Coast Guard Report CG-D-55-79, 1979. 159, 162, 163.

Ramachandran (1982). "A review of mathematical models for assessing fire risk". Fire Prevention No. 149, Fire Protection Association, London, 1982. 127.

Rasbash et al (1956). Rasbash, D.J., Rogowski, Z.W., Stark, G.W.P. Fuel 35 (1) 94, 1956. 157.

Rasmussen, N. (1975). "Reactor safety study - an assessment of accident risks in US commercial nuclear power plants". WASH-1400. Nuclear Regulatory Commission, Washington DC, Oct 1975. 32, 37.

Read, J. (1942). "Explosives". Pelican Books, London, 1942. 141, 216, 454.

Reider et al (1965). Reider, R., Otway, H.J., Knight, H.T. "An unconfined, large volume hydrogen/air explosion". Pyrodynamics 2, Gordon & Breach Scientific Publishers, 1965. 258.

Rice, A.P. (1982). "Seveso accident - dioxin". Chapter in Hazardous Materials Spills Handbook. McGraw-Hill, New York, 1982. 349, 350, 351, 352, 359, 366, 368.

Richardson, L.F. (1920). "The supply of energy to and from atmospheric eddies". Proc. Royal Soc. A97, P.354, 1920. 107.

Richmond et al (1968). Richmond, D.R., Damon, E.G., Fletcher, E.R., Bowen, I.G., White, C.S. "The relationship between selected blast wave parameters and the response of mammals exposed to air blast". Annals of the New York Academy of Sciences, Vol.152, Art.1, 1968. 441.

RMC (1986). "Off-line data base". RM Consultants, Station Road, Didcot, Oxon, UK, 1986. 536.

Robens, A. (1972). "Safety and health at work". Report of the Committee 1970-72, Chairman Lord Robens. HMSO, London, 1972. 395, 442, 443, 445.

Roberts, A.F. (1974). "A survey of certain aspects of the blast damage". A Report given in evidence to the Flixborough court of inquiry, 1974. 295.

Roberts, A.F. (1980). "The collection of data from chemical plant incidents as an aid to the improvement of hazard analyses techniques". 3rd Int. Symposium on Loss Prevention, Basle, Sept 1980. 24.

Roberts, A.F. (1982). "Thermal radiation hazards from releases of LPG from pressurised storage". Fire Safety Journal, Elsevier 6, Amsterdam, 1982. 129, 131, 157.

Roberts, A.F. (1982a). "The effect of conditions prior to loss of containment on fireball behaviour". I.Chem.E, Rugby, Symposium Series No. 71, 1982. 129, 132, 133, 134, 136, 137, 151, 155, 157, 162, 175.

Robertson, R.B. (1976). "Spacing in chemical plant design against loss by fire". I.Chem.E, Rugby, Symposium Series No.47, 1976. 122.

Robinson, C.S. (1944). "Explosives, their anatomy and destructiveness". McGraw-Hill, New York, 1944. 146, 209, 210, 211, 217, 251, 419.

Robinson, J.P. (1968). "Chemistry in warfare". Penguin Science Survey 1968 - Physical Sciences. Penguin Books, 1968. 314.

Rose, S. (1968). [Editor]. "Chemical and biological warfare". Harrap, London, 1968. 314, 343, 345.

Rothschild, Lord (1978). "Risk". Richard Dimbleby Lecture "The Listener" 30/11/78. 403.

Rowe, W.D. (1977). "An anatomy of risk". Wiley, New York, 1977. 39.

Royal Army Medical Corps. (1925). "Training". HMSO, 1925. 314, 342.

Royal Society (1981). "The assessment and perception of risk". Proceedings of a Conference organised by Sir Frederick Warner. The Royal Society, London, 1981. 39, 396, 498.

Royal Society (1983). "Risk assessment - a study group report". The Royal Society, London, 1983. 396, 407, 498.

Royal Society for the Prevention of Accidents (1978). "Care in the home". Royal Society for the Prevention of Accidents, Birmingham, UK. Oct 1978. 44.

RSC (Series). "Chemical hazards in industry". Published monthly. The Royal Society of Chemistry, The Distribution Centre, Blackhorse Road, Letchworth, Herts SG6 1HN. 534, 536.

Rudd, D.F. and Watson, C.C. (1968). "Strategy of process engineering". John Wiley, New York, 1968. 457.

Ruhemann, M. (1949). "The separation of gases". Clarendon Press, Oxford, 2nd Edition, 1949. 8.

Rulkens et al (1984). Rulkens, P.F.M., Bongers, A.J.P., ten Brink, G.P., Guldemond, C.P. "The application of gas curtains for diluting flammable clouds". I.Chem.E. Symposium Series No. 80, 1984. 459.

Sachs, R.G. (1944). "The dependence of blast on ambient pressure and temperature". BRL Report No.466, Aberdeen Proving Ground, Maryland, 1944. 253.

Sachs, P. (1972). "Wind forces in engineering". Pergamon Press, 1972. 89.

Sadee et al (1977). Sadee, C., Samuels, D.E., O'Brien, T.P. "The characteristics of the explosion of cyclohexane at the Nypro (UK) Flixborough plant on 1st June 1974". Journal of Occupational Accidents, P.203 et seq, Elsevier, Amsterdam, 1977, 253, 284, 295, 298, 300, 478.

Safety at Sea (1980). "Collision and fire on the Mississipi". Aug 1980. 147.

Samuels, D.E.J. and O'Brien, T.P. (1974). "An estimate of the energy of the explosion at Nypro". Report made available to the Flixborough court of inquiry, 1974. 294, 295.

Sandermann et al (1957). Sandermann, W., Stockmann, H., Casten, R. "Uber die pyrolyse des pentachlorophenols". Chem.Ber.90, 1957. 351.

Sartori, M. (1939). "The war gases". 1939. 314.

Sax, N.I. (1984). "Dangerous properties of industrial materials". 6th Edition. Reinhold, New York, 1984. 388.

Schardin, H. (1954). Ziviler Luftshutz (Civil Air Defence) No.12, S291-3, 1954. 271.

Schlichting, H. (1968). "Boundary-layer theory". 6th Edition. McGraw-Hill, New York, 1968. 106, 107.

Science Data Book (1971). Oliver & Boyd, Edinburgh, 1971. 326.

Science and Invention Encyclopedia (1977). H.S. Stuttman Inc, Westport, Conn, USA, 1977. 231.

Scilly, N. (1978). Report of incident at "Los Alfaques" camping site, San Carlos de Rapita 11/1/78. HSE (internal), 1978. 147, 184, 186, 187, 194, 198.

Scilly, N. (1982). "View factors and the shape of fireballs". Private communication, 1982. 157.

Seifert et al (1984). Seifert, H., Maurer, B., Giesbrecht, H. "Steam curtains - effectiveness and electrostatic hazards". I.Chem.E, Rugby, Symposium Series No.80, 1984. 459.

Seveso (1978). Final Report of the Italian Parliamentary commission of inquiry into the Seveso disaster. Rome, 1978. 347, 349, 350, 352, 353, 355, 356, 361, 365, 366, 368.

Shaw, P. and Briscoe, F. (1978). "Evaporation from spills of hazardous liquids". UKAEA, SRD R100, Risley, UK, May 1978. 333.

Shell Research (1981). "Refrigerated gas safety, spill tests of LNG and refrigerated liquid propane". Shell Research Ltd, Shell Centre, London, 8/4/81. 104.

Shorter Oxford English Dictionary (1956). 3rd Edition. Clarendon Press, Oxford, 1956. 209.

Shreve, R.N. and Brink, J.A. (1977). "Chemical process industries". McGraw-Hill, New York, 1977. 130, 216, 258.

Simmons et al (1974). Simmons, J.A., Erdmann, R.C., Naft, B.N. "Risk of catastrophic spills of toxic chemicals". Report No. UCLA-Eng-7425, May 1974. 437.

SIPRI (1971). Stockholm International Peace Research Institute. Vol.1. The rise of chemical and biological weapons. Humanities Press, New York, 1971. 323, 326, 337, 343, 345.

Skandia (1985). "Bleve". Skandia International Insurance Corporation, 10350 Stockholm, Sweden. 1985. 202.

Slater, D.H. (1978). "Vapour clouds". Chemistry & Industry, 6th May 1978. 175, 245.

Slater, D.H. (1985). "Risk assessment in practice". Conference at the Administrative Staff College of India, Hyderabad, 27th Oct-1st Nov 1985. Published by Technica, 7 Tavistock Square, London WC1H 9LT, 1985. 373, 378.

Slovic et al (1981). Slovic, P., Fischhoff, B., Lichtenstein, S. "Perceived risk; psychological factors and social implications". In "The assessment and perception of risk". The Royal Society, London, 1981. 398.

Smeaton, J. (1759). "An experimental investigation concerning the natural powers of water and wind". Phil.Trans.Roy.Soc, London, 1759. 89.

Smith, S. (1945). "Forensic medicine". J. & A. Churchill Ltd, London, 1945. 317.

Smith, W. (1982). Private communication, June 1982. 65.

Smith, T.A. (1985). "An analysis of a 100 tonne storage vessel". The Assessment and Control of Major Hazards. I.Chem.E, Rugby, Symposium Series No. 93, 1985. 504.

Smith, T.A. and Warwick, R.G. (1974). "The second survey of defects in pressure vessels . . ." SRD R30, UKAEA, Warrington, UK, 1974. 73.

Smith, T.A. and Warwick, R.G. (1981). "A survey of defects in pressure vessels in the UK for the period 1962-78 and its relevance to nuclear primary circuits". SRD R203, UKAEA, Warrington, UK, 1981. 71.

Spanish Ministry of Justice (1982). Judgement No. OC.8177711, Tarragona 27/1/82. HSE Translation No.2G, May 1982. 184, 186, 187, 188, 195, 196.

Spence, A. (1976). Private communication, Imperial War Museum, 27/9/76. 109.

SRD (1986). Safety & Reliability Directorate. "MHIDAS data base". SRD, UKAEA, Wigshaw Lane, Culcheth, Warrington WA3 4NE, 1986. 536.

Stahl et al (1949). Stahl, W.R., Strassmann, F., Richard, G. "Katastrophe in der BASF vom 28 Juli 1948" (in German). Mediation Court of Land Rheinland-Pfalz, Branch office in Neustadt, Jan 1949. 5, 271, 274.

Steere, N.V. (1977). [Editor]. "CRC handbook of laboratory hazards". CRC Press, Cleveland, Ohio, 1977. 388.

Stehl et al (1973). Stehl, R.H., Papenfuss, R.R., Bredeweg, R.A., Roberts, R.W. "Stability of pentachlorophenol and chlorinated dioxins to sunlight, heat and combustion". In Chlorodioxins, Origin & Fate, American Chemical Society, 1973. 351,

Stern (1978). "Das Inferno". Stern Magazine, Hamburg, Issue of 20th July 1978. 184, 191.

Stevens, J.J. (1980). "What happened at Seveso". Chemistry & Industry, 19th July 1980. 357, 359, 361, 362.

Stinton, H.G. (1978). "Liquid propylene tanker disaster 11/7/78". Hampshire Fire Brigade, 1978. 146, 147, 184, 186, 187, 189, 202.

Stinton, H.G. (1981). "Liquid gas explosions". Hazardous Cargo Bulletin, April 1981. 137.

Strehlow, R.A. (1972). "Unconfined vapour explosions - an overview". 14th Int. Symposium on Combustion, Pennsylvania State University, August 1972. 170, 175, 244, 245, 252, 261.

Strehlow, R.A. and Baker, W.E. (1975). "The characterisation and evaluation of accidental explosions". NASA Report No. CR-134779, Technical Information Service, Springfield, USA, 1975. 211, 245.

Strelzoff, A. (1981). "Technology and manufacture of ammonia". John Wiley, 1981. 12.

Stull, D.R. (1977). "Fundamentals of fire and explosion". A.I.Chem.E, Monograph Series No.10, Vol.73, New York, 1977. 116, 210, 214, 215, 217, 230, 427, 538, 539.

Sunday Star (1979). "The week they closed Mississauga". Special section of edition of Sunday, 18th Nov, 1979. 22, 332.

Sutton, O.G. (1953). "Micrometeorology". McGraw-Hill, New York, 1953. 95, 96.

Sutton, O.G. (1974). "The weather". Teach Yourself Books, London, 1974. 95, 96.

Taylor, H.D. (1975). "Flixborough, the implications for management". Keith Shipton in association with Woodhead Faulkner Ltd, Cambridge, 1975. 284.

Teddington (1974). Teddington Bellows Ltd. "Designers guide to Teddington Bellows". Pontardulais, Swansea, 1974. 85, 289.

Telfer, R.J.G. (1980). "Petrochemical investment in the developing world - its impact on Western Europe". Chemistry & Industry No.1, 1980. 6.

Theophanous, T.G. (1981). "A physicochemical mechanism for the ignition of the Seveso accident". Nature, Vol.291, 25/6/81. 349, 350, 361, 362.

Thielsch, H. (1965). "Defects and failures in pressure vessels". Reinhold Publishing Corp, New York, 1965. 69, 75, 173.

TNO (1985). TNO Department of Industrial Safety. "LPG - a study". P.O.Box 342, 7300 AH Apeldoorn, The Netherlands, 1985. 202.

TNO (1985a). TNO Netherlands organisation for applied scientific research. "A survey of industrial accident data bases". TNO, The Hague, Netherlands, 1985. 533.

TNO (1986). TNO Department of industrial safety. Data base FACTS. TNO, Apeldoorn, The Netherlands, 1986. 536.

Trevethick, R.A. (1973). "Environmental and industrial health hazards". Heinemann Medical Books, London, 1973. 324, 333, 337, 388.

Tucker, D.M. (1975). "The explosion and fire at Nypro (UK) Ltd, Flixborough on 1/6/74". Symposium "The technical lessons of Flixborough" 16/12/75, University of Nottingham. Organised by I.Chem.E. (Nottingham group), 1975. 182.

Tucker, D.M. and Webb, J.S. (1975). "Explosion and fire Nypro (UK) Ltd, Flixborough". Preliminary report, Fire Research Station, Borehamwood, Herts, 1975. 182, 284, 295, 296, 474, 478.

Turner, J.S. (1973). "Buoyancy effects in fluids". Cambridge University Press, 1973. 105, 108.

UEG (1983). Research and Information Group for the Underwater and Offshore engineering industries. "Thermal stress on divers in oxy/helium environment". UEG Technical Note No.25. UEG, 6 Storey's Gate, London SW1P, 1983. 505.

UK Fire Statistics (Series). "UK fire statistics, statistical analysis of reports of fires attended by fire brigades". Home Office, London. 34.

Ulden, van, A.P. (1974). "On the spreading of a heavy gas near the ground". First Int. Loss Prevention Symposium, "Loss prevention and safety promotion in the process industries". The Hague/Delft, May 1974. 94, 97, 99, 102.

Union Carbide (1985). Union Carbide Corporation. Bhopal methyl isocyanate incident investigation. Team report. Danburg, Connecticut, March 1985. 370, 372, 373, 375, 378.

Union Carbide (1985a). Presentation by R. Van Mynen, Corporate Director for Safety and Health for Union Carbide. Press Conference, March 20th 1985. 377.

Uranov, E.B. and Chapman, D.R. (1964). "A dictionary of science". Revised by Isaacs A. Penguin Reference Books, 1964. 209.

URG (1985). Union Research Group. "The role of management practices in the Bhopal gas leak disaster". Bombay, India, 1985. 377, 378.

USIS (1968). United States Information Service. "The age of space". USIS, London, 1968. 132.

Vale, J.A. and Meredith, T.J. (1981). "Poisoning, diagnosis and treatment". Update Books, London, 1981. 314, 339.

Vance, R.W. (1963). "Cryogenic technology". John Wiley, New York, 1963. 56.

van Ulden. See under **Ulden.**

VDI (1979). VDI Richlinien. "Pressure release of dust explosions". VDI Verlag Dusseldorf, 1979. 234.

Ventry, Lord and Kolesnik, E.H. (1976). "Janes pocket book of airships". Collier Books, New York, 1976. 259.

Verkade, M. and Chiotti (1978). "Literature survey of dust explosions in grain handling facilities; causes and prevention". Iowa State University, Iowa, USA 50011, 1978. 230, 234, 235.

Waite et al (1985). Waite, P.J., Sylvester-Evans, R., Hancock, P.L. "Risk assessment; prediction and reality". In "The chemical industry after Bhopal". IBC Technical Service, 56 Holborn Viaduct, London EC1A 2EX, 1985. 372.

Waltham, T. (1978). "Catastrophe, the violent earth". Macmillan, 1978. 22, 89.

Warner, F.E. (1976). "Engineering safety and the environment". Eleventh Maurice Lubbock Memorial Lecture 6/5/76. Exchange Publications, Poland Street, London, 1976. 173, 257.

Warwick, R.G. (1977). "Steam boilers and pressure vessels". In "Industrial safety handbook". McGraw-Hill, Maidenhead, 1977. 408.

Weast, R.C. (1981). [Editor]. "Handbook of chemistry and physics". 62nd Edition. CRC Press, Cleveland, Ohio, 1981. 153, 255, 326.

Weast, R.C. (1985). [Editor]. "Handbook of chemistry and physics". 64th Edition. CRC Press Inc, Boca Raton, Florida, USA, 1985. 326, 370.

Webster, C. (1982). Private communication. Fire Protection Association, April 1982. 112.

Westgate, K. (1975). "Flixborough, the human response". University of Bradford, Disaster Research Unit. Occasional paper No.7, Jan 1975. 284.

Whiddy Island (1980). Report of tribunal of inquiry. "Disaster at Whiddy Island, Bantry, Co. Cork". Government Publication Sales Office, GPO Arcade, Dublin, Eire, 1980. 123, 267.

Whiston, J. (1977). "When you commission a plant". Hydrocarbon Processing 56 (11), 1977, 79.

White, C.S. (1971). "The nature of the problems involved in estimating the immediate casualties from nuclear explosions". Lovelace Foundation for Medical Education and Research, Albuquerque, New Mexico, No. CEX 71.1., 1971. 162.

Whiteside, T. (1978). "The pendulum and the toxic cloud". Yale University Press, 1978. 349, 350, 355.

Whittow, J. (1980). "Disasters, the anatomy of environmental hazards". Pelican Books, 1980. 22.

Wiekema, B.J. (1980). "Vapour cloud explosion model". Journal of Hazardous Materials Vol.3, No.3, Elsevier, Amsterdam, Jan 1980. 253, 254, 255.

Wiekema, B.J. (1984). "Vapour cloud explosions - an analysis based on accidents". Journal of Hazardous Materials Vol.8, No.4, Elsevier, Amsterdam, May 1984. 245, 246, 249, 253, 254, 255, 256, 257, 279, 437, 505.

Wigley, D.A. (1978). "Materials for low temperature use". Oxford University Press, 1978. 75.

Williams, T.I. (1981). "A history of the British gas industry". Oxford University Press, 1981. 237.

Williamson, B.R. and Mann, L.R.B. (1981). "Thermal hazards from propane (LPG) fireballs". Combustion Science & Technology Vol.25, P.141-145, 1981. 151, 163.

Wilson, D.C. (1980). "Flixborough versus Seveso, comparing the hazards". I.Chem.E. Symposium Series No.58, 1980. 440.

Wilson, D.C. (1982). "Lessons from Seveso". Chemistry in Britain, July 1982. 357, 360.

Wingate, P. (1976). "The Penguin medical encyclopedia". 2nd Edition. Penguin Books, London, 1976. 312.

Withers, J. (1985). "The toxic effects of chlorine". A Report of an I.Chem.E. Working party. Loss Prevention Bulletin No.064. I.Chem.E, Rugby, Aug 1985. 109, 319, 337, 435.

Wittcoff, H.A. and Reuben, B.G. (1980). "Industrial organic chemicals in perspective". John Wiley & Sons, 1980. 504.

Wooton, G. (1970). "Interest groups". Prentice-Hall, New Jersey, 1970. 395.

Working Party on Nomenclature (1984). "Nomenclature for hazard and risk assessment in the process industries". Final draft. I.Chem.E, Rugby, 1984. 280.

Worthy, W. (1985). "Methyl isocyanate - the chemistry of a hazard". Chemical & Engineering News, 11th Feb 1985. 370, 373, 377, 378, 379.

Wren, J. (1972). "Great battles of World War 1". Hamlyn Publishers, London, 1972. 109, 327, 328.

Yoshida et al (1985). Yoshida, T., Muranaga, K., Matsunaga, T., Tamura, M. "Evaluation of explosive properties of organic peroxides with a modified Mk 111 ballistic mortar". Journal of Hazardous Materials 12, 1985. 538.

Yusuf, M.F. (1982). "An update on statistics on gas explosions". Private communication from British Gas, 1982. 238, 239.

Zeeuwen et al (1984). Zeeuwen, J.P., van Wingerden, C.J.M., Dauwe, R.M. "Experimental investigation into the blast effect produced by unconfined vapour cloud explosions". I.Chem.E, Rugby, Symposium Series No. 80, 1984. 248.

INDEX